Malaria Parasites

Comparative Genomics, Evolution and Molecular Biology

Edited by

Jane M. Carlton

Center for Genomics and Systems Biology
Department of Biology
New York University
New York, NY
USA

Susan L. Perkins

Sackler Institute for Comparative Genomics
American Museum of Natural History
New York, NY
USA

Kirk W. Deitsch

Department of Microbiology and Immunology
Weill Medical College of Cornell University
New York, NY
USA

Caister Academic Press

Caister Academic Press
Norfolk, UK

www.caister.com

British Library Cataloguing-in-Publication Data
A catalogue record for this book is available from the British Library

ISBN: 978-1-908230-07-2

Cover design adapted from Figure 8.2

Contents

	Contributors	v
	Preface	ix
1	The Diversity of *Plasmodium* and other Haemosporidians: The Intersection of Taxonomy, Phylogenetics and Genomics Ellen S. Martinsen and Susan L. Perkins	1
2	The Apicomplexan Genomic Landscape: The Evolutionary Context of *Plasmodium* Jeremy D. DeBarry, Segun Fatumo and Jessica C. Kissinger	17
3	*Plasmodium* Genomics and the Art of Sequencing Malaria Parasite Genomes Jane M. Carlton, Steven A. Sullivan and Karine G. Le Roch	35
4	Genome Diversity and Applications in Genetic Studies of the Human Malaria Parasites *Plasmodium falciparum* and *Plasmodium vivax* Sittiporn Pattaradilokrat, Jianbing Mu, Philip Awadalla and Xin-zhuan Su	59
5	Functional Genomics of *Plasmodium* Parasites Zbynek Bozdech and Peter R. Preiser	91
6	*Plasmodium* Experimental Genetic Crosses Lisa C. Ranford-Cartwright, Karen L. Hayton and Michael T. Ferdig	127
7	Regulation of Gene Expression Kirk W. Deitsch and Ron Dzikowski	145
8	Invasion of Host Red Blood Cells by Malaria Parasites Amy K. Bei and Manoj T. Duraisingh	169
9	Host Cell Remodelling and Protein Trafficking Silvia Haase, Hayley E. Bullen, Sarah C. Charnaud, Brendan S. Crabb, Paul R. Gilson and Tania F. de Koning-Ward	199

10 Dissecting Mosquito–Parasite Interactions through Molecular Biology and Biochemistry: Genomic, Proteomic and Glycomic Analyses 221

Lindsay A. Parish, Lindsey S. Garver, David R. Colquhoun, Ceereena Ubaida Mohien, Elizabeth Weissbrod and Rhoel R. Dinglasan

11 The Malariologist's Molecular Toolbox 249

Alexander G. Maier

Contributors

Philip Awadalla
Department of Pediatrics
University of Montreal
Montreal, QC
Canada

philip.awadalla@umontreal.ca

Amy K. Bei
Department of Immunology and Infectious Diseases
Harvard School of Public Health
Boston, MA
USA

amy.bei@gmail.com

Zbynek Bozdech
School of Biological Sciences
Nanyang Technological University
Singapore

zbozdech@ntu.edu.sg

Hayley E. Bullen
Burnet Institute
Melbourne, VIC
Australia

bullen@burnet.edu.au

Jane M. Carlton
Center for Genomics and Systems Biology
Department of Biology
New York University
New York, NY
USA

jane.carlton@nyu.edu

Sarah C. Charnaud
Burnet Institute
Melbourne, VIC
Australia

charnaud@burnet.edu.au

David R. Colquhoun
Department of Molecular and Comparative Pathobiology
Johns Hopkins University School of Medicine
Baltimore, MD
USA

dcolquh1@jhmi.edu

Brendan S. Crabb
Burnet Institute
Melbourne, VIC
Australia

crabb@burnet.edu.au

Jeremy D. DeBarry
Center for Tropical and Emerging Global Diseases
Coverdell Center
Athens, GA
USA

jdebarry@uga.edu

Kirk W. Deitsch
Department of Microbiology and Immunology
Weill Medical College of Cornell University
New York, NY
USA

kwd2001@med.cornell.edu

Rhoel R. Dinglasan
Johns Hopkins Bloomberg School of Public Health
Johns Hopkins Malaria Research Institute
W. Harry Feinstone Department of Molecular Microbiology & Immunology
Baltimore, MD
USA

rdinglas@jhsph.edu

Manoj T. Duraisingh
Department of Immunology and Infectious Diseases
Harvard School of Public Health
Boston, MA
USA

mduraisi@hsph.harvard.edu

Ron Dzikowski
Department of Microbiology & Molecular Genetics
The Kuvin Center for the Study of Infectious and Tropical Diseases
The Institute for Medical Research Israel-Canada
The Hebrew University-Hadassah Medical School
Jerusalem
Israel

rond@ekmd.huji.ac.il

Segun Fatumo
Center for Tropical and Emerging Global Diseases
Coverdell Center
Athens, GA
USA

segunfatumo@gmail.com

Michael T. Ferdig
The Eck Institute for Global Health
Department of Biology
University of Notre Dame
Notre Dame, IN
USA

ferdig.1@nd.edu

Lindsey S. Garver
Laboratory of Malaria and Vector Research
National Institute of Allergy and Infectious Diseases
National Institutes of Health
Bethesda, MD
USA

garverls@niaid.nih.gov

Paul R. Gilson
Burnet Institute
Melbourne, VIC
Australia

gilson@burnet.edu.au

Silvia Haase
School of Medicine
Deakin University
Waurn Ponds, VIC
Australia

haase@wehi.edu.au

Karen L. Hayton
Laboratory of Malaria and Vector Research
National Institute of Allergy and Infectious Diseases
National Institutes of Health
Bethesda, MD
USA

khayton1@gmail.com

Jessica C. Kissinger
Department of Genetics
Institute of Bioinformatics and Center for Tropical and Emerging Global Diseases
Coverdell Center
Athens, GA
USA

jkissing@uga.edu

Tania F. de Koning-Ward
School of Medicine
Deakin University
Waurn Ponds, VIC
Australia

taniad@deakin.edu.au

Karine G. Le Roch
Department of Cell Biology & Neuroscience
Institute for Integrative Genome Biology
Center for Disease Vector Research
University of California Riverside
Riverside, CA
USA

karine.leroch@ucr.edu

Alexander G. Maier
Research School of Biology
The Australian National University
Canberra, ACT
Australia

alex.maier@anu.edu.au

Ellen S. Martinsen
Center for Conservation and Evolutionary Genetics
National Zoological Park
National Museum of Natural History
Smithsonian Institution
Washington, DC
USA

ellensarah.martinsen@gmail.com

Ceereena Ubaida Mohien
Department of Molecular and Comparative Pathobiology
Johns Hopkins University School of Medicine
Baltimore, MD
USA

ceereena@jhmi.edu

Jianbing Mu
National Institute of Allergy and Infectious Diseases
National Institutes of Health
Bethesda, MD
USA

jmu@niaid.nih.gov

Lindsay A. Parish
Johns Hopkins Bloomberg School of Public Health
Johns Hopkins Malaria Research Institute
W. Harry Feinstone Department of Molecular Microbiology & Immunology
Baltimore, MD
USA

lindsayannparish@gmail.com

Sittiporn Pattaradilokrat
Laboratory of Malaria and Vector Research
National Institute of Allergy and Infectious Diseases
National Institutes of Health
Bethesda, MD
USA

pattaradilokrats@gmail.com

Susan L. Perkins
Sackler Institute for Comparative Genomics
American Museum of Natural History
New York, NY
USA

perkins@amnh.org

Peter R. Preiser
School of Biological Sciences
Nanyang Technological University
Singapore

prpreiser@ntu.edu.sg

Lisa C. Ranford-Cartwright
Institute of Infection, Immunity and Inflammation
College of Medical, Veterinary and Life Sciences
University of Glasgow
Glasgow
UK

lisa.ranford-cartwright@glasgow.ac.uk

Xin-zhuan Su
National Institute of Allergy and Infectious Diseases
National Institutes of Health
Bethesda, MD
USA

xsu@niaid.nih.gov

Steven A. Sullivan
Department of Biology
Center for Genomics and Systems Biology
New York University
New York, NY
USA

sullis02@nyu.edu

Elizabeth Weissbrod
Johns Hopkins University School of Medicine
Department of Art as Applied to Medicine
Baltimore, MD
USA

leftfieldart@gmail.com

Current books of interest

RNA Editing: Current Research and Future Trends	2013
Microbial Efflux Pumps: Current Research	2013
Cytomegaloviruses: From Molecular Pathogenesis to Intervention	2013
Oral Microbial Ecology: Current Research and New Perspectives	2013
Bionanotechnology: Biological Self-assembly and its Applications	2013
Real-Time PCR in Food Science: Current Technology and Applications	2013
Bacterial Gene Regulation and Transcriptional Networks	2013
Bioremediation of Mercury: Current Research and Industrial Applications	2013
Neurospora: Genomics and Molecular Biology	2013
Rhabdoviruses	2012
Horizontal Gene Transfer in Microorganisms	2012
Two-Component Systems in Bacteria	2012
Foodborne and Waterborne Bacterial Pathogens:	2012
Yersinia: Systems Biology and Control	2012
Stress Response in Microbiology	2012
Bacterial Regulatory Networks	2012
Systems Microbiology: Current Topics and Applications	2012
Quantitative Real-time PCR in Applied Microbiology	2012
Bacterial Spores: Current Research and Applications	2012
Small DNA Tumour Viruses	2012
Extremophiles: Microbiology and Biotechnology	2012
Bacillus: Cellular and Molecular Biology (Second edition)	2012
Microbial Biofilms: Current Research and Applications	2012
Bacterial Glycomics: Current Research, Technology and Applications	2012
Non-coding RNAs and Epigenetic Regulation of Gene Expression	2012
Brucella: Molecular Microbiology and Genomics	2012
Molecular Virology and Control of Flaviviruses	2012
Bacterial Pathogenesis: Molecular and Cellular Mechanisms	2012
Bunyaviridae: Molecular and Cellular Biology	2011
Emerging Trends in Antibacterial Discovery: Answering the Call to Arms	2011
Epigenetics: A Reference Manual	2011
Metagenomics: Current Innovations and Future Trends	2011
Nitrogen Cycling in Bacteria: Molecular Analysis	2011

Full details at www.caister.com

Preface

It has been a decade since the complete genome sequences of the human malaria parasite *Plasmodium falciparum* and a rodent model *Plasmodium yoelii yoelii* were described in a series of landmark papers published in *Nature* in 2002. Two years later this book's predecessor, edited by Andy Waters and Chris Janse, was published. Since then, genome-scale data sets from more species of *Plasmodium* – as well as from *Toxoplasma, Trichomonas, Cryptosporidium, Trypanosoma, Leishmania,* and *Entamoeba* – have been deposited in public databases, offering a cornucopia of data for comparative studies. This book highlights some of the advances in *Plasmodium* research that have resulted. The first six chapters focus on insights gained directly from analyses of the recently available genome sequences. In Chapter 1, recent discoveries regarding the taxonomy and phylogeny of parasites are described, an analysis that is extended in Chapter 2 through a detailed description of the apicomplexan lineage. In Chapters 3 and 4, the 'art' of sequencing *Plasmodium* genomes is presented, as well as what can be gained through both the comparison of different species and through the study of the sequence diversity displayed by different isolates of *P. falciparum*. The application of fully assembled genome sequences to functional genomic techniques and to experimental genetic crosses are described in Chapter 5 and 6, respectively. These powerful tools have contributed to many significant advances in our understanding of parasite biology in recent years.

The remaining chapters of the book focus on many of the molecular advances that have been greatly aided by the availability of parasite genome sequence data. The basic molecular biology of epigenetic modification and transcriptional regulation are highlighted in Chapter 7, while Chapter 8 describes many of the recent advances in our understanding of the pathways used by different *Plasmodium* species to invade host red blood cells. These discoveries have significant implications for many of the vaccine strategies that are being currently being developed. Recently, one area of research that has advanced substantially is our understanding of how proteins are exported and trafficked by malaria parasites into the cytoplasm of their host cells. These advances are described in detail in Chapter 9. While much of the research in the field of malaria molecular biology has been centred on understanding the interactions of malaria parasites with their mammalian hosts, the role of the mosquito vector in disease transmission cannot be ignored. Chapter 10 provides a broad view of recent developments in the study of parasite–mosquito interactions. Finally, the last chapter of the book presents a practical guide to many of the revolutionary new techniques and molecular tools that have become available to researchers over the last decade. These advances promise to further accelerate our ability to dissect and define various molecular and biochemical aspects of parasite biology.

The editors would like to thank the authors for their extensive efforts in putting together clear, concise and informative descriptions of the various research topics covered in each chapter. The overall quality of the book is a credit to their diligent attention to details. We are also indebted to the anonymous reviewers who provided comments and recommendations for improvement for each chapter prior to publication, thereby ensuring the quality and accuracy of the final product. We hope this collection will prove valuable to the malaria research community as we continue to strive to understand the details of this fascinating but deadly organism.

Jane Carlton
Susan Perkins
Kirk Deitsch

The Diversity of *Plasmodium* and other Haemosporidians: The Intersection of Taxonomy, Phylogenetics and Genomics

1

Ellen S. Martinsen and Susan L. Perkins

Abstract

As important agents of disease, a great deal of research has been focused on the malaria parasites. Yet the species that infect humans represent only a small fraction of the diversity of the malaria parasites, and future genomics projects on closely related parasite species with diverse life histories and other key traits will likely serve as important steps to a better understanding of malaria in humans as well as the biology of the group as a whole. Before comparative studies can be performed, however, a robust phylogeny or understanding of the evolutionary history of the group must be in place. The history of the discovery and classification of the malaria parasites has been a long and sometimes circuitous one and while new species surely remain to be discovered, it is important that we continue to adhere to taxonomic principles. Recent advances in molecular systematics have both challenged and enlightened our understanding of the diversity and evolution of these organisms, though the development of new molecular markers still remains a challenge and genome sequencing faces unique hurdles.

Introduction

The malaria parasites produce a staggering amount of morbidity and mortality in mankind, but in terms of the diversity of the group, the five species that commonly infect humans represent a mere fraction of the total, with over 200 species of *Plasmodium* and close to 300 species of closely related parasites formally described in the literature. Within *Plasmodium*, vertebrate hosts include multiple orders of mammals as well as diverse orders of birds and lizards from around the globe (Fig. 1.1), and related genera can be found in these same vertebrate classes. Although all of the haemosporidian parasites share some basic similarities in morphology and life cycle, they also vary greatly in shape, structure, life history, and vertebrate host and insect vector preferences. For over a century, scientists have tried to classify the parasites based on natural groupings of these traits. Recently, molecular studies, particularly of bird and ape populations and species, have suggested a large number of cryptic and undescribed parasite lineages, hinting that the formally described species represent just the tip of the iceberg of the true diversity of these parasites. In addition, molecular studies have also shown that some of the lineages previously classified as other genera due to differences in their life cycle (e.g. *Hepatocystis*) should actually be reclassified as *Plasmodium* (Perkins and Schall, 2002).

To date, complete genomes have been sequenced from six species of *Plasmodium* that infect primates and rodents (*Plasmodium falciparum, Plasmodium vivax, Plasmodium knowlesi, Plasmodium yoelii yoelii, Plasmodium chabaudi, Plasmodium berghei*) with partial sequences available for two other *Plasmodium* species (*Plasmodium reichenowi, Plasmodium gallinaceum*). This first set of species was chosen due to their medical importance or utility as model systems. Although these genomes provide useful information for researchers trying to fight the disease in humans, broadening the scope of genomic studies in a more comparative framework will not only further our understanding

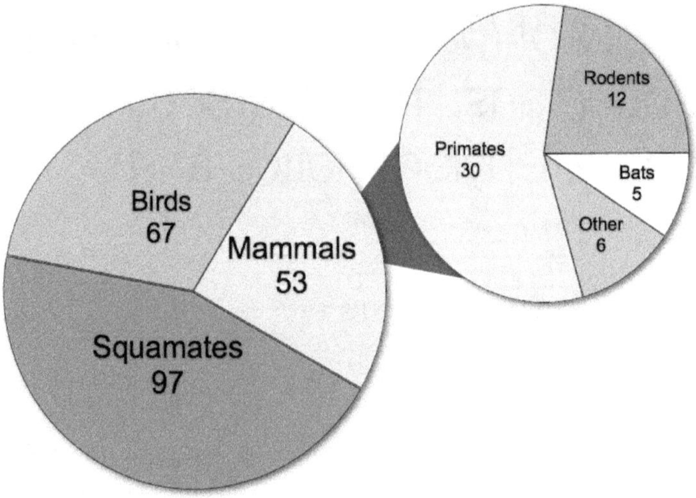

Figure 1.1 Depiction of number of species of *Plasmodium* described from various vertebrate hosts.

of the evolutionary history of the group but also likely unveil new strategies for understanding the parasites' role in disease, including cell invasion mechanisms and traits involved in virulence. In this chapter, we provide a history of the discovery of the *Plasmodium* parasites and the methods to classify their diversity and describe how molecular studies have challenged traditional views of the relationships between groups of taxa, including even the very notion of what constitutes a 'malaria parasite'.

Keeping the 'comparative' in 'comparative genomics'

Perhaps first discussed by Darwin in *On the Origin of Species*, subsequently put into a statistical framework by Felsenstein (1985) and summarized by and advocated for by Harvey and Pagel (1991), the phylogenetic comparative method uses the evolutionary relationships of organisms to compare species. These comparisons may be studies of adaptation, correlation between two or more traits in a set of species, or attempts to discover the ancestral state of a particular trait, for example. The underlying principle is that only with a solid understanding of the evolutionary history of the group that is under investigation, can one make conclusions about the presence of characteristics of a given species or lineage and it is these evolutionary trees that can most fruitfully spawn powerful comparisons about the patterns observed in Nature. For example, in the context of malaria parasites, speculation concerning the high virulence that is seen in *P. falciparum* is often made. However, it is imperative to know whether this species has uniquely acquired phenotypic characteristics or life history traits that render it so virulent or if these are common properties of a larger clade to which it belongs. Comparisons of *P. falciparum* to *P. berghei*, for example, are fairly meaningless without knowing whether or not these two species share a recent shared ancestor. If they do, then a trait that leads to high virulence may be the ancestral condition of this clade and the presence of the trait in both species is a product of their common descent. If they are not recently derived from a common ancestor, however, then these virulence traits have instead arisen independently. This knowledge that the traits are instead convergent can, in turn, spawn more interesting investigations into the pathways that produce these phenotypes or correlations between these traits and others traits. In order to use the phylogenetic comparative method, though, one must have a reasonably well-supported phylogeny for the group and a proper taxonomic understanding of the lineages in question. In the case of the malaria parasites, however, this has not necessarily been an easy task.

The historical context of malaria parasite discovery and description

Uncovering the life cycle of *Plasmodium*

Plasmodium and humans share a long and rich history, one that likely spans the tenure of our species. Within historical texts, symptoms akin to those experienced by current-day malaria patients are recorded from human societies dating back as far as the fourth century BC (Garnham, 1966). Early Greek and Roman physicians sought to understand the cause of 'rigors', the intermittent fevers that struck much of the populace and it was suspected that swamps were connected to these fevers. Although partially correct in this regard, the initial inclination was that noxious gases, or *miasmata*, rising out of them were the agents of illness. In the nineteenth century, upon the discovery of pathogenic microbes, observations by Tomasi-Crudeli hinted that the causative agent of the disease was a bacterium, *Bacillus malariae*. It was not until 1880 that the French army officer, Charles Laveran, pinpointed a protozoan as the culprit of the disease. As no protozoan had previously been recorded from human blood, it took years for the medical community to supplant the notion of a bacterium as the aetiological agent of malaria, but Laveran later received the Nobel Prize for Medicine in 1907 for this work (Cox, 2010). About the same time that Laveran was observing parasites in the blood of his patients suffering from malaria, a Russian physician, Vassily Danilewsky, was noticing a number of similar parasites in the blood of birds and reptiles (Valkiunas, 2005). The discoveries by Laveran and Danilewsky were quite remarkable given that they were limited to the observation of unstained parasites, a state in which parasites are difficult to differentiate from the cells of their hosts.

Perhaps two of the most monumental events in the history of parasitology and in the study of the malaria parasites were the development of the oil immersion microscope objective around 1882 followed by the discovery of an eosin-based blood stain in 1891 (Cox, 2010). A laboratory accident further accelerated the study of malaria parasites, when Romanowsky, an army pathologist, forgot to replace a stopper on one of his bottles of methylene blue and a mould grew upon the solution. When Romanowsky subsequently used this stain along with the typical eosin, he saw that the nuclei of the malaria parasites stained a deep pink whereas the rest of the cell, the cytoplasm, stained blue (Garnham, 1966). This stain has become the basis for all modern differential stains, including Giemsa stain, the most popular blood stain for looking at malaria parasites, and allows the parasites to be visualized under the light microscope and their structures to be differentiated. These discoveries revolutionized the study of the malaria parasites, allowing them to be studied with greater ease from blood samples procured from vertebrate hosts.

One of the most important characters involved in the search for the parasites was the 'malaria pigment,' that is produced and observable within both gametocytes and schizonts of *Plasmodium* species and their close kin (e.g. *Hepatocystis, Haemoproteus* and *Parahaemoproteus*). During the intraerythrocytic stages the parasites may ingest up to 75% of the host cell's haemoglobin, which is cleaved within the parasite's food vacuoles into globin and haem (Jani *et al.*, 2008). While globin is readily degraded into the amino acids necessary for metabolic processes within the parasite, haem, with an iron atom at its core, poses a threat to the parasite cell as it is cytotoxic in its free state. In response, the parasite detoxifies free haem by polymerizing it into an insoluble crystalline material, haemozoin. It was these shiny pigment granules that initially alerted Laveran to the presence of parasites within the red blood cells of his patients. After observing pigment in the spleens of patients infected with malaria, he then sought to find it elsewhere in the body whereupon he discovered it in erythrocytes. Most likely, given his observations, Laveran had observed the schizont or asexual stage of *Plasmodium*.

The cue of haemozoin pigment also sparked subsequent discovery of the different stages of the malaria parasites. In 1897, upon examination of blood from crows infected with *Haemoproteus*, MacCallum and Opie witnessed the emergence of gametocytes from the blood cells and the exflagellation of microgametocytes into microgametes as well as subsequent formation of an ookinete through fertilization of the macrogamete (MacCallum, 1897). These parasite stages, which normally occur within the midgut of the vector host and precede sexual recombination, had been artificially provoked from the blood cells by being exposed to air. This discovery perplexed MacCallum, who postulated that these ookinete forms then went on to reinvade additional tissues within the same host individual. At that point in time, it was still a mystery how the parasites were transmitted from human to human and various hypotheses were put forth, some harkening back to the idea of an airborne route of transmission. Here, Laveran again played an important part in the history of the study of malaria, as he was among the first to suggest mosquitoes as the putative vectors. The role of the mosquito was also strongly suggested by Patrick Manson, a physician who had demonstrated that human filarial worms were transmitted by mosquitoes. Manson had observed similar flagellation of gametocytes and fertilization, as had been previously described by MacCallum, and he directed his colleague and army surgeon, Ronald Ross, who frequented malaria-endemic regions on medical assignments, to 'follow the flagellum' (Cox, 2010). Ross started to experiment with mosquitoes, feeding them on patients infected with malaria; however, a reassignment to India forced Ross to halt his human malaria and mosquito studies. Manson, having learned of the high prevalence of *Plasmodium* species in birds, then suggested to Ross that he instead study transmission of the parasites in a bird system. With *Plasmodium* common in the native birds at Ross's post, he carried out transmission experiments by allowing mosquitoes to feed from infected birds. Ross was then able to document the transmission of the avian malaria parasite *Plasmodium relictum* from chicken to chicken by culicine mosquitoes, thus associating the sexual stages of the parasites to an invertebrate host. Ross observed not only the emergence of the parasites from blood cells and subsequent fertilization within the midgut of the mosquito host but also the development of the oocyst on the wall of the mosquito midgut and subsequent release of the sporozoites. He found these same stages in preparations from the mosquito salivary glands and hypothesized that these were injected into a new bird during the mosquito's next blood meal. Ross then went on to discover the same mode of transmission of human malaria parasites via anopheline mosquitoes in Africa. Ross thus elucidated the complete life cycle of *Plasmodium*, which resulted in his Nobel Prize in Medicine award in 1902. Fifty years later, in 1948, Shortt and Garnham documented the development of malaria parasites within the liver and asexual reproduction in this organ before their entry into the bloodstream, and in 1982 the dormant parasite stages within the liver were conclusively demonstrated by Krotoski (Cox, 2010). Thus, largely through the work of medical doctors, it took close to 100 years to uncover all of the life stages of the malaria parasites, and the milestones achieved along this road of discovery clearly illustrate the importance of the study of parasites in non-human hosts as well.

Discovering the diversity of malaria parasites

With the new methods of staining and microscopy, it became possible for researchers to screen blood films prepared from a variety of vertebrate hosts, resulting in the detection and documentation of a diversity of species of *Plasmodium* from the blood of humans and other mammals, lizards, and birds from all continents except Antarctica. Although sampling of vertebrate hosts would take the primary role in the documentation of new species, in one noted case, the discovery of a species of malaria parasite resulted from the initial detection of sporozoites in a captured *Anopheles dureni* mosquito in western Africa (Garnham, 1966). After tests on the proteins found within the blood meal of these mosquitoes excluded primate hosts, rodents were instead presumed. Finally, after extensive sampling of local rodents, thicket rats (*Thamnomys surdaster*) with *Plasmodium* within their blood were discovered. The rate of discovery of species *Plasmodium* has yet to decrease through time (Fig. 1.2) and it is likely that dozens, if not hundreds, of novel species have yet to be sampled worldwide.

As sampling of vertebrate species continued, so did the wide observation of a diversity of parasites that shared some characteristics with *Plasmodium* but differed in other respects. Depending on various characteristics but also relying largely on

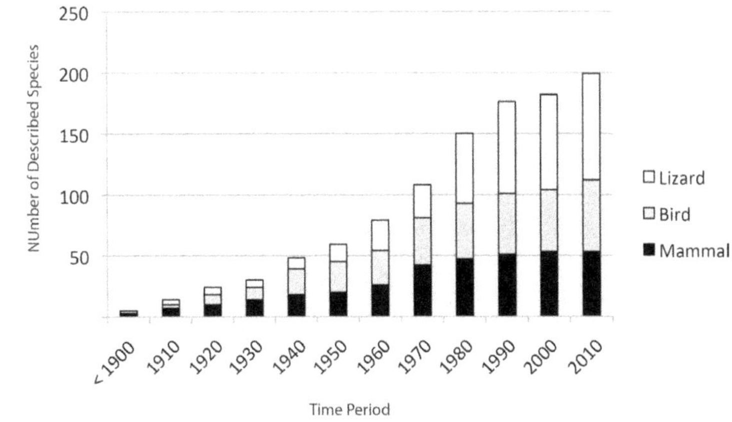

Figure 1.2 Cumulative *Plasmodium* species descriptions through the 20th and 21st centuries, divided by lizard, bird, and mammalian hosts. Only formally described species are included.

combinations of life history traits in certain host groups, these parasites were sometimes classified as distinct genera or families. A total of 18 additional genera of haemosporid blood parasites have now been described. Two of these, *Haemoproteus* and *Leucocytozoon*, are very diverse, geographically widespread, and highly prevalent in host populations. Both of these genera parasitize birds, with the former also found in crocodilians, squamates, and turtles. Further investigation of mammal hosts for blood parasites yielded the discovery of yet another group of morphologically similar parasites, the *Hepatocystis* parasites, which were initially placed within the *Haemamoeba* group but then reassigned to the genus *Haemoproteus* based on the absence of erythrocytic schizogony. But, given that they were documented from primates, bats, and rodents (and even a hippopotamus; Garnham, 1958) and that they produce enormous, distinctive exoerythrocytic schizonts, they were designated to their own genus (Garnham, 1948). As for the other described genera, they have typically been found to be restricted either in a host or geographic sense, or both. Three of the genera (*Bioccala, Biguetiella* and *Dionisia*) are monotypic species each from a single species of host bat; *Billbraya* is likewise a monotypic genus known from just one species of lizard; and *Mesnilium*, the only haemosporid ever reported from a fish, has only been observed once. As of yet, samples for genetic analysis from most of these type hosts and type locations for most of these genera are not available; thus it is uncertain if they are indeed distinct or if, like *Hepatocystis*, the divergent morphological or life history traits are more recently derived and these parasites belong within the *Plasmodium* clade.

Systematics of malaria parasites

The discipline of systematics is composed of two parts. The first component is taxonomy, the hierarchical classification of organisms based on their unique characteristics. The second component is phylogenetics, the study of the evolutionary history of a group of organisms. Cladistics links these two components together – the analysis of the distribution of the same (i.e. homologous) structure across a set of organisms (from an anatomical feature to a nucleotide) is used to create a phylogenetic tree. Likewise, phylogenies can be useful for informing taxonomy by allowing hypotheses of monophyletic groups to be tested. For those studying malaria parasites, systematics has always been a challenge. Before molecular methods were developed for this group, malariologists such as Garnham (1966) were clearly frustrated by the limitations of morphology and the repeated observations of variability of parasite species depending on their hosts or geography or other factors. However, they sought to make as much order out of the diversity of parasites observed as they possibly could.

Taxonomy

Early on in their study, the parasites that would come to comprise the genera *Plasmodium*, *Hepatocystis*, *Haemoproteus*, and *Parahaemoproteus* were all considered members of the same taxonomic category '*Haemamoeba*' (Laveran, 1899). Although parasitologists noted that *Plasmodium* and *Haemoproteus* infections had both morphological and life history similarities, the two genera were kept separated based on a single life history trait difference – the presence or absence of segmenting forms (asexual stages or schizonts) forms in the peripheral blood (Novy and MacNeal, 1904). Garnham's (1966) synthetic opus divided the malaria parasites into three families: Plasmodiidae, Haemoproteidae, and Leucocytozoidae. He considered Plasmodiidae to be monogeneric, but he divided the genus *Plasmodium* into nine subgenera based on host group (mammal, bird, lizard), the size and shape of the schizonts, and the shape of the gametocytes. Levine (1988), in his comprehensive list of all described species of the phylum Apicomplexa (or Sporozoa – there is even debate on what the phylum should be called; Valkiunas, 2005), grouped all haemosporidians into the family Plasmodiidae, but recognized ten different genera in the family. Two life history characters have been traditionally considered most important in the taxonomic structure of malaria parasites. The first of these is whether or not schizogony occurs within erythrocytes or not. Thus, an important distinction between the two common genera, *Plasmodium* and *Haemoproteus*, which have morphologically similar gametocytes, is that *Plasmodium* parasites are able to reproduce asexually within both fixed tissues and blood cells of their vertebrate hosts while *Haemoproteus* parasites are restricted to fixed tissue cells for asexual reproduction. The second trait that was important in differentiating lineages was the presence or absence of stored haemozoin pigment. This character differentiated *Leucocytozoon* from the other parasites, for instance. After these two traits, vertebrate hosts and vectors then often played a role in the classification scheme, but with some inconsistencies. For example, *Saurocytozoon* was erected as a genus for the parasites of lizards that were unpigmented and otherwise resembled *Leucocytozoon* in birds (Lainson and Shaw, 1969). However, the parasite group *Parahaemoproteus*, which solely infects birds and is vectored by biting midges, remained a subgenus within the genus *Haemoproteus*, although the type species for the genus, *Haemoproteus columbae*, is vectored by hippoboscid flies and is restricted to birds of the order Columbiformes.

No one has attempted to amend the taxonomy of the entire order since then, but two other authors have revised bird and lizard malaria parasite taxonomy, respectively. Based on a study of 1200 parasite cells in various stages found to be infecting 32 individuals, Telford (1979) arrived at 18 characteristics that he believed useful for lizard malaria parasite taxonomy. He then used these characteristics and morphometric analyses to classify all saurian malaria parasites, placing them into three genera: *Plasmodium* (7 subgenera), *Fallisia*, and *Saurocytozoon* (Telford, 1988). Valkiunas published a comprehensive summary, which revised the classification of the avian – but only the avian – haemosporid parasites (Valkiunas, 1997; 2005). He proposed four monogeneric families (Haemoproteidae, Plasmodiidae, Garniidae, and Leucocytozoidae) all within the order Haemosporida, with each genus further divided into subgenera. The classification of the remaining genera in Haemosporida, i.e. *Hepatocystis* (bats, primates), *Nycteria* (bats), *Polychromophilus* (bats), *Dionisia* (bats) and *Rayella* (rodents), remains somewhat problematic, though Garnham (1966) placed these genera in the family Haemoproteidae due to a lack of erythrocytic merogony. A summary of the classification of the order as used here is shown in Fig. 1.3.

Challenges of classification: what constitutes a 'malaria parasite'?

The application of a term to a group of organisms should dictate that the group be monophyletic and that all of its members be each other's closest relatives. If such a scheme is applied to the term malaria parasite, then how the term should be applied is not so cut and dry, and thus, not surprisingly, there has been steady debate over this topic in the literature. The application of the term 'malaria parasite' varies widely through history. As the term 'malaria' was originally used to describe the human disease attributed to

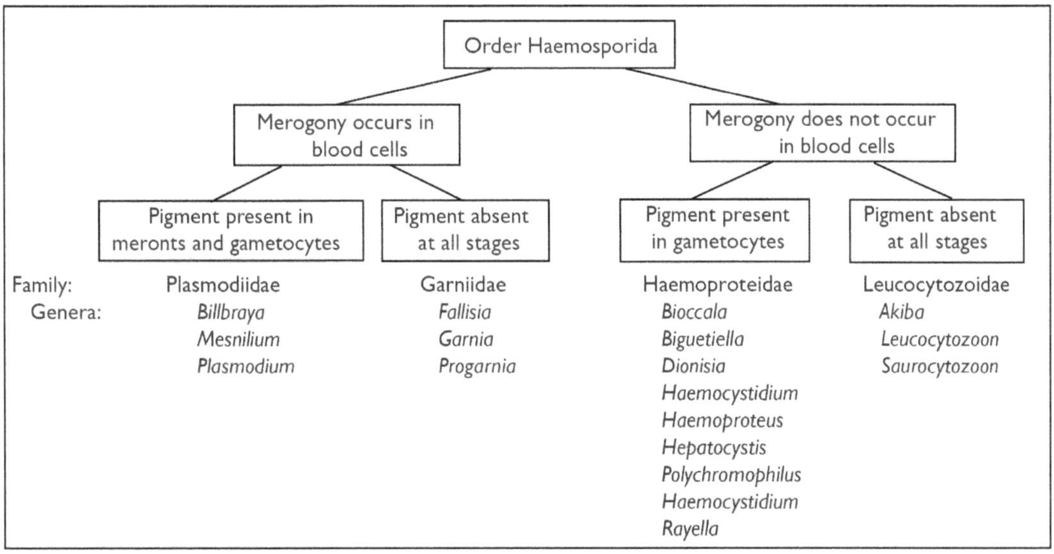

Figure 1.3 Summary of the current taxonomic configuration of the order Haemosporida based on characteristics summarized in Levine (1988), Telford (1988) and Valkiunas *et al.* (2005).

infection by parasites of the genus *Plasmodium*, so it has been argued that in its strictest sense, the term should be applied only to the *Plasmodium* species that infect humans, with such use of the term prominent in medical research, practice, and literature. However, the character of human host association may not be fully justified and certainly, as we now know, the species that infect humans do not comprise a monophyletic group. And, as will be discussed in more detail below, recent studies have suggested the sharing of *Plasmodium* species between humans and closely related primate species including chimpanzees, macaques, and mountain gorillas (Cox-Singh *et al.*, 2008; Liu *et al.*, 2010). If monophyly is to be incorporated, then one must delve back deeper into the evolutionary history of the parasites for more appropriate application of the term.

From a traditional standpoint, the taxonomic classification of a parasite as a 'malaria parasite' is based on assignment to the genus *Plasmodium*, which in turn is based primarily on the presence of erythrocytic schizonts and haemozoin pigment and the use of mosquitoes as the vector, all considered important shared derived and fixed life history traits for the genus. Indeed, it was the presence of haemozoin pigment within the parasite stages present in the red blood cells of the vertebrate hosts that initially led Laveran to their discovery. Occasionally, though, exceptions to the 'rule' have been reported, i.e. species classified as *Plasmodium* that buck the traditional definition. *Plasmodium mexicanum*, a lizard parasite that occurs in western N America, uses not a mosquito but rather a phlebotomine sandfly (*Lutzomyia vexator*) as its vector and *Plasmodium azurophilum* and *Plasmodium leucocytica*, two other lizard parasites that infect Caribbean anoles, do not typically sequester haemozoin pigment (Ayala and Lee, 1970). Thus the boundaries of what constitutes a parasite within the genus *Plasmodium* are not clear-cut even as defined by life history traits, with the discovery of closely related parasite groups such as *Haemoproteus* and *Hepatocystis* further confounding their classification. Many authors now use the term as a common name for any of the parasites in the family Haemosporidia, allowing it to encompass not just *Plasmodium*, but this clade and the related genera that have similar life cycles and that alternate between vertebrates and dipteran vectors.

Phylogeny

Perhaps owing to the paucity of morphological characters with which to construct a character matrix, few scholars attempted to produce a phylogenetic tree or reconstruction of the evolutionary relationships of the Haemosporida prior to the advent of molecular techniques. The two non-molecular phylogenies of malaria parasites that exist in the literature, one in Garnham's tome (1966) and one in Mattingly (1983), depict branches that correspond primarily to the vertebrate host group that the parasites are found to infect. These phylogenies are neither cladistic nor even phenetic and result in numerous cases of paraphyly and polyphyly – with, for example, *Plasmodium* depicted as the tip taxa in each of three clades from mammals, birds, and reptiles (Mattingly, 1983) – and thus are better dismissed as non-cladistic treatments of the phylogeny of the malaria parasites.

The advent of molecular based methods to infer evolutionary relationships shed a whole new light on the phylogeny of the malaria parasites, especially with respect to the relationships between parasites isolated from the different vertebrate host groups. The first molecular phylogenetic study, based on data from 18S ribosomal DNA sequences from six species of *Plasmodium*, suggested that *P. falciparum* was derived from avian malaria parasites and the result of a host switch from chickens to humans (Waters *et al.*, 1991). This study, however, was limited by biased and incomplete taxon sampling and the inclusion of a far too distantly related outgroup taxon. As demonstrated by other studies, choice of outgroup taxa plays an important role in the inference of evolutionary relationships and can affect results (e.g. Pollock *et al.*, 2002; Zwickl and Hillis, 2002). A subsequent study that included 11 species grouped *P. falciparum* with the chimpanzee parasite, *P. reichenowi*, but these two species were grouped with lizard and bird parasites (Escalante and Ayala, 1994). Qari *et al.* (1996) did not find this relationship, but instead placed each of the human *Plasmodium* species as being closely related to a *Plasmodium* species that had been described from a non-human primate. All of these studies were limited largely by their use of 18S genes, which exist as unique and independently evolving loci (i.e. paralogues) across the nuclear genome of the parasites and are differentially expressed during the various life history stages of the parasites, and thus presumably under different forces of selection (Rogers *et al.*, 1995). For these reasons, these molecular loci are not useful as informative phylogenetic markers for haemosporidians. From these earliest sequence-based investigations into the evolutionary associations of *Plasmodium* taxa from different vertebrate groups, it also became very clear that sampling, with respect to both parasite taxa and molecular loci, has a large influence on the reconstruction of evolutionary relationships.

In an effort to avoid paralogy pitfalls associated with the use of the 18S genes and to limit the bias of taxon sampling on phylogenetic inference, later systematic studies of *Plasmodium* began to sequence the cytochrome *b* gene, located in the maternally inherited mitochondrial genome, and to increase the number of parasite taxa represented in the analyses (Escalante *et al.*, 1998). Perkins and Schall (2002) presented a mitochondrial-based phylogeny for over 50 species of malaria parasites and closely related genera, a phylogeny that suggested that all of the mammal-infecting species formed a monophyletic clade, and supported the inclusion of *Hepatocystis* species within this clade of mammalian *Plasmodium* parasites. The phylogeny also indicated the paraphyly of *Haemoproteus* with respect to *Plasmodium*, with *Haemoproteus* parasites found multiple times within the *Plasmodium* clade.

As molecular phylogenies themselves merely represent hypotheses of the relationships between a given set of taxa based on the data available, they are always prone to change. In an attempt to improve our understanding of the evolutionary history of the malaria parasites and closely related genera and to address taxonomic inconsistencies spawned by earlier single-gene phylogenetic studies, Martinsen *et al.* (2008) sequenced four genes including at least one from each of the three genomes of the parasites (nuclear, mitochondrial, and plastid) from a total of 60 parasite taxa (Fig. 1.4). Findings from this study corroborated the findings of Perkins and Schall (2002) as well as some hypotheses put forth by pioneering parasitologists based on traditional study of the parasites. Martinsen *et al.* (2008) found *Plasmodium* to be paraphyletic with respect to *Hepatocystis*, with support from four genes for the placement of *Hepatocystis* parasites within the

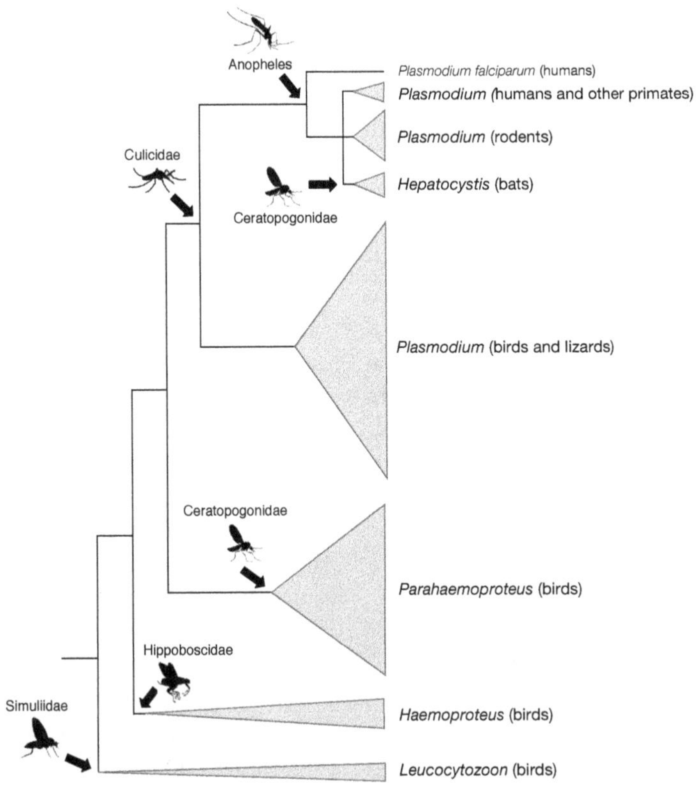

Figure 1.4 Phylogeny for the malaria parasites and closely related genera based on analysis of four genes from three genomes (adapted from Martinsen *et al.*, 2008).

clade of mammal *Plasmodium* parasites (Fig. 1.4). Though the clade still contains the *Hepatocystis* species, the four-gene phylogeny otherwise supported the monophyly of *Plasmodium*, including *Plasmodium* parasites isolated from mammal, lizard, and birds hosts, and thus refuted paraphyly of *Haemoproteus* with respect to *Plasmodium* as previously suggested. As for the genus *Haemoproteus*, despite the overall morphological similarity of the gametocyte stages, Martinsen *et al.* (2008) found the *Haemoproteus* parasites transmitted by hippoboscid flies (subgenus *Haemoproteus*) to be an independent and monophyletic group distinct from parasites of the subgenus *Parahaemoproteus*, which are transmitted by biting midges, thus prompting the authors to elevate the latter subgenus to genus rank. Such a classification system mirrors the one that was proposed decades earlier by Bennett *et al.* (1965) and lends support for some of the traditional groupings of parasites proposed by parasitologists who were restricted to studying them through the light microscope. The insight of parasitologists into the characters of phylogenetic importance was tremendous given the limitations imposed on them. However, these recent sequence-based studies also suggest that some of the traits used to define parasite genera may not be fixed for these parasite groups, but rather are labile and subject to change over evolutionary time, and that parasite morphology, life history traits, and host taxon alone may not be an adequate basis for assigning membership to a genus. Ameliorating the taxonomy of these parasites will surely require further coupling of traditional and molecular characters to infer phylogeny.

Although some clades in these early systematic studies were very strongly supported and consistent, it has been become clear that additional data, i.e. multiple informative loci from the nuclear genome, are needed to resolve the more unstable or tenuous nodes of the tree. Set to this task, Dávalos and Perkins (2008) assembled a dataset of 104 putatively orthologous genes from available *Plasmodium* genomic data and then constructed phylogenies of the eight taxa using both gene-based analyses and analyses of the total concatenated dataset of 169,584 nucleotides. Most genes and the concatenated datasets converged on the same basic (unrooted) topology; however, this study showed that many of the genes that could be identified as orthologues suffered from substitution saturation and base composition biases between taxa, factors that are well known to result in incorrect topologies. These results forewarn that for future systematic study of malaria parasites, discovery of informative nuclear loci will be a difficult task and that dense taxon and locus sampling will be absolutely paramount.

Why molecules are not enough for taxonomy: a cautionary note

Taxonomy is the practice of identifying, describing, and documenting lineages of organisms, at the level of species and above. Practicing taxonomists have created sets of rules to ensure consistency in this field: the various International Codes of Nomenclature (the International Code of Zoological Nomenclature, the International Code of Botanical Nomenclature, the International Code of Nomenclature of Bacteria, and the International Committee on Taxonomy of Viruses). These practices dictate that type specimens be deposited into an accessible collection whenever new species are described in the literature, so that any future taxonomists may compare these species with any putative new ones. Because specimens of Haemosporidia typically exist as thin blood smears that contain multiple individual organisms, a new concept in type specimens – the hapantotype – was created for parasites, which consists of the suite of specimens necessary to show the various stages in the life cycle (Williams, 1986).

Our approach to inferring relationships between parasite taxa, both at the intra- and interspecific level as well as the intra- and intergeneric levels, is changing. The validity of morphology, host taxon, and life history traits, long deemed as key to elucidating shared ancestry for the malaria parasites, are being further evaluated for their phylogenetic utility. For example, *Hepatocystis* parasites, although nested within the *Plasmodium* clade, have lost the trait of erythrocytic schizogony and are thus unable to reproduce asexually within the red blood cells of their mammalian hosts. *Hepatocystis* parasites have also switched to biting midges as their vector hosts, thus displaying major changes in life history traits and host use from their closest *Plasmodium* parasite relatives. Similarly, over evolutionary time, *P. mexicanum* successfully infected and is now transmitted by a different vector group, the sandflies. We are becoming increasingly aware of the ability of key traits to change over evolutionary time in lineages of parasites and of the parasites' ability to host switch between different vertebrate and vector groups (Martinsen *et al.*, 2008). However, taxon names should be representative of monophyletic lineages. We thus need to consider how we can integrate morphological and molecular data into future systematics works and reconsider how to use the colloquial term 'malaria parasite'.

Importantly, we cannot give up the methods of the past as morphological examination via light microscopy is still necessary for a suite of reasons. While the method of PCR has allowed increased accuracy in the screening of infections from putative host species and populations, including individuals with very light or subpatent infections, and has thus allowed better estimates of the true prevalence of infection, it also poses numerous limitations. By PCR methods alone, we are unable to gauge mixed infections and parasitaemia levels. Studies have demonstrated that with mixed infections, the parasite with the greatest parasitaemia is preferentially amplified over the other parasites and that PCR methods may not even indicate the presence of a mixed infection when it is indeed present (Valkiunas *et al.*, 2006; Szollosi, 2008). Also, it is extremely important to verify that a parasite has indeed undergone asexual reproduction within the vertebrate host, but the sensitivity of PCR can make this problematic. Valkiunas *et al.* (2009) revealed that PCR-based methods can detect sporozoites of unknown fate that are circulating in the bloodstream of the vertebrate host. If an individual is bitten by an infected mosquito or other vector that injects sporozoites into its bloodstream, PCR-based methods are able to detect these sporozoites over the course of days to even weeks while they continue to circulate in the bloodstream, even though that animal is not a true host of the parasite. Thus, PCR-based methods may falsely indicate a definitive infection (i.e. one where the parasite has replicated), misinforming our understanding of the distribution of malaria parasites across host taxa. It is clear that studies of the malaria parasites and their close relatives require both traditional and molecular based methods of investigation, with each method informing the other.

Unique challenges for comparative genomics of *Haemosporidia*

As emphasized in this chapter, the fundamental basis for doing comparative biology is a well-supported phylogenetic tree. In order to construct such a robust topology, one must have both enough informative data (Rokas *et al.*, 2003) and an adequately dense level of taxon sampling in order to alleviate the potential biases that can result (Hedtke *et al.*, 2006). In the case of the malaria parasites, there are several challenges to both of these criteria.

Taxon sampling: old species, new species and cryptic species

The first major challenge to reconstructing the evolutionary tree of Haemosporidia is obtaining fresh samples from which to obtain high-quality DNA. Almost all of the species of *Plasmodium* that have been formally described exist in the form of one or a few stained blood smears, many more than 30 years old and with cover slips on them; most of these are stored in just a few museums or universities worldwide. Although some genetic material can be obtained from these samples, the act of doing so destroys the specimen and thus their availability for future study and, because of the DNA-binding properties of the Giemsa stain, typically only small amplicons or DNA segments are obtainable (Vardo and Schall, 2007). Thus, new material must be collected from the field, from the type location (or as close as possible) and ideally from the same type host species. This can be expensive and time-consuming due to permit regulations and travel costs; moreover, in some cases the habitat in the type locality of the species has been so altered that any available hosts have become rare or endangered.

Despite problems associated with the evaluation of previously described type specimens of *Plasmodium* species, there appears to be no limit to the discovery of new parasite lineages, and studies coupling morphological and molecular data have unveiled a vast cryptic diversity. Even within the most well-sampled host taxon, humans, molecular methods have revealed cryptic diversity in the form of genetically divergent lineages masked under a single morphological species, such as with *P. ovale* (Win *et al.*, 2004). In Malaysia, human infections that depicted morphology typical of the common species *P. malariae* under the light microscope, instead were found to be indistinguishable from the macaque monkey malaria parasite *P. knowlesi* upon molecular phylogenetic analysis, suggestive of plasticity in parasite morphology based on host association (Singh *et al.*, 2004; Cox-Singh *et al.*, 2008). Molecular-based scrutiny of species taxa described by traditional means from lizard and bird host taxa has revealed an even greater amount of cryptic diversity, with single *Plasmodium* species found to consist of multiple divergent parasite lineages (Perkins, 2000; Beadell *et al.*, 2006). Indeed, it has been suggested that the diversity of malaria parasites from birds likely approximates the diversity of birds themselves (approximately 10,000 species; Bensch *et al.*, 2004). This indicates a tremendous underestimation of the true diversity of these parasites, given that only roughly 40 species of *Plasmodium* parasites have been formally described to date from avian hosts worldwide (Valkiunas, 2005).

The study of the malaria parasites of non-human primate species has been similarly limited by lack of access to fresh blood samples from the field and limited application of microscopy-based methods. For example, although *Plasmodium reichenowi* was suggested as the sister taxon to *Plasmodium falciparum* by early molecular phylogenetic studies (Escalante *et al.*, 1995; Qari *et al.*, 1996; Perkins and Schall, 2002), it has been extremely difficult to get samples of *Plasmodium reichenowi* (or any other putative *Plasmodium* lineages) from wild chimpanzees, and until very recently, only a single isolate was available. In the last few years, studies of African apes using opportunistically and often non-invasively collected blood samples from the field and even faecal samples have hinted at a much larger diversity of hominid parasites and have challenged previous hypotheses of the origin of *P. falciparum* (Duval *et al.*, 2009; Rich *et al.*, 2009; Liu *et al.*, 2010; Prugnolle *et al.*, 2010). In a study that looked at tissues from 16 deceased chimpanzees from a single national park in Côte d'Ivoire, phylogenetic analysis confirmed the presence of five species of *Plasmodium*: *P. reichenowi*, *P. vivax*, *P. ovale*, *P. malariae* – like strains, and a fifth lineage that was called '*P. gaboni*' (Kaiser *et al.*, 2010). Liu *et al.* (2010) similarly revealed a great diversity of undocumented *Plasmodium* lineages from chimpanzees and gorillas. The phylogenetic analyses from this study also found that the *Plasmodium* parasites isolated from western gorillas (*Gorilla gorilla*) were nearly identical to *P. falciparum* parasites isolated from humans, and suggested western gorillas as the likely origin of this deadly human malaria species. Another study by Krief *et al.* (2010) found parasites of chimpanzees and bonobos to be closely related to the human malaria parasites *P. falciparum*, *P. malariae*, and *P. vivax*, with bonobos hosting the parasite lineages most similar to *P. falciparum*, leading the authors to propose that *P. falciparum* evolved in bonobos and then later switched into humans. On the contrary, Hughes and Verra (2010), in a thorough analysis of sequence data from multiple genes, provide evidence that supports the hypothesis of co-speciation of *P. falciparum* and *P. reichenowi* with their human and chimpanzee hosts, thus refuting the switching of *Plasmodium* parasites into humans. Most of these studies, however, present only molecular-based data, either from blood or faecal samples, and thus are limited by the lack of information on the morphology and life history traits of the parasites. As previously discussed, PCR does not necessarily demonstrate a true infection, and the provenance of a 'positive' PCR result is even more murky when non-invasive samples are used. Overall, these studies make clear the controversy over the origin of the human malaria parasite *Plasmodium falciparum* and also the diversity of previously undescribed *Plasmodium* lineages yielded by molecular investigation of samples from non-human primate species. Thus, with respect to our understanding of the diversity of *Plasmodium* species from vertebrate hosts worldwide, we have likely been presented with only a small percentage of the true diversity of species taxa.

Further surveying and sampling for these parasites, especially from additional mammal, bird, and lizard taxa will likely yield thousands of new species and further our understanding of the distribution and evolution of *Plasmodium* parasites. However, to reiterate, it remains necessary to continue to follow the taxonomic principles that are currently in place. This requires that any new species that are given Latin names are also formally described, and represented by type specimens deposited into a curated and accessible collection. The description of species was initially a long, thorough process that involved detailing every stage from the parasite's complete life cycle. Over time, particularly for many of the avian and squamate parasites where vectors remain unknown, formal species descriptions began to consist solely of documentation of morphological features and measurements from the blood stages observed within the vertebrate hosts only. Some recent species descriptions have argued for using both morphological and molecular characters (Perkins and Austin, 2009; Valkiūnas *et al.*, 2009, 2010), as their close coupling could facilitate linking different stages of the parasite together when samples are obtained from separate hosts or even from candidate vectors. However, any names that are merely given to sequences, or tips on a phylogenetic tree that are just combinations of sequences (e.g. '*Plasmodium gaboni*') are *nomen nudum* according to ICZN – in other words, meaningless, and henceforth should never be used.

Obtaining new loci

Unfortunately, adding additional loci to molecular systematic analysis of the malaria parasites is no easy task as numerous hindrances hamper this process. Although the availability of complete genomes has facilitated primer development in some groups of organisms, this has not been so readily adapted to the malaria parasites as the majority of sequenced genomes have been from isolates of human malaria parasites and other mammalian *Plasmodium* species. In a screening of the 104 primer pairs from the Dávalos and Perkins (2008) study, only 13 of these resulted in any amplified product from any lizard or bird *Plasmodium* isolates and only five of these have been developed into somewhat reliable markers for systematic studies (Perkins, unpublished). Only via additional genomic resources from parasites from non-mammalian hosts will significant progress towards improvement of the size of the data matrix be made and a more thorough phylogenetic study of *Plasmodium* and closely related genera be possible.

The appeal, and challenge, of comparative genomics of malaria parasites

Because of the diversity of life history characteristics and other traits, comparative phylogenetic analyses of *Plasmodium* species that infect birds and lizards as well as representatives of other genera could be a very powerful advance in parasitology. Now that we know that *Hepatocystis* is a more recently derived lineage within the clade of mammalian malaria parasites, it would be interesting to use genomics to better understand the loss of erythrocytic schizogony in this genus or to look for genes that are linked to the production of its macroscopic liver merozoites. Having genome sequences for species of *Plasmodium* that do not sequester haemozoin may offer insight into the importance of this process and uncover new genes to target in an effort to block this process.

Performing genomics on non-mammalian parasites is a particularly challenging task, however. Most important is the fact that the erythrocytes of saurian and avian hosts are nucleated, making it difficult to isolate pure parasite DNA. In addition, the parasitaemia of these infections can be quite low, so, combined with the much larger genome of their hosts, the ratio of host DNA to parasite DNA in a bulk genomic extraction can be as high as a million to one. Although methods such as caesium chloride gradient purifications can be used (McCutchan *et al.*, 1984), unlike the taxa whose genomes have already been sequenced, the parasites of lizards and almost all birds cannot be cultured either *in vitro* or in abundant model hosts. Most of the taxa are quite highly specific to a species of lizard or bird and often these hosts are small-bodied, thus in order to obtain a sufficient quantity of parasites, hundreds of animals would need to be sacrificed. Recently, techniques to isolate infected avian malaria parasite cells through laser microdissection (Palinauskas *et al.*, 2010) and flow cytometry (Omori *et al.*, 2010) have shown promise for obtaining uncontaminated parasite DNA, and perhaps if they are combined with single-cell genomic methodologies (Heywood *et al.*, 2010), the obstacles may soon become surmountable.

Finally, although it may seem foreign to malaria biologists who mainly work with laboratory cultures or model systems, one of the most important tasks remaining for comparative genomics of *Plasmodium* is basic species discovery and natural history. There is undoubtedly a huge diversity of parasites infecting a variety of vertebrate hosts that await discovery and study of their complete life cycles. It seems the better our tools become for investigating the evolutionary history of the malaria parasites and closely related haemosporidians, the more we discover of their diversity and adaptability. This presents challenges to their systematic study but also avenues of study into their tremendous success and complexity, both ecologically and evolutionarily. While the rationale for the choice of malaria parasite species for the first set of genome projects is clear, only through study of non-mammalian *Plasmodium* and other haemosporidian taxa will we be able to uncover the genetic basis for important traits in these pathogens and understand the complex evolutionary patterns that have occurred throughout the diversification of the group.

References

Ayala, S.C., and Lee, D. (1970). Saurian malaria: development of sporozoites in two species of phlebotomine sandflies. Science *167*, 891–892.

Beadell, J.S., Ishtiaq, F., Covas, R., Melo, M., Warren, B.H., Atkinson, C.T., Bensch, S., Graves, B.R., Jhala, Y., Peirce, M.A., Rahmani, A.R., Fonseca, D.M., and Fleischer, R.C. (2006). Global phylogenetic limits of Hawaii's avian malaria. Proc. R. Soc. Lond. B 273, 2935–2944.

Bennett, G.F., Garnham, P.C., and Fallis, A.M. (1965). On the status of the genera *Leucocytozoon* Ziemann, 1893 and *Haemoproteus* Kruse, 1890 (Haemosporidiida: Leucocytozoidae and Haemoproteidae). Can. J. Zool. 43, 927–932.

Bensch, S., Perez-Tris, J., Waldenstrom, J., and Hellgren, O. (2004). Linkage between nuclear and mitochondrial DNA sequences in avian malaria parasites: multiple cases of cryptic speciation? Evolution *58*, 1617–1621.

Cox, F.E.G. (2010). History of the discovery of the malaria parasites and their vectors. Parasites & Vectors *3*, 5.

Cox-Singh, J., Davis, T.M.E., Lee, K.S., Shamsui, S.S.G., Matusop, A., Ratnam, S., Rahman, H.A., Conway, D.J., and Singh, B. (2008). *Plasmodium knowlesi* malaria in humans is widely distributed and potentially life threatening. CID *46*, 165–171.

Dávalos, L.M., and Perkins, S.L. (2008). Saturation and base composition bias explain phylogenomic conflict in *Plasmodium*. Genomics *91*, 433–442.

Duval, L., Nerrienet, E., Rousset, D., Mba, S.A.S., Houze, S., Fourment, M., Le Bras, J., Robert, V., and Ariey, F. (2009). Chimpanzee malaria parasites related to *Plasmodium* ovale in Africa. PLoS One *4*, e5520.

Escalante, A.A., and Ayala, F.J. (1994). Phylogeny of the malarial genus *Plasmodium*, derived from rRNA gene sequences. Proc. Natl. Acad. Sci., USA, *91*, 11373–11377.

Escalante, A.A., Freeland, D.E., Collins, W.E., and Lal, A.A. (1998). The evolution of primate malaria parasites based on the gene encoding cytochrome *b* from the linear mitochondrial genome. Proc. Natl. Acad. Sci., USA *95*, 8124–8129.

Felsenstein, J. (1985). Phylogenies and the comparative method. Am. Nat. *125*, 1–15.

Garnham, P.C.C. (1948). The developmental cycle of *Hepatocystis* (*Plasmodium*) *kopki* in the monkey host. Trans. R. Soc. Trop. Med. Hyg. *41*, 601–616.

Garnham, P.C.C. (1958). A malaria parasite of the hippopotamus. J. Euk. Microbiol. *5*, 149–151.

Garnham, P.C.C. (1966). Malaria Parasites and other Haemosporidia (Oxford: Blackwell Scientific Publications).

Harvey, P.H., and Pagel, M.D. (1991). The Comparative Method in Evolutionary Biology (Oxford: Oxford University Press).

Hedtke, S.M., Townsend, T.M., and Hillis, D.M. (2006). Resolution of phylogenetic conflict in large datasets by increased taxon sampling. Syst. Biol. *55*, 522–529.

Heywood, J.L., Sieracki, M.E., Bellows, W., Poulton, N.J., and Stepanauskas, R. (2011). Capturing diversity of marine heterotrophic protists: one cell at a time. ISME J. *5*, 674–684.

Hughes, A.L., and Verra, F. (2010). Malaria parasites sequences from chimpanzee support the co-speciation hypothesis for the origin of virulent human malaria (*Plasmodium falciparum*). Mol. Phyl. Evol. *57*, 135–143.

Jani, D., Nagarkatti, R., Beatty, W., Angel, R., Slebodnick, C., Andersen, J., Kumar, S., and Rathore, D. (2008). HDP – a novel haem detoxification protein from the malaria parasite. PLoS Pathog. *4*, e1000053.

Kaiserm M., Löwa, A., Ulrich, M., Ellerbrok, H., Goffe, A.S., Zommers, Z., Couacy-Hymann, E., Babweteera, F., Zuberbühler, K., Metzger, S., Geidel, S., Boesch, C., Gillespie, T.R., and Leendertz, F.H. (2010). Wild chimpanzees infected with 5 *Plasmodium* species. Emerg. Infect. Dis. *16*, 1956–1959.

Krief, S., Escalante, A.A., Pacheco, M.A., Mugisha, L., André, C., Halbwax, M., Fischer, A., Krief, J.-M., Kasenene, J.M., Crandfield, M., Cornejo, O.E., Chavatee, J.-M., Lin, C., Letourneur, F., Grüner, C., McCutchan, T.F., Rénia, L., and Snounou, G. (2010). On the diversity of malaria parasites in African apes and the origin of *Plasmodium falciparum* from bonobos. PLoS Pathog. *6*, e1000765.

Lainson, R., and Shaw, J.J. (1969). A new haemosporidian if lizards, *Saurocytozoon tupinambi* gen. nov. sp. nov. in *Tupinambis nigropunctatus* (Teiidae). Parasitology *59*, 159–162.

Laveran, A. (1899). Les hematozoaires endoglobulaires (Haemocytozoa). Cinquantenaire Soc. Biol. 124–133.

Levine N.D. (1988). The Protozoan Phylum Apicomplexa. (Boca Raton, FL: CRC Press).

Liu, W., Li, Y., Learn, G.H., Rudicell, R.S., Robertson, J.D., Keele, B.F., Ndjango, J.B.N., Sanz, C.M., Morgan, D.B., Locatelli, S., Gonder, M.K., Kranzusch, P.J., Walsh, P.D., Delaporte, E., Mpoudi-Ngole, E., Georgiev, A.V., Muller, M.N., Shaw, G.M., and MacCallum, W.G. (1897). On the flagellated form of the malarial parasite. Lancet *2*, 1240–1241.

Peeters, M., Sharp, P.M., Rayner, J.C., and Hahn, B.H. (2010). Origin of the human malaria parasite *Plasmodium falciparum* in gorillas. Nature *467*, 420–427.

Martinsen, E.S., Perkins, S.L., and Schall, J.J. (2008). A three-genome phylogeny of malaria parasites (*Plasmodium* and closely related genera): Evolution of life-history traits and host switches. Mol. Phylogenet. Evol. *47*, 261–273.

Mattingly, P.F. (1983). The paleogeography of mosquito-borne disease. Biol. J. Linn. Soc *19*, 185–210.

McCutchan, T.F., Dame, J.B., Miller, L.H., and Barnwell, J. (1984). Evolutionary relatedness of *Plasmodium* species as determined by the structure of DNA. Science *225*, 808–811.

Novy, F.G., and Macneal, W.J. (1904). Trypanosomes and bird malaria. Am. Med. *8*, 932–934.

Omori, S., Sato, Y., Toda, H., Sasaki, T., Isobe, T., Nakanishi, T., Murata, K., and Yukawa, M. (2010). Use of flow cytometry to separate *Leucocytozoon caulleryi* gametocytes from avian blood. Parasitology *137*, 1899–1903.

Palinauskas, V., Dolnik, O.V., Valkiūnas, G., and Bensch, B. (2010). Laser microdissection microscopy and single cell PCR of avian hemosporidians. J. Parasitol. *96*, 420–424.

Perez-Tris, J., Hasselquist, D., Hellgren, O., Krizanauskiene, A., Waldenstrom, J., and Bensch, S. (2005). What are malaria parasites? Trends Parasitol. *21*, 209–211.

Perkins, S.L. (2000). Species concepts and malaria parasites: detecting a cryptic species of *Plasmodium*. Proc. R. Soc. B *267*, 2345–2350.

Perkins, S.L., and Austin, C.C. (2009). Four new species of *Plasmodium* from New Guinea lizards: Integrating morphology and molecules. J. Parasitol. *95*, 424–433.

Perkins, S.L., and Schall, J.J. (2002). A molecular phylogeny of malarial parasites recovered from cytochrome *b* gene sequences. J. Parasitol. *88*, 972–978.

Pollock, D.D., Zwickl, D.J., McGuire, J.A., and Hillis, D.M. (2002). Increased taxon sampling is advantageous for phylogenetic inference. Syst. Biol. *51*, 664–671.

Prugnolle, F., Durand, P., Neel, C., Ollomo, B., Ayala, F.J., Arnathau, C., Etienne, L., Mpoudi-Ngole, E., Nkoghe, D., Leroy, E., Delaporte, E., Peeters, M., and Renaud, F. (2010). African great apes are natural hosts of multiple related malaria species, including *Plasmodium falciparum*. Proc. Natl. Acad. Sci., USA *107*, 1458–1463.

Qari, S.H., Shi, Y.P., Pieniazek, N J., Collins, W E., and Lal, A.A. (1996). Phylogenetic relationship among the malaria parasites based on small subunit rRNA gene sequences: monophyletic nature of the human malaria parasite, *Plasmodium falciparum*. Mol. Phylogenet. Evol. *6*, 157–165.

Rich, S.M., Leendertz, F.H., Xu, G., LeBreton, M., Djoko, C., Aminake, M.N., Tkang, E.E., Diffo, J.L.D., Pike, B.L., Rosenthal, B.M., Formenty, P., Boesch, C., Ayala, F.J., and Wolfe, N.D. (2009). The origin of malignant malaria. Proc. Natl. Acad. Sci. U.S.A. *106*, 14902–14907.

Rogers, M.J., McConkey, G.A., Li, J., and McCutchan, T.F. (1995). The ribosomal DNA loci in *Plasmodium falciparum* accumulate mutations independently. J. Mol. Biol. *254*, 881–891.

Rokas A., Williams, B.L., King, N., and Carroll, S.B. (2003). Genome-scale approaches to resolving incongruence in molecular phylogenies. Nature *425*, 798–804.

Singh, B., Sung, L.K., Matusop, A., Radhakrishnan, A., Shamsul, S.S.G., Cox-Singh, J., Thomas, A., and Conway, D.J. (2004). A large focus of naturally acquired *Plasmodium knowlesi* infections in human beings. Lancet *363*, 1017–1024.

Szöllsi, E., Hellgren, O., and Hasselquist, D. (2008). A cautionary note on the use of nested PCR for parasite screening: An example from avian blood parasites. J. Parasitol. *94*, 562–564.

Telford, S.R. Jr. (1979). A taxonomic reconsideration of some *Plasmodium* species from iguanid lizards. Ann. Parasitol. Hum. Comp. *54*, 129–144.

Telford, S.R. Jr. (1988). A contribution to the systematics of the reptilian malaria parasites, Family Plasmodiidae (Apicomplexa: Haemospororina). Bull. Florida State Mus., Biol. Sci. *34*, 65–96.

Valkiūnas, G. (2005). Avian Malaria Parasites and Other Haemosporidia. (Boca Raton, FL: CRC Press).

Valkiūnas, G., Anwar, A.M., Atkinson, C.T., Greiner, E.C., Paperna, I., and Peirce, M.A. (2005). What distinguishes malaria parasites from other pigmented haemosporidians? Trends Parasitol. *21*, 357–358.

Valkiūnas, G., Bensch, S., Iezhova, T.A., Križanauskienė, A., Hellgren, O., and Bolshakov, C. (2006). Nested cytochrome *b* polymerase chain reaction diagnostics underestimate mixed infections of avian blood haemosporidian parasites: Microscopy is still essential. J. Parasitol. 92, 418–422.

Valkiūnas, G., Iezhova, T.A., Loiseau, C., and Sehgal, R.N.M. (2009). Nested cytochrome *b* polymerase chain reaction diagnostics detect sporozoites of hemosporidian parasites in peripheral blood of naturally infected birds. J. Parasitol., *95*, 1512–1515.

Valkiūnas, G., Santiago-Alarcon, D., Levin, I., Iezhova, T.A., and Parker, P.G. (2010). A new *Haemoproteus* species (Haemosporidia: Haemoproteidae) from the endemic Galapagos dove *Zenaida galapagoensis*, with remarks on the parasite distribution, vectors, and molecular diagnostics. J. Parasitol. *96*, 783–792.

Vardo, A.M., and Schall, J.J. (2007). Clonal diversity of a lizard malaria parasite, *Plasmodium mexicanum*, in its vertebrate host, the western fence lizard: role of variation in transmission intensity over time and space. Mol. Ecol. *16*, 2712–2720.

Waters, A.P., Higgins, D.G., and McCutchan, T.F. (1991). *Plasmodium falciparum* appears to have arisen as a result of lateral transfer between avian and human hosts. Proc. Natl. Acad. Sci. U.S.A. *88*, 3140–3144.

Williams, R.B. (1986). Hapantotypes: a possible solution to some problems of parasite nomenclature. Parasitol. Today 2, 314–316.

Win, T.T., Jalloh, A., Tanturlar, I.S., Tsuboi, T., Ferreira, M.U., Kimura, M., and Kawamoto, F. (2004). Molecular analysis of *Plasmodium ovale* variants. Emerg. Infect. Dis. 7, 1235–1240.

Zwickl, D.J., and Hillis, D.M. (2002). Increased taxon sampling greatly reduces phylogenetic error. Syst. Biol. *51*, 588–598.

The Apicomplexan Genomic Landscape: The Evolutionary Context of *Plasmodium*

2

Jeremy D. DeBarry, Segun Fatumo and Jessica C. Kissinger

Abstract

The apicomplexan genome is quite different from the typical eukaryotic genome that graces our textbooks and the majority of literature on the topic. It is small (8.5–63 Mb), compact, nearly devoid of transposable elements, and lacking any significant synteny outside of genus- and in some cases, family-level classifications. The nuclear genome has experienced significant gene loss, a characteristic of parasitic organisms. Numerous genes have entered the nuclear genome via *de novo* creation (through recombination, accumulated mutation or gene duplication and subsequent divergence), intracellular gene transfer from organellar and algal endosymbiont genomes, and lateral gene transfer from bacteria and elsewhere. Nuclear genome data are currently available for 18 species within eight genera, namely *Cryptosporidium, Eimeria, Sarcocystis, Neospora, Toxoplasma, Babesia, Theileria* and *Plasmodium*. Analyses have revealed dynamic genomes that share a remarkably small percentage of 'core' genes constituting 11–23% of total gene content, plus large repertoires of lineage- and species-specific genes. Chromosome number and the organization of genes and other genomic features along the chromosome vary across the phylum. *Plasmodium*, the causative agent of malaria, is typical in that genomes from the same genus are highly syntenic, but it differs in that subtelomeric regions host the majority of lineage- and species-specific genes. Tools to perform comparative analyses within *Plasmodium* and across the Apicomplexa are available at EuPathDB.org and other sites.

Introduction

The adage, 'Nothing in biology makes sense except in the light of evolution' (Dobzhanski, 1973) holds as true in this age of genomics as it did when it was first posited. When we compare a genome and its gene content to the genome of a sister taxon we learn the extent to which genes, biological processes, and genomic features are conserved. The power of this comparative approach allows us to ask about genes and biological processes that are shared by all apicomplexan parasites and those that appear to be unique to particular lineages. We can ask questions about the genes and associated processes, such as purine biosynthesis, that have been lost as part of the evolution towards parasitism (Hassan and Coombs, 1988) and we can ask about the origins of new genes that have appeared in the nuclear genome. Were they formed *de novo* or were they acquired from endosymbionts or other sources?

The sheer number of molecular and genomic data that are available from a large number of species make comparative approaches not only possible, but fruitful. Investigations, both comparative and discrete, in related species can have profound impacts on our understanding of the biology of an organism. This is of particular relevance in the study of pathogens. An understanding of host range, tissue preference, life cycle, invasion, evasion of the host's defences, and niche specialization are fundamentally important in our efforts to combat disease. It can be tempting to focus on a single genus or species and remain unaware of important advances in related organisms that can facilitate or enhance our understanding of our favourite species, or place our organismal knowledge in context. Organisms do not evolve in a 'vacuum'; they are changing relative to the species with which they last shared a common ancestor. These changes reflect both stochastic processes and adaptive strategies necessary to support their sometimes very different lifestyles.

As the genomic similarities and differences between organisms, in our case, pathogens, are revealed, they can be used in a variety of ways. Understanding the evolutionary context and history of any biological phenomenon improves our ability to explain and investigate the biology of that phenomenon. In particular, it is beneficial to ascertain why certain pathways or genes are present, or absent, and how genomic landscapes have been shaped. Data are now available not only for the deadliest human malaria parasite *Plasmodium falciparum*, but for an ever-increasing number of lineages within the apicomplexan phylum and free-living sister groups within the Alveolata.

The phylum Apicomplexa contains many pathogens of medical and veterinary importance in addition to *Plasmodium*, including *Cryptosporidium* (now shown to be more closely related to gregarines than coccidians (Leander *et al.*, 2003; Templeton *et al.*, 2009), the primarily Coccidia, *Eimeria*, *Toxoplasma* and *Neospora*, and the tick-borne cattle parasites of the Piroplasmida, *Theileria* and *Babesia*. Owing mainly to their medical and veterinary importance, genome sequences have been generated for many members of the phylum. We now have one or more completed genome sequences for the 12 apicomplexan species shown in Fig. 2.1a, and either complete or partial data for *Cryptosporidium hominis*, *Plasmodium yoelii*, *P. reichenowi*, *P. ovale*, *Sarcocystis neurona* (Howe *et al.*, unpublished), *Neospora caninum*, *Eimeria tenella*, and *E. maxima* (Abrahamsen *et al.*, 2004; Brayton *et al.*, 2007; Carlton *et al.*, 2008; Carlton *et al.*, 2002; *Cryptosporidium muris* Genome Project; Gardner *et al.*, 2005; Gardner *et al.*, 2002; Pain *et al.*, 2008; Pain *et al.*, 2005; Reid *et al.*, 2012; *Toxoplasma gondii* Genome Project; Xu *et al.*, 2004). Apicomplexan genome sizes are relatively small compared to most model eukaryotes, ranging from ~8.5 to ~63 Mb (Fig. 2.1a). The largest known genome size (sequencing and annotation in progress) belongs to *Sarcocystis neurona* and is at least ~122 Mb (Howe *et al.*, unpublished). In addition to small genome sizes, the phylum Apicomplexa is also characterized by massive gene loss within both nuclear and organellar genomes. Gene totals range from ~3700 in *B. bovis* to ~8000 in *T. gondii* (Fig. 2.1a) (Abrahamsen *et al.*, 2004; Brayton *et al.*, 2007; Carlton *et al.*, 2002, 2008; Gardner *et al.*, 2002, 2005). One intriguing example of the effects of gene loss is found in *Cryptosporidium parvum*, where the genes necessary for *de novo* nucleotide synthesis, both purine and pyrimidine, have been lost and nucleotide salvage pathways have been acquired (Striepen *et al.*, 2004).

In addition to the collective differences between apicomplexans and other eukaryotes, there are also many differences between *Plasmodium* species and the other major lineages within the Apicomplexa. These differences extend far beyond broad organismal differences such as host range and pathogenicity and into the genomes themselves. For example, not all apicomplexans have 14 chromosomes and nearly 50% of the genes found in *Plasmodium* species have no orthologue in other apicomplexan genomes (Fig. 2.1) (Kuo and Kissinger, 2008). It is also interesting to note that while all examined apicomplexan species contain genes involved in immune evasion, most do not contain large multi-gene families located in subtelomeric regions as has been observed in most (Carlton *et al.*, 2008; Carlton *et al.*, 2002; Gardner *et al.*, 2002) but not all (Pain *et al.*, 2008) *Plasmodium* species. Discoveries of variations in chromosome number, chromosome organization and gene repertoire have all been possible because of comparative studies. These comparisons provide insight into the biology of *Plasmodium* species (and other apicomplexans) at multiple levels.

Evolutionary relationships

There are ~5000 species in the phylum Apicomplexa and all but one of them are believed to be parasitic (Perkins *et al.*, 2000). The non-parasitic organism, *Nephromyces*, is a symbiont of molgulid ascidian tunicates (Saffo *et al.*, 2010). Parasitism in the phylum is a derived state as the

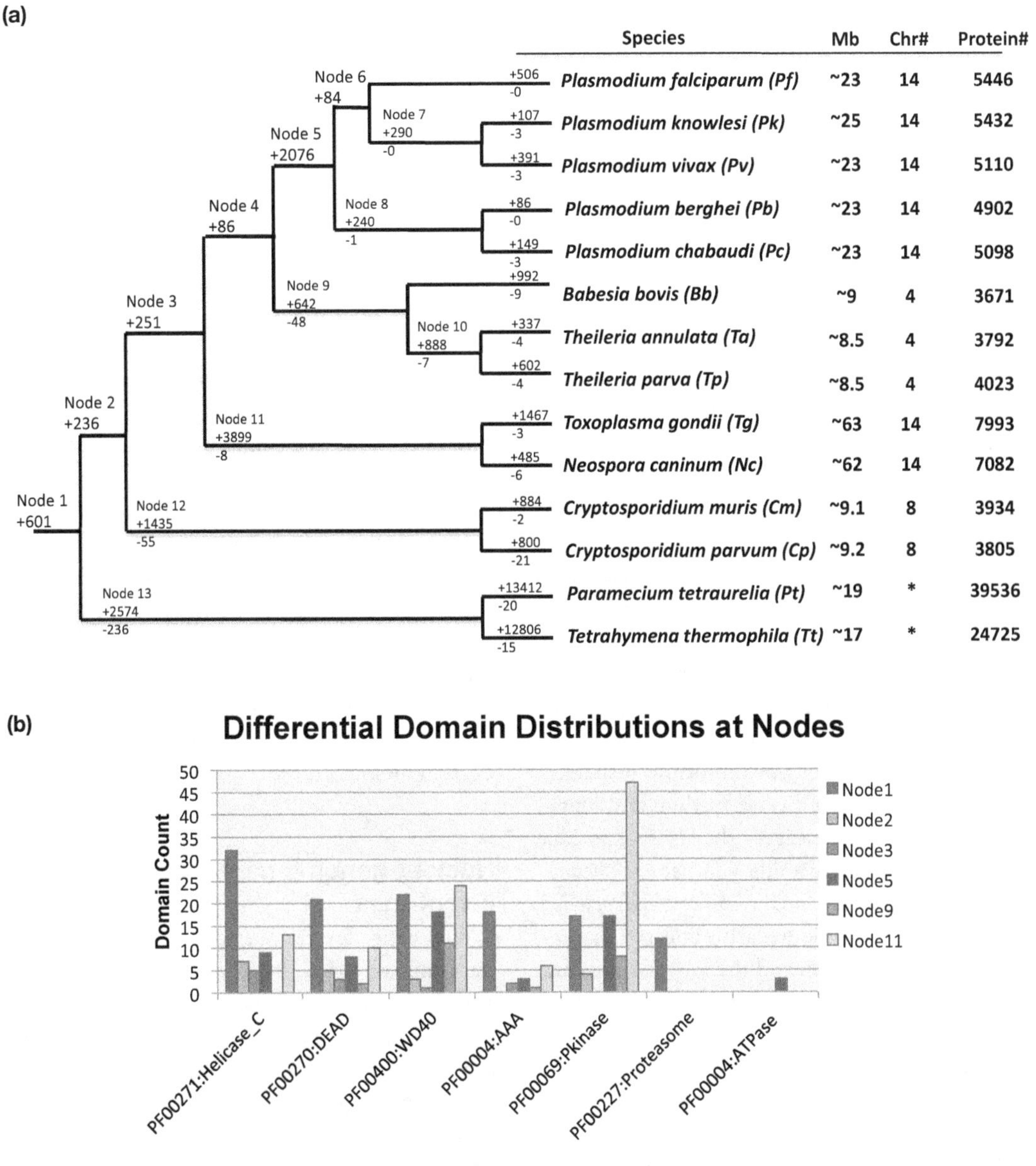

Figure 2.1 Genome characteristics. (a) Phylogenetic distribution of orthologous gene clusters. The cladogram shows the relationships of 12 sequenced apicomplexans. The 2 free-living ciliates *Tetrahymena thermophila* and *Paramecium tetraurelia* are included as outgroups. The genus species abbreviation is shown for each species. Approximate genome sizes are shown in Mb. Sources for the sizes of apicomplexan genome sizes are located in the text. Protein# indicates the number annotated protein-coding genes for each species. * = ~ 225 for *T. thermophila* and > 50 for *P. tetraurelia* macronuclear genomes (Aury *et al.*, 2006; Eisen *et al.*, 2006). The numbers above the branches and proceeded by a '+' sign indicate the number of orthologous groups of protein encoding genes uniquely present in all lineages descending from that branch. The numbers below the branches and proceeded by a '–' indicate the number of orthologous groups of protein encoding genes uniquely absent in all lineages descending from that branch. The node designations refer to the degree of lineage specificity; genes at node 1 are shared by all species on the tree and genes at node 5 are unique to *Plasmodium* species. Data are based on Kuo and Kissinger (2008). (b) Distribution of a sample subset of protein domains by node as determined by the program DAVID (Huang *et al.*, 2008). Note: DAVID has not been optimized for *Plasmodium*. Reference domains are given relative to *P. falciparum*. Node numbers are as in (a). Pfam domain names are: PF00271:Helicase_conserved C-terminal domain; PF00270:DEAD/DEAH box helicase; PF00400:WD40 domain; PF00004:AAA domain; PF00069:Protein kinase domain; PF00227:Proteasome; PF00004:ATPase family associated with various cellular activities.

sister groups (ciliates and dinoflagellates), are primarily free living. Apicomplexans, with the current exception of *Cryptosporidium*, contain a non-photosynthetic plastid of algal origin, called the apicoplast. The plastid arose from a secondary endosymbiosis involving a red alga and another eukaryotic cell. The red (versus green) origin of the apicoplast was originally unclear (Fast *et al.*, 2001; Kohler *et al.*, 1997; Lau *et al.*, 2009), however comparative sequence analysis of the plastid genome of a related, photosynthetic organism, *Chromera velia*, provides the strongest evidence yet of a red algal origin (Janouskovec *et al.*, 2010; Moore *et al.*, 2008). In this case, comparative approaches have helped place apicomplexans and aspects of their biology in a deep evolutionary context.

The last common apicomplexan ancestor is estimated to have existed ~350–824 million years ago (mya) (Escalante and Ayala, 1995). Recent estimates have narrowed the time range to ~420 mya (Berney and Pawlowski, 2006; Okamoto and McFadden, 2008). The 12 apicomplexan species shown in Fig. 2.1a are not uniformly dispersed throughout the highly diverse phylum and represent only four major lineages: The Haemosporidia (*Plasmodium*), Piroplasmida (*Theileria/Babesia*), Coccidia (*Toxoplasma/Neospora/Sarcocystis/Eimeria*) and a probable gregarine lineage (*Cryptosporidium*). *Plasmodium* and the Piroplasmida are estimated to have diverged ~300 mya (Silva *et al.*, 2011).

Apicomplexans have complex life cycles that involve either one or two hosts and several general stages: an invasive haploid sporozoite, an intracellular 'zoite', sexual gametocytes, diploid ookinetes and (depending on the lineage) sometimes oocysts containing haploid meiotic progeny for transmission in the environment. However, there are remarkable differences in the biology of these stages in terms of hosts, mechanisms of transmission and cell types that are infected (reviewed in Wasmuth *et al.*, 2009; see also Perkins *et al.*, 2000 and Roos, 2005) and methods of replication (Striepen *et al.*, 2007). *Plasmodium* and *Piroplasm* species rely on a two-host cycle where the definitive hosts are insect vectors, namely mosquitoes and ticks, respectively. The vertebrate host cell preferences are varied by species; however, human-infecting *Plasmodium* species infect liver and subsequently red blood cells, while *Theileria* infects red and white blood cells and *Babesia* infects red blood cells. The coccidians and *Cryptosporidium* species shown in Fig. 2.1a do not utilize insect vectors and rely instead upon ingested oocysts for infection. Infection occurs in many cell types for the representative Coccidian *T. gondii*, but is limited to gut cells for *Cryptosporidium* species. Each lineage also has specific methods for host interaction and immune evasion. All of the limitations and abilities of these organisms are influenced and determined by the repertoire of genes found in each of their respective genomes. The challenge for those attempting to understand these processes is, can we decipher the underlying mechanisms and complexities of these systems? Comparative evolutionary approaches strike at the heart of this issue, helping us to understand how these organisms arrived at their present state and offer the opportunity to leverage this information and associate it with their current biology.

Nuclear genome gene content inventories

Which genes are unique to a species, shared by the genus or shared by all apicomplexans? There are a variety of methods available to identify groups of genes that share a common evolutionary ancestor (compared in Chen *et al.*, 2007). Orthologous genes arise following speciation events, such that a given gene persists in each descendant species. They are 'copies' of genes that share a common ancestor but are now present in different species. The calculation of orthologous gene groups, or 'clusters', is a powerful tool for investigating the gain and loss of genes within a group of species, such as the apicomplexans. We can ask which genes are shared by all species? Which are species-specific? And, which have either been gained, or lost, in a given lineage? For example, which *Plasmodium* genes have orthologues in all other species of Apicomplexa? As it turns out, the answer is surprisingly few. Only 837 orthologous gene clusters contain genes that are present in all sequenced apicomplexan genomes (independently calculated with multiple groups of organisms in DeBarry and Kissinger, 2011; Kuo and Kissinger, 2008). Depending on the total number of protein-encoding genes, this is only 11–23% of the total gene repertoires of the apicomplexan species shown in Fig. 2.1a. These genes represent 601 ancestral orthologous gene groups shared by the examined alveolates (ciliates and apicomplexans) (node 1 in Fig. 2.1a) and 236 orthologous gene groups shared by all apicomplexans but lacking in the ciliates (node 2 in Fig. 2.1a). The remainder of the genes in these species are lineage- or species-specific and some represent adaptations particular to that lineage or species.

Each node in Fig. 2.1a has been evaluated to determine the number of orthologous groups of genes that are shared by all descendant species from that node to the ends of the cladogram, as indicated by a '+' sign. Numbers proceeded by a '–' sign at interior nodes denote orthologous groups that are shared by all species and groups on the cladogram except for those species that follow the number. These numbers are indicative of the 'ebb, flow and retention' of genes in apicomplexan lineages. There are also numerous more complicated patterns of genes that have been lost or are shared, but these are not shown (Kuo and Kissinger, 2008). The 837 orthologous groups of genes shared by all apicomplexans (601 + 236) can be thought of as a 'core set' of genes that are present and likely required, for all species. Not surprisingly, these genes are primarily involved in fundamental processes like transcription and translation (Wasmuth *et al.*, 2009). However, included in this set of genes will be genes that are necessary for all apicomplexans, an 'inner' core set of genes that are necessary for an apicomplexan to 'be' an apicomplexan. As one moves out towards the tips of the cladogram, the groups of genes that have been differentially retained, generated or acquired by each lineage, or lost from other lineages are indicated. In some ways, the numbers of gene groups can be viewed as an indicator of particular biological factors. For instance, *T. gondii* has the largest number of species-specific groups (Fig. 2.1a, +1467) and also has the largest host range most likely facilitated by the relative abundance of metabolic pathways and processes compared to the other species in the figure (Kim and Weiss, 2004; Perkins *et al.*, 2000). Likewise, several *Plasmodium* species have large numbers of species-specific genes, the majority of which are involved in antigenic variation (Carlton *et al.*, 2002, 2008; Gardner *et al.*, 2002; Pain *et al.*, 2008). While others utilize shared genes in life-cycle or lineage-specific ways, for example sugar energy sources (trehalose, mannitol) located in *Cryptosporidium* and *Eimeria* oocysts, respectively (Schmatz, 1997; Xu *et al.*, 2004) or families of proteins, some conserved and some unique, which are involved in invasion of many different tissues by different apicomplexan parasites (Carruthers and Tomley, 2008; Dowse and Soldati, 2005).

Moving past shared genes, we can then begin to ask questions about the origins of lineage- and species-specific genes. Which genes does *Plasmodium* (or any apicomplexan lineage) not share with other species? Which genes are unique to a species or lineage? In addition to the core set of genes shared by all apicomplexans, there is also a repertoire of genes that are specific to each lineage or species. Within *Plasmodium*, only ~ 85% of genes are orthologous (Kooij *et al.*, 2005). The remainder are specific to *Plasmodium* species and represent innovations and changes that have arisen during the evolution and adaptation of these species to their environments and lifestyles. Lineage- or species-specific genes and multi-copy genes have been shown to be integral to host species interactions (Abrahamsen *et al.*, 2004; Brayton *et al.*, 2007; Carlton *et al.*, 2002, 2008; Gardner *et al.*, 2002, 2005; Kooij *et al.*, 2005; Kuo and Kissinger, 2008; Pain *et al.*, 2005, 2008; Weir *et al.*, 2009; Xu *et al.*, 2004). So where do these lineage- and species-specific or multi-copy genes come from? Many arise via lateral transfer or *de novo* creation (see below). However, others represent genes that have been lost from all other lineages or species except the lineage or species in which they remain. At first, this argument does not seem to be the most parsimonious, but given the extensive amount of gene loss in this phylum it may be the simplest, and most likely, explanation. This argument would imply that the apicomplexan ancestor contained a larger number of genes than is currently present in extant species. Other genes have been differentially acquired or retained from organellar genomes and some have been acquired via horizontal gene transfer from other species. The remainder must have arisen *de novo* via divergence from an ancestral gene, gene duplication and subsequent divergence, recombination or accumulated mutation.

Apicomplexan nuclear genomes have been the recipients of many laterally and intracellularly transferred genes. As a result of their endosymbiotic history, apicomplexans have received genes from the algal nuclear genome, the algal plastid genome (apicoplast) and the mitochondrial genome (Arenas *et al.*, 2010; Dzierszinski *et al.*, 1999; Huang and Kissinger, 2006; Huang *et al.*, 2004a,b; Martens *et al.*, 2008; Nagamune and Sibley, 2006; Striepen *et al.*, 2004; Whitaker *et al.*, 2009; Zhu and Keithly, 2002). The apicomplexan nuclear genome has been greatly affected and its 'landscape' shaped by these transfer events. Genes can also be acquired horizontally from other species. *C. parvum* is the most extensively studied apicomplexan species thus far in terms of horizontal transfer events. There is evidence for at least 31 transfers from other species including seven genes of plastid/endosymbiont origin and 24 of prokaryotic origin (Huang *et al.*, 2004a; Templeton *et al.*, 2004). However, several notable transfers have also been documented in other species. For example, the gene encoding fructose-1,6-bisphosphate in *Eimeria tenella* (a component of the mannitol cycle that is the primary energy source for part of the oocyst life cycle) is of bacterial origin (Huang *et al.*, 2004b; Schmatz, 1997). *Toxoplasma gondii* has many calcium-dependent protein kinases with a diverse phylogenetic distribution including deep relationships with plants and animals (Nagamune and Sibley, 2006). It is possible that genes arising from transfer events may represent potential drug targets since they are unlikely to be present in the host, and may represent essential functions. For example, inhibitors of inosine monophosphate dehydrogenase (a critical component in the pathway for guanine biosynthesis that was acquired from bacteria in *C. parvum*) (Striepen *et al.*, 2002), have recently shown promise as anti-cryptosporidial agents (Sharling *et al.*, 2010; Sullivan *et al.*, 2005; Umejiego *et al.*, 2008).

In addition to acquiring new genes via transfer events, the *de novo* creation of genes also plays an important role. The formation of completely novel genes can occur via recombination and accumulated mutation or more likely, duplication and diversification. The *de novo* generation of genes has played an important role in the virulence of apicomplexans, with species-specific antigen variation genes found in multiple copies, particularly in *Plasmodium* and *Piroplasm* species (al-Khedery *et al.*, 1999; Carlton *et al.*, 2002; Gardner *et al.*, 2002, 2005; Pain *et al.*, 2005). Interestingly, there are species- and lineage-specific patterns of chromosomal distribution associated with these families in the Apicomplexa (Kuo and Kissinger, 2008; DeBarry and Kissinger, 2011).

Significant insight into *Plasmodium* biology has been gained, in particular, from the discovery of the apicoplast organelle and genome (Williamson *et al.*, 1994; Wilson *et al.*, 1996, 1991). Insight into the origin and function of this organelle and how proteins were targeted in the Apicomplexa was gained via comparative studies of plastid-encoded genes, nuclear-encoded plastid targeted genes and comparisons with the plant and cyanobacterial orthologues (Delwiche, 1999; Foth and McFadden, 2003; Foth *et al.*, 2003; Harb *et al.*, 2004; He *et al.*, 2001; McFadden, 1999; Roos *et al.*, 1999; Waller *et al.*, 1998, 2000). Given its plant origin, the likelihood that therapeutic targets could be identified was strong (McFadden and Roos, 1999; Ryall *et al.*, 2003; Soldati, 1999). Studies of plastid metabolism and potential drug targets were undertaken in both *Toxoplasma* and *Plasmodium.* Ciprofloxacin was shown to inhibit plastid replication in *Toxoplasma* (Fichera and Roos, 1997). The non-mevalonate isoprenoid biosynthesis pathway was identified as a potential target (Jomaa *et al.*, 1999; Wiesner and Jomaa, 2007; Wiesner *et al.*, 2008) in *Plasmodium* as was inhibition of FabI with triclosan (McLeod *et al.*, 2001).

One indicator of evolutionary innovation with respect to lineage- or species-specific genes is the distribution of recognizable protein domains throughout the phylum. There are several genus- and species-level domain innovations, e.g. the rifins (al-Khedery *et al.*, 1999; Kyes *et al.*, 1999; Peterson *et al.*, 1995), which are specific to *P. falciparum*, and there are phylum-wide distributions such as the thrombospondin repeat proteins (TRAP and mTRAP) that are involved in parasite adhesion and gliding motility (Baum *et al.*, 2006; Gardner *et al.*, 2005; Harper *et al.*, 2004; Strong and Nelson, 2000; Templeton and Kaslow, 1997) and the micronemal associated proteins, MICs, which are involved in invasion (Carruthers and Tomley, 2008). There are lineage-specific gene family and domain amplifications such as the apetala 2 (AP2) family of transcription factors (Balaji *et al.*, 2005) and novel arrangements of pre-existing protein domains (Templeton *et al.*, 2004; Wasmuth *et al.*, 2009). A preliminary analysis of the distribution of protein domains present in the Apicomplexa at particular nodes using the DAVID software system (Huang *et al.*, 2008) reveals the variable distribution of major domains throughout the phylum. For example, there is a large increase in the number of kinase domains in the Coccidia (Fig. 2.1b). Additional comparative insight has come from studies of motility and diversification of TRAP for gliding motility and merozoite TRAP (mTRAP) for non-gliding motility (Baum *et al.*, 2006). In some cases, orthologous proteins exist but the timing of their expression may vary depending upon the genus of parasite, for example, the membrane occupation and recognition nexus (MORN) domain containing protein, MORN1 is located in the inner membrane complex (IMC) (Ferguson *et al.*, 2008; Gubbels *et al.*, 2006). However, its expression during cell-division depends on the diverse, yet related ways apicomplexan parasites divide and multiply (Striepen *et al.*, 2007). Perhaps the most biologically significant distribution of protein domains is found in the arena of gene regulation. Most common eukaryotic transcription factor proteins are either absent or so divergent as to be undetectable in the Apicomplexa. Instead, the apetala 2 (AP2) family of transcription factors has undergone a lineage-specific expansion within the Apicomplexa leading to large numbers of genes and multiple AP2 domains within these genes (Balaji *et al.*, 2005). The AP2 family of transcription factors have been shown to be expressed and putative binding sites have been found upstream of genes involved in many important developmental pathways and processes ranging from the cell cycle to expression of secreted proteins and their organelles (Behnke *et al.*, 2010; Llinas *et al.*, 2006). Interestingly, orthologous AP2 DNA-binding domains can be identified and their binding specificities are conserved across genera. However, the genes located downstream of these conserved regulatory regions can change (Painter *et al.*, 2011).

The comparative approach provides a glimpse into the pathways and processes that evolution has shown to be important for the parasite because they have been retained or significantly amplified. While existing sequence data have proven highly illuminating, our resolving power varies tremendously within the existing data set. Within the Coccidia, the immune mapped protein 1, IMP-1, is detectable from *Eimeria* to *Neospora* and *Toxoplasma*, but it is not detectable outside this lineage (Blake *et al.*, 2011). However, orthologues of the apical membrane antigen-1, AMA-1 can be detected from *Plasmodium* to the Coccidia, but not *Cryptosporidium* (Blake *et al.*, 2011). *Cryptosporidium* on the other hand, like many eukaryotes contains the E2F/DPI transcription factor which has not been detected in other apicomplexan genomes (Templeton *et al.*, 2004). We are also able to look at fine-scale changes among closely related species and genera like the rodent-infecting *Plasmodium* species (Liew *et al.*, 2010), between the *Theileria/Babesia* (Pain *et al.*, 2005) species and between *C. parvum* and *C. hominis* (Xu *et al.*, 2004), and between strains of *T. gondii* and between *T. gondii* and *N. caninum* (approx. 28 my separation) (Reid *et al.*, 2012). Comparisons among closely related species are quite helpful for correcting genome annotations since most genes are conserved, and for looking at gene family and gene expression differences that may affect pathogenicity and/or other aspects of parasite biology. However, greater amounts of time separate the remaining species in Fig. 2.1a, thus limiting the types of comparisons that can be made. For example, synteny is not conserved over these large distances (DeBarry and Kissinger, 2011), and the number of shared genes and processes are reduced. It also becomes more difficult to correlate life-cycle stages and identify species-specific proteins involved in host interactions.

Genome architectures

Within the nuclear genome, the repertoire of protein-encoding genes has traditionally received the majority of attention and the genomes themselves have been treated mainly as static backgrounds for the transmission of genes from one generation to the next. However, genomes are dynamic landscapes of shifting features that record a significant amount of evolutionary history. Genomic landscapes can be altered by changes to the numbers of chromosomes, centromere location, the state of chromatin remodelling, the numbers and positions of genes (gains and losses as already discussed), and the relocation and rearrangement of chromosomal regions, both small and large. By determining the presence/absence, quantity and distributions of genomic features it is possible to describe, compare and contrast genomic landscapes both within the Apicomplexa and with respect to other eukaryotes (Kissinger and DeBarry, 2011).

At the broadest level one can ask about the number and sizes of chromosomes within a genome and the number and distribution of genes that they contain. Sequenced apicomplexan genomes contain 4–14 linear chromosomes (Fig. 2.1a) with general eukaryotic features such as centromeres and telomeres (though not all centromeres are annotated) (Brooks *et al.*, 2011; Iwanaga *et al.*, 2010). In contrast to most well-studied model eukaryotes, apicomplexan genomes do not contain large transcriptionally inactive and gene poor regions characteristic of centromeric regions. However, there are transcriptionally active and inactive zones spread throughout the genomes that have been examined (Gissot *et al.*, 2007; Deitsch, 2009; Dzikowski and Deitsch, 2009; Epp *et al.*, 2009; Flueck *et al.*, 2009; Tonkin *et al.*, 2009; Voss *et al.*, 2006). Telomere sequences are variable across the phylum. The *Plasmodium* telomere sequence is 5′-GGGTTYA3-′, where Y = T/C (Scherf *et al.*, 2001), *Cryptosporidium* is 5′-GGTTTA3-′ (Liu *et al.*, 1998), *Eimeria* and *Toxoplasma* are 5′-GGGTTTA-3′ (Gajria *et al.*, 2008; Shirley, 2000) and *Theileria* and *Babesia* have variable telomere sequences of the form 5′-$G_{2-3}T_{3-4}$A-3′ (Brayton *et al.*, 2007; Pain *et al.*, 2005). The adjacent subtelomeric regions are relatively gene rich and are particularly important in *Plasmodium* where they have been shown to contain genes with roles in antigen variation, often concentrated in tandem arrays (Gardner *et al.*, 2005; Gardner *et al.*, 2002) or mixed-neighbourhood gene families like the *ves* 1 genes of *B. bovis* (Brayton *et al.*, 2007) or *Tpr* genes of *Theileria parva* (Gardner *et al.*, 2005) and the *Tar* genes of *T. annulata* (Pain *et al.*, 2005).

Transposable elements (TEs) are ubiquitous in all other examined species, but appear to be at least drastically suppressed and possibly completely absent in most members of the Apicomplexa studied to date (A. Barrie, S. Cheng, J. Kissinger and E. Pritham, unpublished observation) (Durand *et al.*, 2006; Templeton *et al.*, 2009). TEs are discrete genomic denizens, able to mobilize and increase their copy numbers within genomes. They can greatly affect genome architectures by mediating genome rearrangements (Bartolome *et al.*, 2002; Bennetzen, 2000; Bohne *et al.*, 2008; Hua-Van *et al.*, 2005; Kent *et al.*, 2003). Genome rearrangement events can be investigated by exploring the relative physical distribution of orthologous protein-encoding genes between genomes. By examining the presence and order of orthologous genes (synteny and collinearity, referred to here as synteny) the extent of rearrangement that has occurred since two species last shared a common ancestor can be determined. Given the absence of rearrangement-mediating TEs, it is logical to think that synteny would have been maintained within in the Apicomplexa. Several studies have shown that synteny has been maintained within genera and at the tips of major lineages (Brayton *et al.*, 2007; DeBarry and Kissinger, 2011; Kooij *et al.*, 2005; Pain *et al.*, 2005; Weir *et al.*, 2009). Extensive synteny is present within the genus *Plasmodium* and even between *Theileria* and *Babesia* species (Fig. 2.2a). There is limited synteny between the *Plasmodium* and *Piroplasm* lineages, indicating that rearrangement has been much more extensive between these groups than within each lineage. However, the same methods were not able to detect any synteny across the entire phylum (Fig. 2.2b) (DeBarry and Kissinger, 2011). This indicates that for the apicomplexans examined to date, rearrangements since the time of divergence from a last common ancestor have

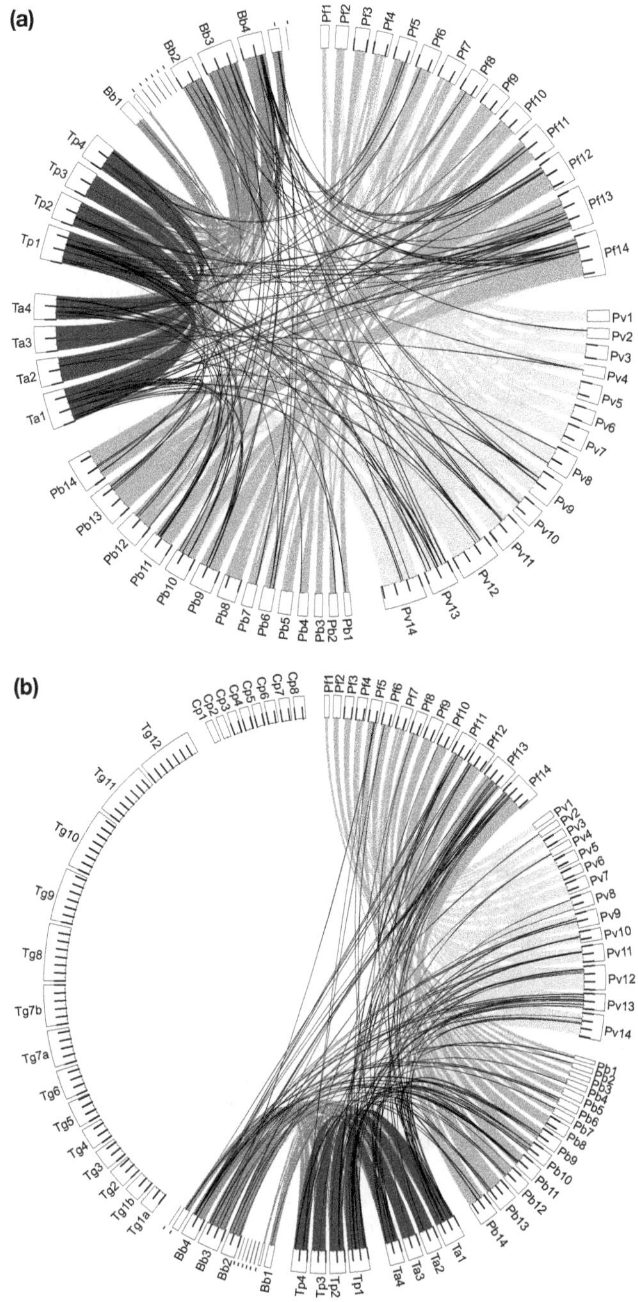

Figure 2.2 Synteny relationships among apicomplexan genomes. (a) Synteny across the *Plasmodium* and Piroplasmida. Figure modified from (DeBarry and Kissinger, 2011). The circle is a graphical representation of the annotated chromosomes and contigs in each genome. Each species' genome is labelled with the genus species abbreviation and chromosome number. Unordered sections of chromosomes are labelled with '–'. Ticks = 1 Mb. Spans that cross the interior of the circle connect syntenic regions identified with protein-coding orthologue markers. 'Twisted' spans represent inversions. Different shades of gray represent synteny between different species. There is a high degree of synteny between all three *Plasmodium* species and (separately) between all three *Piroplasm* species. There is much less synteny between *Plasmodium* and Piroplasm species as indicated by the few, thin spans. (b) Synteny across the apicomplexan phylum – Species and synteny are shown as in (a). There are no spans connecting *T. gondii* or *C. parvum* to each other or to any other species. The results indicate that no synteny was detected between these species and that rearrangement has removed phylum-wide synteny in the Apicomplexa using the detection methods employed.

been so extensive that synteny across the phylum has been completely removed. This level of rearrangement is more extensive than what has been observed in vertebrates or fungal genomes and it has occurred, at least recently, in the absence of TEs.

Comparisons of *Plasmodium* to other apicomplexan species, other alveolates and even their hosts can be extremely useful and provide significant insight into the biology of *Plasmodium*. Many apicomplexan pathogens contain significant data resources and valuable experimental results related to gene function, metabolic pathways, protein–protein interactions, gene regulation and more recently, associations with phenotype (see data sources at EuPathDB). When orthologous genes exist in other species, these resources can be invaluable for suggesting putative functions and enabling testable hypotheses.

Resources and tools for comparative genomics in the Apicomplexa

The *Plasmodium* research community has a large number of useful data sets available to it for comparative analyses both within the genus, with other genera in the phylum Apicomplexa and beyond. Depending on the specific question and the evolutionary distance that needs to be covered, useful data sets range from the entire genome sequence to comparisons of annotated genes, protein motifs, protein localization and RNA or protein expression data sets among others.

The data, depending upon type, can be accessed from a number of different resources (Table 2.1). Genomic sequence and predicted protein data should be accessed from sites where the data are most up to date and reflect any recent assembly or annotation changes. Sites with the most recent

Table 2.1 Major database resources

Resource	URL	Contents
Wellcome Trust Sanger Institute	http://www.sanger.ac.uk/	Genome sequence, data download
Broad Institute	http://www.broadinstitute.org/	Genome sequence, tools for some organisms
Joint Genome Institute (JGI)	http://www.jgi.doe.gov/	Genome sequence, data download, tools
JCVI	http://gsc.jcvi.org/	Genome sequence, data download
IGS	http://gscid.igs.umaryland.edu/	Genome sequence, data download
The genome centre at WashU	http://genome.wustl.edu/	Genome sequence, data download
GeneDB	http://www.genedb.org	BLAST and search tools, gene pages, literature links, curation details
PlasmoDB	http://PlasmoDB.org	BLAST and search tools, gene pages, functional genomics data, synteny and orthologues, community annotation
EuPathDB (CryptoDB, ToxoDB, PiroplasmidaDB) among others	http://EuPathDB.org	BLAST and search tools, gene pages, functional genomics data, synteny and orthologues, community annotation
NCBI	http://www.ncbi.nlm.nih.gov	All types of sequence data, trace archives, BLAST search and download
Ensembl	http://www.ensembl.org	Sequence data, gene pages
OrthoMCL DB	http://www.orthomcl.org/	Precalculated, searchable groups of orthologous genes across the tree of life
ApiLoc	http://apiloc.bio21.unimelb.edu.au	Protein localization
PartiGeneDB	http://www.compsysbio.org/partigene	Sequence data and searches on partial genome data and organized gene sets for Eukaryotes
Galaxy	http://galaxy.psu.edu/	Data integration and analysis tools linked to many different sequence databases

data include the genome sequencing centres (Sanger Institute, JCVI, Broad/MIT, WashU, IGS, JGI, etc.) or laboratories generating the data in-house and community databases where the data are actively deposited and integrated on unified platforms like PlasmoDB (Bahl *et al.*, 2002; Kissinger *et al.*, 2002) and the Eukaryotic Pathogens portal, EuPathDB (Aurrecoechea *et al.*, 2010) where data from PlasmoDB can be compared with other apicomplexan species. Many of the larger sequence repositories like EMBL, NCBI or DDBJ also contain the genome and annotation records, but depending on the record type they are often months to years behind the most recent annotation because of the guidelines associated with updating the original permanent genome records. Reference genome sequences (RefSeq) (Pruitt *et al.*, 2007) or third-party annotation (TPA) records at these resources are often the most up to date. It is important to check the record date and the date of last update and version at all data repository locations prior to using the data. Major resources are described in Table 2.1.

As described above, one of the most valuable data sources for comparative analyses between different species is the examination of orthologous genes. Since orthologous genes are related by evolutionary descent, there is the strong possibility (but not guarantee) that they will have the same or similar function in another species. For example, it is possible that searches of data from the experimentally tractable species *P. falciparum* or *P. berghei* will identify genes with desired properties that may have orthologues in the experimentally intractable *P. vivax*. Searches further abroad are also possible and extend beyond core conserved cellular processes like transcription and translation. Orthologues of many micronemal-, and apicoplast-targeted gene products have been identified in multiple genera. There are two major resources with extensive orthologue tools, OrthoMCL DB (Chen *et al.*, 2006) and EuPathDB (Aurrecoechea *et al.*, 2010).

OrthoMCL DB supports several different types of search strategies. It is possible to begin with a particular protein sequence, ID or protein domain

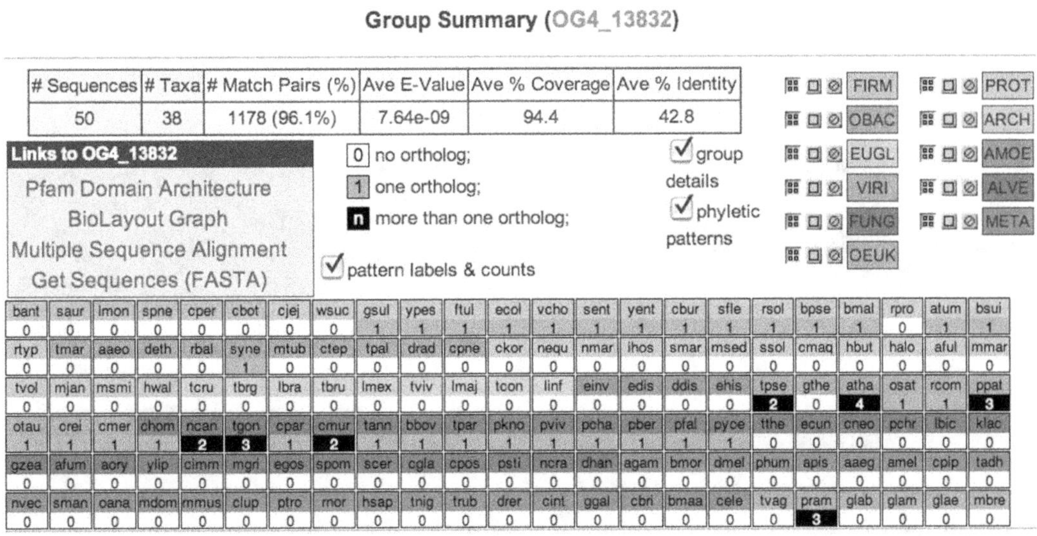

Figure 2.3 OrthoMCL DB result for AminopeptidaseN search. The summary result for orthologous group OG4_13832 provides a quick overview of the phyletic distribution pattern across the tree of life. Species names are four-letter abbreviations. The top half of the rows of species are colour coded according to the key in the upper right corner (FIRM = Firmicutes; PROT = Proteobacteria; OBAC = Other bacteria; ARCH = Archaea; AMOE = Amoebozoa; FUNG = Fungi; VIRI = Viridiplantae; META = Metazoa; EUGL = Euglenozoa; ALVE = Alveolata; OEUK = Other eukaryota). The lower half of each box specifies the number of orthologous genes identified for that species (white = 0; grey = 1 and black = 2 or greater). The overall characteristics of genes in this orthologous cluster are displayed across the top and links to additional views and data download are boxed on the upper left.

name and find other sequences that are related to your initial query. Alternatively, phyletic searches, in which the user specifies particular groups of organisms that should share, or not share, orthologous genes are possible. For example, it is possible to ask for all orthologue groups that are shared by both the Apicomplexa and plants in order to look for endosymbiont derived genes (Fig. 2.3), or cases in which an apicomplexan but not their host(s) contain genes for initially assessing the viability of putative drug or vaccine targets. OrthoMCL DB (Chen *et al.*, 2006) can return groups of orthologous genes that meet a given search criteria. For each orthologous group, it is possible to view the distribution across the tree of life (as represented by ~140 organisms with available genome sequences) (Fig. 2.3), download the orthologous sequences in FASTA format and, depending upon the number of genes in the group, a multiple sequence alignment.

Another route to comparative analysis is via EuPathDB. At this site, researchers can perform the same searches that are performed at component sites, for example, at PlasmoDB, but view results across all eukaryotic pathogens contained in the database (currently: *Babesia, Cryptosporidium, Encephalitozoon, Entamoeba, Enterocytozoon, Giardia, Leishmania, Neospora, Plasmodium, Theileria, Toxoplasma, Trichomonas* and *Trypanosoma* species) or, only a small user-specified subset, e.g. *Plasmodium* and *Toxoplasma,* or all apicomplexans. Fig. 2.4 illustrates the results of a text search for the keyword 'actin' across all organisms in the database. The number of hits found in each database is displayed along the bottom of the top panel as a hyperlink to the exact results. Note: a search for actin will hit 'actin', 'actin-like', 'actin-related' etc., and thus, multiple orthologue groups will be identified in the 'orthologue' box that indicates the number of orthologous gene clusters present within the hits. Alternatively, users can perform a specific search, for example by EC number in the strategy system and then retrieve orthologues of the genes that were identified in the first search from additional species (Fig. 2.5).

Since protein domains are often the most conserved regions of proteins, they can be used to look for related proteins across large distances. The Protein Family Database, PFAM (Bateman *et al.*, 2004), now includes many *Plasmodium* and apicomplexan-specific protein domains. Multiple sequence alignments of protein domains are available as are tools to search user supplied sequences for the presence of protein domains. PFAM names domains with stable identifiers that can be used to look for protein domains in many of the resources

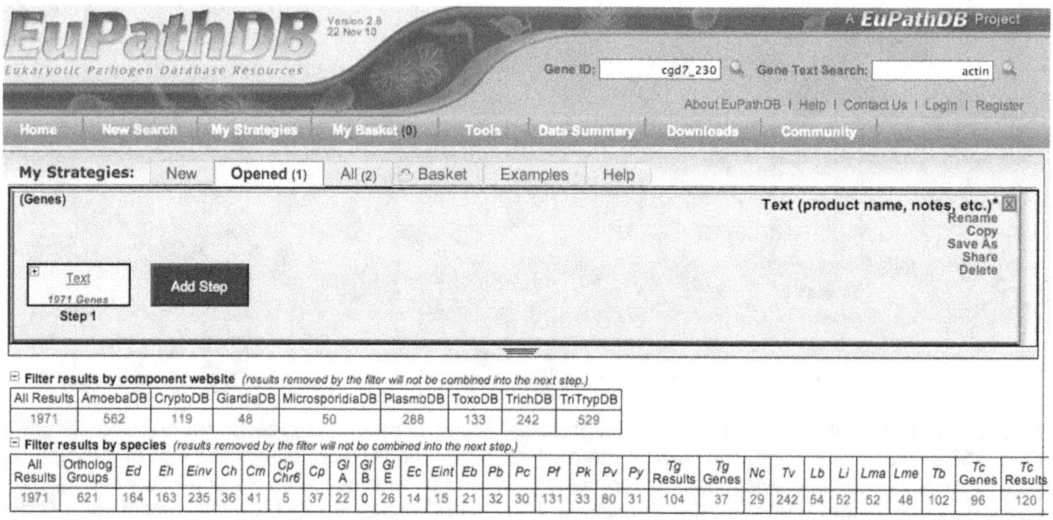

Figure 2.4 EuPathDB Text search with orthologue display. A keyword text search for the word 'actin' was performed and the results are displayed. The row of boxes across the bottom of the panel lists the number of entries found, by species, that contain the word 'actin'. The numbers are hyperlinks that will display the results when selected.

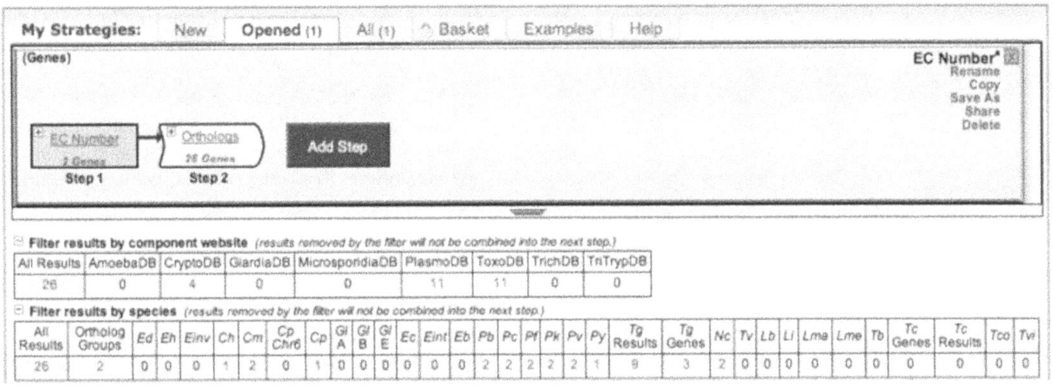

Figure 2.5 EuPathDB EC search with Orthologues. An EC number search for aminopeptidaseN (EC number 3.4.11.2) was performed in the species *P. falciparum*. Two genes were found. A second search across all organisms looked for genes orthologous to the two genes found in the first search. Twenty-four additional genes were found for a total of 26. Note the distribution, by species, where orthologues were found as indicated in the boxes across the bottom of the panel. Orthologues were only found in apicomplexan species.

described above, including OrthoMCL DB, PlasmoDB, EuPathDB and GeneDB.

Drawing conclusions with comparisons

The advantages gained by employing comparative methods to contextualize a phenomenon cannot be overstated. When species, genes, proteins, pathways etc., are placed in an evolutionary context, their characteristics often come into sharper focus. Informative comparisons can be made across vast expanses of evolutionary time. The genome can be regarded as a palimpsest that keeps a record of significant events (Delwiche, 2004). For instance, thanks in part to comparisons with other plastid-containing species spread over the eukaryotic tree of life (Delwiche, 1999; Janouskovec *et al.*, 2010) we now have an enhanced understanding of the roles and mechanisms behind the targeting of the products of nuclear-encoded plastid genes back to the plastid organelle (reviewed in Agrawal and Striepen, 2010). Comparisons within a single phylum can also be greatly informative. For example, discovery of the amplified AP2 family in the Apicomplexa (Balaji *et al.*, 2005) set the stage for comparative studies of different AP2 domains within the *P. falciparum* genome and between different apicomplexan species (Painter *et al.*, 2010).

It should be remembered however, that the phylum Apicomplexa is an old one, at least 400 million years. While there are many common features and processes that unite the Apicomplexa at structural and developmental levels including 837 groups of orthologous genes, there are also intriguing differences that are unique to particular lineages and reflective of their evolutionary history and adaptations to particular hosts and tissue types. The development of tools that facilitate broad comparative analyses has not kept pace with the dramatic increases in sequence generation and the equally dramatic increase in diverse data types (RNA, protein, metabolome, chromatin state, SNP, etc.) that have also scaled to genome-wide analyses. While resources exist that let researchers mine one data type or another, the true power of these data are realized only when they are shared and integrated with other data types and comparisons with other species. This integration allows researchers to gain insight into the basic biology of this most interesting and deadly group of pathogens.

Acknowledgements

We would like to thank the apicomplexan research community for generating and sharing the genome-scale datasets that make comparative analyses possible. This work was supported in part by NIH R01 AI068908 to JCK.

Web resources

Cryptosporidium muris Genome Project http://gsc.jcvi.org/projects/msc/cryptosporidium_muris/.

J Craig Venter Institute. http://www.jcvi.org/.

Sanger Wellcome Trust Pathogens. http://www.sanger.ac.uk/Projects/Pathogens/.

Toxoplasma gondii Genome Project. http://gsc.jcvi.org/projects/msc/toxoplasma_gondii/index.shtml.

References

Abrahamsen, M.S., Templeton, T.J., Enomoto, S., Abrahante, J.E., Zhu, G., Lancto, C.A., Deng, M., Liu, C., Widmer, G., Tzipori, S., *et al.* (2004). Complete genome sequence of the apicomplexan, *Cryptosporidium parvum*. Science *304*, 441–445.

Agrawal, S., and Striepen, B. (2010). More membranes, more proteins: complex protein import mechanisms into secondary plastids. Protist *161*, 672–687.

al-Khedery, B., Barnwell, J.W., and Galinski, M.R. (1999). Antigenic variation in malaria: a 3′ genomic alteration associated with the expression of a *P. knowlesi* variant antigen. Mol. Cell 3, 131–141.

Arenas, A.F., Osorio-Mendez, J.F., Gutierrez, A.J., and Gomez-Marin, J.E. (2010). Genome-wide survey and evolutionary analysis of trypsin proteases in apicomplexan parasites. Genom. Proteom. Bioinform. *8*, 103–112.

Aurrecoechea, C., Brestelli, J., Brunk, B.P., Fischer, S., Gajria, B., Gao, X., Gingle, A., Grant, G., Harb, O.S., Heiges, M., *et al.* (2010). EuPathDB: a portal to eukaryotic pathogen databases. Nucleic Acids Res. *38*, D415–419.

Aury, J.-M., Jaillon, O., Duret, L., Noel, B., Jubin, C., Porcel, B.M., Segurens, B., Daubin, V., Anthouard, V., Aiach, N., *et al.* (2006). Global trends of whole-genome duplications revealed by the ciliate *Paramecium tetraurelia*. Nature *444*, 171–178.

Bahl, A., Brunk, B., Coppel, R.L., Crabtree, J., Diskin, S.J., Fraunholz, M.J., Grant, G.R., Gupta, D., Huestis, R.L., Kissinger, J.C., *et al.* (2002). PlasmoDB: the *Plasmodium* genome resource. An integrated database providing tools for accessing, analysing and mapping expression and sequence data (both finished and unfinished). Nucleic Acids Res. *30*, 87–90.

Balaji, S., Babu, M.M., Iyer, L.M., and Aravind, L. (2005). Discovery of the principal specific transcription factors of Apicomplexa and their implication for the evolution of the AP2-integrase DNA binding domains. Nucleic Acids Res. *33*, 3994–4006.

Bartolome, C., Maside, X., and Charlesworth, B. (2002). On the abundance and distribution of transposable elements in the genome of *Drosophila melanogaster*. Mol. Biol. Evol. *19*, 926–937.

Bateman, A., Coin, L., Durbin, R., Finn, R.D., Hollich, V., Griffiths-Jones, S., Khanna, A., Marshall, M., Moxon, S., Sonnhammer, E.L.L., *et al.* (2004). The Pfam protein families database. Nucleic Acids Res. *32*, D138–D141.

Baum, J., Richard, D., Healer, J., Rug, M., Krnajski, Z., Gilberger, T.W., Green, J.L., Holder, A.A., and Cowman, A.F. (2006). A conserved molecular motor drives cell invasion and gliding motility across malaria life cycle stages and other apicomplexan parasites. J. Biol. Chem. *281*, 5197–5208.

Behnke, M.S., Wootton, J.C., Lehmann, M.M., Radke, J.B., Lucas, O., Nawas, J., Sibley, L.D., and White, M.W. (2010). Coordinated progression through two subtranscriptomes underlies the tachyzoite cycle of *Toxoplasma gondii*. PLoS One 5, e12354.

Bennetzen, J.L. (2000). Transposable element contributions to plant gene and genome evolution. Plant Mol. Biol. *42*, 251–269.

Berney, C.d., and Pawlowski, J. (2006). A molecular time-scale for eukaryote evolution recalibrated with the continuous microfossil record. Proc. Roy. Soc. B: Biol. Sci. *273*, 1867–1872.

Blake, D.P., Billington, K.J., Copestake, S.L., Oakes, R.D., Quail, M.A., Wan, K.L., Shirley, M.W., and Smith, A.L. (2011). Genetic mapping identifies novel highly protective antigens for an apicomplexan parasite. PLoS Pathog. 7, e1001279.

Bohne, A., Brunet, F., Galiana-Arnoux, D., Schultheis, C., and Volff, J.N. (2008). Transposable elements as drivers of genomic and biological diversity in vertebrates. Chromosome Res. *16*, 203–215.

Brayton, K.A., Lau, A.O., Herndon, D.R., Hannick, L., Kappmeyer, L.S., Berens, S.J., Bidwell, S.L., Brown, W.C., Crabtree, J., Fadrosh, D., *et al.* (2007). Genome sequence of *Babesia bovis* and comparative analysis of apicomplexan hemoprotozoa. PLoS Pathog. 3, 1401–1413.

Brooks, C.F., Francia, M.E., Gissot, M., Croken, M.M., Kim, K., and Striepen, B. (2011). *Toxoplasma gondii* sequesters centromeres to a specific nuclear region throughout the cell cycle. Proc. Natl. Acad. Sci. U.S.A. *108* 3767–3772.

Carlton, J.M., Adams, J.H., Silva, J.C., Bidwell, S.L., Lorenzi, H., Caler, E., Crabtree, J., Angiuoli, S.V., Merino, E.F., Amedeo, P., *et al.* (2008). Comparative genomics of the neglected human malaria parasite *Plasmodium vivax*. Nature *455*, 757–763.

Carlton, J.M., Angiuoli, S.V., Suh, B.B., Kooij, T.W., Pertea, M., Silva, J.C., Ermolaeva, M.D., Allen, J.E., Selengut, J.D., Koo, H.L., *et al.* (2002). Genome sequence and comparative analysis of the model rodent malaria parasite *Plasmodium yoelii yoelii*. Nature *419*, 512–519.

Carruthers, V.B., and Tomley, F.M. (2008). Microneme proteins in apicomplexans. Subcell. Biochem. *47*, 33–45.

Chen, F., Mackey, A.J., Stoeckert, C.J., Jr., and Roos, D.S. (2006). OrthoMCL-DB: querying a comprehensive multi-species collection of ortholog groups. Nucleic Acids Res. *34*, D363–368.

Chen, F., Mackey, A.J., Vermunt, J.K., and Roos, D.S. (2007). Assessing performance of orthology detection strategies applied to eukaryotic genomes. PLoS One 2, e383.

DeBarry, J., and Kissinger, J. (2011). Jumbled genomes: missing apicomplexan synteny. Molecular biology and evolution. Mol. Biol. Evol. *27*, 345–354.

Deitsch, K.W. (2009). A mark of silence in malaria parasites. Cell Host Microbe 5, 112–113.

Delwiche, C.F. (1999). Tracing the thread of plastid diversity through the tapestry of life. Am. Nat. *154*, S164-S177.

Delwiche, C.F. (2004). The genomic palimpsest: genomics in evolution and ecology. BioScience *54*, 991–1001.

Dowse, T.J., and Soldati, D. (2005). Rhomboid-like proteins in Apicomplexa: phylogeny and nomenclature. Trends Parasitol. *21*, 254–258.

Durand, P., Oelofse, A., and Coetzer, T. (2006). An analysis of mobile genetic elements in three *Plasmodium* species and their potential impact on the nucleotide composition of the *P. falciparum* genome. BMC Genomics 7, 282.

Dzierszinski, F., Popescu, O., Toursel, C., Slomianny, C., Yahiaoui, B., and Tomavo, S. (1999). The protozoan parasite *Toxoplasma gondii* expresses two functional plant-like glycolytic enzymes. Implications for evolutionary origin of apicomplexans. J. Biol. Chem. *274*, 24888–24895.

Dzikowski, R., and Deitsch, K.W. (2009). Genetics of antigenic variation in *Plasmodium falciparum*. Curr. Genet. 55, 103–110.

Eisen, J.A., Coyne, R.S., Wu, M., Wu, D., Thiagarajan, M., Wortman, J.R., Badger, J.H., Ren, Q., Amedeo, P., Jones, K.M., *et al.* (2006). Macronuclear genome sequence of the ciliate *Tetrahymena thermophila*, a model eukaryote. PLoS Biol. *4*, e286.

Epp, C., Li, F., Howitt, C.A., Chookajorn, T., and Deitsch, K.W. (2009). Chromatin associated sense and antisense noncoding RNAs are transcribed from the *var* gene family of virulence genes of the malaria parasite *Plasmodium falciparum*. RNA *15*, 116–127.

Escalante, A., and Ayala, F. (1995). Evolutionary origin of *Plasmodium* and other Apicomplexa based on rRNA genes. Proc. Natl. Acad. Sci. U.S.A. *92*, 5793–5797.

Fast, N.M., Kissinger, J.C., Roos, D.S., and Keeling, P.J. (2001). Nuclear-encoded, plastid-targeted genes suggest a single common origin for apicomplexan and dinoflagellate plastids. Mol. Biol. Evol. *18*, 418–426.

Ferguson, D.J., Sahoo, N., Pinches, R.A., Bumstead, J.M., Tomley, F.M., and Gubbels, M.J. (2008). MORN1 has a conserved role in asexual and sexual development across the apicomplexa. Eukaryot. Cell 7, 698–711.

Fichera, M.E., and Roos, D.S. (1997). A plastid organelle as a drug target in apicomplexan parasites. Nature *390*, 407–409.

Flueck, C., Bartfai, R., Volz, J., Niederwieser, I., Salcedo-Amaya, A.M., Alako, B.T., Ehlgen, F., Ralph, S.A., Cowman, A.F., Bozdech, Z., *et al.* (2009). *Plasmodium falciparum* heterochromatin protein 1 marks genomic loci linked to phenotypic variation of exported virulence factors. PLoS Pathog. *5*, e1000569.

Foth, B.J., and McFadden, G.I. (2003). The apicoplast: a plastid in *Plasmodium falciparum* and other Apicomplexan parasites. Int. Rev. Cytol. *224*, 57–110.

Foth, B.J., Ralph, S.A., Tonkin, C.J., Struck, N.S., Fraunholz, M., Roos, D.S., Cowman, A.F., and McFadden, G.I. (2003). Dissecting apicoplast targeting in the malaria parasite *Plasmodium falciparum*. Science 299, 705–708.

Gajria, B., Bahl, A., Brestelli, J., Dommer, J., Fischer, S., Gao, X., Heiges, M., Iodice, J., Kissinger, J.C., Mackey, A.J., *et al.* (2008). ToxoDB: an integrated *Toxoplasma gondii* database resource. Nucleic Acids Res. 36, D553–556.

Gardner, M.J., Bishop, R., Shah, T., de Villiers, E.P., Carlton, J.M., Hall, N., Ren, Q., Paulsen, I.T., Pain, A., Berriman, M., *et al.* (2005). Genome sequence of *Theileria parva*, a bovine pathogen that transforms lymphocytes. Science *309*, 134–137.

Gardner, M.J., Hall, N., Fung, E., White, O., Berriman, M., Hyman, R.W., Carlton, J.M., Pain, A., Nelson, K.E., Bowman, S., *et al.* (2002). Genome sequence of the human malaria parasite *Plasmodium falciparum*. Nature *419*, 498–511.

Gissot, M., Kelly, K.A., Ajioka, J.W., Greally, J.M., and Kim, K. (2007). Epigenomic modifications predict active promoters and gene structure in *Toxoplasma gondii*. PLoS Pathog. *3*, e77.

Gubbels, M.J., Vaishnava, S., Boot, N., Dubremetz, J.F., and Striepen, B. (2006). A MORN-repeat protein is a dynamic component of the *Toxoplasma gondii* cell division apparatus. J. Cell Sci. *119*, 2236–2245.

Harb, O.S., Chatterjee, B., Fraunholz, M.J., Crawford, M.J., Nishi, M., and Roos, D.S. (2004). Multiple functionally redundant signals mediate targeting to the apicoplast in the apicomplexan parasite Toxoplasma gondii. Eukaryot. Cell 3, 663–674.

Harper, J.M., Hoff, E.F., and Carruthers, V.B. (2004). Multimerization of the *Toxoplasma gondii* MIC2 integrin-like A-domain is required for binding to heparin and human cells. Mol. Biochem. Parasitol. *134*, 201–212.

Hassan, H.F., and Coombs, G.H. (1988). Purine and pyrimidine metabolism in parasitic protozoa. FEMS Microbiol. Rev. *4*, 47–84.

He, C.Y., Striepen, B., Pletcher, C.H., Murray, J.M., and Roos, D.S. (2001). Targeting and processing of nuclear-encoded apicoplast proteins in plastid segregation mutants of *Toxoplasma gondii*. J. Biol. Chem. *276*, 28436–28442.

Hua-Van, A., Le Rouzic, A., Maisonhaute, C., and Capy, P. (2005). Abundance, distribution and dynamics of retrotransposable elements and transposons: similarities and differences. Cytogenet. Genome Res. *110*, 426–440.

Huang, D.W., Sherman, B.T., and Lempicki, R.A. (2008). Systematic and integrative analysis of large gene lists using DAVID bioinformatics resources. Nat. Protocols *4*, 44–57.

Huang, J., and Kissinger, J. (2006). Horizontal and intracellular gene transfer in the Apicomplexa: The scope and functional consequences. In Genome Evolution in Eukaryotic Microbes, L. Katz, and D. Bhattacharya, eds. (Oxford University Press), p. 256.

Huang, J., Mullapudi, N., Lancto, C.A., Scott, M., Abrahamsen, M.S., and Kissinger, J.C. (2004a). Phylogenomic evidence supports past endosymbiosis, intracellular and horizontal gene transfer in *Cryptosporidium parvum*. Genome Biol. *5*, R88.

Huang, J., Mullapudi, N., Sicheritz-Ponten, T., and Kissinger, J.C. (2004b). A first glimpse into the pattern and scale of gene transfer in Apicomplexa. Int. J. Parasitol. *34*, 265–274.

Iwanaga, S., Khan, S.M., Kaneko, I., Christodoulou, Z., Newbold, C., Yuda, M., Janse, C.J., and Waters, A.P.

(2010). Functional identification of the *Plasmodium* centromere and generation of a *Plasmodium* artificial chromosome. Cell Host Microbe 7, 245–255.

Janouskovec, J., Horak, A., Obornik, M., Lukes, J., and Keeling, P.J. (2010). A common red algal origin of the apicomplexan, dinoflagellate, and heterokont plastids. Proc. Natl. Acad. Sci. U.S.A. *107*, 10949–10954.

Jomaa, H., Wiesner, J., Sanderbrand, S., Altincicek, B., Weidemeyer, C., Hintz, M., Turbachova, I., Eberl, M., Zeidler, J., Lichtenthaler, H.K., *et al.* (1999). Inhibitors of the nonmevalonate pathway of isoprenoid biosynthesis as antimalarial drugs. Science *285*, 1573–1576.

Kent, W.J., Baertsch, R., Hinrichs, A., Miller, W., and Haussler, D. (2003). Evolution's cauldron: duplication, deletion, and rearrangement in the mouse and human genomes. Proc. Natl. Acad. Sci. U.S.A. *100*, 11484–11489.

Kim, K., and Weiss, L.M. (2004). *Toxoplasma gondii*: the model apicomplexan. Int. J. Parasitol. *34*, 423–432.

Kissinger, J.C., Brunk, B.P., Crabtree, J., Fraunholz, M.J., Gajria, B., Milgram, A.J., Pearson, D.S., Schug, J., Bahl, A., Diskin, S.J., *et al.* (2002). The *Plasmodium* genome database. Nature *419*, 490–492.

Kissinger, J.C., and DeBarry, J. (2011). Genome cartography: charting the apicomplexan genome. Trends Parasitol. *27*, 345–354.

Kohler, S., Delwiche, C.F., Denny, P.W., Tilney, L.G., Webster, P., Wilson, R.J., Palmer, J.D., and Roos, D.S. (1997). A plastid of probable green algal origin in Apicomplexan parasites. Science *275*, 1485–1489.

Kooij, T.W., Carlton, J.M., Bidwell, S.L., Hall, N., Ramesar, J., Janse, C.J., and Waters, A.P. (2005). A *Plasmodium* whole-genome synteny map: indels and synteny breakpoints as foci for species-specific genes. PLoS Pathog. *1*, e44.

Kuo, C.H., and Kissinger, J.C. (2008). Consistent and contrasting properties of lineage-specific genes in the apicomplexan parasites *Plasmodium* and *Theileria*. BMC Evol. Biol. *8*, 108.

Kyes, S.A., Rowe, J.A., Kriek, N., and Newbold, C.I. (1999). Rifins: a second family of clonally variant proteins expressed on the surface of red cells infected with *Plasmodium falciparum*. Proc. Natl. Acad. Sci. U.S.A. *96*, 9333–9338.

Lau, A.O., McElwain, T.F., Brayton, K.A., Knowles, D.P., and Roalson, E.H. (2009). *Babesia bovis*: a comprehensive phylogenetic analysis of plastid-encoded genes supports green algal origin of apicoplasts. Exp. Parasitol. *123*, 236–243.

Leander, B.S., Clopton, R.E., and Keeling, P.J. (2003). Phylogeny of gregarines (Apicomplexa) as inferred from small-subunit rDNA and beta-tubulin. Int. J. Syst. Evol. Microbiol. *53*, 345–354.

Liew, K.J., Hu, G., Bozdech, Z., and Peter, P.R. (2010). Defining species specific genome differences in malaria parasites. BMC Genomics *11*, 128.

Liu, C., Schroeder, A.A., Kapur, V., and Abrahamsen, M.S. (1998). Telomeric sequences of *Cryptosporidium parvum*. Mol. Biochem. Parasitol. *94*, 291–296.

Llinas, M., Bozdech, Z., Wong, E.D., Adai, A.T., and DeRisi, J.L. (2006). Comparative whole genome transcriptome analysis of three *Plasmodium falciparum* strains. Nucleic Acids Res. *34*, 1166–1173.

Martens, C., Vandepoele, K., and Van de Peer, Y. (2008). Whole-genome analysis reveals molecular innovations and evolutionary transitions in chromalveolate species. Proc. Natl. Acad. Sci. U.S.A. *105*, 3427–3432.

McClintock, B. (1946). Maize genetics. Year B. Carnegie Inst. Wash. *45*, 176–186.

McFadden, G.I. (1999). Plastids and protein targeting. J. Eukaryot. Microbiol. *46*, 339–346.

McFadden, G.I., and Roos, D.S. (1999). Apicomplexan plastids as drug targets. Trends Microbiol. *7*, 328–333.

McLeod, R., Muench, S.P., Rafferty, J.B., Kyle, D.E., Mui, E.J., Kirisits, M.J., Mack, D.G., Roberts, C.W., Samuel, B.U., Lyons, R.E., *et al.* (2001). Triclosan inhibits the growth of *Plasmodium falciparum* and *Toxoplasma gondii* by inhibition of apicomplexan Fab I. Int. J. Parasitol. *31*, 109–113.

Moore, R.B., Obornik, M., Janouskovec, J., Chrudimsky, T., Vancova, M., Green, D.H., Wright, S.W., Davies, N.W., Bolch, C.J., Heimann, K., *et al.* (2008). A photosynthetic alveolate closely related to apicomplexan parasites. Nature *451*, 959–963.

Nagamune, K., and Sibley, L.D. (2006). Comparative genomic and phylogenetic analyses of calcium ATPases and calcium-regulated proteins in the apicomplexa. Mol. Biol. Evol. *23*, 1613–1627.

Okamoto, N., and McFadden, G.I. (2008). The mother of all parasites. Future Microbiol. *3*, 391–395.

Pain, A., Bohme, U., Berry, A.E., Mungall, K., Finn, R.D., Jackson, A.P., Mourier, T., Mistry, J., Pasini, E.M., Aslett, M.A., *et al.* (2008). The genome of the simian and human malaria parasite *Plasmodium knowlesi*. Nature *455*, 799–803.

Pain, A., Renauld, H., Berriman, M., Murphy, L., Yeats, C.A., Weir, W., Kerhornou, A., Aslett, M., Bishop, R., Bouchier, C., *et al.* (2005). Genome of the host-cell transforming parasite *Theileria annulata* compared with *T. parva*. Science *309*, 131–133.

Painter, H.J., Campbell, T.L., and Llinas, M. (2011). The Apicomplexan AP2 family: integral factors regulating *Plasmodium* development. Mol. Biochem. Parasitol. *176*, 1–7.

Perkins, F.O., Barta, J.R., Clopton, R.E., Peirce, M.A., and Upton, S.J. (2000). Phylum Apicomplexa. In The Illustrated Guide to the Protozoa, Lee, J., Leedale, G., and Bradbury, P., eds. (Lawrence, Allen Press Inc.), pp. 190–369.

Peterson, D.S., Miller, L.H., and Wellems, T.E. (1995). Isolation of multiple sequences from the *Plasmodium falciparum* genome that encode conserved domains homologous to those in erythrocyte-binding proteins. Proc. Natl. Acad. Sci. U.S.A. *92*, 7100–7104.

Pruitt, K.D., Tatusova, T., and Maglott, D.R. (2007). NCBI reference sequences (RefSeq): a curated non-redundant sequence database of genomes, transcripts and proteins. Nucleic Acids Res. *35*, D61–65.

Reid, A.J., Vermont, S.J., Cotton, J.A., Harris, D., Hill-Cawthorne, G.A., Könen-Waisman, S., Latham, S.M., Mourier, T., Norton, R., Quail, M.A., *et al.* (2012). Comparative genomics of the apicomplexan parasites *Toxoplasma gondii* and *Neospora caninum*: Coccidia differing in host range and transmission strategy. PLoS Pathog. *8*, e1002567.

Roos, D.S. (2005). Genetics. Themes and variations in apicomplexan parasite biology. Science *309(5731)*, 72–73.

Roos, D.S., Crawford, M.J., Donald, R.G., Fohl, L.M., Hager, K.M., Kissinger, J.C., Reynolds, M.G., Striepen, B., and Sullivan, W.J., Jr. (1999). Transport and trafficking: *Toxoplasma* as a model for *Plasmodium*. Novartis Found. Symp. *226*, 176–195; discussion 195–178.

Ryall, K., Harper, J.T., and Keeling, P.J. (2003). Plastid-derived Type II fatty acid biosynthetic enzymes in chromists. Gene *313*, 139–148.

Saffo, M.B., McCoy, A.M., Rieken, C., and Slamovits, C.H. (2010). *Nephromyces*, a beneficial apicomplexan symbiont in marine animals. Proc. Natl. Acad. Sci. U.S.A. *107*, 16190–16195.

Scherf, A., Figueiredo, L.M., and Freitas-Junior, L.H. (2001). *Plasmodium* telomeres: a pathogen's perspective. Curr. Opin. Microbiol. *4*, 409–414.

Schmatz, D.M. (1997). The mannitol cycle in *Eimeria*. Parasitology *114* (Suppl.), S81–89.

Sharling, L., Liu, X., Gollapalli, D.R., Maurya, S.K., Hedstrom, L., and Striepen, B. A screening pipeline for antiparasitic agents targeting *Cryptosporidium* inosine monophosphate dehydrogenase. PLoS Negl Trop Dis *4*, e794.

Shirley, M.W. (2000). The genome of *Eimeria* spp., with special reference to *Eimeria tenella* – a coccidium from the chicken. Int. J. Parasitol. *30*, 485–493.

Silva, J.C., Egan, A., Friedman, R., Munro, J.B., Carlton, J.M., and Hughes, A.L. (2011). Genome sequences reveal divergence times of malaria parasite lineages. Parasitology *138*, 1737–1749.

Soldati, D. (1999). The apicoplast as a potential therapeutic target in and other apicomplexan parasites. Parasitol. Today *15*, 5–7.

Striepen, B., Jordan, C.N., Reiff, S., and van Dooren, G.G. (2007). Building the perfect parasite: cell division in apicomplexa. PLoS Pathog. *3*, e78.

Striepen, B., Pruijssers, A.J.P., Huang, J., Li, C., Gubbels, M.J., Umejiego, N.N., Hedstrom, L., and Kissinger, J.C. (2004). Gene transfer in the evolution of parasite nucleotide biosynthesis. Proc. Natl. Acad. Sci. U.S.A. *101*, 3154–3159.

Striepen, B., White, M.W., Li, C., Guerini, M.N., Malik, S.B., Logsdon, J.M., Jr., Liu, C., and Abrahamsen, M.S. (2002). Genetic complementation in apicomplexan parasites. Proc. Natl. Acad. Sci. U.S.A. *99*, 6304–6309.

Strong, W.B., and Nelson, R.G. (2000). Preliminary profile of the *Cryptosporidium parvum* genome: an expressed sequence tag and genome survey sequence analysis. Mol. Biochem. Parasitol. *107*, 1–32.

Sullivan, W.J., Jr., Dixon, S.E., Li, C., Striepen, B., and Queener, S.F. (2005). IMP dehydrogenase from the protozoan parasite *Toxoplasma gondii*. Antimicrob. Agents Chemother. *49*, 2172–2179.

Templeton, T.J., Enomoto, S., Chen, W.J., Huang, C.G., Lancto, C.A., Abrahamsen, M.S., and Zhu, G. (2009). A genome sequence survey for *Ascogregarina taiwanensis* supports evolutionary affiliation, but metabolic diversity between a gregarine and *Cryptosporidium*. Mol. Biol. Evol. 27, 235–248.

Templeton, T.J., Iyer, L.M., Anantharaman, V., Enomoto, S., Abrahante, J.E., Subramanian, G.M., Hoffman, S.L., Abrahamsen, M.S., and Aravind, L. (2004). Comparative analysis of apicomplexa and genomic diversity in eukaryotes. Genome Res. *14*, 1686–1695.

Templeton, T.J., and Kaslow, D.C. (1997). Cloning and cross-species comparison of the thrombospondin-related anonymous protein (TRAP) gene from *Plasmodium knowlesi, Plasmodium vivax* and *Plasmodium gallinaceum*. Mol. Biochem. Parasitol. *84*, 13–24.

Tonkin, C.J., Carret, C.K., Duraisingh, M.T., Voss, T.S., Ralph, S.A., Hommel, M., Duffy, M.F., Silva, L.M., Scherf, A., Ivens, A., *et al.* (2009). Sir2 paralogues cooperate to regulate virulence genes and antigenic variation in *Plasmodium falciparum*. PLoS Biol. *7*, e84.

Umejiego, N.N., Gollapalli, D., Sharling, L., Volftsun, A., Lu, J., Benjamin, N.N., Stroupe, A.H., Riera, T.V., Striepen, B., and Hedstrom, L. (2008). Targeting a prokaryotic protein in a eukaryotic pathogen: identification of lead compounds against cryptosporidiosis. Chem. Biol. *15*, 70–77.

Voss, T.S., Healer, J., Marty, A.J., Duffy, M.F., Thompson, J.K., Beeson, J.G., Reeder, J.C., Crabb, B.S., and Cowman, A.F. (2006). A *var* gene promoter controls allelic exclusion of virulence genes in *Plasmodium falciparum* malaria. Nature *439*, 1004–1008.

Waller, R.F., Keeling, P.J., Donald, R.G., Striepen, B., Handman, E., Lang-Unnasch, N., Cowman, A.F., Besra, G.S., Roos, D.S., and McFadden, G.I. (1998). Nuclear-encoded proteins target to the plastid in *Toxoplasma gondii* and *Plasmodium falciparum*. Proc. Natl. Acad. Sci. U.S.A. *95*, 12352–12357.

Waller, R.F., Reed, M.B., Cowman, A.F., and McFadden, G.I. (2000). Protein trafficking to the plastid of *Plasmodium falciparum* is via the secretory pathway. EMBO J. *19*, 1794–1802.

Wasmuth, J., Daub, J., Peregrin-Alvarez, J.M., Finney, C.A., and Parkinson, J. (2009). The origins of apicomplexan sequence innovation. Genome Res. *19*, 1202–1213.

Weir, W., Sunter, J., Chaussepied, M., Skilton, R., Tait, A., de Villiers, E.P., Bishop, R., Shiels, B., and Langsley, G. (2009). Highly syntenic and yet divergent: A tale of two *Theilerias*. Infect. Gen.Evol. 9, 453–461.

Whitaker, J.W., McConkey, G.A., and Westhead, D.R. (2009). The transferome of metabolic genes explored: analysis of the horizontal transfer of enzyme encoding genes in unicellular eukaryotes. Genome Biol. *10*, R36.

Wiesner, J., and Jomaa, H. (2007). Isoprenoid biosynthesis of the apicoplast as drug target. Curr. Drug Targets *8*, 3–13.

Wiesner, J., Reichenberg, A., Heinrich, S., Schlitzer, M., and Jomaa, H. (2008). The plastid-like organelle of apicomplexan parasites as drug target. Curr. Pharm. Des *14*, 855–871.

Williamson, D.H., Gardner, M.J., Preiser, P., Moore, D.J., Rangachari, K., and Wilson, R.J. (1994). The evolutionary origin of the 35kb circular DNA of *Plasmodium falciparum*: new evidence supports a possible rhodophyte ancestry. Mol. Gen. Genet. *243*, 249–252.

Wilson, R.J., Denny, P.W., Preiser, P.R., Rangachari, K., Roberts, K., Roy, A., Whyte, A., Strath, M., Moore, D.J., Moore, P.W., *et al.* (1996). Complete gene map of the plastid-like DNA of the malaria parasite *Plasmodium falciparum*. J. Mol. Biol. *261*, 155–172.

Wilson, R.J., Gardner, M.J., Feagin, J.E., and Williamson, D.H. (1991). Have malaria parasites three genomes? Parasitol. Today 7, 134–136.

Xu, P., Widmer, G., Wang, Y., Ozaki, L.S., Alves, J.M., Serrano, M.G., Puiu, D., Manque, P., Akiyoshi, D., Mackey, A.J., *et al.* (2004). The genome of *Cryptosporidium hominis*. Nature *431*, 1107–1112.

Zhu, G., and Keithly, J.S. (2002). Alpha-proteobacterial relationship of apicomplexan lactate and malate dehydrogenases. J. Eukaryot. Microbiol. *49*, 255–261.

Plasmodium Genomics and the Art of Sequencing Malaria Parasite Genomes

3

Jane M. Carlton, Steven A. Sullivan and Karine G. Le Roch

Abstract

It may be a cliché to state, but obtaining the genome sequence of an organism is one of the most important – if not *the* most important – step towards interrogating its biology. The first two malaria parasite genome sequences (*Plasmodium falciparum* and the rodent model *Plasmodium yoelii yoelii*) were published in 2002 after more than half a decade of intense sequencing, assembly gap closure, and sequence annotation. Since then, reference genomes of several more *Plasmodium* species have been generated, with an emphasis on malaria parasites that infect humans due to their global health importance. With the recent transformation in technologies available for the rapid and cheap production of genome sequence data, an explosion of *P. falciparum* genomes from a wide variety of geographical locations has started to appear, and with it all of the computational issues of large dataset manipulation, storage and analysis. We begin this chapter with a discussion of sequencing technologies, from Sanger sequencing through to current next generation sequencing platforms, to lay the foundation for many of the studies that are presented in this book. Next we describe the characteristics of a typical *Plasmodium* nuclear genome (with reference to those species of malaria that infect mammals), with a brief mention of the extranuclear apicoplast and mitochondrial genomes also found in the parasite. Finally, we outline how comparative genomics – literally comparing genomes within and between species – has been used as a powerful tool to elucidate malaria parasite biology and evolution.

Platforms and methods for sequencing *Plasmodium* genomes

Sanger sequencing

Over the past three decades, the advent of DNA sequencing technologies has significantly enhanced biomedical research and discovery in general, and considerably improved our understanding of basic malaria parasite biology in particular. In 1977, Fred Sanger and Alan Coulson published a DNA sequencing method that was to be widely used for more than 30 years (Sanger *et al.*, 1977a; Sanger *et al.*, 1977b). The Sanger method, also called dideoxy or 'chain-terminator' sequencing, is based on the use of unmodified and dideoxy nucleotides (NTPs and ddNTPs, respectively) along with DNA polymerase, to synthesize a new DNA strand complementary to a single-stranded DNA 'template' of interest prepared from cloned, amplified duplex DNA. Four sequencing reactions are run in parallel, each containing a mix of all four NTPs plus one radioactively labelled ddNTP. The polymerase synthesizes a growing DNA strand along the template DNA according to Watson–Crick base pairing rules, e.g. if the template strand has a G at the current position, the polymerase must add a C to elongate the copy. In the 'C' reaction mixture, the added C could be either a dCTP, allowing elongation to continue, or a ddCTP, which terminates elongation. When incubation is complete the 'C' reaction yields a set of radiolabelled fragments whose lengths, when arranged in size order, indicate the positions of all the G nucleotides as they occur in order on the template strand. Separation and ordering of these DNA fragments by length is achieved using gel electrophoresis; running the four reactions (ACGT) in adjacent gel lanes allows the nucleotide sequence to be literally 'read' from bottom to top off a gel image captured on X-ray film (Fig. 3.1). Over the years, the Sanger sequencing method underwent numerous improvements (e.g. use of fluorescent rather than radioactive labelling of ddNTPs) that ultimately led to its automation. However, for whole genomes the method was inherently time-consuming and laborious because the template DNAs had to be cloned into small, medium or large insert libraries to amplify it prior to sequencing.

Automated Sanger sequencing using a whole genome or chromosome-by-chromosome shotgun strategy was used to sequence the first human (*P. falciparum*; Gardner *et al.*, 2002) and rodent (*P. yoelii yoelii*; Carlton *et al.*, 2002) malaria parasite genomes in 2002. Using similar methodology, the reference genome sequences of several other malaria parasites, including *Plasmodium vivax* (Carlton *et al.*, 2008), *Plasmodium knowlesi* (Pain *et al.*, 2008) and *Plasmodium chabaudi chabaudi* and *Plasmodium berghei* (Hall *et al.*, 2005), have also been successfully completed, albeit to different levels of coverage and completeness.

Next generation sequencing

The past five years have seen rapid commercialization and adoption of 'next generation sequencing' (NGS) technologies that represent advances beyond Sanger sequencing transforming the field of whole genome sequencing and opening up new applications for sequencing-based approaches. In most NGS methods, sequencing starts with shearing template DNA into fragments and ligating them to platform-specific adaptors (Fig. 3.1). Adapter-ligated fragments are then immobilized on a solid (beads or array) surface. Clonal colonies (called 'polonies' or clusters) are generated by PCR from each immobilized fragment directly on the support surface, eliminating time-consuming cloning into vectors. Clusters are rendered single-stranded, then subjected to multiple sequencing cycles where second-strand synthesis occurs under conditions that generate a signal when one or more nucleotides are added to the end of the synthesized strand. The massively parallel signals generated in each cycle are recorded by highly sensitive detection systems and subsequently decoded in order by base calling software to determine the nucleotide sequence of the DNA. The number of sequencing cycles sharply or loosely limits the length of the 'reads' depending on the chemistry (i.e. whether or not >1 dNTP can be incorporated per cycle). Reads can be aligned to a reference genome or assembled *de novo* based on the experimental plan.

Three main second generation platforms have been commercialized over the past 5 years (Table 3.1): (1) Roche 454 Genome Sequencer (Margulies *et al.*, 2005); (2) Applied Biosystems SOLiD (Shendure and Ji, 2008); and (3) the Illumina® (formerly known as Solexa) Genome Analyser and HiSeq platforms (Bentley *et al.*, 2008).

Roche 454 platforms

The 454 genome sequencer was the first commercially available NGS platform. Each run on the standard 454 GS FLX platform can generate more than 1 million high-quality reads with an average length of > 450 bp in ~ 10 h. The newer GS FLX+ system can generate read lengths up to 1 kb over 23 h. Briefly, the technology is as follows: a single-stranded template DNA library immobilized onto DNA capture beads at a frequency of one template per bead (Fig. 3.1) is amplified *in situ* by 'emulsion PCR', wherein an oil droplet serves as a PCR reaction chamber for each bead. Beads containing clonally amplified templates are loaded into wells of a pico-titre plate for pyrosequencing, with the well diameter allowing only one bead per well. Deoxynucleotides are flowed onto and washed out of the wells cyclically at a rate of one dNTP (dATP or dTTP or dGTP or dCTP) per flow cycle. Polymerase-driven addition of complementary dNTP onto the elongating strand in a well during a flow cycle enzymatically triggers a light signal proportional to the number of dNTPs incorporated, such that homopolymer stretches generate a signal proportional to their lengths. Knowing the order of the dNTP flows and the intensity of signal per cycle allows the DNA sequence in each well to be derived. However, the accuracy of homopolymer pyrosequencing decreases rapidly as homopolymer length increases, making

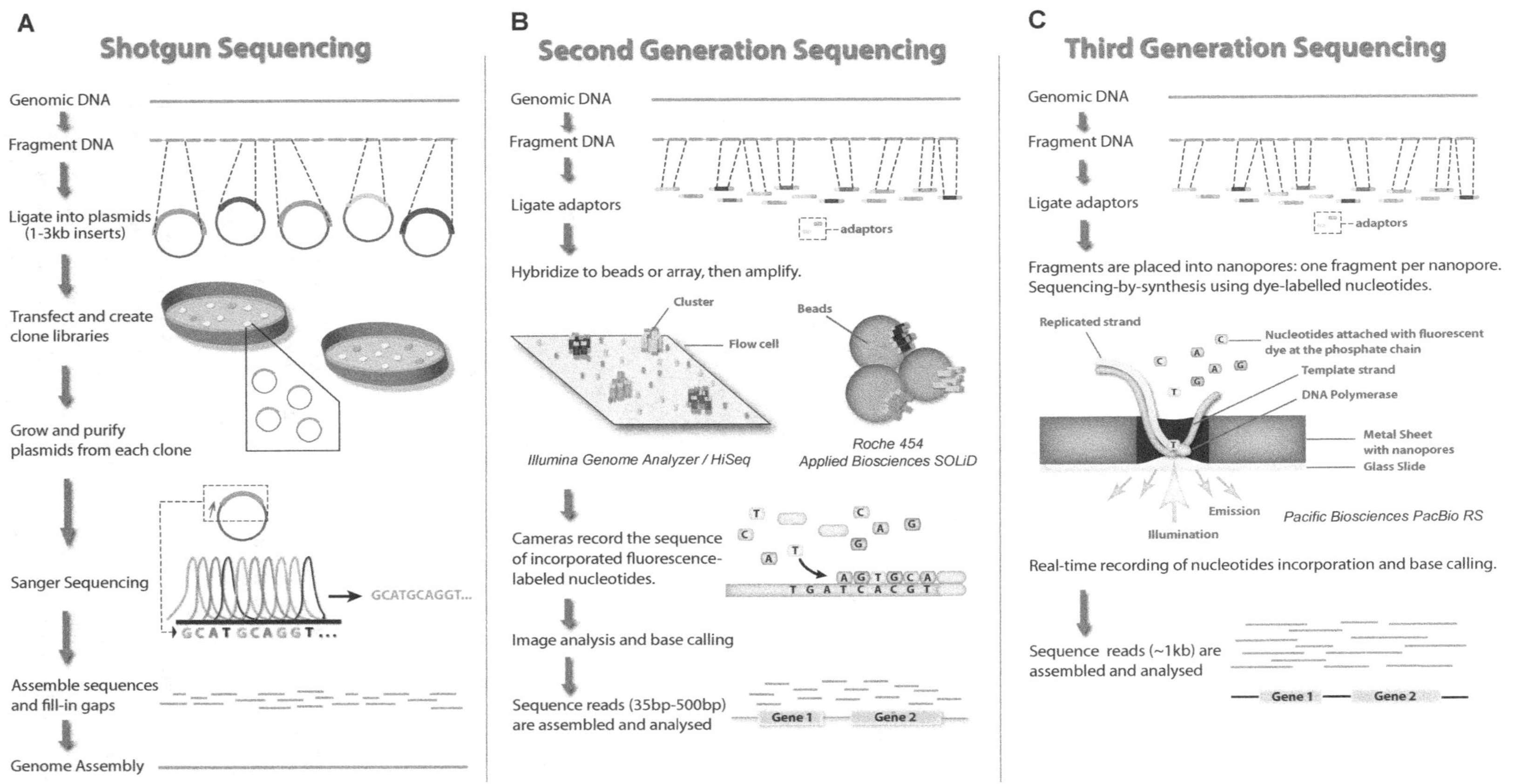

Figure 3.1 Overview of sequencing technologies. A. Overview of the shotgun sequencing method used for sequencing several genomes of *Plasmodium* species e.g. *P. vivax*, *P. y. yoelii*, and *P. knowlesi*. This method is labour-intensive due to generation and processing of insert libraries for Sanger sequencing. B. Overview of the second generation sequencing methodology. There are several different types of this technology touted by different companies, but the outcome in all cases is significantly higher throughput, lower costs and shorter generation time than Sanger sequencing. C. Overview of third generation sequencing (TGS). TGS technology achieves real-time sequencing of single-molecules at a high-throughput level, thus avoiding amplification biases inherent in second generation platforms. In addition, TGS technology is expected to be cheaper and less-time consuming than NGS.

Table 3.1 Summary of the current next generation platforms. The data shown are 'per run' statistics, and assembled from company websites (accessed March 2012), and from Glenn *et al.* (2011)

Company	Platform	Max. read length (bp)[a]	Run time (h or days)	Throughput (Mb)	Usable reads (millions)
Roche/454 Life Sciences	454 FLX	600	10 h	450	1
	454 FLX+	1000	23 h	700	1
	GS Junior	400 (average)	10 h	35	0.1
Applied Biosystems	5500 SOLiD	75+35	8 days	52,000	>700[b]
	5500xl SOLiD[c]	50	8 days	70,000	>1400[b]
	Ion Torrent PGM[d]	400	2 h	10/100/1000	0.1/1/4–8
Illumina	GAIIx	150+150	14 days	95,000	320
	HiSeq2000	100+100	8.5 days	600,000	3000
	MiSeq	150+150	27 h	2000	7
PacBio	RS	>10,000	0.5–2 h	5–10	0.01

[a]Single-end except '+', indicating reads from both ends of fragment; lengths are maxima except where indicated.
[b]Mappable reads; otherwise 'usable' means 'passed quality filters'.
[c]Using 0.75 μm nanobeads.
[d]2012 forecasts for semiconductor chips 314/316/318.

homopolymers one of the major sources of error in 454 sequencing (Gilles *et al.*, 2011). For this reason 454 sequencing of the highly AT-rich genome of *P. falciparum* is problematic when deep coverage across the genome, and especially non-coding regions (which are particularly AT-rich), are required. Roche 454 sequencing is furthermore expensive compared to other NGS platforms (Glenn, 2011). Nevertheless, two *P. falciparum* progeny clones from a genetic cross have been sequenced at more than 30× coverage by 454 methods, from a combination of whole genome shotgun and 3 kb paired-end libraries (Samarakoon *et al.*, 2011). These sequence data were used to identify shared single-nucleotide polymorphisms (SNPs) between the parents and progeny, and discover new crossovers and gene conversions. Recent studies have combined 454 sequencing with cheaper high-throughput methods (see below) to produce reference assemblies for several monkey malaria species, including *Plasmodium cynomolgi* (Tachibana *et al.*, 2012), *Plasmodium coatneyi* and *Plasmodium gonderi* (Jane Carlton, personal communication).

Applied Biosystems platforms

The Applied Biosystems SOLiD platform generates short reads of up to 75 bp at a rate of several million reads per run. As in 454 sequencing, template DNA is immobilized and PCR amplified on beads. Amplified templates are deposited onto glass slides and sequenced by multiple rounds of ligation, detection, and cleavage of labelled oligonucleotides. This approach can produce more accurate sequences than 454 pyrosequencing technology, but the yield in terms of reads is significantly lower compared to the Illumina® platform output, and as a consequence the SOLiD system (and its progeny, the 5500 series) has not been widely used to sequence *Plasmodium* genomes.

Illumina® platforms

The Illumina® suite of sequencers is currently capable of producing up to 3 billion reads per run with a read length of up to 150 bp. Template dsDNA strands are fragmented, ligated to specific adaptors, and PCR amplified to form a clonal library (Fig. 3.1). After hybridization to a flat surface, the DNA fragments are bridge-amplified to create a highly dense sequencing flow cell containing tens of millions of clusters each containing around 1000 copies of the same template fragment. The DNA fragments are sequenced using reversible terminators with removable four-colour fluorescent dyes to generate a base-by-base sequence.

The Illumina sequencing platform has to date been the most widely used NGS method to sequence the *Plasmodium* genome. Since standard Illumina® library preparation and sequencing methods are inadequate for *P. falciparum* AT-rich genomes, procedures have been customized for this species. The first approach, developed by the Wellcome Trust Sanger Institute, proposed an 'amplification-free' library preparation that reduces the incidence of duplicate sequences, and improves read mapping, *de novo* assembly and SNP detection (Kozarewa *et al.*, 2009). Another approach, used by Ponts and colleagues, relies on high-fidelity Taq enzyme mixes and reduced amplification cycles at a low elongation temperature (Ponts *et al.*, 2010). Yet another technique involves linear amplification using a T7 promoter sequence and *in vitro* transcription (Bartfai *et al.*, 2010). Additional experiments have compared a range of optimal enzymes for amplifying *P. falciparum* sequencing libraries (Quail *et al.*, 2012), the effects of reducing PCR extension temperature (Lopez-Barragan *et al.*, 2011b), and the use of recombinase polymerase amplification (RPA) technology, tetramethylammonium chloride (TMAC; a PCR additive that improves thermostability of A-T base pairs) (Oyola *et al.*, 2012). These methods overcome part of the amplification bias and result in more error-free and homogenous libraries.

In addition, Illumina® sequencing has been successfully used in novel functional genomic approaches such as SNP-seq, MAINE-seq/FAIRE-seq, ChIP-seq, and RNA-seq, to study different aspects of *Plasmodium* biology (Fig. 3.2). For example, resequencing (SNP-seq) of different drug resistant lines of the murine malaria parasite *P. c. chabaudi* characterized and linked new mutations to drug resistance (Margulies *et al.*, 2010; Martinelli *et al.*, 2011), and identified polymorphic genes on *P. c. chabaudi* chromosome 8 as potential targets of strain-specific protective immunity in mice (Cheesman *et al.*, 2010). SNP-seq of a single field isolate (IQ07) of the human malaria parasite *P. vivax* from Peru identified thousands of SNPs (Dharia *et al.*, 2010a). Two newly developed methodologies, FAIRE-seq and MAINE-seq, have been used to describe the chromatin landscape of *P. falciparum*. FAIRE (formaldehyde-assisted isolation of regulatory elements) was used to isolate protein-free DNA, and MAINE (MNase-mediated purification of mononucleosomes) to isolate histone-bound DNA. When both techniques were coupled to high-throughput Illumina sequencing, they allowed the characterization of genome-wide maps of nucleosome occupancy in *P. falciparum* across the parasite erythrocytic cycle (Ponts *et al.*, 2011; Ponts *et al.*, 2010). More recently, RNA-seq (deep next-gen sequencing of cDNA to profile transcriptomes) has been used in *P. falciparum* to validate expression profiles previously reported from microarray studies, and characterize new splicing events and transcription start sites (Dimon *et al.*, 2010; Lopez-Barragan *et al.*, 2011a; Otto *et al.*, 2010; Sorber *et al.*, 2011; Tuda *et al.*, 2011). Finally, ChIP-seq (chromatin immunoprecipitation coupled to deep sequencing) selectively enriches for DNA sequences bound to particular protein complexes. This technique has been used to uncover chromatin features in *P. falciparum* (Bartfai *et al.*, 2010). Many of these studies are described in more detail below and in Chapter 5.

Third-generation sequencing: the future

More recently, next-generation genomics has seen the arrival of so-called third generation sequencing (TGS) platforms for sequencing single template molecules (Fig. 3.1). The advantages of directly sequencing single template molecules are that clonal duplication and amplification biases are avoided, and extremely long sequence reads can be obtained. The first TGS platform to be commercialized was Pacific Biosystems' PacBio RS system (Eid *et al.*, 2009), which sequences single molecules in real time in 30 min sequencing runs. A thin 100nm metal film perforated with an array of nanoscopic holes and bound to a glass substrate is anchored to the glass bottom, and a single DNA polymerase molecule added per well. As fluorescently labelled nucleotides are incorporated one-by-one via the polymerase, the fluorescent dye is excited by the laser which transmits a colour signal. The dye is then cleaved off by the DNA polymerase to allow for additional rounds of nucleotide incorporation. The colour fluorescence is detected and decoded, and can be translated to reads averaging >1000 bp, with some reads exceeding 10 kb, although error rates and

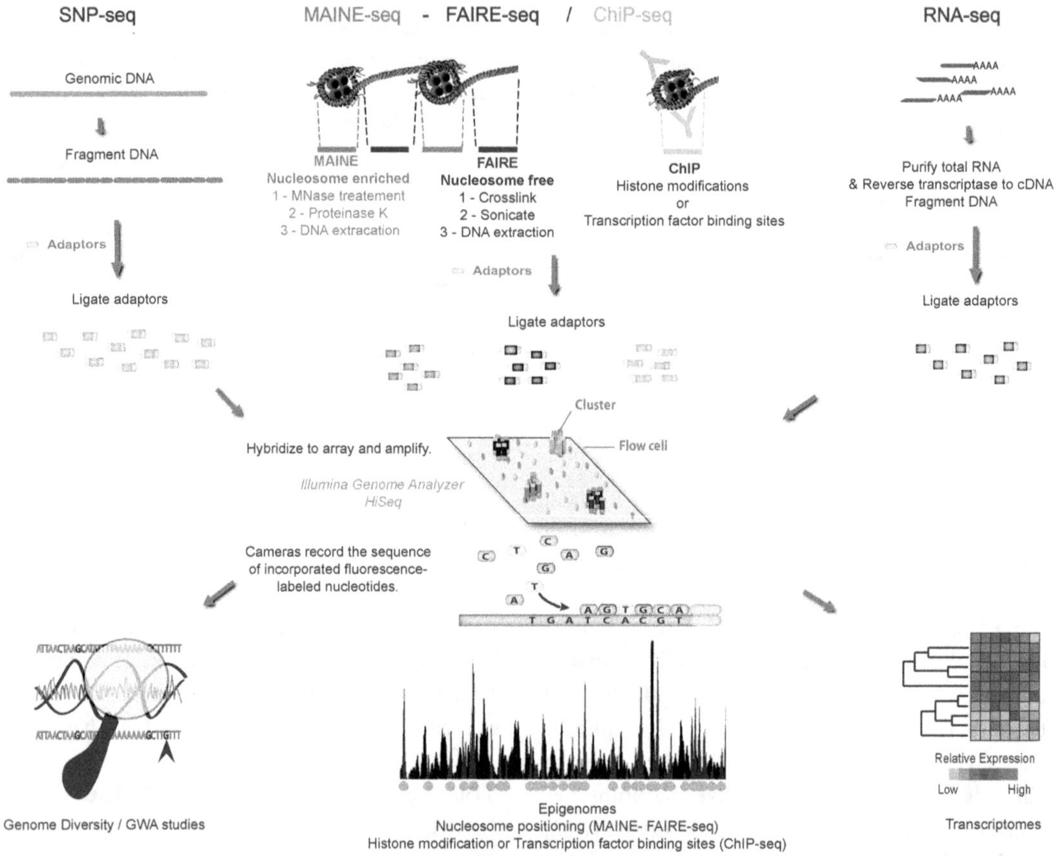

Figure 3.2 Next generation sequencing technologies for novel functional genomic approaches. (1) SNP-seq (single-nucleotide polymorphism sequencing): for re-sequencing parasite genomes to identify DNA variation linked to drug resistance and pathogenesis. (2) FAIRE-seq (formaldehyde-assisted isolation of regulatory elements) and MAINE–seq (MNase-mediated purification of mononucleosomes couple to sequencing): for isolation of protein-free and protein-bound DNA in order to describe the parasite chromatin landscape. (3) ChiP-seq (chromatin immunoprecipitation coupled to sequencing): to analyse protein interactions with DNA. (4) RNA-seq: a sequencing technique to sequence cDNA and measure the parasite transcriptome to detect differential gene expression.

cost-per-run are relatively high. Nevertheless, in addition to eliminating amplification bias, TGS technologies are expected to be ultimately more affordable and less-time consuming than their predecessors. While TGS platforms are gradually becoming accessible to researchers in the field, better algorithms and software are needed to manipulate, store and analyse the millions of long reads generated. Meanwhile, an even newer contender for 'the next next generation' has arisen in the form of Oxford Nanopore's single-molecule sequencing platforms, which rapidly read the sequence of a DNA strand as it passes through a protein (translocase) nanopore embedded in a solid substrate (Clarke *et al.*, 2009).

Procurement of adequate parasite material for sequencing

There are several problems associated with procuring *Plasmodium* genomic material of sufficient quantity and quality for sequencing. First, because malaria parasites are largely intracellular organisms, issues posed by host nucleic acids contaminating the sample are enormous. A *Plasmodium* parasite has a haploid genome size (~30 Mb) approximately 100 times smaller than the diploid human genome (~3 Gb), a disparity that makes even the smallest amount of genomic contamination from host cells (e.g. human leucocytes) confounding for sequencing. Moreover, while human erythrocytes lack nuclei, many *Plasmodium* species (e.g. lizard and bird malaria parasites) infect nucleated erythrocytes, rendering cell filtering strategies ineffective. Second, the parasitaemia (quantity of parasites in the blood) can be extremely low for certain malaria parasite species such as *P. vivax* [and saurian and avian malaria parasites, as discussed in Chapter 1 (Martinsen & Perkins)], making it difficult to procure sufficient quantities of material for sequencing. Finally, many species of *Plasmodium* cannot be cultured *in vitro*, and even for those that can the adaptation process may take many weeks, during which time there can be changes in genetic material (Bourke *et al.*, 1996). By-passing *in vitro* cultivation, a standard technique used for collection of *P. falciparum* or *P. vivax* samples in field settings is 'finger-prick filter paper' whereby blood drops from a finger prick of an infected patient is collected on filter paper, allowed to air-dry, and can be stored at room temperature for months or years.

A number of different techniques have been developed over the past several years to try to alleviate the problems of obtaining sufficient quality and quantity of *Plasmodium* DNA, as outlined below. These methods are particularly under the spotlight now that groups of researchers are attempting to sequence parasite genomes directly from patients without culture adaption of parasites (Auburn *et al.*, 2011; Robinson *et al.*, 2011; Manske *et al.*, 2012).

Host white blood cell depletion

While centrifugation can be used to remove most leucocytes (the 'buffy coat') from a blood sample, it leaves sufficient material behind to interfere with parasite sequence analysis. Inexpensive cellulose powder (CF11)-based methods for clearing residual leucocytes from infected blood date back to the 1950s. The exact basis for their success remains surprisingly unclear, size exclusion and affinity for leucocyte surface molecules being the main suspects. Recent modifications to CF11 depletion protocols (Sriprawat *et al.*, 2009; Venkatesan *et al.*, 2012) largely eliminate earlier concerns about retention of infected red blood cells (IRBCs) by CF11 columns, and offer host contaminant depletion rates comparable to or exceeding that of a commercial depletion filter (Plasmodipur, EuroProxima) at much reduced cost. However, 'home-made' CF11 columns and Plasmodipur filters both require relatively high sample volumes (>1.5 ml) and parasite densities (>5000/µl) in order to recover sufficient parasite DNA for next-generation sequencing.

In the last decade, magnetic separation methods have begun to be used to generate extremely pure samples of parasite DNA from blood. In one method, magnetic beads conjugated to antibodies against common leucocyte surface antigens pull leucocytes from solution when a magnet is applied, leaving purified RBCs in the flow-through (Auburn *et al.*, 2011; Carlton *et al.*, 2001). A converse method exploits the distinctive paramagnetic properties of haemozoin, a metabolic waste produced and stored by malaria parasites, to magnetically harvest IRBCs (Trang *et al.*, 2004). The main drawback of these methods is cost, especially in a field setting.

Parasite selection and amplification

An alternative to cell depletion or selection methods is to perform selection on isolated DNA. This has particular potential for use with parasites that infect nucleated cells (as in reptile and avian *Plasmodium* parasites, although no reference genome exists for species of these hosts yet), but also when facilities to collect and store sufficient blood cells for centrifugation and column-based methods are limited, as can be the case in a field setting. Hybrid capture employs biotinylated RNA probes which hybridize to complementary parasite DNA and are subsequently pulled out of solution using magnetic streptavidin-coated beads (Fig. 3.3). After host DNA is washed away, the parasite DNA is eluted from the beads and processed for sequencing. Probes based either on *P. falciparum* exon sequences or randomly sheared genomic DNA were able to enrich *P. falciparum* DNA as much as ~40-fold in the final eluate compared to the input mixture (Melnikov *et al.*, 2011) and to enrich *P. vivax* DNA from less than 0.5% to a median of 55% (Bright *et al.*, 2012). However, the protocol requires at least 2 µg of DNA (host + parasite) as input, a limiting factor that can be overcome by incorporating a whole genome amplification (WGA) step prior to hybrid selection. Multiple

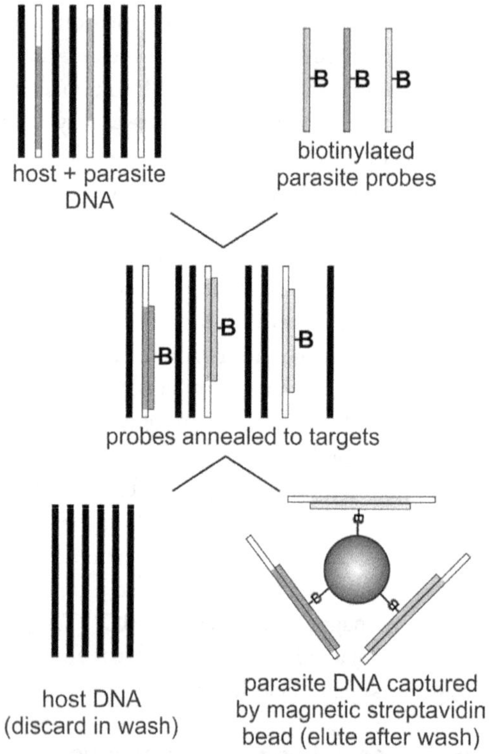

Figure 3.3 Hybrid capture of parasite DNA from a parasite/host DNA mixture. A complex sample containing host and parasite DNA sequences (target) is hybridized in solution with biotinylated oligonucleotide DNA probes complementary to parasite genomic sequence (baits). The annealed target/bait hybrids are captured by streptavidin-coated magnetic beads, which are immobilized on the reaction tube surface by a magnet while the host DNA remaining in solution is washed away. The bound parasite DNA is then eluted from the beads for use in sequencing.

displacement amplification, the WGA method of choice, employs random hexamers and Phi29 polymerase to amplify genomic DNA with minimal position or composition bias. Using WGA on 5–10 ng of DNA from *P. falciparum*-infected whole human blood samples, Wang *et al.* (2009) expanded the amount of DNA approximately 1000-fold, enough to detect SNP, microsatellite, and gene targets from all 14 chromosomes of the parasite's highly AT-rich genome. Starting from >1 year-old blood spot samples whose *P. falciparum* content constituted just ~0.1% of the extracted DNA, WGA followed by hybrid selection with 3D7 genomic probes resulted in a 40 to 70-fold enrichment of parasite DNA, sufficient to allow accurate SNP calling even when the probe and isolate strains were not the same (Melnikov *et al.*, 2011).

Towards single-cell genomics

The ultimate challenge of applying genomics to a single *Plasmodium*-infected RBC – useful for studying expression of *P. falciparum var* or *P. vivax* multi-gene family members, for example – is being met by the development of efficient single-cell capture methods. For example, Miao and Cui (Miao and Cui, 2011) recently described a protocol for using a high-speed cell sorter to quickly isolate individual, singly infected RBCs from a cell mixture. Because no cytotoxic dyes are used in the sorting, the RBCs can be subsequently cultured or, where culture is not an option, WGA could theoretically be used to obtain sufficient DNA for sequencing (although this approach has not yet been tried).

Computational issues involved in *Plasmodium* genome assembly, annotation and analysis

Genome assembly

Assembling a genome, especially *de novo* (i.e. in the absence of a reference sequence), is not for the faint-of-heart (Baker, 2012). *De novo* assembly relies on aligning overlapping reads to produce progressively larger contiguous genomic sequence (contigs). In the ideal case this process would yield one long 'supercontig' per chromosome, but in practice, eukaryote chromosomes contain regions that are difficult to sequence and/or assemble, and first-pass sequencing yields a set of non-overlapping contigs whose positional relationships to each other in the genome are unknown. The next level of assembly is achieved by using a subset of reads designed to occur a known distance apart (paired-end or mate-pair reads). These serve as markers within contigs for assembling them in correct order into larger constructs (scaffolds), separated by unknown sequence of known (to a good approximation) length. A fully sequenced chromosome is thus essentially a scaffold where the inter-contig sequence (and sequence at the ends of the chromosome) has been determined more or less laboriously in the 'closure' stage of the sequencing project.

At the time when the *P. falciparum* genome was being sequenced via Sanger sequencing and early assembly software, two main sources of difficulty were its large ~23 Mb size (it was the largest microbial genome to be attempted) and extraordinarily high >80% AT compositional bias (Gardner, 2001). The latter factor impeded not only the cloning of genomic template fragments into bacterial and yeast vectors then in common use for generating sequencing libraries, but also the sequencing itself, with extensive secondary structure and long homopolymer tracts posing obstacles to the progress and fidelity of sequencing polymerases. Even in more compositionally balanced genomes, regional variability in base composition results in some regions being sequenced less or with a higher error rate than others. Furthermore, while assembling single-chromosome bacterial genomes was relatively straightforward by the turn of the 21st century, the computational complexity of assembling the numerous shotgun reads generated from megabase-sized eukaryote genomes posed an unprecedented computational challenge, and remains a significant consideration today (Salzberg *et al.*, 2012).

A third challenge, the presence of long regions of highly similar ('repeat') sequences dispersed within eukaryotic genomes, remains the bane of nearly all sequencing and assembly methods developed to date. Briefly stated, while repeat regions A and A′ may lie far apart or even on separate chromosomes, without significant differences in the sequences the assembly software is unable to differentiate them and will assemble reads from both of them into a single contig. The only surefire way to correctly assemble such regions is to generate reads long enough to sequence through the repeat region, and into the flanking unique sequence. For massively repetitive genomes, conventional first- and second-generation sequencing yields many instances of 'pileup' of reads into small contigs that are impossible to assemble further, resulting in a highly fragmented, weakly scaffolded genome assembly. The ultimate salvation for such assemblies is technology that can produce extremely long high-quality reads, the so-called third generation sequencing described above, currently represented by bulky, expensive, and error-prone single molecule platforms (e.g. Pacific Biosystem's PacBio), and nanopore technology (e.g. Oxford Nanopore's GridION and MinION systems).

Other persistent obstacles to *Plasmodium* genome assembly include (1) read (basecalling) errors such as mismatches and insertions/deletions, especially length inaccuracies produced while reading through stretches of homopolymeric sequence; and (2) the presence of contaminating host DNA sequence in samples (see above). Read errors confound correct alignment of reads, while sequencing and assembly of contaminant reads wastes valuable chemical and computational resources, reduces coverage of the target genome, and can produce false chimerical contigs.

De novo assemblers of eukaryotic genomes faced the daunting task of performing all-against-all pairwise comparison of millions of reads comprising gigabytes of data, and this spurred development of more powerful assembly software at the turn of the 21st century. Celera Assembler (Myers *et al.*, 2000), which was based on the earlier TIGR Assembler, was used in TIGR's *P. falciparum*, *P. y. yoelii*, and *P. vivax* projects as well as Celera's high-profile *Homo sapiens* genome project. Researchers at the Wellcome Trust Sanger Institute, where *P. falciparum* chromosomes 1, 3–9 and 13 (Hall *et al.*, 2002), and the genomes of the rodent malaria species *P. c. chabaudi* and *P. berghei* (Hall *et al.*, 2005) and the monkey species *P. knowlesi* (Pain *et al.*, 2008b) were sequenced, favoured Phrap (P. Green, http://www.phrap.org/) and its offshoot Phusion (Mullikin and Ning, 2003), which use base quality score information to inform final sequence determination for *de novo* assemblies.

The advent of next-generation sequencing required *de novo* assemblers that could handle reads much shorter and more numerous than those generated by Sanger sequencing, representing a huge increase in computational complexity. Three strains of the monkey malaria species *P. cynomolgi* (Berok, strain B, and Cambodian) were sequenced primarily on Roche 454 and

Illumina platforms, the first reference *Plasmodium* genomes to be generated from next-generation reads (Tachibana *et al.*, 2012). *P. cynomolgi* Berok was assembled *de novo* from decontaminated (primate sequence-depleted) 454 reads, using GS De Novo Assembler ('Newbler'), the proprietary software supplied by Roche. *P. cynomolgi* strain B was assembled from a combination of Roche 454, Illumina, and Sanger reads, using Newbler for *de novo* assembly. Four more *P. vivax* genomes from Brazil, Mauritania, North Korea and India, were assembled *de novo* using the ALLPATHS LG assembly algorithm (Gnerre *et al.*, 2011). Assembly quality was significantly higher for *P. vivax* than for a *P. falciparum* assembly generated using the same approach due to the more moderate AT base composition of the *P. vivax* genome (Neafsey *et al.*, 2012). *De novo* assembly today is a rapidly developing field where new software packages proliferate, each offering multiple customizable run parameters. The result is an emerging, urgent need for assembler performance evaluations to help new users make sense of the welter of options available to them. Tentative movement in this direction is evidenced by comparative evaluations such as GAGE (Salzberg *et al.*, 2012) and 'Assemblathon' competitions (Earl *et al.*, 2011), but getting the many players in the assembly software arena to agree on standards seems unlikely, especially as the nascent next generation of sequencing technology could in short order render today's *de novo* assembly problems and solutions obsolete.

Reference mapping – alignment of reads to a reference genome – is the assembly method of choice when a genome is being re-sequenced or is closely related to a sequenced genome, as it avoids most of the computational burden associated with *de novo* assembly. However, a 'reference' genome may not be an entirely reliable model of the genome (or transcriptome) being sequenced; significant genomic deletions, duplications, and rearrangements can occur between isolates of the same species (Tachibana *et al.*, 2012), as well as between species. This is either a pitfall or, if the intent is to explore sequence copy number variation, a boon. The number of read mapping applications exceeds that for *de novo* assembly and appears to be growing monthly due to the crucial role read mapping plays in transcriptome sequencing (RNA-Seq). Mapping assemblers (see Bao *et al.*, 2011, for an overview) can be broadly categorized into two groups: (1) programs that use a memory-intensive BLAST-like sequence similarity search strategy, e.g. SSAHA2 (Ning *et al.*, 2001) and 454's GS Reference Mapper; and (2) a newer class based on the Burrows–Wheeler Transform, a memory-light algorithm originally designed for data compression, e.g. Bowtie2 (Langmead and Salzberg, 2012), BWA (Li and Durbin, 2009), SOAP2 (Li *et al.*, 2009b). In addition, applications have been developed which leverage these mapping assemblers to, for example, identify RNA splice junctions (TopHat; Trapnell *et al.*, 2009), estimate transcript abundance, and test for differential expression (Cufflinks; Trapnell *et al.*, 2012).

Reference mapping has been used to survey the transcriptome of *P. falciparum* (e.g. Otto *et al.*, 2010, which used SSAHA) and was employed in the *P. cynomolgi* strain B sequencing project to map large *de novo*-assembled contigs to a *P. vivax* reference genome and detect gene copy number variants between strains Berok and B (Tachibana *et al.*, 2012). Read mapping is also a crucial tool for detecting fine-resolution sequence variation, i.e. SNPs between samples and even, in the case of infection with multiple clones of a *Plasmodium* species, within a single sample (Bailey *et al.*, 2012; Juliano *et al.*, 2010). Mapping assemblers with SNP-calling extensions include SAMtools (Li *et al.*, 2009a), which was used to call SNPs in the *P. cynomolgi* project, SOAP, and SSAHA. With many more SNP-detection sequencing projects of *P. falciparum* and *P. vivax* field isolates expected in the next few years, read mapping will undoubtedly become one of the most-used genomic tools in malaria research.

Lastly, an exciting and largely unexplored frontier in the study of *Plasmodium* genome assembly and the study of variation is represented by the work of Riley *et al.* (2011), who have applied cutting edge optical mapping techniques to the problem. Gardner *et al* (2002) employed an early version of optical mapping in the assembly of the reference *P. falciparum* 3D7 genome. Briefly, chromosomes are separated and immobilized on a substrate, digested with a restriction enzyme(s), and imaged. In each immobilized chromosome, restriction cuts are revealed as spaces between fragments arranged in a line. The sizes of these fragments are calculated and compared to the sizes generated by a 'virtual' digest of the same chromosome from a reference genome. In this way, fragments can be ordered and orientated onto chromosomes. In addition, differences in fragment size between the reference genome and test genome(s) represent possible large-scale variations, e.g. copy number variants. Thus, multiple chromosomes can be compared simultaneously, allowing for rapid whole-genome comparison of multiple strains at comparatively low cost (<$1000/genome).

Gene finding and genome annotation

Until recently, the gene finding and genomic feature annotation stages of *Plasmodium* genome projects (e.g. for the *P. falciparum*, *P. y. yoelii*, *P. vivax* and *P. knowlesi* reference genomes) relied on workflows that were highly customized to the genome centre running the project, and typically involved a large manual curation component (Gardner *et al.*, 2001). This has begun to change, starting with the *P. cynomolgi* strain B reference genome, which was annotated using a portable and easily configurable genome annotation pipeline called MAKER (Cantarel *et al.*, 2008). Such tools will be vital to support the decentralization of whole-genome sequencing from large sequencing cores to individual labs of all sizes that are taking advantage of the compact, relatively inexpensive high-throughput sequencing platforms now available.

Computational issues with manipulation of large datasets

Manipulation and storage of enormous and ever-growing datasets is now widely recognized as the most critical problem facing next-generation sequencing. Data sizes can now easily overwhelm the processing power of typical computing clusters, and tax the storage capacities of sophisticated core facilities, requiring terabytes of RAM and petabytes of disc space. In response, attention is increasingly turning to large-scale distributed computing and storage solutions, including cloud-based systems (Wall *et al.*, 2010), using, for example, an open-source framework such as Hadoop to manage them (Niemenmaa *et al.*, 2012; Taylor, 2010). Such solutions almost certainly reflect the future, though issues with data transfer speeds and data security will need to be resolved first.

The *Plasmodium* nuclear and organellar genomes and comparative genomics

A snapshot of current malaria parasite genome sequencing initiatives is shown in Table 3.2. Draft reference assemblies of seven different *Plasmodium* species have been published as of Fall 2012, all from species that infect mammals and with a particular focus on human malaria species. The *P. falciparum* and *P. y. yoelii* genomes were first published in 2002 (Carlton *et al.*, 2002; Gardner *et al.*, 2002), followed by two more rodent malaria genomes *Plasmodium berghei* and *P. c. chabaudi* in 2005 (Hall *et al.*, 2005), the *P. vivax* and *P. knowlesi* genomes in 2008 (Carlton *et al.*, 2008; Pain *et al.*, 2008), and one reference assembly and two SNP-discovery sequences of the monkey malaria species *P. cynomolgi* in 2012 (Tachibana *et al.*, 2012). SNP-discovery projects for many *P. falciparum* and *P. vivax* field isolates are under way, in particular two large-scale sequencing efforts at the Wellcome Trust Sanger Malaria Program and at the Broad Institute. Whole genome sequence data are available from various sources including DDBJ/EMBL/GenBank, the centralized malaria sequence database PlasmoDB (Aurrecoechea *et al.*, 2009), and specialized websites at several genome sequencing centres.

Plasmodium nuclear genome size, base composition and architecture

The current set of draft genome assemblies are all from mammalian malaria parasite species, which biases our understanding of the variety of genomes in the *Plasmodium* genus. However, based upon this limited number from species that infect rodents, humans and monkeys, an interesting finding is the apparent conservation of features such as size, composition and architecture (Table 3.3). The mammalian *Plasmodium* genome is haploid, with an average size of ~25 Mb, distributed

Table 3.2 *Plasmodium* whole-genome sequencing projects. Reference genomes now exist for several *Plasmodium* species, and SNP-discovery projects for many field isolates are under way. This table does not aim to be comprehensive but instead provide a snapshot of *Plasmodium* genome projects that have been published. For more up-to-date information, two large-scale sequencing efforts in particular should be viewed at (1) the Sanger Malaria Program http://www.sanger.ac.uk/research/projects/malariaprogramme-kwiatkowski/sequencing.html; and (2) the Broad Institute http://www.broadinstitute.org/science/projects/gscid/white-papers.

Primary host	Species, strain or clone	Technology	Reference	Comments
Human	*P. falciparum*, 3D7	Sanger	Gardner *et al.* (2002a)	Reference genome
	P. falciparum, Dd2, HB3	Sanger	Volkman *et al.* (2007)	For SNP detection
	P. falciparum IT, Ghanaian, PFCLIN	Sanger	Jeffares *et al.* (2007)	For SNP detection
	P. falciparum, 7C126, SC05	Roche 454	Samarakoon *et al.* (2011)	Assembled
	P. falciparum, 227 isolates Asia, Africa, Oceania	Illumina	Manske *et al.* (2012)	For SNP detection
	P. vivax, Salvador I	Sanger	Carlton *et al.* (2008)	Reference genome
	P. vivax, India VII, North Korea, Brazil I, Mauritania	Illumina	Neafsey *et al.* (2012)	Reference assemblies for diverse strains
	P. vivax, Peru strains	Illumina	Bright *et al.* (2012); Dharia *et al.* (2010a)	For SNP detection
	P. vivax, Belem, Madagascar and Cambodia strains	Illumina	Chan *et al.* (2012)	For SNP detection
Chimpanzee	*P. reichenowi*, Dennis strain	Sanger	Jeffares *et al.* (2007)	Low coverage for SNP detection
Human/monkey	*P. knowlesi*, H strain	Sanger	Pain *et al.* (2008)	Reference genome
Monkey	*P. cynomolgi*, strains B, Berok and Cambodian	Sanger, Roche 454, Illumina	Tachibana *et al.* (2012)	Reference genome strain B; SNP/CNV detection Berok and Cambodian
Rodent	*P. yoelii yoelii*, NS	Sanger	Carlton *et al.* (2002)	Low-coverage draft assembly
	P. chabaudi chabaudi, AS	Sanger	Hall *et al.* (2005)	Low-coverage draft assembly
	P. berghei, ANKA	Sanger	Hall *et al.* (2005)	Low-coverage draft assembly
	P. chabaudi chabaudi, AS	Illumina	Hunt *et al.* (2010)	Resequencing of drug-resistant clones

among 14 linear chromosomes that range in size from 500 kb to >3 Mb (Carlton *et al.*, 1999; Janse *et al.*, 1994; Kemp *et al.*, 1987). Base composition varies from species to species, from the extreme 81% AT in *P. falciparum* (Gardner *et al.*, 2002) to 57.7% AT in *P. vivax* (Carlton *et al.*, 2008), but this variation is not specific to the host lineage. Intriguingly, an 'isochore' structure exists in some species (*P. cynomolgi, P. vivax*) whereby regions of the genome differ in their density (GC content) and are separable by CsCl centrifugation (McCutchan *et al.*, 1984). In contrast, *P. falciparum*, rodent malaria parasites and *P. knowlesi* have a more uniform base composition along the length of their nuclear genomes.

Several features of chromosome structure are conserved in all sequenced *Plasmodium* species. For example, short regions of extremely (>97%) AT- and repeat-rich sequences containing the centromeres are conserved in position in all species. In *P. falciparum*, the 14 centromeres have been found to conform to a strictly conserved

Table 3.3 Comparison of genome features between five *Plasmodium* genomes: the human malaria parasite *P. falciparum*; three species of the monkey malaria clade *P. cynomolgi*, *P. vivax* and *P. knowlesi*; and a representative of the rodent malaria clade, *P. y. yoelii*

Feature	*P. falciparum*	*P. cynomolgi*	*P. vivax*	*P. knowlesi*	*P. y. yoelii*
Reference	Gardner *et al.* (2002)	Tachibana *et al.* (2012)	Carlton *et al.* (2008)	Pain *et al.* (2008)	Carlton *et al.* (2002)
Assembly					
Size (Mb)	23.3	26.2	26.9	23.7	23.1
No. scaffolds[a]	14	14 (1649)	14 (2547)	14 (67)	5687
Coverage (fold)	14.5	161	10	8	5
G+C content (%)	19.4	40.4	42.3	38.8	22.6
Genes					
No. genes	5403	5722	5432	5197	5878
Mean gene length	2283	2240	2164	2180	1298
Gene density (bp per gene)	4312	4428.2	4950.5	4416.1	2566
Coding (%)	52.6	51.0	47.1	49.0	50.6
Structural RNAs					
No. tRNA genes	43	43	44	41	39
No. 5S rRNA genes	3	3	3	0[b]	3
No. 5.8S/18S/28S rRNA units	7	7	7	5	4
Nuclear genome					
No. chromosomes	14	14	14	14	14
No. centromeres	14	14	14	14	14
Isochore structure	No	Yes	Yes	No	No
Mitochondrial genome					
Size (bp)	5967	5986	5990	5958	5957
G+C content (%)	31.6	30.3	30.5	30.5	30.9
Apicoplast genome					
Size (bp)	29,430	29,297	29,093	Not available	29,227
G+C content (%)	13.1	13.0	13.2	Not available	13.9

[a]Small unassigned contigs indicated in parentheses.
[b]Not present in *P. knowlesi* assembly version 4.0.

4–4.5 kb size and structure due to the occupation by histone H3 variant CENH3 (Hoeijmakers *et al.*, 2012). All *Plasmodium* chromosomes possess telomeres consisting of degenerate, canonical, tandem repeats (Scherf *et al.*, 2001), with the mean length of the telomeric array (~800 to ~6700 bp) varying among species but remaining constant within species (Figueiredo *et al.*, 2002). Only in *P. knowlesi* and *P. cynomolgi* have single or tandemly repeated intrachromosomal telomeric sequences (ITSs, containing the heptad sequence GGGTT[T/C]A) been found at multiple internal sites within chromosomes, although whether their role is in facilitating recombination or transcriptional control is unclear (Tachibana *et al.*, 2012; Pain *et al.*, 2008). The subtelomeric regions of several *Plasmodium* species chromosomes consist of a variable number of species-specific repeats that extend 10–40 kb towards the centre of chromosomes, and which have extensive large-scale similarity between chromosomes, indicative of intra-chromosomal exchange (Carlton *et al.*, 2002; Gardner *et al.*, 2002). Since species-specific gene families coding for immunodominant antigens and proteins known to be involved in antigenic variation are found within these regions in several species (*P. falciparum*, *P. y. yoelii* and *P. vivax*), this ectopic recombination among heterologous subtelomeric chromosome regions provides a mechanism for the generation of different repertoires of antigen genes (Freitas-Junior *et al.*, 2000).

Plasmodium gene content: orthologues and paralogues

All *Plasmodium* genomes analysed to date have a predicted content of ~5400–5800 genes (Table 3.3). The difference in gene number between species is primarily due to gene family expansion in different lineages, and species-specific antigen and cell-surface protein-coding genes. The number of orthologous genes (genes that have evolved from a common ancestor by speciation) shared between different *Plasmodium* species is directly correlated to the evolutionary distance between them. For example, the three species belonging to the monkey malaria clade (*P. vivax*, *P. cynomolgi* and *P. knowlesi*) share ~90% of genes (4613; Tachibana *et al.*, 2012), whereas *P. falciparum* and *P. y. yoelii* have ~60% reciprocal best match orthologues (Carlton *et al.*, 2002). Other characteristics of coding and non-coding regions suggests more similarities than differences (Table 3.3). For example, the per cent coding and gene density is similar between all species, with the exception of *P. y. yoelii*, although statistics for this genome may be complicated by the fragmented nature of this assembly. A possible exception appears to be the mean length of genes; *P. falciparum* orthologues are slightly longer than other *Plasmodium* orthologues due to the presence of more low complexity regions (LCRs). LCRs constitute a larger proportion of *P. falciparum* proteins on average (60%) than *P. vivax* proteins (39%), and this may be linked to different immune evasion mechanisms deployed by the parasites (Carlton *et al.*, 2008).

The largest differences between *Plasmodium* genomes is seen in their gene families. For example, repeated lineage-specific gene duplication and/or gene deletion in multigene families appears to have occurred within the three monkey malaria clade species (Tachibana *et al.*, 2012). Moreover, the copy number of multigene families is larger in the *P. cynomolgi*/*P. vivax* lineage than in *P. knowlesi*, suggesting repeated gene duplication in the ancestral lineage of *P. cynomolgi*/*P. vivax* (or repeated gene deletion in the *P. knowlesi* lineage). Intriguingly therefore, since so many other features are conserved, the genomes of *P. cynomolgi*, *P. vivax* and *P. knowlesi* can almost be completely characterized by variations in the copy number of multigene families.

There are many fascinating gene families in the *Plasmodium* genus, and several of these appear in other chapters in this book. We focus here on one example that provides a snapshot of some of the comparative genomics research that can be done now that multiple whole *Plasmodium* genomes are available. First, the *pir* (plasmodium interspersed repeat) superfamily first described in *P. vivax* (the *vir* family, 319 copies; del Portillo *et al.*, 2001), and latterly in *P. y. yoelii* (the *yir* family, 838 copies; Carlton *et al.*, 2002), *P. berghei* (the *bir* family, 180 copies; Hall *et al.*, 2005) and *P. chabaudi* (the *cir* family, 138 copies; Janssen *et al.*, 2002), *P. knowlesi* (the *kir* family, 70 copies; Pain *et al.*, 2008b), and more recently in *P. cynomolgi* (the *cyir* family, 256 copies; Tachibana *et al.*, 2012), is the largest gene family in *Plasmodium*. PIR proteins reside on the surface of infected erythrocytes and play an important role in immune evasion through antigenic variation (Cunningham *et al.*, 2010). The majority of *pir* genes are located in subtelomeric regions of *Plasmodium* chromosomes, except for 11 *cyirs* that have sequence similarity to *P. knowlesi kirs* and occur more distantly from telomeres – as many of the *kirs* do in *P. knowlesi*. An intriguing finding from the *P. knowlesi* genome project was the identification of host monkey sequences in the KIRs, in particular a ~36 amino acid sequence coding for the immunoregulatory molecule CD99 (involved in the regulation of T-cell function) within the predicted KIR extracellular domain (Pain *et al.*, 2008a). This example of 'molecular mimicry' has also recently been found in one *P. cynomolgi* CYIR protein

(PCYB_032250) containing a 56-amino acid region highly similar to the extracellular domain of primate CD99. These observations indicate acquisition of host peptide sequences in several malaria parasite proteins across different species, an acquisition that may enhance the immunomodulatory role of the parasite.

Genome synteny

Plasmodium synteny maps are extraordinarily useful for a number of reasons, including (1) as a means to infer the evolution of the genus, i.e. tracking the gross chromosomal rearrangements that gave rise to the karyotype of current species seen today; (2) as a method for identifying orthologues between species which cannot be identified through sequence similarity alone; (3) for refinement of gene predictions through simultaneous annotation of multiple *Plasmodium* genomes; (4) for identification of conserved non-coding regions of the *Plasmodium* genome; and (5) as a means for the classification of genes under different evolutionary pressures. The first multispecies malaria parasite synteny map (Carlton *et al.*, 2008) was generated between *P. falciparum*, *P. vivax*, *P. knowlesi*, and the rodent malaria parasites *P. y. yoelii*, *P. berghei* and *P. chabaudi* (considered as a single lineage due to their almost complete synteny). Of the 3336 orthologues on the map between all six species, 3305 (99%) were found to be positionally conserved. In addition, it was possible to reconstruct the ancestral karyotype of the six species, which turned out to correspond to the karyotypes of *P. vivax* and *P. knowlesi*; the karyotypes of *P. falciparum* and the rodent malaria parasites can be reconstructed from this form through nine and six chromosomal rearrangements, respectively. No synteny breakage hot-spots were identified, indicating that intersyntenic breakpoints were unique during the divergence of the species. More recently, the synteny map was expanded to include *P. cynomolgi* (Tachibana *et al.*, 2012), and this identified two apparent errors in existing public sequence databases: an inversion in chromosome 3 of *P. knowlesi*, and an inversion in chromosome 6 of *P. vivax*. This underlies the importance of such maps for assembly and annotation checking.

Whole genome evolutionary analyses

Several *Plasmodium* whole genome selective constraint analyses have been carried out over the past several years. For example, maximum likelihood estimates of the rate of substitution in synonymous (d_S) and non-synonymous (d_N) sites, as well as the ratio d_N/d_S (ω) were calculated for 3322 orthologue pairs between *P. vivax* and *P. knowlesi* (Carlton *et al.*, 2008). Interestingly, genes in more G + C-rich, centromere-proximal *P. vivax* locations appear to evolve faster, while genes located in AT-rich isochores evolve slower. Genes coding for GPI-anchored proteins, cell adhesion genes, genes of the exportome and genes containing transmembrane or signal peptides motifs, all of which have partial extra-cellular presence, were found to evolve significantly faster than other groups of genes involved in, for example, carbohydrate metabolism, enzyme regulation and structural molecules, perhaps indicating how the host immune system has shaped the pattern of evolutionary rate across different classes of genes. More recently, ~4500 pairs of orthologues within *P. cynomolgi* strain B and Berok, and ~4600 orthologs between these two *P. cynomolgi* genomes and *P. vivax* Salvador I were analysed (Tachibana *et al.*, 2012). Less than 70 genes were found with $d_N > d_S$ within the two *P. cynomolgi* strains, and 3265 genes with $d_S > d_N$, providing evidence of extensive purifying selection across the *P. cynomolgi* genome. As far as evolution of orthologues between the two species, 83 genes were found under possible accelerated evolution but 3739 genes under possible purifying selection. This indicates that overall the genome of *P. cynomolgi* is highly conserved in single locus genes when compared to *P. vivax*, and emphasizes the value of *P. cynomolgi* as a biomedical and evolutionary model for studying *P. vivax*.

Plasmodium extranuclear genomes

This chapter has focused almost exclusively on the nuclear genome of malaria parasites. However, *Plasmodium* has two extranuclear organelles that contain genomes, the apicoplast (a non-photosynthetic plastid descended from a secondary endosymbiosis of algal origin) and the mitochondrion. Genomes of both are highly conserved in *Plasmodium* parasites (Table 3.3). The *Plasmodium* mitochondrial genome is a linear ~6 kb element and contains only three protein-coding genes, cytochrome c oxidase subunit I and III, cytochrome b, and large subunit and small subunit ribosomal RNA (rRNA) genes. Comparison of 23 complete genomes from different *Plasmodium* species revealed complete conservation of the arrangement of rRNA gene fragments in the genomes (Hikosaka *et al.*, 2011). The 35 kb circular apicoplast genome possesses translation- and transcription-related protein genes, rRNAs, tRNAs, and several other genes such as caseinolytic protease C, sulphur mobilizing protein B, and seven open reading frames of unknown function. A comparison of the apicoplast genomes from eight *Plasmodium* species showed almost complete conservation of gene content, gene arrangement, and AT content (Arisue *et al.*, 2012).

Applied *Plasmodium* genomics

Ideally, anti-malaria strategies comprise a balanced use of mosquito control, anti-parasite drug treatment and prophylaxis, and a general improvement of public health sanitation and awareness. Using these three methodologies malaria was successfully eradicated from North America and Europe after the Second World War. Unfortunately, malaria still remains a major concern in most tropical regions and there is a continuous need to develop new anti-malaria drugs to keep up with constant emergence of drug resistances.

The availability of the complete sequences of *Plasmodium* genomes coupled with the exponential development of bioinformatics tools has revolutionized malaria research. Genome-wide studies have already delivered an expansion of biological knowledge relevant to new drug design and vaccine interventions. For example, large-scale analyses of *P. falciparum* transcripts and protein levels using microarray technologies and mass spectrometry, respectively, have traced expression profiles of genes and their proteins, and allowed functional association with various stages of the parasite (Bozdech *et al.*, 2003; Florens *et al.*, 2002; Lasonder *et al.*, 2002; Le Roch *et al.*, 2003). Moreover, transcriptome analysis of *P. falciparum* field isolates can identify factors involved in pathogenesis (Daily *et al.*, 2007; Siau *et al.*, 2007). Genome-wide methods also identified various regulatory elements and changes in chromatin structure that in turn could possibly regulate gene expression in the parasite genome (Elemento *et al.*, 2007; Gonzales *et al.*, 2008; Ponts *et al.*, 2010, 2011; Westenberger *et al.*, 2009). The recent application of RNA-seq methodology to *P. falciparum* shows promising results for the prediction of non-annotated flanking regions and alternative splicing events (Otto *et al.*, 2010). Furthermore, genomic analysis is fostering new leads to understand the importance of both post-transcriptional and post-translational events in malaria parasite development (Foth *et al.*, 2003; Ganesan *et al.*, 2008; Le Roch *et al.*, 2004, 2008; Margulies *et al.*, 2005; Ponts *et al.*, 2010, 2011). The following paragraphs summarize major advances generated by genomics and system biology studies for the discovery of new anti-malaria strategies.

Genomics and antigen discovery for vaccine design

For an effective vaccine to be developed and prevent *Plasmodium* infection or the disease, antigens must be selected according to their role in protective immunity. For example, the circumsporozoite protein (CSP) was shown decades ago to be a protective antigen against malaria, and is still today the most promising malaria vaccine candidate against *P. falciparum*. CSP is the major surface antigen of the sporozoite, the parasite stage transmitted by the *Anopheles* mosquito. Once sporozoites enter the bloodstream, they are rapidly sequestered in the liver through the specific interaction of parasite surface proteins, primarily CSP, with hepatic cell heparan sulphate proteoglycans. The RTS,S candidate malaria vaccine (or Mosquirix from GlaxoSmithKline) consists of a recombinant subunit vaccine that combines the hepatitis B surface antigen and epitopes from CSP (see Ballou, 2009, for a review). Though the RTS,S vaccine recently entered phase III testing in clinical trials and was shown to be relatively safe while conferring significant protection against severe *P. falciparum* infection, the mechanisms behind how RTS,S vaccination actually protects against malaria are poorly understood, and the potential risk of the loss of natural immunity because of mass vaccination are still unknown. These unknown parameters raise serious concerns that could compromise the vaccine submission initially scheduled for 2012, and represent major obstacles to vaccine development and implementation. Thus, a better understanding of mechanisms involved in innate immunity as well as the discovery of new vaccine strategies are essential for a successful malaria eradication campaign. Genome-wide proteomics analyses have now surveyed all parasite intracellular and surface proteins to identify new antigens (Florens *et al.*, 2002, 2004; Lasonder *et al.*, 2002; Sam-Yellowe *et al.*, 2004). These antigens can now be used for the rational development of vaccines. Furthermore, the availability of sequenced parasite genomes and proteomes has facilitated the development of genome-wide protein microarrays (Doolan *et al.*, 2008). These protein microarrays can now be used to probe human plasma from individuals before and after the malaria transmission season. This novel genome-wide method has already uncovered important insights into parasite proteins associated with immunoreactivity in an unbiased manner (Crompton *et al.*, 2010; Doolan *et al.*, 2008). It is likely that studies of this type will improve our understanding of the molecular basis of immunity, identify immune responses that correlate with significant protection and finally usher in the discovery of new, more efficient vaccine strategies.

Genomics and drug discovery and drug resistance

Genomic studies are extremely powerful tools for identifying novel drugs and drug targets, and to understand the mechanism of drug action as well as drug resistance.

Drug discovery

Many of the current drug therapies are based on chemically engineered variants of already known anti-malarial compounds (e.g. aminoquinolines and/or peroxides). Intensive exploration of the *Plasmodium* genome as well as comparative genomic studies lead to the identification of essential genes or metabolic bottlenecks that can be targeted for rational designs of new drugs (Daily *et al.*, 2005, 2007; Llinas *et al.*, 2006; Margulies *et al.*, 2005). For example, a fosmidomycin-sensitive mevalonate-independent pathway of isoprenoid biosynthesis, which is absent in higher eukaryotes and located in the plant-like parasite plastid, was identified in *Plasmodium* (Jomaa *et al.*, 1999). Similarly, an apicoplast-associated type II fatty acid biosynthesis pathway (usually found in plants, fungi and certain bacteria) has been identified as a potential target for drugs such as triclosan (Goodman and McFadden, 2007; Surolia *et al.*, 2002). More recently, the polyamine biosynthetic pathway was identified as a possible target for anti-malarial drugs (Becker *et al.*, 2010; Dharia *et al.*, 2010b). This essential pathway is responsible for the metabolism of important amines such as spermidine that are involved in cellular metabolism. A striking example of the power of genomics in drug discovery is the identification of a potent drug by cell proliferation-based compound screening (Plouffe *et al.*, 2008). The anti-malarial properties of the spiroindolone NITD609 were discovered and its target was identified in a 2-year time span using high-density microarrays and sequencing (Rottmann *et al.*, 2010). Without the advent of genomics, such a process would have required many years. In the long run, identifying multiple drug targets in various metabolic pathways could minimize the risk of the emergence of drug resistance.

Elucidating mechanisms of drug resistance

Typically, mechanisms of drug resistance are determined by examining the genetic differences between sensitive and resistant strains. Genes carrying differences are then correlated with the mode of action. The advent of whole-genome analyses (deep sequencing, microarrays) offers the possibility of rapidly analysing the genetic changes such as single nucleotide polymorphisms (SNPs), copy number variants (CNVs), or insertions/deletions (indels) that occur in the genome of a resistant strain. Whole-genome scanning using tiling microarrays has been widely used for this purpose. For example, point mutations in the apicoplast genome were recently associated with resistance to clindamycin, a drug used in combination with quinine for the treatment of malaria in pregnant women and infants (Dharia et al., 2010b). Tiling microarrays also revealed that resistance to the spiroindolone NITD609 involves the P-type cation-transporter ATPase4 (Rottmann *et al.*, 2010).

Resistance to chloroquine (CQR) is probably the best-studied drug resistance in *P. falciparum*. CQR is mediated by the CQ transporter gene (*Pfcrt*) and by the multi-drug resistance gene (*Pfmdr1*), also involved in resistance to other anti-malarials, potentially including artemisinin. As a consequence, the genetic variants of the *Pfcrt* and *Pfmdr1* loci have been the subject of intense analysis, providing a wide range of information in the area of drug resistance. For example, a recent large-scale (460 blood samples) analysis of microsatellite polymorphisms at these two loci proposed that different mechanisms led to the rise of these mutations in Africa and Asia, which may have important consequences for the future evolution of the parasite (Mehlotra *et al.*, 2008). Whole-genome scanning for the presence of evolutionary signatures could, indeed, provide critical information for understanding the mechanisms of resistance to drugs and identify new potential targets for drug discovery [these considerations have been briefly reviewed in (Gupta *et al.*, 2010)]. For example, a genome-wide SNP analysis of the genomes from almost 200 different *P. falciparum* isolates identified regions of high and low recombination frequencies (*hotspots* and *coldspots*) and found that loci associated with resistance to known drugs, e.g. chloroquine, are likely found in hotspots (Mu *et al.*, 2007). A practical rule to abstract from these observations could be that drug targets should be proteins encoded by genes located in coldspots rather than hotspots. Beyond SNP analyses, tiling microarray analyses found relatively abundant CNVs in *P. falciparum* that may be involved in drug resistance (Kidgell *et al.*, 2006). For example, CNVs have been identified surrounding *Pfmdr1* and the first gene involved in the folate pathway (as a result of widespread use of pyrimethamine). Recently, *in vitro*-derived fosmidomycin resistance was attributed to a CNV in the *Pfdxr* gene that suppresses the drug-induced inhibition of isoprenoid biosynthesis in the parasite (Dharia *et al.*, 2009). Despite their abundance and importance, CNVs are however understudied in the malaria parasite (Anderson *et al.*, 2009). Advances in deep sequencing technologies should allow the field to develop rapidly in the near future. Moreover, novel classes of drugs can rise from genomics drug design with the development of AT-specific DNA-binding agents that target the AT-rich genome of the malaria parasite (Woynarowski *et al.*, 2007; Yanow *et al.*, 2007).

Genomics and development of novel diagnostics

Diagnostic PCR using *Plasmodium* 18S rRNA genes as targets long predates genome sequencing projects, and remains the most widely used molecular diagnostic method for malaria (MalERA, 2011). Preliminary studies coupling PCR amplification of *Plasmodium* 18S rRNA genes with pyrosequencing produced mixed results (Lulitanond *et al.*, 2012). In addition, the sensitivity and accuracy of PCR diagnosis is important for appropriate malaria treatment, and in *Plasmodium* species 18s rRNA genes are present as just 4–8 relatively divergent copies per genome, reducing their sensitivity as PCR targets. At the same time the high sequence similarity of the orthologues between species reduces the efficiency of multiplex (multiple-target) PCR for detection of mixed infections. Having complete *Plasmodium* genome sequences now allows scanning across genomes for new, more sensitive and efficient diagnostic PCR targets. Ideal targets are high copy number, low divergence, non-tandem repeat sequences that are species-specific. A bioinformatic search of *P. falciparum* and *P. vivax* genomes for such targets (Demas *et al.*, 2011) identified several promising candidates in each species. The two chosen for testing, Pfr363 and Pv47, present as 41 and 14 copies per genome, respectively, proved notably more sensitive and specific than 18S rRNA in single and multiplex PCR.

Future diagnostics may rely on sequencing whole genomes rather than gene targets. Direct NGS of *Plasmodium* genomes from patient blood samples can detect multi-clonal infections and polymorphisms associated with drug resistance (Robinson *et al.*, 2011). Coupled with the advent of exciting new (though as yet untested in the field) 'portable' nanopore sequencing technologies (http://www.nytimes.com/2012/02/18/health/

oxford-nanopore-unveils-tiny-dna-sequencing-device.html), there is now a huge potential to make diagnoses highly information-rich, and thus drive development of more specific and effective malaria treatments.

Summary and future trends

Plasmodium genomics has undoubtedly changed the way that malaria research is undertaken today. Within the short period of a decade, we have progressed from a gene-by-gene approach to undertaking basic research, to a whole genome approach, a leap forward that has enabled fundamental questions regarding the parasite's biological processes, biochemical pathways, genetic diversity and evolution to be answered. Research can now be undertaken based upon genotyping hundreds of parasites and their vertebrate hosts to detect and analyse genetic variants, and identify molecular mechanisms involved in pathogenicity and disease resistance – all the while improving our understanding of the disease and developing efficient preventative and curative strategies (Volkman *et al.*, 2012). These approaches have been driven by the rapid evolution of high-throughput genomics and sequencing technologies. A major challenge in the coming years will be the curation, storage and access to *Plasmodium* sequence data, and fortunately the *Plasmodium* Genomics Resource PlasmoDB (http://www.plasmodb.org) provides a centralized resource to support such endeavours. In addition, the establishment of large multicentre collaborations such as the Malaria Genomic Epidemiology Network (MalariaGEN; http://www.malariagen.net/), a network of researchers in more than 20 countries working to understand how genome variation in human, *Plasmodium* and *Anopheles* populations affects the biology and epidemiology of malaria, have established policies for data-sharing that are focused on the development of large-scale collaborative projects across multiple populations, bringing together detailed clinical and epidemiological data with state-of-the art technologies for genome sequencing and web tools for data analysis. The future is bright for *Plasmodium* genomics.

References

Anderson, T.J.C., Patel, J., and Ferdig, M.T. (2009). Gene copy number and malaria biology. Trends Parasitol. *25*, 336–343.

Arisue, N., Hashimoto, T., Mitsui, H., Palacpac, N.M., Kaneko, A., Kawai, S., Hasegawa, M., Tanabe, K., and Horii, T. (2012). The Plasmodium apicoplast genome: conserved structure and close relationship of *P. ovale* to rodent malaria parasites. Mol. Biol. Evol. *29*, 2095–2099.

Auburn, S., Campino, S., Clark, T.G., Djimde, A.A., Zongo, I., Pinches, R., Manske, M., Mangano, V., Alcock, D., Anastasi, E., *et al.* (2011). An effective method to purify *Plasmodium falciparum* DNA directly from clinical blood samples for whole genome high-throughput sequencing. PLoS One *6*, e22213.

Aurrecoechea, C., Brestelli, J., Brunk, B.P., Dommer, J., Fischer, S., Gajria, B., Gao, X., Gingle, A., Grant, G., Harb, O.S., *et al.* (2009). PlasmoDB: a functional genomic database for malaria parasites. Nucleic Acids Res. *37*, D539–543.

Bailey, J.A., Mvalo, T., Aragam, N., Weiser, M., Congdon, S., Kamwendo, D., Martinson, F., Hoffman, I., Meshnick, S.R., and Juliano, J.J. (2012). Diversity of and Selection on *Plasmodium falciparum* csp T cell epitopes determined by massively parallel pyrosequencing in Lilongwe, Malawi. J. Infect. Dis. *206*, 580–587.

Baker, M. (2012). De novo genome assembly: what every biologist should know. Nat. Methods *9*, 333–337.

Ballou, W.R. (2009). The development of the RTS, S malaria vaccine candidate: challenges and lessons. Parasite Immunol. *31*, 492–500.

Bao, S., Jiang, R., Kwan, W., Wang, B., Ma, X., and Song, Y.Q. (2011). Evaluation of next-generation sequencing software in mapping and assembly. J. Hum. Genet. *56*, 406–414.

Bartfai, R., Hoeijmakers, W.A.M., Salcedo-Amaya, A.M., Smits, A.H., Janssen-Megens, E., Kaan, A., Treeck, M., Gilberger, T.-W., Francoijs, K.-J., and Stunnenberg, H.G. (2010). H2A.Z demarcates intergenic regions of the *Plasmodium falciparum* epigenome that are dynamically marked by H3K9ac and H3K4me3. PLoS Pathog. *6*.

Becker, J.V., Mtwisha, L., Crampton, B.G., Stoychev, S., van Brummelen, A.C., Reeksting, S., Louw, A.I., Birkholtz, L.M., and Mancama, D.T. (2010). *Plasmodium falciparum* spermidine synthase inhibition results in unique perturbation-specific effects observed on transcript, protein and metabolite levels. BMC Genomics *11*, 235.

Bentley, D.R., Balasubramanian, S., Swerdlow, H.P., Smith, G.P., Milton, J., Brown, C.G., Hall, K.P., Evers, D.J., Barnes, C.L., Bignell, H.R., *et al.* (2008). Accurate whole human genome sequencing using reversible terminator chemistry. Nature *456*, 53–59.

Bourke, P.F., Holt, D.C., Sutherland, C.J., Currie, B., and Kemp, D.J. (1996). Positional cloning of a sequence from the breakpoint of chromosome 9 commonly associated with the loss of cytoadherence. Ann. Trop. Med. Parasitol. *90*, 353–357.

Bozdech, Z., Llinas, M., Pulliam, B.L., Wong, E.D., Zhu, J., and DeRisi, J.L. (2003). The transcriptome of the intraerythrocytic developmental cycle of *Plasmodium falciparum*. PLoS Biol. *1*, E5.

Bright, A.T., Tewhey, R., Abeles, S., Chuquiyauri, R., Llanos-Cuentas, A., Ferreira, M.U., Schork, N.J., Vinetz, J.M., and Winzeler, E.A. (2012). Whole genome sequencing analysis of Plasmodium vivax using whole genome capture. BMC Genomics *13*, 262.

Cantarel, B.L., Korf, I., Robb, S.M., Parra, G., Ross, E., Moore, B., Holt, C., Sanchez Alvarado, A., and Yandell, M. (2008). MAKER: an easy-to-use annotation pipeline designed for emerging model organism genomes. Genome Res. *18*, 188–196.

Carlton, J.M., Galinski, M.R., Barnwell, J.W., and Dame, J.B. (1999). Karyotype and synteny among the chromosomes of all four species of human malaria parasite. Mol. Biochem. Parasitol. *101*, 23–32.

Carlton, J.M., Yowell, C.A., Sturrock, K.A., and Dame, J.B. (2001). Biomagnetic separation of contaminating host leukocytes from plasmodium-infected erythrocytes. Exp. Parasitol. *97*, 111–114.

Carlton, J.M., Angiuoli, S.V., Suh, B.B., Kooij, T.W., Pertea, M., Silva, J.C., Ermolaeva, M.D., Allen, J.E., Selengut, J.D., Koo, H.L., *et al.* (2002). Genome sequence and comparative analysis of the model rodent malaria parasite *Plasmodium yoelii yoelii*. Nature *419*, 512–519.

Carlton, J.M., Adams, J.H., Silva, J.C., Bidwell, S.L., Lorenzi, H., Caler, E., Crabtree, J., Angiuoli, S.V., Merino, E.F., Amedeo, P., *et al.* (2008). Comparative genomics of the neglected human malaria parasite *Plasmodium vivax*. Nature 455, 757–763.

Chan, E.R., Menard, D., David, P.H., Ratsimbasoa, A., Kim, S., Chim, P., Do, C., Witkowski, B., Mercereau-Puijalon, O., Zimmerman, P.A., *et al.* (2012). Whole genome sequencing of field isolates provides robust characterization of genetic diversity in *Plasmodium vivax*. PLoS Negl. Trop. Dis. *6*, e1811.

Cheesman, S., O'Mahony, E., Pattaradilokrat, S., Degnan, K., Knott, S., and Carter, R. (2010). A single parasite gene determines strain-specific protective immunity against malaria: the role of the merozoite surface protein I. Int. J. Parasitol. *40*, 951–961.

Clarke, J., Wu, H.C., Jayasinghe, L., Patel, A., Reid, S., and Bayley, H. (2009). Continuous base identification for single-molecule nanopore DNA sequencing. Nat. Nanotechnol. *4*, 265–270.

Crompton, P.D., Kayala, M.A., Traore, B., Kayentao, K., Ongoiba, A., Weiss, G.E., Molina, D.M., Burk, C.R., Waisberg, M., Jasinskas, A., *et al.* (2010). A prospective analysis of the Ab response to *Plasmodium falciparum* before and after a malaria season by protein microarray. Proc. Natl. Acad. Sci. U.S.A. *107*, 6958–6963.

Cunningham, D., Lawton, J., Jarra, W., Preiser, P., and Langhorne, J. (2010). The pir multigene family of *Plasmodium*: antigenic variation and beyond. Mol. Biochem. Parasitol. *170*, 65–73.

Daily, J.P., Le Roch, K.G., Sarr, O., Ndiaye, D., Lukens, A., Zhou, Y., Ndir, O., Mboup, S., Sultan, A., Winzeler, E.A., *et al.* (2005). In vivo transcriptome of *Plasmodium falciparum* reveals overexpression of transcripts that encode surface proteins. J. Infect. Dis. *191*, 1196–1203.

Daily, J.P., Scanfeld, D., Pochet, N., Le Roch, K., Plouffe, D., Kamal, M., Sarr, O., Mboup, S., Ndir, O., Wypij, D., *et al.* (2007). Distinct physiological states of *Plasmodium falciparum* in malaria-infected patients. Nature *450*, 1091–1095.

del Portillo, H.A., Fernandez-Becerra, C., Bowman, S., Oliver, K., Preuss, M., Sanchez, C.P., Schneider, N.K., Villalobos, J.M., Rajandream, M.A., Harris, D., *et al.* (2001). A superfamily of variant genes encoded in the subtelomeric region of *Plasmodium vivax*. Nature *410*, 839–842.

Demas, A., Oberstaller, J., DeBarry, J., Lucchi, N.W., Srinivasamoorthy, G., Sumari, D., Kabanywanyi, A.M., Villegas, L., Escalante, A.A., Kachur, S.P., *et al.* (2011). Applied genomics: data mining reveals species-specific malaria diagnostic targets more sensitive than 18S rRNA. J. Clin. Microbiol. *49*, 2411–2418.

Dharia, N.V., Sidhu, A.B., Cassera, M.B., Westenberger, S.J., Bopp, S.E., Eastman, R.T., Plouffe, D., Batalov, S., Park, D.J., Volkman, S.K., *et al.* (2009). Use of high-density tiling microarrays to identify mutations globally and elucidate mechanisms of drug resistance in *Plasmodium falciparum*. Genome Biol.*10*, R21.

Dharia, N.V., Bright, A.T., Westenberger, S.J., Barnes, S.W., Batalov, S., Kuhen, K., Borboa, R., Federe, G.C., McClean, C.M., Vinetz, J.M., *et al.* (2010a). Whole-genome sequencing and microarray analysis of ex vivo *Plasmodium vivax* reveal selective pressure on putative drug resistance genes. Proc. Natl. Acad. Sci. U.S.A. *107*, 20045–20050.

Dharia, N.V., Plouffe, D., Bopp, S.E., Gonzalez-Paez, G.E., Lucas, C., Salas, C., Soberon, V., Bursulaya, B., Kochel, T.J., Bacon, D.J., *et al.* (2010b). Genome scanning of Amazonian *Plasmodium falciparum* shows subtelomeric instability and clindamycin-resistant parasites. Genome Res. *20*, 1534–1544.

Dimon, M.T., Sorber, K., and DeRisi, J.L. (2010). HMMSplicer: a tool for efficient and sensitive discovery of known and novel splice junctions in RNA-Seq data. PLoS One 5.

Doolan, D.L., Mu, Y., Unal, B., Sundaresh, S., Hirst, S., Valdez, C., Randall, A., Molina, D., Liang, X., Freilich, D.A., *et al.* (2008). Profiling humoral immune responses to *P. falciparum* infection with protein microarrays. Proteomics *8*, 4680–4694.

Earl, D., Bradnam, K., St John, J., Darling, A., Lin, D., Fass, J., Yu, H.O., Buffalo, V., Zerbino, D.R., Diekhans, M., *et al.* (2011). Assemblathon 1: a competitive assessment of de novo short read assembly methods. Genome Res. *21*, 2224–2241.

Eid, J., Fehr, A., Gray, J., Luong, K., Lyle, J., Otto, G., Peluso, P., Rank, D., Baybayan, P., Bettman, B., *et al.* (2009). Real-time DNA sequencing from single polymerase molecules. Science *323*, 133–138.

Elemento, O., Slonim, N., and Tavazoie, S. (2007). A universal framework for regulatory element discovery across all genomes and data types. Mol. Cell *28*, 337–350.

Figueiredo, L.M., Freitas-Junior, L.H., Bottius, E., Olivo-Marin, J.C., and Scherf, A. (2002). A central role for

Plasmodium falciparum subtelomeric regions in spatial positioning and telomere length regulation. EMBO J. *21*, 815–824.

Florens, L., Washburn, M.P., Raine, J.D., Anthony, R.M., Grainger, M., Haynes, J.D., Moch, J.K., Muster, N., Sacci, J.B., Tabb, D.L., *et al.* (2002). A proteomic view of the *Plasmodium falciparum* life cycle. Nature *419*, 520–526.

Florens, L., Liu, X., Wang, Y., Yang, S., Schwartz, O., Peglar, M., Carucci, D.J., Yates, J.R., 3rd, and Wub, Y. (2004). Proteomics approach reveals novel proteins on the surface of malaria-infected erythrocytes. Mol. Biochem. Parasitol. *135*, 1–11.

Foth, B.J., Ralph, S.A., Tonkin, C.J., Struck, N.S., Fraunholz, M., Roos, D.S., Cowman, A.F., and McFadden, G.I. (2003). Dissecting apicoplast targeting in the malaria parasite *Plasmodium falciparum*. Science *299*, 705–708.

Freitas-Junior, L.H., Bottius, E., Pirrit, L.A., Deitsch, K.W., Scheidig, C., Guinet, F., Nehrbass, U., Wellems, T.E., and Scherf, A. (2000). Frequent ectopic recombination of virulence factor genes in telomeric chromosome clusters of *P. falciparum*. Nature *407*, 1018–1022.

Ganesan, K., Ponmee, N., Jiang, L., Fowble, J.W., White, J., Kamchonwongpaisan, S., Yuthavong, Y., Wilairat, P., and Rathod, P.K. (2008). A genetically hard-wired metabolic transcriptome in *Plasmodium falciparum* fails to mount protective responses to lethal antifolates. PLoS pathogens *4*, e1000214.

Gardner, M.J. (2001). A status report on the sequencing and annotation of the *P. falciparum* genome. Mol. Biochem. Parasitol. *118*, 133–138.

Gardner, M.J., Hall, N., Fung, E., White, O., Berriman, M., Hyman, R.W., Carlton, J.M., Pain, A., Nelson, K.E., Bowman, S., *et al.* (2002). Genome sequence of the human malaria parasite *Plasmodium falciparum*. Nature *419*, 498–511.

Gilles, A., Meglecz, E., Pech, N., Ferreira, S., Malausa, T., and Martin, J.F. (2011). Accuracy and quality assessment of 454 GS-FLX Titanium pyrosequencing. BMC Genomics *12*, 245.

Glenn, T.C. (2011). Field guide to next-generation DNA sequencers. Mol. Ecol. Resour. *11*, 759–769.

Gnerre, S., Maccallum, I., Przybylski, D., Ribeiro, F.J., Burton, J.N., Walker, B.J., Sharpe, T., Hall, G., Shea, T.P., Sykes, S., *et al.* (2011). High-quality draft assemblies of mammalian genomes from massively parallel sequence data. Proc. Natl. Acad. Sci. U.S.A. *108*, 1513–1518.

Gonzales, J.M., Patel, J.J., Ponmee, N., Jiang, L., Tan, A., Maher, S.P., Wuchty, S., Rathod, P.K., and Ferdig, M.T. (2008). Regulatory hotspots in the malaria parasite genome dictate transcriptional variation. PLoS Biol. *6*, e238.

Goodman, C.D., and McFadden, G.I. (2007). Fatty acid biosynthesis as a drug target in apicomplexan parasites. Current Drug Targets *8*, 15–30.

Gupta, B., Awasthi, G., and Das, A. (2010). Malaria parasite genome scan: insights into antimalarial resistance. Parasitology Res. *107*, 495–499.

Hall, N., Pain, A., Berriman, M., Churcher, C., Harris, B., Harris, D., Mungall, K., Bowman, S., Atkin, R., Baker, S., *et al.* (2002). Sequence of *Plasmodium falciparum* chromosomes 1, 3–9 and 13. Nature *419*, 527–531.

Hall, N., Karras, M., Raine, J.D., Carlton, J.M., Kooij, T.W.A., Berriman, M., Florens, L., Janssen, C.S., Pain, A., Christophides, G.K., *et al.* (2005). A comprehensive survey of the Plasmodium life cycle by genomic, transcriptomic, and proteomic analyses. Science *307*, 82–86.

Hikosaka, K., Watanabe, Y., Kobayashi, F., Waki, S., Kita, K., and Tanabe, K. (2011). Highly conserved gene arrangement of the mitochondrial genomes of 23 *Plasmodium* species. Parasitol. Int. *60*, 175–180.

Hoeijmakers, W.A., Flueck, C., Francoijs, K.J., Smits, A.H., Wetzel, J., Volz, J.C., Cowman, A.F., Voss, T., Stunnenberg, H.G., and Bartfai, R. (2012). *Plasmodium falciparum* centromeres display a unique epigenetic makeup and cluster prior to and during schizogony. Cell. Microbiol. *14*, 1391–1401.

Hunt, P., Martinelli, A., Modrzynska, K., Borges, S., Creasey, A., Rodrigues, L., Beraldi, D., Loewe, L., Fawcett, R., Kumar, S., *et al.* (2010). Experimental evolution, genetic analysis and genome re-sequencing reveal the mutation conferring artemisinin resistance in an isogenic lineage of malaria parasites. BMC Genomics *11*, 499.

Janse, C.J., Carlton, J.M., Walliker, D., and Waters, A.P. (1994). Conserved location of genes on polymorphic chromosomes of four species of malaria parasites. Mol. Biochem. Parasitol. *68*, 285–296.

Janssen, C.S., Barrett, M.P., Turner, C.M., and Phillips, R.S. (2002). A large gene family for putative variant antigens shared by human and rodent malaria parasites. Proc. R. Soc. Lond. B Biol. Sci. *269*, 431–436.

Jeffares, D.C., Pain, A., Berry, A., Cox, A.V., Stalker, J., Ingle, C.E., Thomas, A., Quail, M.A., Siebenthall, K., Uhlemann, A.-C., *et al.* (2007). Genome variation and evolution of the malaria parasite *Plasmodium falciparum*. Nat. Genet. *39*, 120–125.

Jomaa, H., Wiesner, J., Sanderbrand, S., Altincicek, B., Weidemeyer, C., Hintz, M., Turbachova, I., Eberl, M., Zeidler, J., Lichtenthaler, H.K., *et al.* (1999). Inhibitors of the nonmevalonate pathway of isoprenoid biosynthesis as antimalarial drugs. Science *285*, 1573–1576.

Juliano, J.J., Porter, K., Mwapasa, V., Sem, R., Rogers, W.O., Ariey, F., Wongsrichanalai, C., Read, A., and Meshnick, S.R. (2010). Exposing malaria in-host diversity and estimating population diversity by capture-recapture using massively parallel pyrosequencing. Proc. Natl. Acad. Sci. U.S.A. *107*, 20138–20143.

Kemp, D.J., Thompson, J.K., Walliker, D., and Corcoran, L.M. (1987). Molecular karyotype of *Plasmodium falciparum*: conserved linkage groups and expendable histidine-rich protein genes. Proc. Natl. Acad. Sci. U.S.A. *84*, 7672–7676.

Kidgell, C., Volkman, S.K., Daily, J., Borevitz, J.O., Plouffe, D., Zhou, Y., Johnson, J.R., Le Roch, K., Sarr, O., Ndir, O., *et al.* (2006). A systematic map of genetic variation in *Plasmodium falciparum*. PLoS Pathogens *2*, e57.

Kozarewa, I., Ning, Z., Quail, M.A., Sanders, M.J., Berriman, M., and Turner, D.J. (2009). Amplification-free Illumina sequencing-library preparation facilitates improved mapping and assembly of (G + C)-biased genomes. Nat. Methods *6*, 291–295.

Langmead, B., and Salzberg, S.L. (2012). Fast gapped-read alignment with Bowtie 2. Nat. Methods *9*, 357–359.

Lasonder, E., Ishihama, Y., Andersen, J.S., Vermunt, A.M., Pain, A., Sauerwein, R.W., Eling, W.M., Hall, N., Waters, A.P., Stunnenberg, H.G., *et al.* (2002). Analysis of the *Plasmodium falciparum* proteome by high-accuracy mass spectrometry. Nature *419*, 537–542.

Le Roch, K.G., Zhou, Y., Blair, P.L., Grainger, M., Moch, J.K., Haynes, J.D., De La Vega, P., Holder, A.A., Batalov, S., Carucci, D.J., *et al.* (2003). Discovery of gene function by expression profiling of the malaria parasite life cycle. Science *301*, 1503–1508.

Le Roch, K.G., Johnson, J.R., Florens, L., Zhou, Y., Santrosyan, A., Grainger, M., Yan, S.F., Williamson, K.C., Holder, A.A., Carucci, D.J., *et al.* (2004). Global analysis of transcript and protein levels across the *Plasmodium falciparum* life cycle. Genome Res. *14*, 2308–2318.

Le Roch, K.G., Johnson, J.R., Ahiboh, H., Chung, D.W., Prudhomme, J., Plouffe, D., Henson, K., Zhou, Y., Witola, W., Yates, J.R., *et al.* (2008). A systematic approach to understand the mechanism of action of the bisthiazolium compound T4 on the human malaria parasite, *Plasmodium falciparum*. BMC genomics *9*, 513.

Li, H., and Durbin, R. (2009). Fast and accurate short read alignment with Burrows–Wheeler transform. Bioinformatics *25*, 1754–1760.

Li, H., Handsaker, B., Wysoker, A., Fennell, T., Ruan, J., Homer, N., Marth, G., Abecasis, G., and Durbin, R. (2009a). The Sequence Alignment/Map format and SAMtools. Bioinformatics *25*, 2078–2079.

Li, R., Yu, C., Li, Y., Lam, T.W., Yiu, S.M., Kristiansen, K., and Wang, J. (2009b). SOAP2: an improved ultrafast tool for short read alignment. Bioinformatics *25*, 1966–1967.

Llinas, M., Bozdech, Z., Wong, E.D., Adai, A.T., and DeRisi, J.L. (2006). Comparative whole genome transcriptome analysis of three *Plasmodium falciparum* strains. Nucleic Acids Res. *34*, 1166–1173.

Lopez-Barragan, M.J., Lemieux, J., Quinones, M., Williamson, K.C., Molina-Cruz, A., Cui, K., Barillas-Mury, C., Zhao, K., and Su, X.Z. (2011a). Directional gene expression and antisense transcripts in sexual and asexual stages of *Plasmodium falciparum*. BMC Genomics *12*, 587.

Lopez-Barragan, M.J., Quinones, M., Cui, K., Lemieux, J., Zhao, K., and Su, X.Z. (2011b). Effect of PCR extension temperature on high-throughput sequencing. Mol. Biochem. Parasitol. *176*, 64–67.

Lulitanond, V., Intapan, P.M., Tantrawatpan, C., Sankuntaw, N., Sanpool, O., Janwan, P., and Maleewong, W. (2012). Molecular markers for detection and differentiation of *Plasmodium falciparum* and *Plasmodium vivax* in Human blood samples by pyrosequencing. J. Clin. Microbiol. *50*, 1455–1457.

MalERA (2011). A research agenda for malaria eradication: diagnoses and diagnostics. PLoS Med. *8*, e1000396.

Manske, M., Miotto, O., Campino, S., Auburn, S., Almagro-Garcia, J., Maslen, G., O'Brien, J., Djimde, A., Doumbo, O., Zongo, I., *et al.* (2012). Analysis of Plasmodium falciparum diversity in natural infections by deep sequencing. Nature *487(7407)*, 375–379.

Margulies, M., Egholm, M., Altman, W.E., Attiya, S., Bader, J.S., Bemben, L.A., Berka, J., Braverman, M.S., Chen, Y.-J., Chen, Z., *et al.* (2005). Genome sequencing in microfabricated high-density picolitre reactors. Nature *437*, 376–380.

Martinelli, A., Henriques, G., Cravo, P., and Hunt, P. (2011). Whole genome re-sequencing identifies a mutation in an ABC transporter (mdr2) in a *Plasmodium chabaudi* clone with altered susceptibility to antifolate drugs. Int. J. Parasitol. *41*, 165–171.

McCutchan, T.F., Dame, J.B., Miller, L.H., and Barnwell, J. (1984). Evolutionary relatedness of Plasmodium species as determined by the structure of DNA. Science (New York, NY) *225*, 808–811.

Mehlotra, R.K., Mattera, G., Bockarie, M.J., Maguire, J.D., Baird, J.K., Sharma, Y.D., Alifrangis, M., Dorsey, G., Rosenthal, P.J., Fryauff, D.J., *et al.* (2008). Discordant patterns of genetic variation at two chloroquine resistance loci in worldwide populations of the malaria parasite *Plasmodium falciparum*. Antimicrob. Agents Chemother. *52*, 2212–2222.

Melnikov, A., Galinsky, K., Rogov, P., Fennell, T., Van Tyne, D., Russ, C., Daniels, R., Barnes, K.G., Bochicchio, J., Ndiaye, D., *et al.* (2011). Hybrid selection for sequencing pathogen genomes from clinical samples. Genome Biol. *12*, R73.

Miao, J., and Cui, L. (2011). Rapid isolation of single malaria parasite-infected red blood cells by cell sorting. Nat. Protoc. *6*, 140–146.

Mu, J., Awadalla, P., Duan, J., McGee, K.M., Keebler, J., Seydel, K., McVean, G.A.T., and Su, X.-Z. (2007). Genome-wide variation and identification of vaccine targets in the *Plasmodium falciparum* genome. Nat. Genet. *39*, 126–130.

Mullikin, J.C., and Ning, Z. (2003). The phusion assembler. Genome Res. *13*, 81–90.

Myers, E.W., Sutton, G.G., Delcher, A.L., Dew, I.M., Fasulo, D.P., Flanigan, M.J., Kravitz, S.A., Mobarry, C.M., Reinert, K.H., Remington, K.A., *et al.* (2000). A whole-genome assembly of *Drosophila*. Science *287*, 2196–2204.

Neafsey, D.E., Galinsky, K., Jiang, R.H.Y., Young, L., Sykes, S.M., Saif, S., Gujja, S., Goldberg, J.M., Young, S., Zeng, Q., *et al.* (2012). The malaria parasite *Plasmodium vivax* exhibits greater genomic diversity than *Plasmodium falciparum*. Nature Genet. *44*, 1046–1050.

Niemenmaa, M., Kallio, A., Schumacher, A., Klemela, P., Korpelainen, E., and Heljanko, K. (2012). Hadoop-BAM: directly manipulating next generation sequencing data in the cloud. Bioinformatics *28*, 876–877.

Ning, Z., Cox, A.J., and Mullikin, J.C. (2001). SSAHA: a fast search method for large DNA databases. Genome Res. *11*, 1725–1729.

Otto, T.D., Wilinski, D., Assefa, S., Keane, T.M., Sarry, L.R., Bohme, U., Lemieux, J., Barrell, B., Pain, A., Berriman, M., *et al.* (2010). New insights into the blood-stage transcriptome of *Plasmodium falciparum* using RNA-Seq. Mol. Microbiol. *76*, 12–24.

Oyola, S.O., Otto, T.D., Gu, Y., Maslen, G., Manske, M., Campino, S., Turner, D.J., Macinnis, B., Kwiatkowski, D.P., Swerdlow, H.P., *et al.* (2012). Optimizing illumina next-generation sequencing library preparation for extremely at-biased genomes. BMC Genomics *13*, 1.

Pain, A., Bohme, U., Berry, A.E., Mungall, K., Finn, R.D., Jackson, A.P., Mourier, T., Mistry, J., Pasini, E.M., Aslett, M.A., *et al.* (2008). The genome of the simian and human malaria parasite *Plasmodium knowlesi*. Nature *455*, 799–803.

Plouffe, D., Brinker, A., McNamara, C., Henson, K., Kato, N., Kuhen, K., Nagle, A., Adrian, F., Matzen, J.T., Anderson, P., *et al.* (2008). In silico activity profiling reveals the mechanism of action of antimalarials discovered in a high-throughput screen. Proc. Natl. Acad. Sci. U.S.A. *105*, 9059–9064.

Ponts, N., Harris, E.Y., Prudhomme, J., Wick, I., Eckhardt-Ludka, C., Hicks, G.R., Hardiman, G., Lonardi, S., and Le Roch, K.G. (2010). Nucleosome landscape and control of transcription in the human malaria parasite. Genome Res. *20*, 228–238.

Ponts, N., Harris, E.Y., Lonardi, S., and Le Roch, K.G. (2011). Nucleosome occupancy at transcription start sites in the human malaria parasite: a hard-wired evolution of virulence? Infect. Genet. Evol. *11*, 716–724.

Quail, M.A., Otto, T.D., Gu, Y., Harris, S.R., Skelly, T.F., McQuillan, J.A., Swerdlow, H.P., and Oyola, S.O. (2012). Optimal enzymes for amplifying sequencing libraries. Nature Methods *9*, 10–11.

Riley, M.C., Kirkup, B.C., Jr., Johnson, J.D., Lesho, E.P., and Ockenhouse, C.F. (2011). Rapid whole genome optical mapping of *Plasmodium falciparum*. Malar. J. *10*, 252.

Robinson, T., Campino, S.G., Auburn, S., Assefa, S.A., Polley, S.D., Manske, M., MacInnis, B., Rockett, K.A., Maslen, G.L., Sanders, M., *et al.* (2011). Drug-resistant genotypes and multi-clonality in *Plasmodium falciparum* analysed by direct genome sequencing from peripheral blood of malaria patients. PLoS One *6*, e23204.

Rottmann, M., McNamara, C., Yeung, B.K., Lee, M.C., Zou, B., Russell, B., Seitz, P., Plouffe, D.M., Dharia, N.V., Tan, J., *et al.* (2010). Spiroindolones, a potent compound class for the treatment of malaria. Science *329*, 1175–1180.

Salzberg, S.L., Phillippy, A.M., Zimin, A., Puiu, D., Magoc, T., Koren, S., Treangen, T.J., Schatz, M.C., Delcher, A.L., Roberts, M., *et al.* (2012). GAGE: A critical evaluation of genome assemblies and assembly algorithms. Genome Res. *22*, 557–567.

Sam-Yellowe, T.Y., Florens, L., Johnson, J.R., Wang, T., Drazba, J.A., Le Roch, K.G., Zhou, Y., Batalov, S., Carucci, D.J., Winzeler, E.A., *et al.* (2004). A Plasmodium gene family encoding Maurer's cleft membrane proteins: structural properties and expression profiling. Genome Res. *14*, 1052–1059.

Samarakoon, U., Regier, A., Tan, A., Desany, B.A., Collins, B., Tan, J.C., Emrich, S.J., and Ferdig, M.T. (2011). High-throughput 454 resequencing for allele discovery and recombination mapping in *Plasmodium falciparum*. BMC Genomics *12*, 116.

Sanger, F., Air, G.M., Barrell, B.G., Brown, N.L., Coulson, A.R., Fiddes, C.A., Hutchison, C.A., Slocombe, P.M., and Smith, M. (1977a). Nucleotide sequence of bacteriophage phi X174 DNA. Nature *265*, 687–695.

Sanger, F., Nicklen, S., and Coulson, A.R. (1977b). DNA sequencing with chain-terminating inhibitors. Proc. Natl. Acad. Sci. U.S.A. *74*, 5463–5467.

Scherf, A., Figueiredo, L.M., and Freitas-Junior, L.H. (2001). Plasmodium telomeres: a pathogen's perspective. Curr. Opin. Microbiol. *4*, 409–414.

Shendure, J., and Ji, H. (2008). Next-generation DNA sequencing. Nature Biotechnol. *26*, 1135–1145.

Siau, A., Toure, F.S., Ouwe-Missi-Oukem-Boyer, O., Ciceron, L., Mahmoudi, N., Vaquero, C., Froissard, P., Bisvigou, U., Bisser, S., Coppee, J.Y., *et al.* (2007). Whole-transcriptome analysis of *Plasmodium falciparum* field isolates: identification of new pathogenicity factors. J. Infect. Dis. *196*, 1603–1612.

Sorber, K., Dimon, M.T., and DeRisi, J.L. (2011). RNA-Seq analysis of splicing in *Plasmodium falciparum* uncovers new splice junctions, alternative splicing and splicing of antisense transcripts. Nucleic Acids Res. *39*, 3820–3835.

Sriprawat, K., Kaewpongsri, S., Suwanarusk, R., Leimanis, M.L., Lek-Uthai, U., Phyo, A.P., Snounou, G., Russell, B., Renia, L., and Nosten, F. (2009). Effective and cheap removal of leukocytes and platelets from *Plasmodium vivax* infected blood. Malar. J. *8*, 115.

Surolia, N., RamachandraRao, S.P., and Surolia, A. (2002). Paradigm shifts in malaria parasite biochemistry and anti-malarial chemotherapy. BioEssays *24*, 192–196.

Tachibana, S.-I., Sullivan, S.A., Kawai, S., Nakamura, S., Kim, H.R., Goto, N., Arisue, N., Palacpac, N.M.Q., Honma, H., Yagi, M., *et al.* (2012). *Plasmodium cynomolgi* genome sequences provide insight into *Plasmodium vivax* and the monkey malaria clade. Nature Genet. *44*, 1051–1055.

Taylor, R.C. (2010). An overview of the Hadoop/MapReduce/HBase framework and its current applications in bioinformatics. BMC Bioinformatics *11* (Suppl. 12), S1.

Trang, D.T., Huy, N.T., Kariu, T., Tajima, K., and Kamei, K. (2004). One-step concentration of malarial parasite-infected red blood cells and removal of contaminating white blood cells. Malar. J. 3, 7.

Trapnell, C., Pachter, L., and Salzberg, S.L. (2009). TopHat: discovering splice junctions with RNA-Seq. Bioinformatics *25*, 1105–1111.

Trapnell, C., Roberts, A., Goff, L., Pertea, G., Kim, D., Kelley, D.R., Pimentel, H., Salzberg, S.L., Rinn, J.L., and Pachter, L. (2012). Differential gene and transcript expression analysis of RNA-seq experiments with TopHat and Cufflinks. Nat Protoc. *7*, 562–578.

Tuda, J., Mongan, A.E., Tolba, M.E.M., Imada, M., Yamagishi, J., Xuan, X., Wakaguri, H., Sugano, S., Sugimoto, C., and Suzuki, Y. (2011). Full-parasites: database of full-length cDNAs of apicomplexa parasites, 2010 update. Nucleic Acids Res. *39*, 625–631.

Venkatesan, M., Amaratunga, C., Campino, S., Auburn, S., Koch, O., Lim, P., Uk, S., Socheat, D., Kwiatkowski, D.P., Fairhurst, R.M., *et al.* (2012). Using CF11 cellulose columns to inexpensively and effectively remove human DNA from *Plasmodium falciparum*-infected whole blood samples. Malar. J. *11*, 41.

Volkman, S.K., Sabeti, P.C., DeCaprio, D., Neafsey, D.E., Schaffner, S.F., Milner, D.A., Daily, J.P., Sarr, O., Ndiaye, D., Ndir, O., *et al.* (2007). A genome-wide map of diversity in *Plasmodium falciparum*. Nat. Genet. *39*, 113–119.

Volkman, S.K., Neafsey, D.E., Schaffner, S.F., Park, D.J., and Wirth, D.F. (2012). Harnessing genomics and genome biology to understand malaria biology. Nat. Rev. Genet. *13*, 315–328.

Wall, D.P., Kudtarkar, P., Fusaro, V.A., Pivovarov, R., Patil, P., and Tonellato, P.J. (2010). Cloud computing for comparative genomics. BMC Bioinformatics *11*, 259.

Wang, Y., Nair, S., Nosten, F., and Anderson, T.J. (2009). Multiple displacement amplification for malaria parasite DNA. J. Parasitol. *95*, 253–255.

Westenberger, S.J., Cui, L., Dharia, N., and Winzeler, E. (2009). Genome-wide nucleosome mapping of *Plasmodium falciparum* reveals histone-rich coding and histone-poor intergenic regions and chromatin remodelling of core and subtelomeric genes. BMC Genomics *10*, 610.

Woynarowski, J.M., Krugliak, M., and Ginsburg, H. (2007). Pharmacogenomic analyses of targeting the AT-rich malaria parasite genome with AT-specific alkylating drugs. Mol. Biochem. Parasitol. *154*, 70–81.

Yanow, S.K., Purcell, L.A., Lee, M., and Spithill, T.W. (2007). Genomics-based drug design targets the AT-rich malaria parasite: implications for antiparasite chemotherapy. Pharmacogenomics *8*, 1267–1272.

Genome Diversity and Applications in Genetic Studies of the Human Malaria Parasites *Plasmodium falciparum* and *Plasmodium vivax*

4

Sittiporn Pattaradilokrat, Jianbing Mu, Philip Awadalla and Xin-zhuan Su

Abstract

The publication of the genomes of the human malaria parasites *Plasmodium falciparum* and *Plasmodium vivax* has provided the foundation for developing high-throughput methods for systematic investigation into genomic variability in parasite populations. Various tools for discovering genome-wide polymorphisms and genotyping, including DNA microarrays and high-throughput sequencing, are readily available for the malaria research community. Elucidating the genetic diversity within and between populations forms the basis of population genetic analysis that not only will aid in deciphering the evolution and adaptation of the parasites but also has important implications for the development and implementation of therapeutic interventions. Here we summarize the recent advances in the studies of genome variation, population genetics, and the development of high-throughput genetic tools for investigating gene function in *P. falciparum* and *P. vivax*.

Introduction

Genetic variation forms the molecular basis of phenotypic variation and the driving force underpinning the evolution of an organism. Genetic variation is manifested in many forms, including single nucleotide polymorphism (SNP), insertion or deletion of a section of DNA (indel), copy number variation (CNV), and various modifications of DNA and DNA binding proteins. Mutations can occur when DNA is replicated during cell division. In malaria parasites, asexual multiplication takes place during the pre-erythrocytic and erythrocytic stages in vertebrate hosts and during sporogony in *Anopheline* mosquitoes. In addition, the existence of a sexual phase during the parasite's life cycle in the mosquito presents an opportunity for genetic exchange between genetically distinct malaria parasites. These variations lead to genetically and phenotypically diverse populations of malaria parasites.

Genetic diversity in *Plasmodium falciparum* and *Plasmodium vivax* is of biological and medical importance and presents a great challenge to controlling malaria. The high levels of genetic diversity endow the parasite with the evolutionary agility to survive and propagate within a highly selective and hostile host environment. Such diversity also allows the parasite to develop resistance to antimalarial drugs rapidly (Rathod *et al.*, 1997) and to evade host immune responses through antigenic variation (Bull *et al.*, 2005, 2008; Ferreira *et al.*, 2004), impeding efforts to develop an effective vaccine. Understanding the extent and basis of genetic variation and its molecular mechanism is fundamental for designing strategies for disease control and new therapies.

Malaria parasite genomes include three components: a 23- to 30-Mb nuclear genome depending on the species of malaria parasite, a 6-kilobase pair (kb) mitochondrial genome, and a 35-kb plastid genome associated with an organelle known as an apicoplast. The linear mitochondrial genome contains genes encoding many truncated ribosomal RNAs and three components of cellular respiration (cytochrome *c* oxidase subunits I and III, and cytochrome *b*) (Feagin, 1992; Vaidya and Mather,

2009). There are approximately 20 copies of mitochondrial genome per cell in *P. falciparum* (Preiser *et al.*, 1996). The apicoplast genome contains ~30 genes encoding RNA polymerase, ribosomal proteins, and non-coding RNA (tRNA and rRNA) (Ralph *et al.*, 2004; Wilson *et al.*, 1996). Proteins translocated to the apicoplast are thought to have important roles in lipid metabolism (Lim and McFadden, 2010). Both the mitochondrial and apicoplast genomes harbour many genes essential for parasite development that may provide new potential targets for chemotherapies, and genetic variations in these genomes can be informative for studies of evolutionary history, including the age and origin of malaria parasites (Conway, 2003; Conway *et al.*, 2000b; Joy *et al.*, 2003; Roy and Irimia, 2008) (see also Chapter 1).

The nuclear genome of *P. falciparum* has 14 chromosomes ranging from 0.64 Mb for chromosome 1 to 3.29 Mb for chromosome 14. The first genome sequence of *P. falciparum* (3D7 clone) was published in 2002 (Gardner *et al.*, 2002). It contains a high proportion of A+T (~80.6%) (McCutchan *et al.*, 1984; Pollack *et al.*, 1982) and carries ~5500 predicted genes. The proportion of A+T rises to ~90% in introns and intergenic regions, and to >97% in putative centromeric regions (Gardner *et al.*, 2002; Kelly *et al.*, 2006). The sequence information is freely available through public databases such as the *PlasmoDB* (http://plasmodb.org/plasmo/) and the National Center for Biotechnology Information (NCBI; http://www.ncbi.nlm.nih.gov/sites/genomeprj?Db= genomeprj&cmd=ShowDetailView&TermToSearch=9538) websites. Since publication of the 3D7 genome sequence, genomes of many more laboratory clones and field isolates of *P. falciparum* have been or are being sequenced (Carlton *et al.*, 2008b). The genomic information provides unprecedented opportunities for studying parasite genetic diversity, molecular evolution, drug resistance, parasite biology, and the diseases it causes.

Genomic diversity in *P. falciparum*

The genome of *P. falciparum* has been shown to be highly diverse (Jeffares *et al.*, 2007; Mu *et al.*, 2007; Volkman *et al.*, 2007), although there have been debates on its age, origin, and diversity (Su *et al.*, 2003). For the past four decades, there has been continuous development of molecular tools to evaluate the extent of genetic variation within natural populations of human malaria parasites. Beginning in the early 1970s, detection of genetic variation depended initially upon the observation of size differences in isoenzyme profiles using starch gel electrophoresis. The studies revealed a large number of electrophoretic enzyme variants among natural isolates of *P. falciparum* (Carter and McGregor, 1973; Carter and Voller, 1975). The variation could also be revealed by two-dimensional protein gel electrophoresis, in which variant forms of a protein were differentiated by their isoelectric point and molecular size (Fenton *et al.*, 1985). Serological methods, especially those using monoclonal antibodies, have also proved useful for differentiating antigenic variant forms in field isolates (McBride *et al.*, 1982, 1984, 1985).

Subsequently, DNA-based methodologies became the choice for detection of genetic variation in natural populations of human malaria parasites. The DNA-based techniques have a number of advantages over isoenzyme electrophoresis, including requiring a much smaller quantity of material than protein electrophoresis. Restriction fragment length polymorphism (RFLP), random amplified polymorphic DNA (RAPD), and amplified fragment length polymorphism (AFLP) have enabled us to assess genetic polymorphisms without prior sequence knowledge (Dolan *et al.*, 1993; Howard *et al.*, 1996; Rubio *et al.*, 2001; Triglia *et al.*, 1992; Walker-Jonah *et al.*, 1992); however, because these techniques rely on detection of polymorphism within or between defined endonuclease restriction sites, these methods can only detect a small fraction of the nucleotide substitutions in genomes. The relatively high cost and intensive laboratory work associated with these methods have also limited their applications for larger-scale functional studies. Nevertheless, RFLP has facilitated development of genetic markers for the initial mapping of the determinant conferring chloroquine resistance in progeny of the Dd2 and HB3 cross (Wellems *et al.*, 1990, 1991) and subsequently led to the construction of the first genetic linkage map of *P. falciparum* (Sinnis and Wellems, 1988; Walker-Jonah *et al.*, 1992) (see Chapter 6). The map was established with ~90 parasite clone-specific RFLP markers that were allocated to 14 inferred linkage groups corresponding to the 14 nuclear chromosomes using pulse field gel electrophoresis (PFGE) and Southern hybridizations (Walker-Jonah *et al.*, 1992). Crossover rates obtained from RFLP markers mapped on *P. falciparum* chromosomes allowed calculation of the map unit for this malaria parasite. An average map unit in the RFLP-based linkage map of *P. falciparum* was estimated to be 15–30 kb per centimorgan (kb/cM). Following this work, there are continued demands for techniques for large-scale genotyping and for developing high-resolution genetic linkage maps. In the 1990s, microsatellite (MS) markers were introduced into genetic studies and trait mapping of *P. falciparum* (Su *et al.*, 2007, 2009). Here, we summarize the use of MS and SNP to study genome diversity, population genetics, parasite evolution, and trait mapping in *P. falciparum*.

MS polymorphism and applications

MS are short (typically 1–5 base pairs in length) tandem repeats. Different MS alleles are characterized by variation in the number of repeating units (Ferdig and Su, 2000; Tan *et al.*, 2010). MS markers offer many advantages over RFLP. Typing MS is based on PCR and requires a small amount of DNA. PCR products can be separated in agarose or polyacrylamide gels, and the results can be obtained within hours instead of days for RFLP. MS can also be incorporated into multiplex and automated assays, an important advantage in the analysis of large numbers of parasite clones (Joy *et al.*, 2004).

Because they are highly polymorphic and generally have multiple alleles, MS can contribute greatly to genome diversity. MS in the form of $(T \text{ or } A)_n$, $(TA)_n$, or $(TAA)_n$ are present abundantly in the AT-rich *P. falciparum* genome, at a frequency of about 2–3 kb (Su and Wellems, 1996). In one study, an average of 15 alleles was obtained from 87 worldwide isolates typed with 342 MS markers (Wootton *et al.*, 2002). MS are apparently much less abundant in the human malaria parasite *Plasmodium vivax* and the rodent malaria parasite *Plasmodium yoelii*, likely due to the higher GC content of these parasite genomes (Feng *et al.*, 2003; Li *et al.*, 2009). Additionally, MS mutation rates were estimated to be 6.98×10^{-5} to 3.7×10^{-4} based on genotypes from 35 cloned progeny of the Dd2 × HB3 cross (Anderson *et al.*, 2000b). The highly polymorphic MS markers provide sensitive genetic tools for differentiating recrudescence from new infections in clinical studies (Mwangi *et al.*, 2006).

MS have also been proven extremely useful in genetic mapping in malaria parasites (see Chapter 6). In one study, more than 800 polymorphic MS were identified between Dd2 and HB3 by searching available GenBank databases and screening a *Dra*I genome library. Thirty-five progeny from the Dd2 × HB3 cross were then typed with the MS markers, leading to the construction of a high-density genetic linkage map (Su *et al.*, 1999). MS markers were also assigned to chromosome locations by PFGE analysis (Su and Wellems, 1999). The MS markers played a critical role in localizing the *P. falciparum* chloroquine resistance transporter (*pfcrt*) gene to a 36-kb region in *P. falciparum* chromosome 7 (Fidock *et al.*, 2000; Su *et al.*, 1997). Likewise, the genetic determinant of a male gametocytogenesis defect was mapped to *P. falciparum* chromosome 12 containing *P. falciparum* male development gene 1 (*pfmdv-1*) (Furuya *et al.*, 2005) and the determinant of sulfadoxine resistance to chromosome 4 (Wang *et al.*, 1997). The MS-based linkage map of the Dd2 × HB3 cross was also employed in quantitative trait loci (QTL) analysis to identify genes associated with parasite response to quinine and artemisinin (Beez *et al.*, 2010; Ferdig *et al.*, 2004), expression level polymorphisms (ELPs) (Gonzales *et al.*, 2008), and variation in growth rate/cell cycle duration (Reilly Ayala *et al.*, 2010). More recently, MS markers have been utilized to construct a genetic linkage map for another genetic cross between *P. falciparum* clones GB4 and 7G8. Using this map, Hayton and others (2008) identified the *pfRh5* gene on chromosome 4 that determines specific red blood cell invasion pathways and virulence of *P. falciparum* in the *Aotus* monkey (Hayton *et al.*, 2008). The genetic maps from the two crosses have also facilitated mapping genes that control differences in parasite responses to many compounds from chemical libraries (Nguitragool *et al.*, 2011; Yuan *et al.*, 2011; Yuan *et al.*, 2009), demonstrating the utility of high-throughput chemical screening and genome-wide MS variation for studying malaria traits. Genetic recombination hotspots and motifs enriched in the hotspots were also identified from the progeny of the GB4 and 7G8 cross, providing critical information for studying mechanism of genetic recombination in the parasite (Jiang *et al.*, 2011). Furthermore, in conjunction with an optical map (Lai *et al.*, 1999) and the yeast artificial chromosome (YAC) tiling contigs (Thompson *et al.*, 1999; Thompson and Cowman, 1997), the MS genetic linkage map also contributed to contig assembly of the *P. falciparum* genome sequencing project (Gardner *et al.*, 2002). Recently, a MS-based linkage map was constructed using multiple genetic crosses of the rodent malaria parasite *P. yoelii*. The map has not only enabled us to identify loci linked to growth related virulence phenotype but also proved useful for assembly of orphan contigs in the *P. yoelii* genome (Li *et al.*, 2009, 2011).

In addition to their usefulness in genetic mapping, MS loci are informative for population genetic studies (Anderson *et al.*, 2000a; Joy *et al.*, 2004) because they are highly polymorphic. Multiple alleles of a MS marker make it valuable for searching genetic variation and signature of natural selection within and between parasite populations (Anderson, 2004). Indeed, signatures of drug selection have been extensively investigated since the first study on the *pfcrt* locus for chloroquine selective sweeps (Wootton *et al.*, 2002). Evidence of selective sweeps at the *pfdhfr* and *pfdhps* loci that confer resistance to pyrimethamine and sulfadoxine, respectively, were also detected (Lumb *et al.*, 2009; McCollum *et al.*, 2006, 2007, 2008; Mixson-Hayden *et al.*, 2010; Nair *et al.*, 2003; Pearce *et al.*, 2009; Roper *et al.*, 2003; Salgueiro *et al.*, 2010).

Copy number variation (CNV)

CNV is another important source of genetic variation and appears to be associated with important phenotypes in *P. falciparum*. Large-scale chromosomal rearrangement, amplification, and deletion events frequently occur at the subtelomeric compartments containing genes involved in antigenic variation (*var*, *rifin*, and *stevor*) (Biggs *et al.*, 1989; Mattei and Scherf, 1994; Shirley *et al.*, 1990). Likewise, genes in other parts of the *P. falciparum* genome are found to vary widely in copy numbers (Cheeseman *et al.*, 2009; Jiang *et al.*, 2008b; Nair *et al.*, 2008). PFGE and Southern hybridization are very useful for broad-scale detection of the translocation, amplification, or deletion of specific genes across the parasite genomes (Corcoran *et al.*, 1986); however, this method is low-throughput, time-consuming, and cannot yield accurate measurement of gene copy number. Thanks to the availability of whole genome sequences of *P. falciparum* (Gardner *et al.*, 2002), several high-throughput methods such as DNA microarrays have been employed for the detection of genome-wide CNV (Carret *et al.*, 2005; Dharia *et al.*, 2009; Jiang *et al.*, 2008b; Kidgell *et al.*, 2006; Mu *et al.*, 2010; Neafsey *et al.*, 2008; Ribacke *et al.*, 2007; Tan *et al.*, 2009). Real-time quantitative PCR and fluorescence *in situ* hybridization are also useful for fine-scale detection and validation of CNV.

Various studies have shown that CNV may be associated with important phenotypes such as growth rate and virulence factors (Day *et al.*, 1993; Nair *et al.*, 2010) or parasite responses to antimalarial drugs (Barnes *et al.*, 1992; Cowman *et al.*, 1994; Gonzales *et al.*, 2008; Preechapornkul *et al.*, 2009; Price *et al.*, 2004; Sidhu *et al.*, 2006; Suwanarusk *et al.*, 2008; Triglia *et al.*, 1991). Recently, studies using comparative genome hybridization (CGH) and high-density oligonucleotide arrays have shown that CNV in the DNA segments containing *P. falciparum* multidrug resistant gene-1 (*pfmdr*-1), GTP cyclohydrolase I (*pfgch*1), and 1-deoxy-D-xylulose 5-phosphate reductoisomerase are associated with decreased susceptibility to mefloquine, antifolate drugs, and fosmidomycin *in vitro* and *in vivo* (Dharia *et al.*, 2009; Kidgell *et al.*, 2006). Because of direct effects of gene dosage, CNV is thought to alter levels of gene expression and the parasite's transcriptional network. Indeed, the expression level of the *pfgch*1 gene is increased in the clinical parasite samples with higher half-maximum inhibitory concentration (IC_{50}) values to antifolate drugs (Nair *et al.*, 2008). Likewise, progeny inheriting multiple copies of *pfmdr-1* from the multidrug-resistant parent Dd2 showed higher expression levels of the gene than the progeny inheriting a single copy from the drug-sensitive parent HB3 (Gonzales *et al.*, 2008). More recently, using *P. falciparum* field isolates from Mali, MacKinnon *et al.* (2009) revealed both positive and negative correlations between expression level and DNA copy number (Mackinnon *et al.*, 2009). The positive association appeared to be due to gene amplification. In contrast, the negative association between DNA copy number and expression level was suggested to be due to release of gene suppression. Deletions of large DNA segments are also frequently seen in culture-adapted *P. falciparum* laboratory isolates and in field isolates after short-term cultures (Hinterberg *et al.*, 1995; Kemp *et al.*, 1985). The deletions on the left arm of chromosome 2 and right arm of chromosome 9 have been well documented and appear to be associated with the loss of parasites' ability to produce gametocytes and the loss of cytoadherence phenotypes (Biggs *et al.*, 1989; Day *et al.*, 1993; Tan-Ariya *et al.*, 1988). These findings indicate that CNV and gene deletion are an important source of polymorphisms that underline important phenotypes and may determine parasite survival, virulence, and transmission.

Single nucleotide polymorphism (SNP)

SNP represents a difference in a single nucleotide (A, T, G, or C) in the DNA sequence. A SNP within a coding sequence does not necessarily cause the alternation of amino acid in a protein sequence, because the genetic code is degenerate. A SNP that does not lead to a change in amino acid is termed a 'synonymous SNP.' On the other hand, a SNP that causes a change in amino acid is termed a 'non-synonymous SNP.' A SNP within non-coding regions is termed a 'non-coding SNP'; this type of SNP could play an essential role in transcriptional regulation by affecting the binding of transcriptional factors to DNA and in gene splicing. In humans, variations in DNA sequences can affect responses to pathogens, drugs, and chemicals and can lead to genetic disorders (Shastry, 2007, 2009).

SNP analysis is easily automated, relatively cheap to assay, and is becoming the marker of choice for genetic studies. A number of molecular methods are readily available for rapid SNP detection, including dot-blot hybridization, multiple-PCR RFLP, molecular beacon, PCR-enzyme linked immunosorbent assay, and pyrosequencing (Abdel-Muhsin *et al.*, 2002; Alifrangis *et al.*, 2005; Bin Dajem *et al.*, 2011; Durand *et al.*, 2000; Veiga *et al.*, 2006); however, these techniques are suitable for relatively small-scale studies. Larger-scale high-throughput SNP analysis is routinely performed using two major methods: DNA sequencing-based and DNA hybridization-based approaches. Here we summarize the current advances in SNP discovery and genotyping technologies that are now available for genetic studies in *P. falciparum*. Methods of high-throughput SNP detection in other organisms have been published elsewhere (Kim and Misra, 2007; Komar, 2009; Maresso and Broeckel, 2008; Shah, 2008; Voisey and Morris, 2008).

SNP discovery and analysis using DNA sequencing-based approaches

A traditional method for SNP discovery is to sequence DNA samples from different individuals (isolates and clones of the malaria parasites) and align the DNA sequences to search for sequence variation. A project to systematically identify SNP was initiated by sequencing portions of 204 genes on *P. falciparum* chromosome 3 from four geographically divergent isolates (Dd2 from South East Asia, D10 from Papua New Guinea, HB3 from Honduras and 7G8 from Brazil) (Mu *et al.*, 2002). *P. falciparum* chromosome 3 was one of the first two chromosomes sequenced in the *P. falciparum* genome sequence project (Bowman *et al.*, 1999; Gardner *et al.*, 1998). The study generated ~218-kb coding and non-coding sequences from each of the four laboratory clones. After alignment of the sequences to the corresponding genome sequence of clone 3D7, ~240 SNPs were identified from ~400 polymorphic sites, giving approximately one SNP in every ~0.9-kb DNA sequence [average pairwise nucleotide diversity $(\pi) = 4.9 \times 10^{-4}$] (Mu *et al.*, 2002). Following this study, DNA fragments spanning 183 SNP positions on chromosome 3 from an additional 94 worldwide *P. falciparum* isolates were determined, which resulted in identification of 185 new SNPs (Mu *et al.*, 2005a). Among the 293 SNP in the 99 parasite isolates, 108 were common SNPs, with a minor allele frequency (MAF) equal to or greater than 0.05. The study also identified sets of alleles of closely linked loci on chromosome 3 that were conserved and likely to be inherited together in the parasite populations, thereby establishing a chromosome-wide SNP haplotype map in *P. falciparum*. SNP-typing results also showed that the parasites were mainly clustered according to their geographic origins, and different parasite population clusters exhibit wide variations in recombination rates, linkage disequilibrium, and size of haplotype blocks (Mu *et al.*, 2005a). Subsequently, approximately 4000 genome-wide SNPs were identified in ~3500 annotated genes (~3.5-Mb sequence analysed in each of four *P. falciparum* laboratory isolates [HB3, Dd2, 7G8, and D10]) (Mu *et al.*, 2007). The SNP and MS markers constitute a high-resolution map of genome-wide diversity in *P. falciparum*, with an average of one polymorphic marker per ~4 kb for the genome; however, analysis of genome-wide SNP diversity in the 99 *P. falciparum* isolates indicated that survey of the five laboratory clones might capture only ~45% of the total common SNP. Based on this observation, the frequency of SNP in the *P. falciparum* genome was estimated to be approximately one SNP per 842 bp (Mu *et al.*, 2005a), which is consistent with the previous estimate (0.9 kb per SNP) based on comparison of the chromosome 3 sequences (Mu *et al.*, 2002).

With coordinating efforts from the Wellcome Trust Sanger Institute and the Broad Institute, more *P. falciparum* isolates have been sequenced to search for polymorphisms. The two sequencing centres used the whole-genome shotgun (WGS) approach and *in silico* analysis to detect SNP, and a large number of SNP from an assortment of *P. falciparum* isolates of different geographic origins have been identified and deposited in PlasmoDB. A study by Volkman and others (2007) identified ~47,000 SNPs and ~37,000 indels from culture-adapted clinical isolates (Volkman *et al.*, 2007). In another study, ~27,000 SNPs and ~28,000 indels were identified between 3D7, one clinical isolate of *P. falciparum* from Ghana (PFCLIN), and one culture-adapted *P. falciparum* clone IT (Jeffares *et al.*, 2007). The PFCLIN parasite was isolated directly from the blood of an infected female patient at St Georges Hospital (London), which has not been subjected to repeated laboratory passages that may generate unwanted changes in the parasite genome. The Sanger sequencing team also presented a low-coverage genome sequence of the chimpanzee malaria parasite *Plasmodium reichenowi* and identified more than 200,000 SNP and 37,000 indels after comparison with the *P. falciparum* 3D7 genome sequence (Jeffares *et al.*, 2007). The results indicated that divergence between *P. falciparum* and *P. reichenowi* is ~10-fold greater than within-species polymorphism. So far, ~180,000 SNPs have been collectively identified from the genomes of 19 full or partially sequenced strains of *P. falciparum* (see NCBI SNP database http://www.ncbi.nlm.nih.gov/projects/SNP/and NCBI Genome project of *P. falciparum* http://www.ncbi.nlm.nih.gov/sites/entrez?Db=genomeprj& cmd=ShowDetailView &TermToSearch=9538).

In addition to the traditional Sanger sequencing, next-generation sequencing (NGS) methods have been developed recently (Ansorge, 2009; Hall, 2007; MacLean *et al.*, 2009; Metzker, 2010). NGS methods employ massive parallel approaches that are capable of sequencing millions of nucleotides (nt) simultaneously, thereby allowing inexpensive production of large volumes of sequence data. Genome projects that take several years using Sanger sequencing methods may now be completed in a matter of days or weeks. SNP and CNV discovery will undoubtedly benefit from the enormous data sets generated by these technologies (see Chapter 3).

SNP analysis using microarray

The availability of the 3D7 reference genome sequence has facilitated the development of DNA microarrays. DNA microarray (DNA chip) is a multiplex technology that consists of thousands to millions of microscopic spots of DNA oligonucleotides also called 'features' or 'probes.' The probe can be a short oligonucleotide, 25 to 83 nt in length, representing genes or regions in the genomes of interest. Hybridization of probes to RNA or cDNA samples (targets) is normally quantified by detection of fluorescent or *chemiluminescent*-labelled targets. Although DNA microarrays were originally designed to measure gene expression level, they are now used for applications such as detection of SNP, CNV, and chromatin modifications. DNA microarray-based assays allow detection of tens of thousands of single-feature polymorphisms at the same time. Hence, DNA microarrays serve as rapid high-throughput molecular analysis platforms and are now commonplace in many research laboratories. Generally, DNA microarray technologies can be categorized into two formats: spotted microarrays and *in situ* synthesized arrays. Each type of microarray platform can be manufactured in different ways, depending on probe numbers and applications. Indeed, various malaria microarrays have been developed for monitoring gene expression, SNP detection, genotyping, and chromatin modification in malaria parasites in recent years (Table 4.1).

Spotted microarrays

The spotted microarray is a platform in which probes are synthesized prior to being 'spotted' onto glass slides (usually poly-L-lysine-coated glass slides). Probes can be cDNA, genomic DNA fragments (generated by PCR amplification of a genomic library), gene sequence tag (GST), expressed sequence tag (EST), or oligonucleotides (Ben Mamoun *et al.*, 2001; Gissot *et al.*, 2004; Hayward *et al.*, 2000; Kappe *et al.*, 2001; Miller *et al.*, 2002; Rathod *et al.*, 2002). Spotted arrays provide flexibility in array design and can be easily customized and produced in-house for specific experiments. *P. falciparum* spotted arrays were pioneered by Drs. Pradipsinh Rathod and Joseph DeRisi. The first *P. falciparum*-specific spotted array consisted of 3648 cDNA probes from a mung bean nuclease-generated genomic DNA library of *P. falciparum* clone 7G8 (Hayward *et al.*, 2000), and the first synthetic oligonucleotide spotted array (DeRisi array) had 7462 70-mer oligonucleotide probes representing ~6000 predicted coding sequences or open reading frames (also known as the Malaria Oligo Set, Qiagen-Operon) (Bozdech *et al.*, 2003a,b). Using the synthetic oligoarray, Bozdech *et al.* (2003a,b) examined the transcriptome of asexual stages of *P. falciparum* clone HB3, providing the first comprehensive view of the timing and level of transcription throughout the asexual erythrocytic stages (Bozdech *et al.*, 2003a) (see also Chapter 5). A second generation of the spotted array (Bozdech 10K spotted array) comprises 10,166 oligonucleotide probes representing 5363 conserved regions of the predicted genes (1–2 probes per gene, but no probes representing variant antigens (Hu *et al.*, 2007; MacKinnon, M., personal communication). Using this array, MacKinnon *et al.* (2009) showed that transcriptional variations were prevalent in among culture-adapted field isolates of *P. falciparum*, which appeared to be associated with deletion of chromosomal segments or gene copy number variation (Mackinnon *et al.*, 2009). The array was also used to study the genome-wide transcriptional regulations during growth arrests induced by chemical treatment and to study protein interaction network of the intraerythrocytic development cycle (Hu *et al.*, 2010). Additionally, the Malaria Oligo Set, combined with new sets of oligonucleotides representing *var* genes, was also employed to study antigenic variation and pregnancy malaria-associated gene expression in field isolates (Ralph *et al.*, 2005; Tuikue Ndam *et al.*, 2008). Spotted oligonucleotide arrays are also useful for detecting genome-wide CNV in conjunction with fluorescence *in situ* hybridization (FISH) and real-time quantitative PCR. A study by Ribacke *et al.* (2007) showed that gene amplification and deletion events were located at specific locations in the *P. falciparum* genome, suggesting the presence of genomic hot spots where CNV are prone to occur.

There are also small-scale spotted arrays developed to target specific sets of SNPs such as ResMalChip (developed by Swiss Tropical Institute, Switzerland) and FlexiChip (developed by Pasteur Institute, France). The ResMalChip probes (20 to 35 bp) were designed to interrogate 32 SNPs in five *P. falciparum* genes (*pfdhfr*, *pfdhps*, *pfcrt*, *pfmdr-1* and *pfATPase*) reported to be associated with resistance to antimalarial drugs (Crameri *et al.*, 2007). This chip has been recently used to study molecular epidemiology of drug resistance in *P. falciparum* in Papua New

Table 4.1 Microarrays for the detection of polymorphism, genotyping, and gene expression analysis in the human malaria parasite *Plasmodium falciparum*

Microarray	Probe length (bp)	No. of probes	No. of genes or SNP	Density (probes/ SNP or gene)	Application	Selected references
Spotted arrays						
Rathod cDNA array	1–2 kb	3648	3648 cDNA	1/gene	Gene expression	Hayward *et al.* (2000)
DeRisi array	70	7462	4488 genes	1+/gene	Gene expression, eQTL, CNV	Bozdech *et al.* (2003a,b), Gonzalez *et al.* (2008)
Bozdech 10K array	70	10,417	5577 genes	1–2/gene	CNV, Gene expression	Hu *et al* (2007), MacKinnon *et al.* (2009), Hu *et al.* (2010)
Wahlgren array	70	6850	~4000 genes	1–13/gene	CNV	Ribacke *et al.* (2007)
ResMalChip	20–35	32	32 SNP (5 genes)	1/SNP	SNP	Crameri *et al.* (2007)
Guo Qin Zhang array (China)	35–54	45	21 SNP (5 genes)	1+/SNP	SNP	Zhang *et al.* (2008)
FlexiChip	>24	Up to 96	Up to 96 SNP	1/SNP	SNP	Steenkeste *et al.* (2009)
In situ *synthesized oligonucleotide arrays*						
Winzeler Affymetrix Ch2 array	25	4167	210 genes	2 – 117/ gene	SNP, CNV	Volkman *et al.* (2002)
Winzeler Affymetrix 300K array	25	327,989	5159 genes	1+/SNP	SNP, CNV, Gene expression	Le Roch *et al.* (2003), Kidgell *et al.* (2006)
Affymetrix GeneChip *P. falciparum/Anopheles*	25	>500,000	4700 *Pf* genes	11/gene	SNP	Turner *et al.* (2005), Jiang *et al.* (2008a)
Affymetrix scrMalaria arrays	25	260,596	5159 genes	1–20/gene	CNV	Carret *et al.* (2005)
PFSANGER tiling array	25	2,560,000	5486 genes	1–45/SNP	SNP, CNV, Gene expression	Cortes *et al.* (2007), Jiang *et al.* (2008b); Cheeseman *et al.* (2009)
Broad Affymetrix 3K SNP array	25	165,000	Up to 3K SNP	56 probes/ SNP	SNP	Neafsey *et al.* (2008), Park, personal communication
Winzeler 4.8M tiling array	25	4,800,000	5159 genes	9–13/SNP	SNP, CNV	Dharia *et al.* (2009)
Ferdig NimbleGen 385K array	45–83	385,585	5272 gene	1–710/gene	SNP, CNV	Tan *et al.* (2009), Nair *et al.* (2010)
NIAID 3K and 8k MIP SNP arrays*	–	3354	3K SNP or 8k SNP	1/SNP	SNP	Mu *et al.* (2010), Su, unpublished
Broad Affymetrix 75K SNP array	25	4,400,000	Up to 75K SNP	12–84 probes/SNP (1+/gene)	SNP, CNV	Van Tyne *et al.* (2011), Park, personal communication
Ferdig CNV-SNP array (NimbleGen 3-plex)	29–41 (SNP), 50–75 (CGH)	719,995	5183 genes, 45,524 SNP	1–600/ gene,8/SNP	SNP, CNV	Tan *et al.*(2011)

Microarray	Probe length (bp)	No. of probes	No. of genes or SNP	Density (probes/ SNP or gene)	Application	Selected references
Ferdig Falciparum Exon array (Nimble gene 12-plex)	50–75	128,099	5681 genes	1–296/gene	Gene expression, alternative splicing	Ferdig and Tan, personal communication
Llinas Agilent 15K array	60	14,826	5789 genes	1+/gene	Gene expression	Llinas, personal communication

Note: Asterisk (*) indicates that an updated NIAID MIP array with ~8000 SNPs is available, but not published; the probes are anchored oligonucleotides, not real detection targets; CGH, comparative genomic hybridization, CNV, copy number variation; SNP, single nucleotide polymorphism. We apologize for not including other custom spotted arrays (see text for references) as they are applied for gene expression studies only. More information on microarrays can be found in Chapter 5.

Guinea and Niger (Crameri *et al.*, 2007; Ibrahim *et al.*, 2009). The FlexiChip consists of 96 universal probes each containing a 24-nucleotide artificial sequence (so-called 'zip-code') (Gerry *et al.*, 1999) and an amino C-7 nucleotide linker that facilitates attachment of probes to the aldehyde-coated slide. Because the PCR primers that amplify genomic regions carrying SNP of interest also contain sequences complementary to the zip-codes, various SNP assays can be performed using a single protocol. Unlike the ResMalChip, SNP positions in the FlexiChip can be modified. Currently, up to 96 SNP sites can be interrogated simultaneously. Steenkeste and others (2009) have recently utilized the FlexiChips to scan 32 SNP associated with antimalarial drug resistance in uncomplicated malaria cases in Cambodia (Steenkeste *et al.*, 2009). Another spotted array platform containing 45 oligonucleotides (35 to 54-mer probes) has been developed for detection of 21 SNP in the *pfhdfr*, *pfdhps*, *pfcrt*, *pfmdr-1*, and *pfATPase* genes (Zhang *et al.*, 2008). The targeted SNP genotyping spotted arrays can become a new tool for monitoring *P. falciparum* drug resistance markers on an epidemiologic scale. Although there has been substantial progress towards developing spotted microarrays with higher probe density for malaria research, which can now accommodate up to 10,000 probes (Mackinnon *et al.*, 2009), their use in genome-wide SNP discovery has been limited, because the number of probes in the microarrays is relatively small compared with the other types of DNA microarrays.

In situ synthesized oligonucleotide arrays

An alternative DNA microarray platform is the *in situ* synthesized oligonucleotide array, which can be effectively used to measure changes in gene expression level, detect SNP, or even resequence a genome (Daily *et al.*, 2007; Le Roch *et al.*, 2003; Palacios *et al.*, 2007; Volkman *et al.*, 2002; Wurtz *et al.*, 2009) (Table 4.1). The array is made by *in situ* synthesizing short oligonucleotides onto the array surface (silicon chip) directly using photolithographic chemistry. The size of the oligonucleotide probes range from 25 nt (Affymetrix, California) to 83 nt long (the Nimblegen array). The advantage of *in situ* synthesized oligonucleotides is that the probes may be synthesized to very high density (now up to ~5 million probes).

For many years, Affymetrix has facilitated production of high-density short-oligonucleotide arrays for transcriptomic analysis and for screening for genome-wide polymorphism in malaria parasites (Daily *et al.*, 2007; Le Roch *et al.*, 2003; Volkman *et al.*, 2002). For SNP detection, a reference DNA sample (usually from *P. falciparum* clone 3D7) and a test sample are fluorescently labelled before hybridizing to the array. The strength of hybridization (signals) between a probe and its target sequence depends largely on sequence similarity between the probe and the target. An equivalent signal ratio indicates uniform hybridization and signifies nucleotide identity. On the other hand, a single base change in a 25-mer probe and its genome targets may result in a reproducible loss of hybridization signal compared with that of the reference 3D7 strain. Because the specific location of the nucleotide substitution is unknown, the change is simply called single-feature polymorphism (SFP). It should be noted that accuracy of SFP calls can be influenced by many factors, such as the number and location of mismatches between the target and probe sequences, the G+C content of the probe, and algorithm for SFP call (Jiang *et al.*, 2008b; Tan *et al.*, 2011). These factors require optimization and standardization.

To evaluate the potential use of Affymetrix oligonucleotide arrays for elucidating genetic diversity in *P. falciparum*, a microarray consisting of 4167 unique 25-mer oligonucleotide probes was initially developed to detect polymorphism on chromosome 2 (Volkman *et al.*, 2002). The *P. falciparum* probes were produced on a prototype array that also included ~400,000 probes for ~80,000 different cDNA from human tissues, so it permitted detection of cross-contaminations of host mRNA. In this investigation, the microarray detected 981 putative SNPs among four laboratory clones (W2 from South East Asia, D6 from Sierra Leone, 7G8, and HB3) relative to clone 3D7. The results showed that most SNPs were concentrated on the subtelomeric region within 100 kb of the chromosome ends. The greatest variation occurred in genes encoding parasite surface antigens and proteins exported to the infected erythrocyte membrane. A likely explanation for this observation is that these proteins have a high likelihood of interacting with host immune systems and may be subjected to immune selection pressure. In addition, the array confirmed a 42-kb segmental deletion on the left arm of chromosome 2 in the W2 parasite (Volkman *et al.*, 2002). These results demonstrate that the high-density oligonucleotide array is applicable for detection of SNP and gene deletions in *P. falciparum*.

Consequently, an expanded version of Affymetrix higher-density oligonucleotide microarray was developed for identification of genetic variation in the entire genome (Kidgell *et al.*, 2006). The array contains 327,989 (300K) non-overlapping 25-mer probes covering approximately 55% of the coding sequences of 5179 predicted genes and some intergenic regions. This 300K array was used to detect genome-wide variation in 13 culture-adapted isolates of *P. falciparum* and recombination events in one progeny from the Dd2 × HB3 cross. Results from the study using the 300K array identified 23,653 SFPs across all the strains and various gene amplifications and deletions (Kidgell *et al.*, 2006). The majority of the variable genes are predicted to have roles in evading host immune responses and/or in conferring resistance to antimalarial drugs, which is in agreement with the view that the human immune system and drug therapy are major driving forces of malaria parasite evolution.

At almost the same time, a higher-density tiling array with 2.56 million probes (PfSANGER GeneChips; http://www.sanger.ac.uk/Projects/P_falciparum/news.shtml) designed by researchers at the Wellcome Trust Sanger Institute became commercially available (Cortes *et al.*, 2007; Jiang *et al.*, 2008b; Mourier *et al.*, 2008). The tiling array comprised overlapping probes covering both DNA strands that systematically tile through the chromosomes, including approximately 1.2 million probes covering predicted gene-coding sequences. The PfSANGER array has been used to identify novel structural RNA such as non-coding RNAs in *P. falciparum* (Mourier *et al.*, 2008). It has also been used for SNP detection (Jiang *et al.*, 2008b). Using this technology, more than 120,000 mSFPs (SFP identified from multiple probes) from five *P. falciparum* isolates were detected, giving an average of one unique SFP per 570 bp (Jiang *et al.*, 2008b), a frequency that is consistent to those of previous studies (Mu *et al.*, 2002; Volkman *et al.*, 2007). Like other Affymetrix arrays, the PfSANGER array can be used to detect CNV in the *P. falciparum* genome (Cheeseman *et al.*, 2009; Jiang *et al.*, 2008b). More recently, another even higher-density Affymetrix tiling array with 4.8 million 25-mer probes was developed and validated for accurate SFP calling (~ 10% false discovery rate) (Dharia *et al.*, 2009). This array has enabled detection of *pfdxr-1* gene amplification in the Dd2 parasite that has decreased *in vitro* susceptibility to fosmidomycin, which has been shown to target the parasite isoprenoid synthesis pathway. These studies demonstrated that the Affymetrix tiling array is an effective tool for detecting novel SNP and CNV in the *P. falciparum* genome.

In addition to the high-density tiling arrays that interrogate SFP, two microarray platforms have been developed specifically for high-throughput SNP genotyping and mapping genes associated with phenotypes of interest in genome-wide association studies (GWAS). One of the SNP arrays used molecular inversion probe (MIP) technology to interrogate 3354 SNPs (Mu *et al.*, 2010). The MIP technology was initially developed at Stanford University (Hardenbol *et al.*, 2003) and has been further developed and commercialized by ParAllele BioScience, currently part of Affymetrix. This array was used to genotype 189 *P. falciparum* isolates with a high call rate (97%) and high accuracy, as 99.9% of SNPs called matched the reference genome of *P. falciparum* clone 3D7 (Mu *et al.*, 2010). Because of PCR amplification of DNA templates before hybridization, this array requires a relatively small amount of DNA (100–300 ng), providing a platform for direct detection of SNP from clinical isolates. The other is the Broad Affymetrix 3K SNP array platform that employed a prototype of the human 500K SNP array to detect ~3000 SNPs (Neafsey *et al.*, 2008). This method utilizes a set of 56 25-mer probes arranged in 14 quartets to interrogate each SNP, with each polymorphic site having 'perfect match' and 'mismatch' probes for each of the two alleles. The mismatch probes are included so that the effectiveness of the SNP calling algorithm can be systematically evaluated. Approximately half of the assays (54.6%) yield high-quality signals that provide correct genotype calls. Currently, the MIP chip and the Affymetrix 3000 SNP array have been scaled up to interrogate up to ~8000 and ~75000 SNP, respectively (Van Tyne *et al.*, 2011; Jiangbing Mu and Xin-zhuan Su, unpublished). Furthermore, in conjunction with whole-genome amplification (WGA) methods such as multiple displacement amplification (MDA), primer extension pre-amplification (PEP), degenerate oligonucleotide primed PCR (DOP-PCR), or ligation-mediated PCR (LMP) (Carret *et al.*, 2005), these SNP arrays can be used for large-scale genotyping clinical isolates from patient blood without *in vitro* culture adaption.

In addition, Affymetrix also launched Affymetrix GeneChip *Plasmodium/Anopheles* (http://www.affymetrix.com/support/technical/byproduct.affx?product=plasmodium; http://www.osa.sunysb.edu/udmf/ArraySheets/plasmodium datasheet.pdf), which is the first microarray to provide comprehensive coverage of the two organisms on a single array. The array, developed in collaboration with the Malaria Research Institute at the Johns Hopkins University, can interrogate over 4300 *P. falciparum* transcripts and approximately 14,900 *Anopheles gambiae* transcripts. The inclusion of both organisms on a single array allows researchers to study the molecular dynamics involved in the host–parasite relationship. Using this microarray, Turner and others (2005) identified three genomic loci on chromosomes 2L, 2R and X, also known as 'speciation islands', that differentiate two molecular forms (taxa), M and S, of *An. gambiae*, the major vector of human malaria diseases in Africa (Turner *et al.*, 2005). The array has also been used to study global gene expression in *P. falciparum* (Jiang *et al.*, 2008a) and in the *Anopheles* mosquitoes under various conditions (Aguilar *et al.*, 2005; Kumar *et al.*, 2010; Marinotti *et al.*, 2005, 2006; Neira Oviedo *et al.*, 2009; Pinto *et al.*, 2009; Rogers *et al.*, 2008). More recently, another gene expression array (Agilent) comprising 60-mer 15K probes targeting 5789 genes has been developed in the laboratory of Dr Manuel Llinás's laboratory at Princeton university and will soon be available to the malaria community (Table 4.1) (M. Llinas, personal communication)

Like Affymetrix arrays, the NimbleGen array (Roche NimbleGen) is an alternative hybridization-based platform on which photolithographic chemistry is utilized to synthesize short probes on a silicon chip *in situ*. A NimbleGen array for genome-wide CNV and SNP detection in *P. falciparum* has been constructed with more than 385,000 unique tiling probes of variable length, ranging from 45 nt to 83 nt, that cover both coding and intergenic regions of the genome (Tan *et al.*, 2009). Unlike the standard Affymetrix array in which different samples are subjected to separated single-dye hybridizations and detections, the NimbleGen platform has competitive-base assays in which two samples of interest (i.e. one reference and one test sample) are labelled with two different fluorophores, commonly Cy3 and Cy5, and then analysed in pairs. A relative Log_2 signal ratio of a test sample over a reference is used to estimate differences in copy number between the two samples. Because NimbleGen probes are longer than standard Affymetrix 25-mer probes, the array yields good quality data for comparative genomic analysis, in particular CNV detection, but may not be sensitive to nucleotide changes at the middle of a probe. Thus, preliminary analysis showed that the NimbleGen array could be used to detect large-scale gene amplification and deletion events in the genomes of Dd2 and HB3 with respective to 3D7 (Tan *et al.*, 2009). More recently, another NimbleGen array with higher probe density (720K probes) has been developed and optimized for detection of genome-wide SNP and CNV in samples from *P. falciparum* lab strains with 95% accuracy in a single hybridization. This new microarray is also capable of defining SNP and CNV in *P. falciparum* genomic DNA from field isolates even with contamination of some host DNA (Tan *et al.*, 2011). [Note that a new high-density Nimblegen exon array for estimating gene expression in *P. falciparum* (12-plex of ~130K probes) is currently being developed at Dr Mike Ferdig's laboratory; M. Ferdig and J. Tan, personal communication]. Together with genome sequencing, microarrays will continue to play an important role in studies of genome diversity, population structure, inbreeding, recombination frequency, natural selection, and biology of malaria parasites.

P. falciparum population genetics and functional genomics studies using SNPs

An overall estimate of genetic diversity in the *P. falciparum* genome has proven difficult to obtain, with different markers showing different levels of polymorphism. For the past few decades, investigations into genetic diversity and population structure of malaria parasites have derived mostly from a limited number of loci, including genes encoding malaria surface antigens such as merozoite surface protein-1 (MSP-1), MSP-2, apical membrane antigen-1 (AMA-1), and circumsporozoite protein (CSP) (Da Silveira *et al.*, 2001; Escalante *et al.*, 2001; Farooq *et al.*, 2009; Ferreira and Hartl, 2007; Ferreira *et al.*, 1998, 2003; Heidari *et al.*, 2007; Mamillapalli *et al.*, 2007; Rajesh *et al.*, 2008; Umar *et al.*, 2006). Although these genes have exhibited high levels of genetic diversity within and between parasite populations, such polymorphisms may not reflect global relationships among the parasite populations because these molecules are generally under immune selection for antigenic polymorphism. By contrast, the use of antimalarial drugs has imposed additional selective forces on its target genes. In this case, an antimalarial drug may lead to eradication of parasites with susceptible alleles while selecting parasites with resistant alleles. As a result, population genetic studies based on these genetic markers known to be under selection pressure may be problematic. Current advances in high-throughput molecular techniques, such as microarrays and whole genome sequencing, can be readily used to investigate the extent of the parasite's genomic diversity. These technologies will transform our approaches in studying parasite evolution, population structure, geographic origin, host–parasite interaction, and gene function. The information obtained from these studies will have a significant impact on vaccine and drug development.

Origin and recent expansion of *P. falciparum*

The possibility that *P. falciparum* might have a clonal population structure was first described by Tibayrenc and others in 1990 (Tibayrenc *et al.*, 1990), and the subject has been debated ever since (Conway and Baum, 2002; Dye, 1991; Hartl, 2004; Su *et al.*, 2003). Evidence for a clonal population structure in *P. falciparum* was obtained from two preliminary observations. First, a lack of correlation between linkage disequilibrium (LD) and nucleotide distances was observed in the *csp* gene (Rich *et al.*, 1997). Additionally, analyses of ten protein-coding genes (~16.5 kb), including genes encoding metabolic enzymes and highly polymorphic antigens deposited in GenBank, identified no synonymous mutations (Rich *et al.*, 1998). To explain the observed low level of polymorphisms, the authors proposed that *P. falciparum* had undergone a severe population bottleneck ~5000 year ago, which forms the 'Malaria's Eve' hypothesis. Additional evidence supporting the Malaria's Eve hypothesis came from analyses of mitochondrial DNA haplotypes of a worldwide collection of isolates (Conway *et al.*, 2000b) and from intron sequences of ten housekeeping genes on *P. falciparum* chromosomes 2 and 3 (Volkman *et al.*, 2001). In both cases, very low levels of polymorphisms were observed among parasites from different continents, supporting the view that current *P. falciparum* populations originate from a relatively young ancestor.

Other studies, however, suggested a more diverse parasite genome and older evolutionary age of *P. falciparum*. A survey of 204 predicted genes on chromosome 3 for five *P. falciparum* isolates identified 403 polymorphic sites (165 polymorphic MS sites and 238 SNP) (Mu *et al.*, 2002). Of the resulting SNPs, 207 occurred in the coding regions of the genes, 62 of which were synonymous. On the basis of synonymous and non-coding SNPs, the time to the most recent common ancestor was estimated to be ~ 100,000–180,000 years (Mu *et al.*, 2002), significantly older than the date previously proposed in the Malaria's Eve hypothesis (Rich *et al.*, 1998). Similarly, Hughes and Verra (2001) found that *P. falciparum* has maintained a large effective population size (10^5) for the last 300,000–400,000 years on the basis of variation in 23 nuclear loci coding for proteins (Hughes and Verra, 2001). A similar conclusion was also drawn from a subsequent study of 100 mitochondrial genome sequences from worldwide isolates (from Africa, Papua New Guinea, and Asia) (Joy *et al.*, 2003). Large numbers of SNPs obtained from searching genome-wide polymorphisms also supported the conclusion that *P. falciparum* has a relatively long evolutionary history and a large effective population size (Jeffares *et al.*, 2007; Mu *et al.*, 2007; Volkman *et al.*, 2007). Although the discrepancy in age estimates could result from the use of different calculation methods, the technical quality of the data set, the use of closely related parasite isolates, and the selection of genes used in the analyses (Conway and Baum, 2002; Hartl, 2004; Su *et al.*, 2003), most studies agree that the African *P. falciparum* population is ancestral to those from other continents. Sudden expansion of the parasite population in Africa corroborates a major prediction of Coluzzi's hypothesis that elevated rates of malaria transmission accompanied speciation of *An. gambiae* (Coluzzi, 1999; Coluzzi *et al.*, 2002).

Population structure

A number of local epidemiologic and demographic factors can influence diversity within parasite populations, such as the multiplicity of infection (clonal or mixed-clone infections), the level of vector transmission intensity, and the density and distribution of human hosts and vectors (Conway, 2007). Early studies that attempted to infer the population structure of *P. falciparum* were mostly based on SNP data from highly polymorphic genes encoding antigens (Conway, 1997; Drakeley *et al.*, 1996; Walliker *et al.*, 1998). Caution should be taken in interpreting population structure results with these markers, because polymorphisms in these genes may be due to selection mediated by the immune responses of human hosts (Day and Marsh, 1991). In a recent metapopulation genetic analysis of the 12 genes coding pre-erythrocytic stage antigens (CSP, TRAP, and LSA1), erythrocytic-stage antigens (MSP-1, -2, -3, and -5, EBA-175, and GLURP), gametocyte antigen (*pfs*48/45), and antigens expressing at multiple stages (MSP-4, AMA-1, and GLURP), Barry *et al.* (2009) showed that merozoite antigens had high levels of diversity globally and a worldwide distribution of each subgroup. In contrast, the diversity of the non-merozoite antigens showed geographically variable levels of diversity and geographic restriction of specific subgroups, which could be organized by physical and location-specific barriers restricting gene flow (Barry *et al.*, 2009). Therefore, results of population structure analysis will be influenced by the type and nature of genetic markers used.

Genome-wide MS markers have been employed to study *P. falciparum* populations. As MS loci are mostly located in non-coding regions and are not expected to be under strong selection, they offer relatively neutral genetic markers for population genetic studies in *P. falciparum*. Anderson and others (2000a) conducted a worldwide survey of variations using 12 MS in 465 infections from nine geographic populations and showed that genetic diversity in *P. falciparum* varied greatly among populations from different continents (Anderson *et al.*, 2000a). Diversity was greatest in parasite populations in Africa, where high genetic recombination rates with intense parasite transmission were observed (Conway *et al.*, 1999). The high diversity observed in Africa may be due to the older age of this population, providing support for the argument that malaria originated from Africa. Intermediate levels of genetic diversity were detected in *P. falciparum* populations in Asia, while South American parasites had the lowest diversity. The low level of diversity in South American populations could be due to chloroquine selection and rapid expansion of drug-resistant parasites. Consistent with the above study, significant MS-allele sharing was found among the South American population but not among those from Africa (Wootton *et al.*, 2002).

Likewise, analysis of 183 SNPs on chromosome 3 in 99 worldwide isolates of *P. falciparum* supported the view that the parasites could be categorized into five genetic clusters according to major geographic boundaries, regardless of their time and locations of isolation (Mu *et al.*, 2005a) (Fig. 4.1A). Additional supporting evidence came from two independent studies based on analysis of 76 worldwide isolates using the Broad Affymetrix 3K SNP array (Neafsey *et al.*, 2008), and the genome-wide SNP profiles in 189 isolates of *P. falciparum* using the 3K MIP chip (Mu *et al.*, 2010) (Fig. 4.1B). Both studies confirmed the genetic differentiation between parasite populations from different continental boundaries. In addition, there is some evidence of subpopulation structures within the continents, including the separation of Southeast Asian parasites from Papua New Guinea parasites (Mu *et al.*, 2005a) (Fig. 4.1B), the separation between the two Central American parasites (Honduras and El Salvador) and Brazilian parasites (Neafsey *et al.*, 2008) and, more recently, the separation of Cambodian parasites into two populations (Mu *et al.*, 2010). The clustering of parasites into different subgroups suggested either recent population admixtures or the presence of SNPs that are selected for advantageous phenotypes. Together, the studies also indicated that worldwide *P. falciparum* parasites cluster primarily according to geographic origin. However, population structure clustering is depended on the types of genetic markers used. Parasites from Africa were clustered into two groups, one resistant to chloroquine and another sensitive to chloroquine, based on SNPs from 37 putative transporters (Fig. 4.1C). If SNPs from an immune target were used, no geographical clustering was found; instead, similar clusters were found in each geographic population. In two recent studies, SNPs in the gene encoding PfAMA-1, a target of protective immunity, grouped parasites into subpopulations that were independent of geographic origin (Duan *et al.*, 2008; Ouattara *et al.*, 2010) (Fig. 4.1D).

Recombination hotspots and linkage disequilibrium (LD)

It is well established that sexual recombination in *P. falciparum* occurs at high frequency during laboratory crosses, with a 1% possibility of crossover (1 centimorgan) in every 10- to 30-kb DNA segment (Jiang *et al.*, 2011; Su *et al.*, 1999; Walker-Jonah *et al.*, 1992). Even though the overall population recombination rate is highly variable among parasite populations and along different chromosomes of *P. falciparum*, the recombination hotspots appear to be conserved among populations. Many recombination hotspots are located towards subtelomeric regions of chromosomes that contain multigene families (*var*, *rifin*, and *stevor*) (Mu *et al.*, 2005a, 2010), suggesting the effect of balancing selection that perpetuates allelic variability in these multigene families. Interestingly, high recombination rates were detected in a large segment of chromosome 7 containing the *pfcrt* gene but not in the small ~100-kb fragment flanking *pfcrt*, which showed significantly reduced recombination activity and genetic diversity near the *pfcrt* gene. The observation suggested positive selection and a recent selective sweep due to chloroquine pressure (Wootton *et al.*, 2002).

Using progeny from the GB4×7G8 genetic cross, Jiang *et al.* (2011) studied recombination hotspots in *P. falciparum* (Jiang *et al.*, 2011). From 54 hotpots detected after analysing genome-wide SNP and MS polymorphisms among 32 progeny from the GB4×7G8 cross, several motifs were

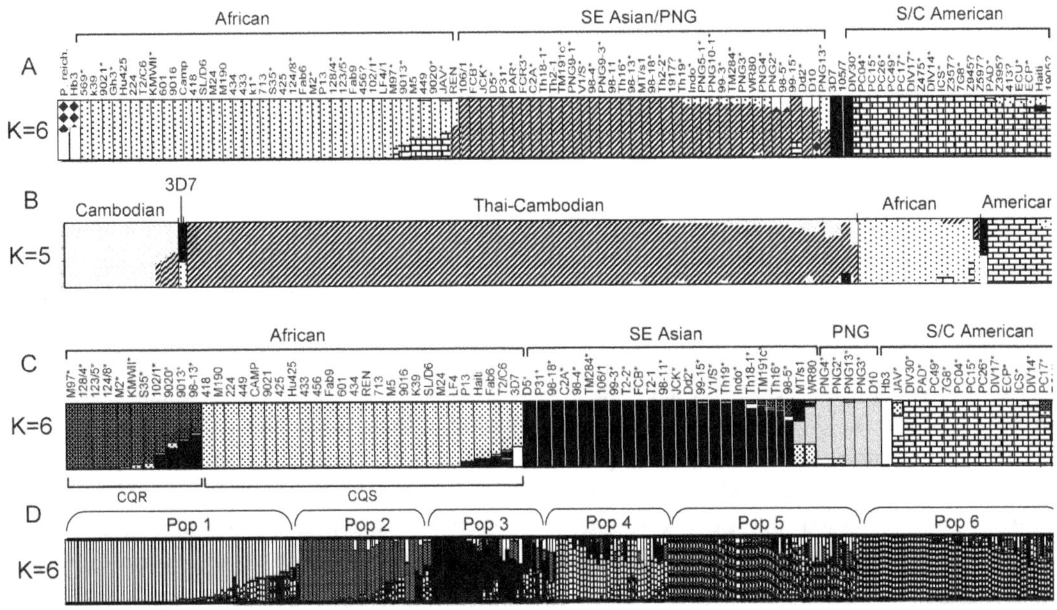

Figure 4.1 Inferred population structures of worldwide *Plasmodium falciparum* parasites. Population partition was assessed using a Bayesian clustering method implemented in structure 2.2. Each parasite is represented by a vertical bar/line and the patterns of different textures indicate their proportional membership in the different populations. *P. falciparum* isolates are largely clustered according to their geographical origins using either 183 chromosome-wide (A) or 3354 genome-wide SNP (B). Note although there are some overlaps in the parasite isolates used in the two studies, the sample sizes (99 isolates in A and 189 isolates in B) are different; the Cambodian parasites were only present in B, and Papua New Guinea (PNG) parasites were only analysed in A. However, when the same 99 parasites in A were clustered using 102 SNP from 37 putative transporters, the African parasites were further partitioned into chloroquine-resistant and -susceptible parasites (C). Population structures analysis based on coding SNPs of the apical membrane antigen-1 gene from 315 isolates (the parasites in A plus field isolates from Mali, west Africa) clustered parasites into six populations with each population containing parasites from different continental regions (D). These results show that different clustering patterns can be obtained using different genes/ SNP; genome-wide or chromosome-wide SNP will cluster parasites mostly according to their geographical origins, whereas the use of genes involved in drug resistance or immune targets will cluster the parasites differently. These figures were modified from the original coloured plots published in Mu *et al.* (2005, 2010), Duan *et al.* (2008), Ouattara *et al.* (2010).

identified, including a 12bp G/C-rich motif with 3-bp periodicity that may interact with a protein containing 11 predicted zinc finger arrays. The putative sequences of *P. falciparum* centromeres (Kelly *et al.*, 2006) were also found in chromosome regions largely devoid of recombination activity, which is consistent with the observation that centromeres generally have reduced recombination rates.

LD, the non-random association of genetic markers in the genome, has been extensively studied in *Plasmodium* parasites in different populations using SNP and MS markers. Comparison among several populations of different transmission intensities showed that variation in recombination rate in different parasite populations could affect the size of LD and haplotype blocks in the genome. In regions where transmission of malaria is high with frequent occurrence of mixed-genotype infections, genetic recombination is present at extremely high rates (Babiker *et al.*, 1994, 1997; Conway *et al.*, 1999). Strong LD only occurs between closely linked nucleotides. Early studies of the antigen-encoding genes, *msp*-1 and *ama*-1 in *P. falciparum* showed that LD between the two polymorphic sites rarely extends beyond 500 bp in African populations (Conway *et al.*, 1999; Da Silveira *et al.*, 2001; Polley and

Conway, 2001; Sakihama *et al.*, 1999). Similarly, an expanded analysis of LD using genome-wide 3000 SNPs revealed zero LD in the parasite samples from Senegal and Malawi (Neafsey *et al.*, 2008). In contrast, in the areas of low transmission, parasites with the same genotypes are likely to be present in populations, suggesting clonal structure (Arez *et al.*, 1999). The size of LD was extended to ~20kb and ~100kb in parasite populations in Thailand and South America, respectively, where transmission intensity and multiplicity of infection were lower than those of Africa (Neafsey *et al.*, 2008). Analysis of LD and haplotype block using chromosome-wide (Mu *et al.*, 2005a) and genome-wide SNP markers (Mu *et al.*, 2010) also showed that LD decayed more rapidly in the African parasite population than in Southeast Asian and American populations. For *P. falciparum* populations in Africa, more than one marker for every kb of DNA (>25,000–30,000 SNP) will likely be required for successful association mapping of novel genes that are involved in longstanding phenotypes (Neafsey *et al.*, 2008). In contrast, a density of one marker for every 5–15kb might provide sufficient resolution for Asian and South American populations (see *Genome-wide association studies* section).

Signatures of selection

Analysis of single nucleotide and MS polymorphisms can reveal signatures of selection in the parasite's genomes. There are two major types of selections: balancing selection and directional selection. Balancing selection perpetuates different alleles in a population (Charlesworth, 2006). Signature of balancing selection can be exploited to identify novel vaccine targets. Conway *et al.* proposed that polymorphism in genes encoding target malarial antigens of protective immunity is maintained due to negative frequency-dependent selection (Conway, 1997; Conway and Polley, 2002). In this hypothesis, a parasite expressing a dominant allele is more likely to be eliminated by protective immunity than a parasite with a rare allelic type. Thus, high prevalence of a frequent antigenic type should lead to its reduction in prevalence, while prevalence of a rare antigenic type is, in turn, increased under antiparasitic allele-specific immune selection. These genes are thought to be a target of strain- or allele-specific protective immune responses and could be explored for malaria vaccine development.

Evaluation for signatures of balancing selection has been performed using a number of population genetic and statistical analyses (Conway, 1997; Conway and Polley, 2002; Weedall and Conway, 2010). Conway and colleagues measured allele frequencies on different blocks of the *pfmsp*-1 gene in parasite populations from different geographic regions in Africa and Asia (Conway *et al.*, 2000a). Their results showed that *pfmsp-1* domain 2 is evolving under balancing selection, as would be expected of a target of protective immunity against a malarial infection. Studies in the rodent malaria parasite *P. chabaudi chabaudi* also indicated involvement of MSP-1 as a target of strain-specific protective immunity (Cheesman *et al.*, 2010; Martinelli *et al.*, 2005; Pattaradilokrat *et al.*, 2007). Subsequently, several other genes encoding merozoite surface antigens, such as *msp-3*, *ama-1*, *eba-175*, *sera5*, *trap*, and PF10_0348 (hypothetical protein), have been identified as being under strong balancing selection and are targets of allele-specific antibodies (Baum *et al.*, 2003; Conway *et al.*, 2000a; Ochola *et al.*, 2011; Polley *et al.*, 2003, 2007; Polley and Conway, 2001; Tetteh *et al.*, 2009; Verra *et al.*, 2006; Weedall *et al.*, 2007).

Directional selection can cause loss or fixation of an allele in a population (Anderson, 2004; Biswas and Akey, 2006). Selection that favours an allele that is advantageous for survival is termed positive directional selection (Hide and Isokpehi, 2004), whereas selection against an allele that has a deleterious effect on fitness is called negative selection. A typical example of positive selection in malaria parasites is the rapid spread of chloroquine-resistant parasites, leading to the 'selective sweeps' in which genetic diversity surrounding the *P. falciparum pfcrt* on chromosome 7 is significantly reduced due to removal of wild-type alleles (Wootton *et al.*, 2002). Subsequent studies of resistance to pyrimethamine and sulfadoxine in *P. falciparum* field isolates also revealed signatures of positive selection and selective sweeps at the *pfdhfr* and *pfdhps* loci that are the targets of pyrimethamine and sulfadoxine, respectively (Lumb *et al.*, 2009; McCollum *et al.*,

2006, 2007, 2008; Mixson-Hayden *et al.*, 2010; Nair *et al.*, 2003; Pearce *et al.*, 2009; Roper *et al.*, 2003; Salgueiro *et al.*, 2010). Although the genetic determinants of chloroquine, pyrimethamine, and sulfadoxine resistance in *P. falciparum* are already known, the ability to identify the signature of positive selection and selective sweeps may provide a promising approach to identify genes conferring resistance to other antimalarial drugs.

A systematic search for polymorphisms in ~3500 genes from five *P. falciparum* isolates produced a map of genomic diversity in *P. falciparum* (Fig. 4.2) and led to the identification of various genomic regions under balancing selection (Mu *et al.*, 2007). Further investigation of 52 candidate genes under balancing selection identified 11 proteins with predicted signal peptide and/or transmembrane domains that were recognized by pooled human immune sera but not by pooled non-immune sera, suggesting that some of the proteins could be novel antigens (Mu *et al.*, 2007). More recently, a genome-wide SNP map was developed to search for signals of positive selection in parasite populations of Asia, Africa, and America (Mu *et al.*, 2010). Many loci potentially under positive selection were mapped using three methods: (i) relatively extended haplotype homozygosity (REHH) (Sabeti *et al.*, 2002, 2007); (ii) integrated haplotype score (Voight *et al.*, 2006), and (iii) cross-population extended haplotype homozygosity (XP-EHH) (Pickrell *et al.*, 2009). A total of 11 loci under significant positive selection were detected by all of the three methods. These loci include the well characterized loci on chromosomes 7, 11, and 13 that contain *pfcrt*, *pfama*-1, and PF03_0271 (encoding ABC transporter MDR-6), respectively (see Fig. 4.3 for loci under positive selection detected by the REHH method). The *pfcrt* gene was already known to be under chloroquine selection (Wootton *et al.*, 2002). PF13_0271, known as *pfmdr-6*, is thought to have a role in iron transport in mitochondria.

Figure 4.2 A map of genome-wide polymorphisms across 14 *Plasmodium falciparum* chromosomes. Approximately 4000 SNP were identified from the comparison of nucleotide sequences of 3539 genes or fragments from four *P. falciparum* cloned isolates (Dd2, Hb3, D10 and 7G8) and the reference genome of *P. falciparum* 3D7 strain (Gardner *et al.*, 2002). SNPs are not distributed evenly across the genome or chromosomes; rather, some regions have consecutive genes without any SNPs, while other DNA segments contain clusters of genes with multiple SNPs. SNPs have been deposited at PlasmoDB. The horizontal bars represent the 14 *P. falciparum* chromosomes. The vertical lines above the horizontal bars indicate the positions of SNPs (synonymous, non-synonymous and non-coding SNPs). Only one line is shown if a gene contains more than one SNP. The vertical lines under the chromosomal bars show the locations of polymorphic microsatellites. SNPs in repetitive regions and in gene families, such as *var*, *rifin* and *stevor*, were excluded in the analysis. This figure is modified from the original work by Mu *et al.* (2007).

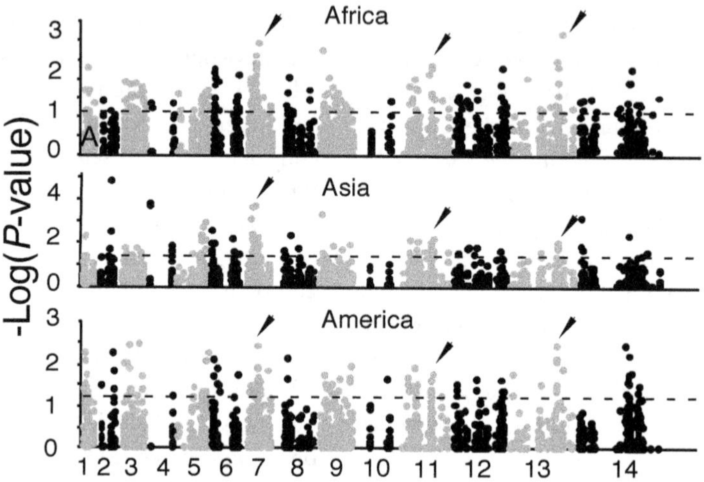

Figure 4.3 Loci under recent positive selection in *Plasmodium falciparum* populations detected by using relative extended haplotype homozygosity (REHH). Log *P* value of each locus (SNP) was plotted along the 14 chromosomes. The horizontal dashed lines indicate significant levels of selection. Arrows indicate loci under positive selection detected in more than one population (Asia, Africa and America), such as the loci containing the gene encoding chloroquine resistance transporter on chromosome 7, the gene encoding apical membrane antigen-1 on chromosome 11 and the gene for an ABC transporter on chromosome 13 (MDR6). The original figure was published by Mu *et al.* (2010)

A substitution in *pfmdr-6* was shown to be significantly associated with decreased sensitivity to chloroquine in culture-adapted parasites (Mu *et al.*, 2003), which was later shown to be associated with responses to quinine and artesunate in fresh isolates (Anderson *et al.*, 2005). These results imply that *pfmdr-6* is a target of drug selection. In addition, signals of positive selection on *pfama*-1, PFA0655w (*surfin* gene) (Winter *et al.*, 2005), and PF14_0260 (encoding a putative metabolite and drug transporter) on chromosomes 1, 11, and 14, respectively, were detected. Many of the proteins encoded by these genes have predicted transmembrane domains and are likely to be antigens in *P. falciparum*. Further functional investigations will be required to determine their roles in parasite development and survival.

Genome-wide association studies (GWAS)

One of the goals of population genetic studies in malaria parasites is to identify genes of medical and biological importance without the need to undertake a genetic cross that is very expensive and time consuming. The success of GWAS depends on a number of factors including the nature of the phenotypes, selection forces, LD in the parasite population, parasite transmission dynamics, and the number of markers used. GWAS studies have been carried out in human populations to identify polymorphisms associated with various human diseases (Hardy and Singleton, 2009; Hindorff *et al.*, 2009; Rosenberg *et al.*, 2010). Similar approaches have been applied to identify genes associated with drug resistance in the human malaria parasite *P. falciparum*. In one study, the parasite half-maximum inhibitory concentrations (IC_{50}) of seven different drugs (chloroquine, mefloquine, sulfadoxine–pyrimethamine, dihydroartemisinin, amodiaquine, and piperaquine) were measured, and SNP genotypes of 3354 SNPs were collected from 185 culture-adapted isolates (Mu *et al.*, 2010). Strong associations between chloroquine resistance and *pfcrt* and between quinine and *pfmdr-1* were detected. This study, therefore, provides proof-of-concept evidence to suggest that using a population genetics approach could help locate regions in the *P. falciparum* genome associated with drug resistance. Additional candidate genes, PFA0665w, PF11_0079, and PFC0460w, that had not previously been associated with the parasite response to any antimalarial drugs, were also identified. PFA0665w is known to be co-transported with *Pf*EMP-1 and RIFIN to the infected erythrocyte surface (Winter *et al.*, 2005) and could be part of a protein complex involved in binding or transporting chemical compounds. Because it is possible that the associations of some of these candidates are due to linkage to genes nearby, the functions of all the genes in the loci identified by GWAS and their contributions to antimalarial drug resistance will require further investigation. In another recent study, Van Tyne *et al.* (2011) used a high-density genotyping array containing over 17,000 SNP to type 57 culture-adapted parasites from three continents and identified 11 loci that were significantly associated with parasite responses to five antimalarial drugs, including a gene (PF10_0355) encoding a hypothetical protein that was functionally shown to contribute to response to halofantrine (Van Tyne *et al.*, 2011). Genome-wide genetic diversity within and between populations was also investigated. More recently, GWAS has been effectively used in combination with high throughput chemical screening to identify three common genomic regions associated with different responses to drugs approved for human and animal use (Yuan *et al.*, 2011).

Genomic diversity and population structure of *P. vivax*

Most of the markers used for population genetic studies in *P. vivax* were derived from known candidate genes for vaccine development (e.g. *pvcsp*, *pydpb* and *msps*; Cui *et al.*, 2003; Garzon-Ospina *et al.*, 2011; Hernandez-Martinez *et al.*, 2011; Leclerc *et al.*, 2004b; Lim *et al.*, 2005; Patil *et al.*, 2011; Prajapati *et al.*, 2010; Putaporntip *et al.*, 2002, 2009, 2011; Sousa *et al.*, 2010; Yang *et al.*, 2006; Zeyrek *et al.*, 2010) and potential targets of antimalarial drugs (e.g. *pvdhfr*, *pvdhps*, *pvcrt-o*, and *pvmdr1* (Barnadas *et al.*, 2008; Hawkins *et al.*, 2008; Orjuela-Sanchez *et al.*, 2009). However, interpretation of genetic diversity based on these markers requires great caution because antigen-coding genes may be under strong immune selection that promotes genetic diversity, while targets of antimalarial drugs may be under directional/purifying selection that promotes allelic homogeneity in parasite populations.

Evaluation of genetic variation can be performed using more neutral genetic markers such as non-coding SNPs, tandem repeats (TR) and microsatellite markers. Prior to the publication of the genome of *P. vivax*, population genetic studies in *P. vivax* relied primarily upon DNA sequence data generated from multiple sources, including the 200-kb contiguous DNA sequence from one *P. vivax* chromosome 3 (Tchavtchitch *et al.*, 2001), the 1.5-Mb *P. vivax* whole-genome yeast artificial chromosome library (Camargo *et al.*, 1997), the collections of *P. vivax* gene sequence tags from genomic DNA libraries of two monkey adapted strains Salvador I (from El Salvador) and Belem (from Brazil) of *P. vivax* (Carlton *et al.*, 2001; Cui *et al.*, 2005), and the sequences of the *P. vivax* multigene family (*vir*) (del Portillo *et al.*, 2001). As a result, only a limited number of MS loci were available for the evaluation of genetic diversity in this species (Feng *et al.*, 2003; Gomez *et al.*, 2003). In one study, Feng *et al.* (2003) compared the frequency of SNPs and microsatellite/tandem repeats (MS/TR) in a contiguous 100 kb chromosome segment homologous to a region on *P. falciparum* chromosome 3 from five *P. vivax* isolates (Feng *et al.*, 2003). Four *P. vivax* isolates were derived from laboratory-adapted lines from different localities (India, El Salvador, Brazil and Thailand). The analysis revealed that SNP frequency is much greater in *P. vivax* than in *P. falciparum*, while MS/TR becomes less abundant in *P. vivax*. The low levels of MS variation are probably due to the lower AT content in the *P. vivax* genome (A + T content of 58% in *P. vivax* and 81 in *P. falciparum*). In a follow up study, Leclerc *et al.* (2004a) genotyped 108 *P. vivax* isolates from Asia, Africa, the Middle East, and Central and South America with 13 MS loci and 96 isolates with 96 tandem repeats and found only few MS variants from one of the 13 MS loci. The low level of microsatellite polymorphism in the *P. vivax* isolates led the authors to conclude that *P. vivax* had undergone a demographic sweep or a recent expansion <10,000 years ago (Leclerc *et al.*, 2004a). This hypothesis was supported by evidence from a lack of synonymous substitutions within and between two forms of *csp* genes (Lim *et al.*, 2005).

However, a subsequent study by Imwong *et al.* (2006) demonstrated that MS repeat array size can affect the ability to detect polymorphism (Imwong *et al.*, 2006) (see also a review by Russell *et al.*, 2006). In this study, MS markers with longer repeat length (range = 13–18) produced multiple allelic variants in single clone *P. vivax* infections from various isolates from India, Thailand and Columbia. Therefore, the low level of MS polymorphism in *P. vivax* shown by Leclerc *et al.* (2004a) may be explained by the use of MS loci with fewer repeat units. Following this work, thanks to the availability of the *P. vivax* genome sequence (Carlton *et al.*, 2008a), multiple sets of MS and other polymorphic genes have been developed by a number of research groups. Several studies revealed extensive genetic variation in *P. vivax* from different localities, including South Asia and South East Asia (Sri Lanka, Vietnam, Myanmar and Papua New Guinea), East Africa (Ethiopia and Madagascar) and South America (Mexico, Peru, Brazil) (Alam *et al.*, 2007; Ferreira *et al.*, 2007; Gunawardena *et al.*, 2010; Imwong *et al.*, 2007; Joy *et al.*, 2008; Karunaweera *et al.*, 2007; Karunaweera *et al.*, 2008; Koepfli *et al.*, 2009; Menard *et al.*, 2010; Rezende *et al.*, 2009, 2010; Van den Eede *et al.*, 2010a,b). These studies suggest that a spectrum of population structures exist in this species.

MS typing analysis in 345 *P. vivax* isolates from Thailand, Laos, India and Columbia revealed that parasites clustered according to their geographical origins, although subpopulation structures do exist within the same continent (Imwong *et al.*, 2007), an observation similar to that found in *P. falciparum* (Mu *et al.*, 2005a). The similar conclusion was obtained from two recent studies based on analyses using 12 MS loci in 425 *P. vivax* isolates from Sri Lanka, Myanmar, and Ethiopia and using SNP in a 100-kb region on chromosome 8 in 238 *P. vivax* isolates from the Amazon Basin of Brazil, Sri Lanka and Southeast Asia countries (Gunawardena *et al.*, 2010; Orjuela-Sanchez *et al.*, 2010). These studies provide evidence of the strong differentiation in *P. vivax* populations between the two continents (South America and Asia). In *P. falciparum*, population structure and the extent of genetic diversity tends to be positively correlated with malaria transmission intensity (Anderson *et al.*, 2000a; Mu *et al.*, 2005a; Neafsey *et al.*, 2007). However, the contrasting patterns were found in *P. vivax*, for which the highest SNP diversity was detected in populations where the *P. vivax* transmission is lowest (Karunaweera *et al.*, 2008; Orjuela-Sanchez *et al.*, 2010). Significant LD over long chromosome segments and conserved haplotype blocks were also present in *P. vivax* populations in areas of low-to-moderate transmission in Columbia and Brazil, but absent in populations from areas of high endemicity in Southeast Asia (Thailand, Lao, Cambodia or Vietnam) (Imwong *et al.*, 2007; Orjuela-Sanchez *et al.*, 2010). These results support the feasibility of conducting GWAS studies with relatively low marker density in South American *P. vivax* populations.

More recently, genetic studies in *P. vivax* have also expanded from limited MS analysis to genome-wide markers. Two microarray platforms that were developed for studying gene expressions in *P. vivax* have been employed to scan en masse SNP and CNV variation in *P. vivax* genomes (Bozdech *et al.*, 2008; Westenberger *et al.*, 2010). One of these platforms is a tiling array contains 4.2 million 25-mer probes, 1.8 million of which represent sense and anti-sense strands of 5419 predicted genes at 6-bp spacing from the genome assembly (2809 contigs) of *P. vivax* Salvador I. This microarray includes 1.7 and 2.3 million probes uniquely mapped to coding regions and non-coding regions, respectively, and will be made available for purchase from Affymetrix (Westenberger *et al.*, 2010) (part number PvivaxLi520507). Using this array, Dharia *et al.* (2010) identified thousands of polymorphic loci in uncultured *P. vivax* clinical isolates from Peru. The study also showed that the genes containing the largest ratio of non-synonymous-to-synonymous SNPs include two AP2 transcription factors and the *P. vivax* multidrug resistance-associated protein (PvMRP1) (Dharia *et al.*, 2010), a homologue of an ABC transporter shown to be associated with quinoline and antifolate tolerance in *Plasmodium falciparum* (Raj *et al.*, 2009). The other platform is the 60-mer spotted array platform developed by Dr Zbynek Bozdech at Nanyang Technological University (Singapore) in collaboration with Dr

Jane Carlton at New York University. The microarray contained 5727 oligonucleotide probes that represented a total of 5335 predicted genes in the *P. vivax* genome. From these, 5063 are represented by a single oligonucleotide, whereas 272 genes (typically long ORFs) are represented by two or more oligonucleotide elements. The *P. vivax* long-oligo array is also available through the Pathogen Functional Genomics Resource Centre (http://pfgrc.jcvi.org). Currently, the platform is being used in CNV studies in *P. vivax* isolates (J. Carlton, personal communication). These tools will provide important resources for genome-wide comparative analysis with important potentials for identifying markers for global parasite diversity and drug resistance mapping studies.

In addition to the extent of genetic diversity in nuclear genomes, the polymorphisms in the mitochondrial genomes of *P. vivax* can also be used to infer evolutionary ages of parasites. By comparing the mitochondrial genome sequences of *P. vivax* and *P. falciparum*, Mu and colleagues estimated the evolutionary age of *P. vivax* to be 58,000–265,000 years (Mu *et al.*, 2005b). This is in agreement with estimates of 46,000–82,000 and 200,000–300,000 years obtained from two independent analyses based on two nuclear genes (β-tubulin and cell division cycle 2), a gene from plastid genome (the elongation factor TU), and the mitochondrial genomes (Escalante *et al.*, 2005; Jongwutiwes *et al.*, 2005). Therefore, *Plasmodium vivax* MS/TR sequences and mitochondrial genomes consistently show levels of variation comparable to those seen in *P. falciparum*, providing strong support for relatively ancient origins in this species.

Conclusion and future research perspectives

The development of molecular tools has widened our ability to study and understand genetic diversity, population structure, transmission dynamics, gene function, and biology of malaria parasites. The advances in microarray and genome sequencing will allow us to generate data from a larger number of samples at lower cost and shorter time. There is an urgent need, however, to develop assays for phenotyping malaria parasites. So far, there are only a limited number of methods for systematic phenotyping malaria parasites such as high-throughput *in vitro* chemical screening for asexual blood stages (Brown *et al.*, 2011; Gamo *et al.*, 2010; Guiguemde *et al.*, 2010; Ndiaye *et al.*, 2010; Plouffe *et al.*, 2008; Yuan *et al.*, 2009) and for gametocyte stages (Buchholz *et al.*, 2011), *in vitro* growth rate (Reilly Ayala *et al.*, 2010; Reilly *et al.*, 2007) and metabolite characterization (Olszewski *et al.*, 2009). Our main challenge is how to translate the genetic and genomic data into effective strategies for combating malaria.

Acknowledgments

This work was supported by the Intramural Research Program of the Division of Intramural Research, National Institute of Allergy and Infectious Diseases (NIAID), National Institutes of Health (NIH). We thank Drs John Tan, Michael Ferdig, Danny Park, Pardis Sabeti, Elizabeth Winzeler, Manuel Llinás and Margaret MacKinnon for sharing information on current microarray platforms in their laboratories. We are grateful to Dr Wang Nguitragool and Verma Walker for critical reading of the manuscript and to NIAID intramural editor Brenda Rae Marshall for editorial assistance.

References

Abdel-Muhsin, A.M., Ranford-Cartwright, L.C., Medani, A.R., Ahmed, S., Suleiman, S., Khan, B., Hunt, P., Walliker, D., and Babiker, H.A. (2002). Detection of mutations in the *Plasmodium falciparum* dihydrofolate reductase (dhfr) gene by dot-blot hybridization. Am. J. Trop. Med. Hyg. *67*, 24–27.

Aguilar, R., Jedlicka, A.E., Mintz, M., Mahairaki, V., Scott, A.L., and Dimopoulos, G. (2005). Global gene expression analysis of *Anopheles gambiae* responses to microbial challenge. Insect Biochem. Mol. Biol. *35*, 709–719.

Alam, M.T., Agarwal, R., and Sharma, Y.D. (2007). Extensive heterozygosity at four microsatellite loci flanking *Plasmodium vivax* dihydrofolate reductase gene. Mol. Biochem. Parasitol. *153*, 178–185.

Alifrangis, M., Enosse, S., Pearce, R., Drakeley, C., Roper, C., Khalil, I.F., Nkya, W.M., Ronn, A.M., Theander, T.G., and Bygbjerg, I.C. (2005). A simple, high-throughput method to detect *Plasmodium falciparum* single nucleotide polymorphisms in the dihydrofolate reductase, dihydropteroate synthase, and *P. falciparum* chloroquine resistance transporter genes using polymerase chain reaction- and enzyme-linked immunosorbent assay-based technology. Am. J. Trop. Med. Hyg. 72, 155–162.

Anderson, T.J. (2004). Mapping drug resistance genes in *Plasmodium falciparum* by genome-wide association. Curr. Drug Targets Infect. Disord. *4*, 65–78.

Anderson, T.J., Haubold, B., Williams, J.T., Estrada-Franco, J.G., Richardson, L., Mollinedo, R., Bockarie, M., Mokili, J., Mharakurwa, S., French, N., *et al.* (2000a). Microsatellite markers reveal a spectrum of population structures in the malaria parasite *Plasmodium falciparum*. Mol. Biol. Evol. *17*, 1467–1482.

Anderson, T.J., Nair, S., Qin, H., Singlam, S., Brockman, A., Paiphun, L., and Nosten, F. (2005). Are transporter genes other than the chloroquine resistance locus (pfcrt) and multidrug resistance gene (pfmdr) associated with antimalarial drug resistance? Antimicrob. Agents Chemother. *49*, 2180–2188.

Anderson, T.J., Su, X.Z., Roddam, A., and Day, K.P. (2000b). Complex mutations in a high proportion of microsatellite loci from the protozoan parasite *Plasmodium falciparum*. Mol. Ecol. *9*, 1599–1608.

Ansorge, W.J. (2009). Next-generation DNA sequencing techniques. Nat. Biotechnol. *25*, 195–203.

Arez, A.P., Snounou, G., Pinto, J., Sousa, C.A., Modiano, D., Ribeiro, H., Franco, A.S., Alves, J., and do Rosario, V.E. (1999). A clonal *Plasmodium falciparum* population in an isolated outbreak of malaria in the Republic of Cabo Verde. Parasitology *118*, 347–355.

Babiker, H.A., Lines, J., Hill, W.G., and Walliker, D. (1997). Population structure of *Plasmodium falciparum* in villages with different malaria endemicity in east Africa. Am. J. Trop. Med. Hyg. *56*, 141–147.

Babiker, H.A., Ranford-Cartwright, L.C., Currie, D., Charlwood, J.D., Billingsley, P., Teuscher, T., and Walliker, D. (1994). Random mating in a natural population of the malaria parasite *Plasmodium falciparum*. Parasitology *109*, 413–421.

Barnadas, C., Tichit, M., Bouchier, C., Ratsimbasoa, A., Randrianasolo, L., Raherinjafy, R., Jahevitra, M., Picot, S., and Menard, D. (2008). *Plasmodium vivax* dhfr and dhps mutations in isolates from Madagascar and therapeutic response to sulphadoxine-pyrimethamine. Malar. J. *7*, 35.

Barnes, D.A., Foote, S.J., Galatis, D., Kemp, D.J., and Cowman, A.F. (1992). Selection for high-level chloroquine resistance results in deamplification of the pfmdr1 gene and increased sensitivity to mefloquine in *Plasmodium falciparum*. EMBO J. *11*, 3067–3075.

Barry, A.E., Schultz, L., Buckee, C.O., and Reeder, J.C. (2009). Contrasting population structures of the genes encoding ten leading vaccine-candidate antigens of the human malaria parasite, *Plasmodium falciparum*. PLoS One *4*, e8497.

Baum, J., Thomas, A.W., and Conway, D.J. (2003). Evidence for diversifying selection on erythrocyte-binding antigens of *Plasmodium falciparum* and P. vivax. Genetics *163*, 1327–1336.

Beez, D., Sanchez, C.P., Stein, W.D., and Lanzer, M. (2010). Genetic predisposition favors the acquisition of stable artemisinin resistance in malaria parasites. Antimicrob. Agents Chemother. *55*, 50–55.

Ben Mamoun, C., Gluzman, I.Y., Hott, C., MacMillan, S.K., Amarakone, A.S., Anderson, D.L., Carlton, J.M., Dame, J.B., Chakrabarti, D., Martin, R.K., *et al.* (2001). Co-ordinated programme of gene expression during asexual intraerythrocytic development of the human malaria parasite *Plasmodium falciparum* revealed by microarray analysis. Mol. Microbiol. *39*, 26–36.

Biggs, B.A., Kemp, D.J., and Brown, G.V. (1989). Subtelomeric chromosome deletions in field isolates of *Plasmodium falciparum* and their relationship to loss of cytoadherence in vitro. Proc. Natl. Acad. Sci. U.S.A. *86*, 2428–2432.

Bin Dajem, S.M., Al-Sheikh, A.A., Bohol, M.F., Alhawi, M., Al-Ahdal, M.N., and Al-Qahtani, A. (2011). Detecting mutations in PfCRT and PfMDR1 genes among *Plasmodium falciparum* isolates from Saudi Arabia by pyrosequencing. Parasitol. Res. *108*, 6775–6780.

Biswas, S., and Akey, J.M. (2006). Genomic insights into positive selection. Trends Genet. *22*, 437–446.

Bowman, S., Lawson, D., Basham, D., Brown, D., Chillingworth, T., Churcher, C.M., Craig, A., Davies, R.M., Devlin, K., Feltwell, T., *et al.* (1999). The complete nucleotide sequence of chromosome 3 of *Plasmodium falciparum*. Nature *400*, 532–538.

Bozdech, Z., Llinas, M., Pulliam, B.L., Wong, E.D., Zhu, J., and DeRisi, J.L. (2003a). The transcriptome of the intraerythrocytic developmental cycle of *Plasmodium falciparum*. PLoS Biol. *1*, E5.

Bozdech, Z., Mok, S., Hu, G., Imwong, M., Jaidee, A., Russell, B., Ginsburg, H., Nosten, F., Day, N.P., White, N.J., *et al.* (2008). The transcriptome of *Plasmodium vivax* reveals divergence and diversity of transcriptional regulation in malaria parasites. Proc. Natl. Acad. Sci. U.S.A. *105*, 16290–16295.

Bozdech, Z., Zhu, J., Joachimiak, M.P., Cohen, F.E., Pulliam, B., and DeRisi, J.L. (2003b). Expression profiling of the schizont and trophozoite stages of *Plasmodium falciparum* with a long-oligonucleotide microarray. Genome Biol. *4*, R9.

Brown, L.E., Chih-Chien Cheng, K., Wei, W.G., Yuan, P., Dai, P., Trilles, R., Ni, F., Yuan, J., Macarthur, R., Guha, R., *et al.* (2011). Discovery of new antimalarial chemotypes through chemical methodology and library development. Proc. Natl. Acad. Sci. U.S.A. *109*, 291–296.

Buchholz, K., Burke, T.A., Williamson, K.C., Wiegand, R.C., Wirth, D.F., and Marti, M. (2011). A high-throughput screen targeting malaria transmission stages opens new avenues for drug development. J. Infect. Dis. *203*, 1445–1453.

Bull, P.C., Buckee, C.O., Kyes, S., Kortok, M.M., Thathy, V., Guyah, B., Stoute, J.A., Newbold, C.I., and Marsh, K. (2008). *Plasmodium falciparum* antigenic variation. Mapping mosaic var gene sequences onto a network of shared, highly polymorphic sequence blocks. Mol. Microbiol. *68*, 1519–1534.

Bull, P.C., Pain, A., Ndungu, F.M., Kinyanjui, S.M., Roberts, D.J., Newbold, C.I., and Marsh, K. (2005). *Plasmodium falciparum* antigenic variation: relationships between in vivo selection, acquired antibody response, and disease severity. J. Infect. Dis. *192*, 1119–1126.

Camargo, A.A., Fischer, K., Lanzer, M., and del Portillo, H.A. (1997). Construction and characterization of a *Plasmodium vivax* genomic library in yeast artificial chromosomes. Genomics *42*, 467–473.

Carlton, J.M., Adams, J.H., Silva, J.C., Bidwell, S.L., Lorenzi, H., Caler, E., Crabtree, J., Angiuoli, S.V., Merino, E.F., Amedeo, P., *et al.* (2008a). Comparative genomics of the neglected human malaria parasite *Plasmodium vivax*. Nature *455*, 757–763.

Carlton, J.M., Escalante, A.A., Neafsey, D., and Volkman, S.K. (2008b). Comparative evolutionary genomics of human malaria parasites. Trends Parasitol. *24*, 545–550.

Carlton, J.M., Muller, R., Yowell, C.A., Fluegge, M.R., Sturrock, K.A., Pritt, J.R., Vargas-Serrato, E., Galinski, M.R., Barnwell, J.W., Mulder, N., *et al.* (2001). Profiling the malaria genome: a gene survey of three species of malaria parasite with comparison to other apicomplexan species. Mol. Biochem. Parasitol. *118*, 201–210.

Carret, C.K., Horrocks, P., Konfortov, B., Winzeler, E., Qureshi, M., Newbold, C., and Ivens, A. (2005). Microarray-based comparative genomic analyses of the human malaria parasite *Plasmodium falciparum* using Affymetrix arrays. Mol. Biochem. Parasitol. *144*, 177–186.

Carter, R., and McGregor, I.A. (1973). Enzyme variation in *Plasmodium falciparum* in the Gambia. Trans. R. Soc. Trop. Med. Hyg. *67*, 830–837.

Carter, R., and Voller, A. (1975). The distribution of enzyme variation in populations of *Plasmodium falciparum* in Africa. Trans. R. Soc. Trop. Med. Hyg. *69*, 371–376.

Charlesworth, D. (2006). Balancing selection and its effects on sequences in nearby genome regions. PLoS Genet. *2*, e64.

Cheeseman, I.H., Gomez-Escobar, N., Carret, C.K., Ivens, A., Stewart, L.B., Tetteh, K.K., and Conway, D.J. (2009). Gene copy number variation throughout the *Plasmodium falciparum* genome. BMC Genomics *10*, 353.

Cheesman, S., O'Mahony, E., Pattaradilokrat, S., Degnan, K., Knott, S., and Carter, R. (2010). A single parasite gene determines strain-specific protective immunity against malaria: the role of the merozoite surface protein I. Int. J. Parasitol. *40*, 951–961.

Coluzzi, M. (1999). The clay feet of the malaria giant and its African roots: hypotheses and inferences about origin, spread and control of *Plasmodium falciparum*. Parassitologia *41*, 277–283.

Coluzzi, M., Sabatini, A., della Torre, A., Di Deco, M.A., and Petrarca, V. (2002). A polytene chromosome analysis of the *Anopheles gambiae* species complex. Science *298*, 1415–1418.

Conway, D.J. (1997). Natural selection on polymorphic malaria antigens and the search for a vaccine. Parasitol. Today *13*, 26–29.

Conway, D.J. (2003). Tracing the dawn of *Plasmodium falciparum* with mitochondrial genome sequences. Trends Genet. *19*, 671–674.

Conway, D.J. (2007). Molecular epidemiology of malaria. Clin. Microbiol. Rev. *20*, 188–204.

Conway, D.J., and Baum, J. (2002). In the blood – the remarkable ancestry of *Plasmodium falciparum*. Trends Parasitol. *18*, 351–355.

Conway, D.J., Cavanagh, D.R., Tanabe, K., Roper, C., Mikes, Z.S., Sakihama, N., Bojang, K.A., Oduola, A.M., Kremsner, P.G., Arnot, D.E., *et al.* (2000a). A principal target of human immunity to malaria identified by molecular population genetic and immunological analyses. Nat. Med. *6*, 689–692.

Conway, D.J., Fanello, C., Lloyd, J.M., Al-Joubori, B.M., Baloch, A.H., Somanath, S.D., Roper, C., Oduola, A.M., Mulder, B., Povoa, M.M., *et al.* (2000b). Origin of *Plasmodium falciparum* malaria is traced by mitochondrial DNA. Mol. Biochem. Parasitol. *111*, 163–171.

Conway, D.J., and Polley, S.D. (2002). Measuring immune selection. Parasitology *125* (Suppl.), S3–16.

Conway, D.J., Roper, C., Oduola, A.M., Arnot, D.E., Kremsner, P.G., Grobusch, M.P., Curtis, C.F., and Greenwood, B.M. (1999). High recombination rate in natural populations of *Plasmodium falciparum*. Proc. Natl. Acad. Sci. U.S.A. *96*, 4506–4511.

Corcoran, L.M., Forsyth, K.P., Bianco, A.E., Brown, G.V., and Kemp, D.J. (1986). Chromosome size polymorphisms in *Plasmodium falciparum* can involve deletions and are frequent in natural parasite populations. Cell *44*, 87–95.

Cortes, A., Carret, C., Kaneko, O., Yim Lim, B.Y., Ivens, A., and Holder, A.A. (2007). Epigenetic silencing of *Plasmodium falciparum* genes linked to erythrocyte invasion. PLoS Pathog. 3, e107.

Cowman, A.F., Galatis, D., and Thompson, J.K. (1994). Selection for mefloquine resistance in *Plasmodium falciparum* is linked to amplification of the pfmdr1 gene and cross-resistance to halofantrine and quinine. Proc. Natl. Acad. Sci. U.S.A. *91*, 1143–1147.

Crameri, A., Marfurt, J., Mugittu, K., Maire, N., Regos, A., Coppee, J.Y., Sismeiro, O., Burki, R., Huber, E., Laubscher, D., *et al.* (2007). Rapid microarray-based method for monitoring of all currently known single-nucleotide polymorphisms associated with parasite resistance to antimalaria drugs. J. Clin. Microbiol. *45*, 3685–3691.

Cui, L., Escalante, A.A., Imwong, M., and Snounou, G. (2003). The genetic diversity of *Plasmodium vivax* populations. Trends Parasitol. *19*, 220–226.

Cui, L., Fan, Q., Hu, Y., Karamycheva, S.A., Quackenbush, J., Khuntirat, B., Sattabongkot, J., and Carlton, J.M. (2005). Gene discovery in *Plasmodium vivax* through sequencing of ESTs from mixed blood stages. Mol. Biochem. Parasitol. *144*, 1–9.

Da Silveira, L.A., Ribeiro, W.L., Kirchgatter, K., Wunderlich, G., Matsuoka, H., Tanabe, K., and Ferreira, M.U. (2001). Sequence diversity and linkage disequilibrium within the merozoite surface protein-1 (Msp-1) locus of *Plasmodium falciparum*: a longitudinal study in Brazil. J. Eukaryot. Microbiol. *48*, 433–439.

Daily, J.P., Scanfeld, D., Pochet, N., Le Roch, K., Plouffe, D., Kamal, M., Sarr, O., Mboup, S., Ndir, O., Wypij, D., *et al.* (2007). Distinct physiological states of *Plasmodium falciparum* in malaria-infected patients. Nature *450*, 1091–1095.

Day, K.P., Karamalis, F., Thompson, J., Barnes, D.A., Peterson, C., Brown, H., Brown, G.V., and Kemp, D.J. (1993). Genes necessary for expression of a virulence determinant and for transmission of *Plasmodium falciparum* are located on a 0.3-megabase region of chromosome 9. Proc. Natl. Acad. Sci. U.S.A. *90*, 8292–8296.

Day, K.P., and Marsh, K. (1991). Naturally acquired immunity to *Plasmodium falciparum*. Immunol. Today *12*, A68–71.

del Portillo, H.A., Fernandez-Becerra, C., Bowman, S., Oliver, K., Preuss, M., Sanchez, C.P., Schneider, N.K., Villalobos, J.M., Rajandream, M.A., Harris, D., *et al.* (2001). A superfamily of variant genes encoded in the subtelomeric region of *Plasmodium vivax*. Nature *410*, 839–842.

Dharia, N.V., Bright, A.T., Westenberger, S.J., Barnes, S.W., Batalov, S., Kuhen, K., Borboa, R., Federe, G.C., McClean, C.M., Vinetz, J.M., *et al.* (2010). Whole-genome sequencing and microarray analysis of ex vivo *Plasmodium vivax* reveal selective pressure on putative drug resistance genes. Proc. Natl. Acad. Sci. U.S.A. *107*, 20045–20050.

Dharia, N.V., Sidhu, A.B., Cassera, M.B., Westenberger, S.J., Bopp, S.E., Eastman, R.T., Plouffe, D., Batalov, S., Park, D.J., Volkman, S.K., *et al.* (2009). Use of high-density tiling microarrays to identify mutations globally and elucidate mechanisms of drug resistance in *Plasmodium falciparum*. Genome Biol. *10*, R21.

Dolan, S.A., Herrfeldt, J.A., and Wellems, T.E. (1993). Restriction polymorphisms and fingerprint patterns from an interspersed repetitive element of *Plasmodium falciparum* DNA. Mol. Biochem. Parasitol. *61*, 137–142.

Drakeley, C.J., Duraisingh, M.T., Povoa, M., Conway, D.J., Targett, G.A., and Baker, D.A. (1996). Geographical distribution of a variant epitope of Pfs48/45, a *Plasmodium falciparum* transmission-blocking vaccine candidate. Mol. Biochem. Parasitol. *81*, 253–257.

Duan, J., Mu, J., Thera, M.A., Joy, D., Kosakovsky Pond, S.L., Diemert, D., Long, C., Zhou, H., Miura, K., Ouattara, A., *et al.* (2008). Population structure of the genes encoding the polymorphic *Plasmodium falciparum* apical membrane antigen 1: implications for vaccine design. Proc. Natl. Acad. Sci. U.S.A. *105*, 7857–7862.

Durand, R., Eslahpazire, J., Jafari, S., Delabre, J.F., Marmorat-Khuong, A., di Piazza, J.P., and Le Bras, J. (2000). Use of molecular beacons to detect an antifolate resistance-associated mutation in *Plasmodium falciparum*. Antimicrob. Agents Chemother. *44*, 3461–3464.

Dye, C. (1991). Population genetics of nonclonal, nonrandomly mating malaria parasites. Parasitol. Today *7*, 236–240.

Escalante, A.A., Cornejo, O.E., Freeland, D.E., Poe, A.C., Durrego, E., Collins, W.E., and Lal, A.A. (2005). A monkey's tale: the origin of *Plasmodium vivax* as a human malaria parasite. Proc. Natl. Acad. Sci. U.S.A. *102*, 1980–1985.

Escalante, A.A., Grebert, H.M., Chaiyaroj, S.C., Magris, M., Biswas, S., Nahlen, B.L., and Lal, A.A. (2001). Polymorphism in the gene encoding the apical membrane antigen-1 (AMA-1) of *Plasmodium falciparum*. X. Asembo Bay Cohort Project. Mol. Biochem. Parasitol. *113*, 279–287.

Farooq, U., Malla, N., and Dubey, M.L. (2009). Genetic polymorphism in msp-2, ama-1 and csp genes in *Plasmodium falciparum* field isolates from north and north-western India. J. Vector Borne Dis. *46*, 109–116.

Feagin, J.E. (1992). The 6-kb element of *Plasmodium falciparum* encodes mitochondrial cytochrome genes. Mol. Biochem. Parasitol. *52*, 145–1488.

Feng, X., Carlton, J.M., Joy, D.A., Mu, J., Furuya, T., Suh, B.B., Wang, Y., Barnwell, J.W., and Su, X.Z. (2003). Single-nucleotide polymorphisms and genome diversity in *Plasmodium vivax*. Proc. Natl. Acad. Sci. U.S.A. *100*, 8502–8507.

Fenton, B., Walker, A., and Walliker, D. (1985). Protein variation in clones of *Plasmodium falciparum* detected by two dimensional electrophoresis. Mol. Biochem. Parasitol. *16*, 173–183.

Ferdig, M.T., Cooper, R.A., Mu, J., Deng, B., Joy, D.A., Su, X.Z., and Wellems, T.E. (2004). Dissecting the loci of low-level quinine resistance in malaria parasites. Mol. Microbiol. *52*, 985–997.

Ferdig, M.T., and Su, X.Z. (2000). Microsatellite markers and genetic mapping in *Plasmodium falciparum*. Parasitol. Today *16*, 307–312.

Ferreira, M.U., da Silva Nunes, M., and Wunderlich, G. (2004). Antigenic diversity and immune evasion by malaria parasites. Clin. Diagn. Lab. Immunol. *11*, 987–995.

Ferreira, M.U., and Hartl, D.L. (2007). *Plasmodium falciparum*: worldwide sequence diversity and evolution of the malaria vaccine candidate merozoite surface protein-2 (MSP-2). Exp. Parasitol. *115*, 32–40.

Ferreira, M.U., Karunaweera, N.D., da Silva-Nunes, M., da Silva, N.S., Wirth, D.F., and Hartl, D.L. (2007). Population structure and transmission dynamics of *Plasmodium vivax* in rural Amazonia. J. Infect. Dis. *195*, 1218–1226.

Ferreira, M.U., Liu, Q., Zhou, M., Kimura, M., Kaneko, O., Van Thien, H., Isomura, S., Tanabe, K., and Kawamoto, F. (1998). Stable patterns of allelic diversity at the Merozoite surface protein-1 locus of *Plasmodium falciparum* in clinical isolates from southern Vietnam. J. Eukaryot. Microbiol. *45*, 131–136.

Ferreira, M.U., Ribeiro, W.L., Tonon, A.P., Kawamoto, F., and Rich, S.M. (2003). Sequence diversity and evolution of the malaria vaccine candidate merozoite surface protein-1 (MSP-1) of *Plasmodium falciparum*. Gene *304*, 65–75.

Fidock, D.A., Nomura, T., Talley, A.K., Cooper, R.A., Dzekunov, S.M., Ferdig, M.T., Ursos, L.M., Sidhu, A.B., Naude, B., Deitsch, K.W., *et al.* (2000). Mutations in the *P. falciparum* digestive vacuole transmembrane protein PfCRT and evidence for their role in chloroquine resistance. Mol. Cell *6*, 861–871.

Furuya, T., Mu, J., Hayton, K., Liu, A., Duan, J., Nkrumah, L., Joy, D.A., Fidock, D.A., Fujioka, H., Vaidya, A.B., *et al.* (2005). Disruption of a *Plasmodium falciparum* gene linked to male sexual development causes early arrest in gametocytogenesis. Proc. Natl. Acad. Sci. U.S.A. *102*, 16813–16818.

Gamo, F.J., Sanz, L.M., Vidal, J., de Cozar, C., Alvarez, E., Lavandera, J.L., Vanderwall, D.E., Green, D.V., Kumar, V., Hasan, S., *et al.* (2010). Thousands of chemical starting points for antimalarial lead identification. Nature *465*, 305–310.

Gardner, M.J., Hall, N., Fung, E., White, O., Berriman, M., Hyman, R.W., Carlton, J.M., Pain, A., Nelson, K.E., Bowman, S., *et al.* (2002). Genome sequence of the human malaria parasite *Plasmodium falciparum*. Nature *419*, 498–511.

Gardner, M.J., Tettelin, H., Carucci, D.J., Cummings, L.M., Aravind, L., Koonin, E.V., Shallom, S., Mason, T., Yu, K., Fujii, C., *et al.* (1998). Chromosome 2 sequence of the human malaria parasite *Plasmodium falciparum*. Science *282*, 1126–1132.

Garzon-Ospina, D., Romero-Murillo, L., Tobon, L.F., and Patarroyo, M.A. (2011). Low genetic polymorphism of merozoite surface proteins 7 and 10 in Colombian *Plasmodium vivax* isolates. Infect. Genet. Evol. *11*, 528–531.

Gerry, N.P., Witowski, N.E., Day, J., Hammer, R.P., Barany, G., and Barany, F. (1999). Universal DNA microarray method for multiplex detection of low abundance point mutations. J. Mol. Biol. *292*, 251–262.

Gissot, M., Refour, P., Briquet, S., Boschet, C., Coupe, S., Mazier, D., and Vaquero, C. (2004). Transcriptome of 3D7 and its gametocyte-less derivative F12 *Plasmodium falciparum* clones during erythrocytic development using a gene-specific microarray assigned to gene regulation, cell cycle and transcription factors. Gene *341*, 267–277.

Gomez, J.C., McNamara, D.T., Bockarie, M.J., Baird, J.K., Carlton, J.M., and Zimmerman, P.A. (2003). Identification of a polymorphic *Plasmodium vivax* microsatellite marker. Am. J. Trop. Med. Hyg. *69*, 377–379.

Gonzales, J.M., Patel, J.J., Ponmee, N., Jiang, L., Tan, A., Maher, S.P., Wuchty, S., Rathod, P.K., and Ferdig, M.T. (2008). Regulatory hotspots in the malaria parasite genome dictate transcriptional variation. PLoS Biol. *6*, e238.

Guiguemde, W.A., Shelat, A.A., Bouck, D., Duffy, S., Crowther, G.J., Davis, P.H., Smithson, D.C., Connelly, M., Clark, J., Zhu, F., *et al.* (2010). Chemical genetics of *Plasmodium falciparum*. Nature *465*, 311–315.

Gunawardena, S., Karunaweera, N.D., Ferreira, M.U., Phone-Kyaw, M., Pollack, R.J., Alifrangis, M., Rajakaruna, R.S., Konradsen, F., Amerasinghe, P.H., Schousboe, M.L., *et al.* (2010). Geographic structure of *Plasmodium vivax*: microsatellite analysis of parasite populations from Sri Lanka, Myanmar, and Ethiopia. Am. J. Trop. Med. Hyg. *82*, 235–242.

Hall, N. (2007). Advanced sequencing technologies and their wider impact in microbiology. J. Exp. Biol. *210*, 1518–1525.

Hardenbol, P., Baner, J., Jain, M., Nilsson, M., Namsaraev, E.A., Karlin-Neumann, G.A., Fakhrai-Rad, H., Ronaghi, M., Willis, T.D., Landegren, U., *et al.* (2003). Multiplexed genotyping with sequence-tagged molecular inversion probes. Nat. Biotechnol. *21*, 673–678.

Hardy, J., and Singleton, A. (2009). Genomewide association studies and human disease. N. Engl. J. Med. *360*, 1759–1768.

Hartl, D.L. (2004). The origin of malaria: mixed messages from genetic diversity. Nat. Rev. Microbiol. 2, 15–22.

Hawkins, V.N., Auliff, A., Prajapati, S.K., Rungsihirunrat, K., Hapuarachchi, H.C., Maestre, A., O'Neil, M.T., Cheng, Q., Joshi, H., Na-Bangchang, K., *et al.* (2008). Multiple origins of resistance-conferring mutations in *Plasmodium vivax* dihydrofolate reductase. Malar. J. *7*, 72.

Hayton, K., Gaur, D., Liu, A., Takahashi, J., Henschen, B., Singh, S., Lambert, L., Furuya, T., Bouttenot, R., Doll, M., *et al.* (2008). Erythrocyte binding protein PfRH5 polymorphisms determine species-specific pathways of *Plasmodium falciparum* invasion. Cell Host Microbe *4*, 40–51.

Hayward, R.E., Derisi, J.L., Alfadhli, S., Kaslow, D.C., Brown, P.O., and Rathod, P.K. (2000). Shotgun DNA microarrays and stage-specific gene expression in *Plasmodium falciparum* malaria. Mol. Microbiol. *35*, 6–14.

Heidari, A., Keshavarz, H., Rokni, M.B., and Jelinek, T. (2007). Genetic diversity in merozoite surface protein (MSP)-1 and MSP-2 genes of *Plasmodium falciparum* in a major endemic region of Iran. Korean J. Parasitol. *45*, 59–63.

Hernandez-Martinez, M.A., Escalante, A.A., Arevalo-Herrera, M., and Herrera, S. (2011). Antigenic diversity of the *Plasmodium vivax* circumsporozoite protein in parasite isolates of Western Colombia. Am. J. Trop. Med. Hyg. *84*, 51–57.

Hide, W.A., and Isokpehi, R.D. (2004). Positive selection scanning of parasite DNA sequences. Methods Mol. Biol. *270*, 127–150.

Hindorff, L.A., Sethupathy, P., Junkins, H.A., Ramos, E.M., Mehta, J.P., Collins, F.S., and Manolio, T.A. (2009). Potential etiologic and functional implications of genome-wide association loci for human diseases and traits. Proc. Natl. Acad. Sci. U.S.A. *106*, 9362–9367.

Hinterberg, K., Muanza, K., Hernandez-Rivas, R., Gay, F., Gysin, J., Mattei, D., and Scherf, A. (1995). Karyotype analysis of virulent *Plasmodium falciparum* strains propagated in *Saimiri sciureus*: strain adaptation leads to deletion of the RESA gene. Infect. Immun. *63*, 693–695.

Howard, J., Carlton, J.M., Walliker, D., and Jensen, J.B. (1996). Use of random amplified polymorphic DNA (RAPD) technique in inheritance studies of *Plasmodium falciparum*. J. Parasitol. *82*, 941–946.

Hu, G., Llinas, M., Li, J., Preiser, P.R., and Bozdech, Z. (2007). Selection of long oligonucleotides for gene expression microarrays using weighted rank-sum strategy. BMC Bioinformatics *8*, 350.

Hu, G., Cabrera, A., Kono, M., Mok, S., Chaal, B.K., Haase, S., Engelberg, K., Cheemadan, S., Spielmann, T., Preiser, P.R., *et al.* (2010). Transcriptional profiling of growth perturbations of the human malaria parasite *Plasmodium falciparum*. Nat. Biotechnol. *28*, 91–98.

Hughes, A.L., and Verra, F. (2001). Very large long-term effective population size in the virulent human malaria parasite *Plasmodium falciparum*. Proc. Biol. Sci. *268*, 1855–1860.

Ibrahim, M.L., Steenkeste, N., Khim, N., Adam, H.H., Konate, L., Coppee, J.Y., Ariey, F., and Duchemin, J.B. (2009). Field-based evidence of fast and global increase of *Plasmodium falciparum* drug-resistance by DNA-microarrays and PCR/RFLP in Niger. Malar. J. *8*, 32.

Imwong, M., Nair, S., Pukrittayakamee, S., Sudimack, D., Williams, J.T., Mayxay, M., Newton, P.N., Kim, J.R., Nandy, A., Osorio, L., *et al.* (2007). Contrasting genetic structure in *Plasmodium vivax* populations from Asia and South America. Int. J. Parasitol. *37*, 1013–1022.

Imwong, M., Sudimack, D., Pukrittayakamee, S., Osorio, L., Carlton, J.M., Day, N.P., White, N.J., and Anderson, T.J. (2006). Microsatellite variation, repeat array length, and population history of *Plasmodium vivax*. Mol. Biol. Evol. 23, 1016–1018.

Jeffares, D.C., Pain, A., Berry, A., Cox, A.V., Stalker, J., Ingle, C.E., Thomas, A., Quail, M.A., Siebenthall, K., Uhlemann, A.C., *et al.* (2007). Genome variation and evolution of the malaria parasite *Plasmodium falciparum*. Nat. Genet. *39*, 120–125.

Jiang, H., Li, N., Gopalan, V., M.M., Z., Varma, S., Nagarajan, V., Li, J., Mu, J., Hayton, K., Henschen, B., *et al.* (2011). High recombination rates and hotspots in a *Plasmodium falciparum* genetic cross. Genome Biol. *12*. R33.

Jiang, H., Patel, J.J., Yi, M., Mu, J., Ding, J., Stephens, R., Cooper, R.A., Ferdig, M.T., and Su, X.Z. (2008a). Genome-wide compensatory changes accompany drug- selected mutations in the *Plasmodium falciparum* crt gene. PLoS One 3, e2484.

Jiang, H., Yi, M., Mu, J., Zhang, L., Ivens, A., Klimczak, L.J., Huyen, Y., Stephens, R.M., and Su, X.Z. (2008b). Detection of genome-wide polymorphisms in the AT-rich *Plasmodium falciparum* genome using a high-density microarray. BMC Genomics 9, 398.

Jongwutiwes, S., Putaporntip, C., Iwasaki, T., Ferreira, M.U., Kanbara, H., and Hughes, A.L. (2005). Mitochondrial genome sequences support ancient population expansion in *Plasmodium vivax*. Mol. Biol. Evol. 22, 1733–1739.

Joy, D.A., Feng, X., Mu, J., Furuya, T., Chotivanich, K., Krettli, A.U., Ho, M., Wang, A., White, N.J., Suh, E., *et al.* (2003). Early origin and recent expansion of *Plasmodium falciparum*. Science *300*, 318–321.

Joy, D.A., Gonzalez-Ceron, L., Carlton, J.M., Gueye, A., Fay, M., McCutchan, T.F., and Su, X.Z. (2008). Local adaptation and vector-mediated population structure in *Plasmodium vivax* malaria. Mol. Biol. Evol. *25*, 1245–1252.

Joy, D.A., Mu, J., and Su, X.Z. (2004). Microsatellite Markers and Population Genetics in *Plasmodium falciparum*. In Malaria Parasites: Genomes and Molecular Biology, A. Waters, and C.J. Janse, eds. (Norfolk, Caister Academic Press), pp. 167–186.

Kappe, S.H., Gardner, M.J., Brown, S.M., Ross, J., Matuschewski, K., Ribeiro, J.M., Adams, J.H., Quackenbush, J., Cho, J., Carucci, D.J., *et al.* (2001). Exploring the transcriptome of the malaria sporozoite stage. Proc. Natl. Acad. Sci. U.S.A. *98*, 9895–9900.

Karunaweera, N., Ferreira, M., Hartl, D., and Wirth, D. (2007). Fourteen polymorphic microsatellite DNA markers for the human malaria parasite *Plasmodium vivax*. Mol. Ecol. Notes 7, 172–175.

Karunaweera, N.D., Ferreira, M.U., Munasinghe, A., Barnwell, J.W., Collins, W.E., King, C.L., Kawamoto, F., Hartl, D.L., and Wirth, D.F. (2008). Extensive microsatellite diversity in the human malaria parasite *Plasmodium vivax*. Gene *410*, 105–112.

Kelly, J.M., McRobert, L., and Baker, D.A. (2006). Evidence on the chromosomal location of centromeric DNA in *Plasmodium falciparum* from etoposide-mediated topoisomerase-II cleavage. Proc. Natl. Acad. Sci. U.S.A. *103*, 6706–6711.

Kemp, D.J., Corcoran, L.M., Coppel, R.L., Stahl, H.D., Bianco, A.E., Brown, G.V., and Anders, R.F. (1985). Size variation in chromosomes from independent cultured isolates of *Plasmodium falciparum*. Nature *315*, 347–350.

Kidgell, C., Volkman, S.K., Daily, J., Borevitz, J.O., Plouffe, D., Zhou, Y., Johnson, J.R., Le Roch, K., Sarr, O., Ndir, O., *et al.* (2006). A systematic map of genetic variation in *Plasmodium falciparum*. PLoS Pathog. *2*, e57.

Kim, S., and Misra, A. (2007). SNP genotyping: technologies and biomedical applications. Annu. Rev. Biomed. Eng. *9*, 289–320.

Koepfli, C., Mueller, I., Marfurt, J., Goroti, M., Sie, A., Oa, O., Genton, B., Beck, H.P., and Felger, I. (2009). Evaluation of *Plasmodium vivax* genotyping markers for molecular monitoring in clinical trials. J. Infect. Dis. *199*, 1074–1080.

Komar, A.A. (2009). Single Nucleotide Polymorphisms: Methods and Protocols, Vol. 578, 2nd edn (New York: Humana Press).

Kumar, S., Molina-Cruz, A., Gupta, L., Rodrigues, J., and Barillas-Mury, C. (2010). A peroxidase/dual oxidase system modulates midgut epithelial immunity in *Anopheles gambiae*. Science *327*, 1644–1648.

Lai, Z., Jing, J., Aston, C., Clarke, V., Apodaca, J., Dimalanta, E.T., Carucci, D.J., Gardner, M.J., Mishra, B., Anantharaman, T.S., *et al.* (1999). A shotgun optical map of the entire *Plasmodium falciparum* genome. Nat. Genet. *23*, 309–313.

Le Roch, K.G., Zhou, Y., Blair, P.L., Grainger, M., Moch, J.K., Haynes, J.D., De La Vega, P., Holder, A.A., Batalov, S., Carucci, D.J., *et al.* (2003). Discovery of gene function by expression profiling of the malaria parasite life cycle. Science *301*, 1503–1508.

Leclerc, M.C., Durand, P., Gauthier, C., Patot, S., Billotte, N., Menegon, M., Severini, C., Ayala, F.J., and Renaud, F. (2004a). Meager genetic variability of the human malaria agent *Plasmodium vivax*. Proc. Natl. Acad. Sci. U.S.A. *101*, 14455–14460.

Leclerc, M.C., Menegon, M., Cligny, A., Noyer, J.L., Mammadov, S., Aliyev, N., Gasimov, E., Majori, G., and

Severini, C. (2004b). Genetic diversity of *Plasmodium vivax* isolates from Azerbaijan. Malar. J. 3, 40.

Li, J., Pattaradilokrat, S., Zhu, F., Jiang, H., Liu, S., Hong, L., Fu, Y., Koo, L., Xu, W., Pan, W., *et al.* (2011). Linkage maps from multiple genetic crosses and loci linked to growth-related virulent phenotype in *Plasmodium yoelii*. Proc. Natl. Acad. Sci. U.S.A. *108*, E374–382.

Li, J., Zhang, Y., Liu, S., Hong, L., Sullivan, M., McCutchan, T.F., Carlton, J.M., and Su, X.Z. (2009). Hundreds of microsatellites for genotyping *Plasmodium yoelii* parasites. Mol. Biochem. Parasitol. *166*, 153–158.

Lim, C.S., Tazi, L., and Ayala, F.J. (2005). *Plasmodium vivax*: recent world expansion and genetic identity to *Plasmodium simium*. Proc. Natl. Acad. Sci. U.S.A. *102*, 15523–15528.

Lim, L., and McFadden, G.I. (2010). The evolution, metabolism and functions of the apicoplast. Phil. Trans. R. Soc. Lond. B Biol. Sci. *365*, 749–763.

Lumb, V., Das, M.K., Singh, N., Dev, V., Wajihullah, and Sharma, Y.D. (2009). Characteristics of genetic hitchhiking around dihydrofolate reductase gene associated with pyrimethamine resistance in *Plasmodium falciparum* isolates from India. Antimicrob. Agents Chemother. *53*, 5173–5180.

Mackinnon, M.J., Li, J., Mok, S., Kortok, M.M., Marsh, K., Preiser, P.R., and Bozdech, Z. (2009). Comparative transcriptional and genomic analysis of *Plasmodium falciparum* field isolates. PLoS Pathog. *5*, e1000644.

MacLean, D., Jones, J.D., and Studholme, D.J. (2009). Application of 'next-generation' sequencing technologies to microbial genetics. Nat. Rev. Microbiol. *7*, 287–296.

Mamillapalli, A., Sunil, S., Diwan, S.S., Sharma, S.K., Tyagi, P.K., Adak, T., Joshi, H., and Malhotra, P. (2007). Polymorphism and epitope sharing between the alleles of merozoite surface protein-1 of *Plasmodium falciparum* among Indian isolates. Malar. J. *6*, 95.

Maresso, K., and Broeckel, U. (2008). Genotyping platforms for mass-throughput genotyping with SNPs, including human genome-wide scans. Adv. Genet. *60*, 107–139.

Marinotti, O., Calvo, E., Nguyen, Q.K., Dissanayake, S., Ribeiro, J.M., and James, A.A. (2006). Genome-wide analysis of gene expression in adult *Anopheles gambiae*. Insect Mol. Biol. *15*, 1–12.

Marinotti, O., Nguyen, Q.K., Calvo, E., James, A.A., and Ribeiro, J.M. (2005). Microarray analysis of genes showing variable expression following a blood meal in *Anopheles gambiae*. Insect Mol. Biol. *14*, 365–373.

Martinelli, A., Cheesman, S., Hunt, P., Culleton, R., Raza, A., Mackinnon, M., and Carter, R. (2005). A genetic approach to the de novo identification of targets of strain-specific immunity in malaria parasites. Proc. Natl. Acad. Sci. U.S.A. *102*, 814–819.

Mattei, D., and Scherf, A. (1994). Subtelomeric chromosome instability in *Plasmodium falciparum*: short telomere-like sequence motifs found frequently at healed chromosome breakpoints. Mutat. Res. *324*, 115–120.

McBride, J.S., Newbold, C.I., and Anand, R. (1985). Polymorphism of a high molecular weight schizont antigen of the human malaria parasite *Plasmodium falciparum*. J. Exp. Med. *161*, 160–180.

McBride, J.S., Walliker, D., and Morgan, G. (1982). Antigenic diversity in the human malaria parasite *Plasmodium falciparum*. Science *217*, 254–257.

McBride, J.S., Welsby, P.D., and Walliker, D. (1984). Serotyping *Plasmodium falciparum* from acute human infections using monoclonal antibodies. Trans. R. Soc. Trop. Med. Hyg. *78*, 32–34.

McCollum, A.M., Basco, L.K., Tahar, R., Udhayakumar, V., and Escalante, A.A. (2008). Hitchhiking and selective sweeps of *Plasmodium falciparum* sulfadoxine and pyrimethamine resistance alleles in a population from central Africa. Antimicrob. Agents Chemother. *52*, 4089–4097.

McCollum, A.M., Mueller, K., Villegas, L., Udhayakumar, V., and Escalante, A.A. (2007). Common origin and fixation of *Plasmodium falciparum* dhfr and dhps mutations associated with sulfadoxine-pyrimethamine resistance in a low-transmission area in South America. Antimicrob. Agents Chemother. *51*, 2085–2091.

McCollum, A.M., Poe, A.C., Hamel, M., Huber, C., Zhou, Z., Shi, Y.P., Ouma, P., Vulule, J., Bloland, P., Slutsker, L., *et al.* (2006). Antifolate resistance in *Plasmodium falciparum*: multiple origins and identification of novel dhfr alleles. J. Infect. Dis. *194*, 189–197.

McCutchan, T.F., Dame, J.B., Miller, L.H., and Barnwell, J. (1984). Evolutionary relatedness of Plasmodium species as determined by the structure of DNA. Science *225*, 808–811.

Menard, D., Barnadas, C., Bouchier, C., Henry-Halldin, C., Gray, L.R., Ratsimbasoa, A., Thonier, V., Carod, J.F., Domarle, O., Colin, Y., *et al.* (2010). *Plasmodium vivax* clinical malaria is commonly observed in Duffy-negative Malagasy people. Proc. Natl. Acad. Sci. U.S.A. *107*, 5967–5971.

Metzker, M.L. (2010). Sequencing technologies – the next generation. Nat. Rev. Genet. *11*, 31–46.

Miller, S.K., Good, R.T., Drew, D.R., Delorenzi, M., Sanders, P.R., Hodder, A.N., Speed, T.P., Cowman, A.F., de Koning-Ward, T.F., and Crabb, B.S. (2002). A subset of *Plasmodium falciparum* SERA genes are expressed and appear to play an important role in the erythrocytic cycle. J. Biol. Chem. 277, 47524–47532.

Mixson-Hayden, T., Jain, V., McCollum, A.M., Poe, A., Nagpal, A.C., Dash, A.P., Stiles, J.K., Udhayakumar, V., and Singh, N. (2010). Evidence of selective sweeps in genes conferring resistance to chloroquine and pyrimethamine in *Plasmodium falciparum* isolates in India. Antimicrob. Agents Chemother. *54*, 997–1006.

Mourier, T., Carret, C., Kyes, S., Christodoulou, Z., Gardner, P.P., Jeffares, D.C., Pinches, R., Barrell, B., Berriman, M., Griffiths-Jones, S., *et al.* (2008). Genome-wide discovery and verification of novel structured RNAs in *Plasmodium falciparum*. Genome Res. *18*, 281–292.

Mu, J., Awadalla, P., Duan, J., McGee, K.M., Joy, D.A., McVean, G.A., and Su, X.Z. (2005a). Recombination hotspots and population structure in *Plasmodium falciparum*. PLoS Biol. 3, e335.

Mu, J., Awadalla, P., Duan, J., McGee, K.M., Keebler, J., Seydel, K., McVean, G.A., and Su, X.Z. (2007). Genome-wide variation and identification of vaccine targets in the *Plasmodium falciparum* genome. Nat. Genet. *39*, 126–130.

Mu, J., Duan, J., Makova, K.D., Joy, D.A., Huynh, C.Q., Branch, O.H., Li, W.H., and Su, X.Z. (2002). Chromosome-wide SNPs reveal an ancient origin for *Plasmodium falciparum*. Nature *418*, 323–326.

Mu, J., Ferdig, M.T., Feng, X., Joy, D.A., Duan, J., Furuya, T., Subramanian, G., Aravind, L., Cooper, R.A., Wootton, J.C., *et al.* (2003). Multiple transporters associated with malaria parasite responses to chloroquine and quinine. Mol. Microbiol. *49*, 977–989.

Mu, J., Joy, D.A., Duan, J., Huang, Y., Carlton, J., Walker, J., Barnwell, J., Beerli, P., Charleston, M.A., Pybus, O.G., *et al.* (2005b). Host switch leads to emergence of *Plasmodium vivax* malaria in humans. Mol. Biol. Evol. *22*, 1686–1693.

Mu, J., Myers, R.A., Jiang, H., Liu, S., Ricklefs, S., Waisberg, M., Chotivanich, K., Wilairatana, P., Krudsood, S., White, N.J., *et al.* (2010). *Plasmodium falciparum* genome-wide scans for positive selection, recombination hot spots and resistance to antimalarial drugs. Nat. Genet. *42*, 268–271.

Mwangi, J.M., Omar, S.A., and Ranford-Cartwright, L.C. (2006). Comparison of microsatellite and antigen-coding loci for differentiating recrudescing *Plasmodium falciparum* infections from reinfections in Kenya. Int. J. Parasitol. *36*, 329–336.

Nair, S., Miller, B., Barends, M., Jaidee, A., Patel, J., Mayxay, M., Newton, P., Nosten, F., Ferdig, M.T., and Anderson, T.J. (2008). Adaptive copy number evolution in malaria parasites. PLoS Genet. *4*, e1000243.

Nair, S., Nkhoma, S., Nosten, F., Mayxay, M., French, N., Whitworth, J., and Anderson, T. (2010). Genetic changes during laboratory propagation: copy number At the reticulocyte-binding protein 1 locus of *Plasmodium falciparum*. Mol. Biochem. Parasitol. *172*, 145–148.

Nair, S., Williams, J.T., Brockman, A., Paiphun, L., Mayxay, M., Newton, P.N., Guthmann, J.P., Smithuis, F.M., Hien, T.T., White, N.J., *et al.* (2003). A selective sweep driven by pyrimethamine treatment in Southeast Asian malaria parasites. Mol. Biol. Evol. *20*, 1526–1536.

Ndiaye, D., Patel, V., Demas, A., LeRoux, M., Ndir, O., Mboup, S., Clardy, J., Lakshmanan, V., Daily, J.P., and Wirth, D.F. (2010). A non-radioactive DAPI-based high-throughput in vitro assay to assess *Plasmodium falciparum* responsiveness to antimalarials – increased sensitivity of *P. falciparum* to chloroquine in Senegal. Am. J. Trop. Med. Hyg. *82*, 228–230.

Neafsey, D.E., Schaffner, S.F., Volkman, S.K., Park, D., Montgomery, P., Milner, D.A., Jr., Lukens, A., Rosen, D., Daniels, R., Houde, N., *et al.* (2008). Genome-wide SNP genotyping highlights the role of natural selection in *Plasmodium falciparum* population divergence. Genome Biol. *9*, R171.

Neira Oviedo, M., Ribeiro, J.M., Heyland, A., VanEkeris, L., Moroz, T., and Linser, P.J. (2009). The salivary transcriptome of *Anopheles gambiae* (Diptera: Culicidae) larvae: A microarray-based analysis. Insect Biochem. Mol. Biol. *39*, 382–394.

Nguitragool, W., Bokhari, A.A., Pillai, A.D., Rayavara, K., Sharma, P., Turpin, B., Aravind, L., and Desai, S.A. (2011). Malaria parasite clag3 genes determine channel-mediated nutrient uptake by infected red blood cells. Cell *145*, 665–677.

Ochola, L.I., Tetteh, K.K., Stewart, L.B., Riitho, V., Marsh, K., and Conway, D.J. (2011). Allele frequency-based and polymorphism-versus-divergence indices of balancing selection in a new filtered set of polymorphic genes in *Plasmodium falciparum*. Mol. Biol. Evol. *27*, 2344–2351.

Olszewski, K.L., Morrisey, J.M., Wilinski, D., Burns, J.M., Vaidya, A.B., Rabinowitz, J.D., and Llinas, M. (2009). Host–parasite interactions revealed by *Plasmodium falciparum* metabolomics. Cell Host Microbe *5*, 191–199.

Orjuela-Sanchez, P., de Santana Filho, F.S., Machado-Lima, A., Chehuan, Y.F., Costa, M.R., Alecrim, M.G., and del Portillo, H.A. (2009). Analysis of single-nucleotide polymorphisms in the crt-o and mdr1 genes of *Plasmodium vivax* among chloroquine-resistant isolates from the Brazilian Amazon region. Antimicrob. Agents Chemother. *53*, 3561–3564.

Orjuela-Sanchez, P., Karunaweera, N.D., da Silva-Nunes, M., da Silva, N.S., Scopel, K.K., Goncalves, R.M., Amaratunga, C., Sa, J.M., Socheat, D., Fairhust, R.M., *et al.* (2010). Single-nucleotide polymorphism, linkage disequilibrium and geographic structure in the malaria parasite *Plasmodium vivax*: prospects for genome-wide association studies. BMC Genet. *11*, 65.

Ouattara, A., Mu, J., Takala-Harrison, S., Saye, R., Sagara, I., Dicko, A., Niangaly, A., Duan, J., Ellis, R.D., Miller, L.H., *et al.* (2010). Lack of allele-specific efficacy of a bivalent AMA1 malaria vaccine. Malar. J. *9*, 175.

Palacios, G., Quan, P.L., Jabado, O.J., Conlan, S., Hirschberg, D.L., Liu, Y., Zhai, J., Renwick, N., Hui, J., Hegyi, H., *et al.* (2007). Panmicrobial oligonucleotide array for diagnosis of infectious diseases. Emerg. Infect. Dis. *13*, 73–81.

Patil, A., Orjuela-Sanchez, P., da Silva-Nunes, M., and Ferreira, M.U. (2011). Evolutionary dynamics of the immunodominant repeats of the *Plasmodium vivax* malaria-vaccine candidate circumsporozoite protein (CSP). Infect. Genet. Evol. *10*, 298–303.

Pattaradilokrat, S., Cheesman, S.J., and Carter, R. (2007). Linkage group selection: towards identifying genes controlling strain specific protective immunity in malaria. PLoS One 2, e857.

Pearce, R.J., Pota, H., Evehe, M.S., Ba el, H., Mombo-Ngoma, G., Malisa, A.L., Ord, R., Inojosa, W., Matondo, A., Diallo, D.A., *et al.* (2009). Multiple origins and regional dispersal of resistant dhps in African *Plasmodium falciparum* malaria. PLoS Med. *6*, e1000055.

Pickrell, J.K., Coop, G., Novembre, J., Kudaravalli, S., Li, J.Z., Absher, D., Srinivasan, B.S., Barsh, G.S., Myers, R.M., Feldman, M.W., *et al.* (2009). Signals of recent positive selection in a worldwide sample of human populations. Genome Res. *19*, 826–837.

Pinto, S.B., Lombardo, F., Koutsos, A.C., Waterhouse, R.M., McKay, K., An, C., Ramakrishnan, C., Kafatos, F.C., and Michel, K. (2009). Discovery of Plasmodium modulators by genome-wide analysis of circulating hemocytes in *Anopheles gambiae*. Proc. Natl. Acad. Sci. U.S.A. *106*, 21270–21275.

Plouffe, D., Brinker, A., McNamara, C., Henson, K., Kato, N., Kuhen, K., Nagle, A., Adrian, F., Matzen, J.T., Anderson, P., *et al.* (2008). In silico activity profiling reveals the mechanism of action of antimalarials discovered in a high-throughput screen. Proc. Natl. Acad. Sci. U.S.A. *105*, 9059–9064.

Pollack, Y., Katzen, A.L., Spira, D.T., and Golenser, J. (1982). The genome of *Plasmodium falciparum*. I: DNA base composition. Nucleic Acids Res. *10*, 539–546.

Polley, S.D., Chokejindachai, W., and Conway, D.J. (2003). Allele frequency-based analyses robustly map sequence sites under balancing selection in a malaria vaccine candidate antigen. Genetics *165*, 555–561.

Polley, S.D., and Conway, D.J. (2001). Strong diversifying selection on domains of the *Plasmodium falciparum* apical membrane antigen 1 gene. Genetics *158*, 1505–1512.

Polley, S.D., Tetteh, K.K., Lloyd, J.M., Akpogheneta, O.J., Greenwood, B.M., Bojang, K.A., and Conway, D.J. (2007). *Plasmodium falciparum* merozoite surface protein 3 is a target of allele-specific immunity and alleles are maintained by natural selection. J. Infect. Dis. *195*, 279–287.

Prajapati, S.K., Joshi, H., and Valecha, N. (2010). *Plasmodium vivax* merozoite surface protein-3 alpha: a high-resolution marker for genetic diversity studies. J. Vector Borne Dis. *47*, 85–90.

Preechapornkul, P., Imwong, M., Chotivanich, K., Pongtavornpinyo, W., Dondorp, A.M., Day, N.P., White, N.J., and Pukrittayakamee, S. (2009). *Plasmodium falciparum* pfmdr1 amplification, mefloquine resistance, and parasite fitness. Antimicrob. Agents Chemother. *53*, 1509–1515.

Preiser, P.R., Wilson, R.J., Moore, P.W., McCready, S., Hajibagheri, M.A., Blight, K.J., Strath, M., and Williamson, D.H. (1996). Recombination associated with replication of malarial mitochondrial DNA. EMBO J. *15*, 684–693.

Price, R.N., Uhlemann, A.C., Brockman, A., McGready, R., Ashley, E., Phaipun, L., Patel, R., Laing, K., Looareesuwan, S., White, N.J., *et al.* (2004). Mefloquine resistance in *Plasmodium falciparum* and increased pfmdr1 gene copy number. Lancet *364*, 438–447.

Putaporntip, C., Jongwutiwes, S., Ferreira, M.U., Kanbara, H., Udomsangpetch, R., and Cui, L. (2009). Limited global diversity of the *Plasmodium vivax* merozoite surface protein 4 gene. Infect. Genet. Evol. 9, 821–826.

Putaporntip, C., Jongwutiwes, S., Sakihama, N., Ferreira, M.U., Kho, W.G., Kaneko, A., Kanbara, H., Hattori, T., and Tanabe, K. (2002). Mosaic organization and heterogeneity in frequency of allelic recombination of the *Plasmodium vivax* merozoite surface protein-1 locus. Proc. Natl. Acad. Sci. U.S.A. *99*, 16348–16353.

Putaporntip, C., Udomsangpetch, R., Pattanawong, U., Cui, L., and Jongwutiwes, S. (2011). Genetic diversity of the *Plasmodium vivax* merozoite surface protein-5 locus from diverse geographic origins. Gene *456*, 24–35.

Raj, D.K., Mu, J., Jiang, H., Kabat, J., Singh, S., Sullivan, M., Fay, M.P., McCutchan, T.F., and Su, X.Z. (2009). Disruption of a *Plasmodium falciparum* multidrug resistance-associated protein (PfMRP) alters its fitness and transport of antimalarial drugs and glutathione. J. Biol. Chem. *284*, 7687–7696.

Rajesh, V., Singamsetti, V.K., Vidya, S., Gowrishankar, M., Elamaran, M., Tripathi, J., Radhika, N.B., Kochar, D., Ranjan, A., Roy, S.K., *et al.* (2008). *Plasmodium falciparum*: genetic polymorphism in apical membrane antigen-1 gene from Indian isolates. Exp. Parasitol. *119*, 144–151.

Ralph, S.A., van Dooren, G.G., Waller, R.F., Crawford, M.J., Fraunholz, M.J., Foth, B.J., Tonkin, C.J., Roos, D.S., and McFadden, G.I. (2004). Tropical infectious diseases: metabolic maps and functions of the *Plasmodium falciparum* apicoplast. Nat. Rev. Microbiol. 2, 203–216.

Ralph, S.A., Bischoff, E., Mattei, D., Sismeiro, O., Dillies, M.A., Guigon, G., Coppee, J.Y., David, P.H, and Scherf, A. (2005). Transcriptome analysis of antigenic variation in *Plasmodium falciparum* – var silencing is not dependent on antisense RNA. Genome Biol. *6*, R93.

Rathod, P.K., Ganesan, K., Hayward, R.E., Bozdech, Z., and DeRisi, J.L. (2002). DNA microarrays for malaria. Trends Parasitol. *18*, 39–45.

Rathod, P.K., McErlean, T., and Lee, P.C. (1997). Variations in frequencies of drug resistance in *Plasmodium falciparum*. Proc. Natl. Acad. Sci. U.S.A. *94*, 9389–9393.

Reilly Ayala, H.B., Wacker, M.A., Siwo, G., and Ferdig, M.T. (2010). Quantitative trait loci mapping reveals candidate pathways regulating cell cycle duration in *Plasmodium falciparum*. BMC Genomics *11*, 577.

Reilly, H.B., Wang, H., Steuter, J.A., Marx, A.M., and Ferdig, M.T. (2007). Quantitative dissection of clone-specific growth rates in cultured malaria parasites. Int. J. Parasitol. *37*, 1599–1607.

Rezende, A.M., Tarazona-Santos, E., Couto, A.D., Fontes, C.J., De Souza, J.M., Carvalho, L.H., and Brito, C.F. (2009). Analysis of genetic variability of *Plasmodium vivax* isolates from different Brazilian Amazon areas using tandem repeats. Am. J. Trop. Med. Hyg. *80*, 729–733.

Rezende, A.M., Tarazona-Santos, E., Fontes, C.J., Souza, J.M., Couto, A.D., Carvalho, L.H., and Brito, C.F. (2010). Microsatellite loci: determining the genetic variability of *Plasmodium vivax*. Trop. Med. Int. Health *15*, 718–726.

Ribacke, U., Mok, B.W., Wirta, V., Normark, J., Lundeberg, J., Kironde, F., Egwang, T.G., Nilsson, P., and Wahlgren, M. (2007). Genome-wide gene amplifications and deletions in *Plasmodium falciparum*. Mol. Biochem. Parasitol. *155*, 33–44.

Rich, S.M., Hudson, R.R., and Ayala, F.J. (1997). *Plasmodium falciparum* antigenic diversity: evidence of clonal population structure. Proc. Natl. Acad. Sci. U.S.A. *94*, 13040–13045.

Rich, S.M., Licht, M.C., Hudson, R.R., and Ayala, F.J. (1998). Malaria's Eve: evidence of a recent population bottleneck throughout the world populations of *Plasmodium falciparum*. Proc. Natl. Acad. Sci. U.S.A. *95*, 4425–4430.

Rogers, D.W., Whitten, M.M., Thailayil, J., Soichot, J., Levashina, E.A., and Catteruccia, F. (2008). Molecular and cellular components of the mating machinery in *Anopheles gambiae* females. Proc. Natl. Acad. Sci. U.S.A. *105*, 19390–19395.

Roper, C., Pearce, R., Bredenkamp, B., Gumede, J., Drakeley, C., Mosha, F., Chandramohan, D., and Sharp, B. (2003). Antifolate antimalarial resistance in southeast Africa: a population-based analysis. Lancet *361*, 1174–1181.

Rosenberg, N.A., Huang, L., Jewett, E.M., Szpiech, Z.A., Jankovic, I., and Boehnke, M. (2010). Genome-wide association studies in diverse populations. Nat. Rev. Genet. *11*, 356–366.

Roy, S.W., and Irimia, M. (2008). Origins of human malaria: rare genomic changes and full mitochondrial genomes confirm the relationship of *Plasmodium falciparum* to other mammalian parasites but complicate the origins of *Plasmodium vivax*. Mol. Biol. Evol. *25*, 1192–1198.

Rubio, J.M., Berzosa, P.J., and Benito, A. (2001). Amplified fragment length polymorphism (AFLP) protocol for genotyping the malarial parasite *Plasmodium falciparum*. Parasitology *123*, 331–336.

Russell, B., Suwanarusk, R., and Lek-Uthai, U. (2006). *Plasmodium vivax* genetic diversity: microsatellite length matters. Trends Parasitol. *22*, 399–401.

Sabeti, P.C., Reich, D.E., Higgins, J.M., Levine, H.Z., Richter, D.J., Schaffner, S.F., Gabriel, S.B., Platko, J.V., Patterson, N.J., McDonald, G.J., *et al.* (2002). Detecting recent positive selection in the human genome from haplotype structure. Nature *419*, 832–837.

Sabeti, P.C., Varilly, P., Fry, B., Lohmueller, J., Hostetter, E., Cotsapas, C., Xie, X., Byrne, E.H., McCarroll, S.A., Gaudet, R., *et al.* (2007). Genome-wide detection and characterization of positive selection in human populations. Nature *449*, 913–918.

Sakihama, N., Kimura, M., Hirayama, K., Kanda, T., Na-Bangchang, K., Jongwutiwes, S., Conway, D., and Tanabe, K. (1999). Allelic recombination and linkage disequilibrium within Msp-1 of *Plasmodium falciparum*, the malignant human malaria parasite. Gene *230*, 47–54.

Salgueiro, P., Vicente, J.L., Ferreira, C., Teofilo, V., Galvao, A., do Rosario, V.E., Cravo, P., and Pinto, J. (2010). Tracing the origins and signatures of selection of antifolate resistance in island populations of *Plasmodium falciparum*. BMC Infect. Dis. *10*, 163.

Shah, N. (2008). Single Nucleotide Polymorphisms, 2nd eedn (Totowa, NJ: Humana Press).

Shastry, B.S. (2007). SNPs in disease gene mapping, medicinal drug development and evolution. J. Hum. Genet. *52*, 871–880.

Shastry, B.S. (2009). SNPs: impact on gene function and phenotype. Methods Mol. Biol. *578*, 3–22.

Shirley, M.W., Biggs, B.A., Forsyth, K.P., Brown, H.J., Thompson, J.K., Brown, G.V., and Kemp, D.J. (1990). Chromosome 9 from independent clones and isolates of *Plasmodium falciparum* undergoes subtelomeric deletions with similar breakpoints in vitro. Mol. Biochem. Parasitol. *40*, 137–145.

Sidhu, A.B., Uhlemann, A.C., Valderramos, S.G., Valderramos, J.C., Krishna, S., and Fidock, D.A. (2006). Decreasing pfmdrl copy number in *plasmodium falciparum* malaria heightens susceptibility to mefloquine, lumefantrine, halofantrine, quinine, and artemisinin. J. Infect. Dis. *194*, 528–535.

Sinnis, P., and Wellems, T.E. (1988). Long-range restriction maps of *Plasmodium falciparum* chromosomes: crossingover and size variation among geographically distant isolates. Genomics 3, 287–295.

Sousa, T.N., Tarazona-Santos, E.M., Wilson, D.J., Madureira, A.P., Falcao, P.R., Fontes, C.J., Gil, L.H., Ferreira, M.U., Carvalho, L.H., and Brito, C.F. (2010). Genetic variability and natural selection at the ligand domain of the Duffy binding protein in Brazilian *Plasmodium vivax* populations. Malar. J. *9*, 334.

Steenkeste, N., Dillies, M.A., Khim, N., Sismeiro, O., Chy, S., Lim, P., Crameri, A., Bouchier, C., Mercereau-Puijalon, O., Beck, H.P., *et al.* (2009). FlexiChip package: an universal microarray with a dedicated analysis software for high-throughput SNPs detection linked to anti-malarial drug resistance. Malar. J. *8*, 229.

Su, X., Ferdig, M.T., Huang, Y., Huynh, C.Q., Liu, A., You, J., Wootton, J.C., and Wellems, T.E. (1999). A genetic map and recombination parameters of the human malaria parasite *Plasmodium falciparum*. Science *286*, 1351–1353.

Su, X., Hayton, K., and Wellems, T.E. (2007). Genetic linkage and association analyses for trait mapping in *Plasmodium falciparum*. Nat. Rev. Genet. *8*, 497–506.

Su, X., Kirkman, L.A., Fujioka, H., and Wellems, T.E. (1997). Complex polymorphisms in an approximately 330 kDa protein are linked to chloroquine-resistant *P. falciparum* in Southeast Asia and Africa. Cell *91*, 593–603.

Su, X., and Wellems, T.E. (1996). Toward a high-resolution *Plasmodium falciparum* linkage map: polymorphic markers from hundreds of simple sequence repeats. Genomics 33, 430–444.

Su, X.Z., Jiang, H., Yi, M., Mu, J., and Stephens, R.M. (2009). Large-scale genotyping and genetic mapping in Plasmodium parasites. Korean J. Parasitol. *47*, 83–91.

Su, X.Z., Mu, J., and Joy, D.A. (2003). The 'Malaria's Eve' hypothesis and the debate concerning the origin of the human malaria parasite *Plasmodium falciparum*. Microbes Infect 5, 891–896.

Su, X.Z., and Wellems, T.E. (1999). *Plasmodium falciparum*: assignment of microsatellite markers to chromosomes by PFG-PCR. Exp. Parasitol. *91*, 367–369.

Suwanarusk, R., Chavchich, M., Russell, B., Jaidee, A., Chalfein, F., Barends, M., Prasetyorini, B.,

Kenangalem, E., Piera, K.A., Lek-Uthai, U., *et al.* (2008). Amplification of pvmdr1 associated with multidrug-resistant *Plasmodium vivax*. J. Infect. Dis. *198*, 1558–1564.

Tan, J.C., Miller, B.A., Tan, A., Patel, J.J., Cheeseman, I.H., Anderson, T.J., Manske, M., Maslen, G., Kwiatkowski, D.P., and Ferdig, M.T. (2011). An optimized microarray platform for assaying genomic variation in *Plasmodium falciparum* field populations. Genome Biol. *12*, R35.

Tan, J.C., Patel, J.J., Tan, A., Blain, J.C., Albert, T.J., Lobo, N.F., and Ferdig, M.T. (2009). Optimizing comparative genomic hybridization probes for genotyping and SNP detection in *Plasmodium falciparum*. Genomics *93*, 543–550.

Tan, J.C., Tan, A., Checkley, L., Honsa, C.M., and Ferdig, M.T. (2010). Variable numbers of tandem repeats in *Plasmodium falciparum* genes. J. Mol. Evol. *71*, 268–278.

Tan-Ariya, P., Yang, Y.F., and Kilejian, A. (1988). *Plasmodium falciparum*: comparison of the genomic organization of the knob protein gene in knobby and knobless variants. Exp. Parasitol. *67*, 129–136.

Tchavtchitch, M., Fischer, K., Huestis, R., and Saul, A. (2001). The sequence of a 200 kb portion of a *Plasmodium vivax* chromosome reveals a high degree of conservation with *Plasmodium falciparum* chromosome 3. Mol. Biochem. Parasitol. *118*, 211–222.

Tetteh, K.K., Stewart, L.B., Ochola, L.I., Amambua-Ngwa, A., Thomas, A.W., Marsh, K., Weedall, G.D., and Conway, D.J. (2009). Prospective identification of malaria parasite genes under balancing selection. PLoS One 4, e5568.

Thompson, J.K., Caruana, S.R., and Cowman, A.F. (1999). YAC contigs and restriction maps of chromosomes 4 and 5 from the cloned line 3D7 of *Plasmodium falciparum*. Mol. Biochem. Parasitol. *102*, 197–204.

Thompson, J.K., and Cowman, A.F. (1997). A YAC contig and high resolution restriction map of chromosome 3 from *Plasmodium falciparum*. Mol. Biochem. Parasitol. *90*, 537–542.

Tibayrenc, M., Kjellberg, F., and Ayala, F.J. (1990). A clonal theory of parasitic protozoa: the population structures of *Entamoeba, Giardia, Leishmania, Naegleria, Plasmodium, Trichomonas*, and *Trypanosoma* and their medical and taxonomical consequences. Proc. Natl. Acad. Sci. U.S.A. *87*, 2414–2418.

Triglia, T., Foote, S.J., Kemp, D.J., and Cowman, A.F. (1991). Amplification of the multidrug resistance gene pfmdr1 in *Plasmodium falciparum* has arisen as multiple independent events. Mol. Cell Biol. *11*, 5244–5250.

Triglia, T., Wellems, T.E., and Kemp, D.J. (1992). Towards a high-resolution map of the *Plasmodium falciparum* genome. Parasitol. Today *8*, 225–229.

Tuikue Ndam, N., Bischoff, E., Proux, C., Lavstsen, T., Salanti, A., Guitard, J., Nielsen, M.A., Coppee, J.Y., Gaye, A., Theander, T., *et al.* (2008). *Plasmodium falciparum* transcriptome analysis reveals pregnancy malaria associated gene expression. PLoS One 3, e1855.

Turner, T.L., Hahn, M.W., and Nuzhdin, S.V. (2005). Genomic islands of speciation in *Anopheles gambiae*. PLoS Biol. 3, e285.

Umar, F., Dubey, M.L., Malla, N., and Mahajan, R.C. (2006). *Plasmodium falciparum*: polymorphism in the MSP-1 gene in Indian isolates and predominance of certain alleles in cerebral malaria. Exp. Parasitol. *112*, 139–143.

Vaidya, A.B., and Mather, M.W. (2009). Mitochondrial evolution and functions in malaria parasites. Annu. Rev. Microbiol. *63*, 249–267.

Van den Eede, P., Erhart, A., Van der Auwera, G., Van Overmeir, C., Thang, N.D., Hung le, X., Anne, J., and D'Alessandro, U. (2010a). High complexity of *Plasmodium vivax* infections in symptomatic patients from a rural community in central Vietnam detected by microsatellite genotyping. Am. J. Trop. Med. Hyg. *82*, 223–227.

Van den Eede, P., Van der Auwera, G., Delgado, C., Huyse, T., Soto-Calle, V.E., Gamboa, D., Grande, T., Rodriguez, H., Llanos, A., Anne, J., *et al.* (2010b). Multilocus genotyping reveals high heterogeneity and strong local population structure of the *Plasmodium vivax* population in the Peruvian Amazon. Malar. J. 9, 151.

Van Tyne, D., Park, D.J., Schaffner, S.F., Neafsey, D.E., Angelino, E., Cortese, J.F., Barnes, K.G., Rosen, D.M., Lukens, A.K., Daniels, R.F., *et al.* (2011). Identification and functional validation of the novel antimalarial resistance locus PF10_0355 in *Plasmodium falciparum*. PLoS Genet. 7, e1001383.

Veiga, M.I., Ferreira, P.E., Bjorkman, A., and Gil, J.P. (2006). Multiplex PCR-RFLP methods for pfcrt, pfmdr1 and pfdhfr mutations in *Plasmodium falciparum*. Mol. Cell Probes *20*, 100–104.

Verra, F., Chokejindachai, W., Weedall, G.D., Polley, S.D., Mwangi, T.W., Marsh, K., and Conway, D.J. (2006). Contrasting signatures of selection on the *Plasmodium falciparum* erythrocyte binding antigen gene family. Mol. Biochem. Parasitol. *149*, 182–190.

Voight, B.F., Kudaravalli, S., Wen, X., and Pritchard, J.K. (2006). A map of recent positive selection in the human genome. PLoS Biol. 4, e72.

Voisey, J., and Morris, C.P. (2008). SNP technologies for drug discovery: a current review. Curr. Drug Discov. Technol. 5, 230–235.

Volkman, S.K., Barry, A.E., Lyons, E.J., Nielsen, K.M., Thomas, S.M., Choi, M., Thakore, S.S., Day, K.P., Wirth, D.F., and Hartl, D.L. (2001). Recent origin of *Plasmodium falciparum* from a single progenitor. Science 293, 482–484.

Volkman, S.K., Hartl, D.L., Wirth, D.F., Nielsen, K.M., Choi, M., Batalov, S., Zhou, Y., Plouffe, D., Le Roch, K.G., Abagyan, R., *et al.* (2002). Excess polymorphisms in genes for membrane proteins in *Plasmodium falciparum*. Science *298*, 216–218.

Volkman, S.K., Sabeti, P.C., DeCaprio, D., Neafsey, D.E., Schaffner, S.F., Milner, D.A., Jr., Daily, J.P., Sarr, O., Ndiaye, D., Ndir, O., *et al.* (2007). A genome-wide map of diversity in *Plasmodium falciparum*. Nat. Genet. 39, 113–119.

Walker-Jonah, A., Dolan, S.A., Gwadz, R.W., Panton, L.J., and Wellems, T.E. (1992). An RFLP map of the *Plasmodium falciparum* genome, recombination rates and favored linkage groups in a genetic cross. Mol. Biochem. Parasitol. *51*, 313–320.

Walliker, D., Babiker, H., and Ranford-Cartwright, L. (1998). The Genetic Structure of Malaria Parasite Populations. In Malaria: Parasite Biology, Pathogenesis, and Protection, I.W. Sherman, ed. (Washington D.C., ASM Press), pp. 235–252.

Wang, P., Read, M., Sims, P.F., and Hyde, J.E. (1997). Sulfadoxine resistance in the human malaria parasite *Plasmodium falciparum* is determined by mutations in dihydropteroate synthetase and an additional factor associated with folate utilization. Mol. Microbiol. *23*, 979–986.

Weedall, G.D., and Conway, D.J. (2010). Detecting signatures of balancing selection to identify targets of anti-parasite immunity. Trends Parasitol. *26*, 363–369.

Weedall, G.D., Preston, B.M., Thomas, A.W., Sutherland, C.J., and Conway, D.J. (2007). Differential evidence of natural selection on two leading sporozoite stage malaria vaccine candidate antigens. Int. J. Parasitol. *37*, 77–85.

Wellems, T.E., Panton, L.J., Gluzman, I.Y., do Rosario, V.E., Gwadz, R.W., Walker-Jonah, A., and Krogstad, D.J. (1990). Chloroquine resistance not linked to mdr-like genes in a *Plasmodium falciparum* cross. Nature *345*, 253–255.

Wellems, T.E., Walker-Jonah, A., and Panton, L.J. (1991). Genetic mapping of the chloroquine-resistance locus on *Plasmodium falciparum* chromosome 7. Proc. Natl. Acad. Sci. U.S.A. *88*, 3382–3386.

Westenberger, S.J., McClean, C.M., Chattopadhyay, R., Dharia, N.V., Carlton, J.M., Barnwell, J.W., Collins, W.E., Hoffman, S.L., Zhou, Y., Vinetz, J.M., *et al.* (2010). A systems-based analysis of *Plasmodium vivax* life cycle transcription from human to mosquito. PLoS Negl. Trop. Dis. *4*, e653.

Wilson, R.J., Denny, P.W., Preiser, P.R., Rangachari, K., Roberts, K., Roy, A., Whyte, A., Strath, M., Moore, D.J., Moore, P.W., *et al.* (1996). Complete gene map of the plastid-like DNA of the malaria parasite *Plasmodium falciparum*. J. Mol. Biol. *261*, 155–172.

Winter, G., Kawai, S., Haeggstrom, M., Kaneko, O., von Euler, A., Kawazu, S., Palm, D., Fernandez, V., and Wahlgren, M. (2005). SURFIN is a polymorphic antigen expressed on *Plasmodium falciparum* merozoites and infected erythrocytes. J. Exp. Med. *201*, 1853–1863.

Wootton, J.C., Feng, X., Ferdig, M.T., Cooper, R.A., Mu, J., Baruch, D.I., Magill, A.J., and Su, X.Z. (2002). Genetic diversity and chloroquine selective sweeps in *Plasmodium falciparum*. Nature *418*, 320–323.

Wurtz, N., Desplans, J., and Parzy, D. (2009). Phenotypic and transcriptomic analyses of *Plasmodium falciparum* protein kinase A catalytic subunit inhibition. Parasitol. Res. *105*, 1691–1699.

Yang, Z., Miao, J., Huang, Y., Li, X., Putaporntip, C., Jongwutiwes, S., Gao, Q., Udomsangpetch, R., Sattabongkot, J., and Cui, L. (2006). Genetic structures of geographically distinct *Plasmodium vivax* populations assessed by PCR/RFLP analysis of the merozoite surface protein 3beta gene. Acta Trop. *100*, 205–212.

Yuan, J., Cheng, K.C., Johnson, R.L., Huang, R., Pattaradilokrat, S., Liu, A., Guha, R., Fidock, D.A., Inglese, J., Wellems, T.E., *et al.* (2011). Chemical genomic profiling for antimalarial therapies, response signatures, and molecular targets. Science *333*, 724–729.

Yuan, J., Johnson, R.L., Huang, R., Wichterman, J., Jiang, H., Hayton, K., Fidock, D.A., Wellems, T.E., Inglese, J., Austin, C.P., *et al.* (2009). Genetic mapping of targets mediating differential chemical phenotypes in *Plasmodium falciparum*. Nat. Chem. Biol. *5*, 765–771.

Zeyrek, F.Y., Tachibana, S., Yuksel, F., Doni, N., Palacpac, N., Arisue, N., Horii, T., Coban, C., and Tanabe, K. (2010). Limited polymorphism of the *Plasmodium vivax* merozoite surface protein 1 gene in isolates from Turkey. Am. J. Trop. Med. Hyg. *83*, 1230–1237.

Zhang, G.Q., Guan, Y.Y., Sheng, H.H., Zheng, B., Wu, S., Xiao, H.S., and Tang, L.H. (2008). Multiplex PCR and oligonucleotide microarray for detection of single-nucleotide polymorphisms associated with *Plasmodium falciparum* drug resistance. J. Clin. Microbiol. *46*, 2167–2174.

Functional Genomics of *Plasmodium* Parasites

5

Zbynek Bozdech and Peter R. Preiser

Abstract

Over the last decade, functional genomics of *Plasmodium* species uncovered many new insights into gene and protein expression that characterizes both the growth and multiplication of malaria parasites in their natural hosts. Genome- and proteome-wide technologies provided new insights into the physiology of all major developmental stages as well as many regulatory mechanisms that control progression of the complex *Plasmodium* life cycle. It is now clear that the morphological transition between the developmental stages is linked to broad changes in gene expression that are regulated by several types of regulatory mechanisms including specific transcription factors and chromatin remodelling machinery as well as global DNA/RNA metabolic processes. At the protein level, translation rate, protein turnover and posttranslational modifications appeared to be important contributing factors to the life cycle regulation. In addition, it is clear that *Plasmodium* parasites are able to transcriptionally and post-transcriptionally respond to some external stimuli. Both the broad regulation of gene/protein expression during the life cycle and the transcriptional/translational plasticity likely reflect a robust adaptation of malaria parasites to their natural hosts and thus their ability to spread amongst many populations throughput the world. Better understanding for these dynamic properties of the *Plasmodium* genome will provide the tools for the design and development of new malaria control strategies.

Introduction

By definition, functional genomics is a scientific field that utilizes the wealth of information obtained from genome sequencing to study functions of genes and proteins. Unlike genomics, which describes the sequence or sequence variations, functional genomics focuses on the dynamic properties of genes such as transcription and translation, posttranslational modifications, and protein–protein interactions. Ultimately, functional genomics strives to reconstruct the overall system of biochemical and biological functionalities within a cell or a whole organism, in order to capture their intricate interactions and relationships. For this, functional genomics utilizes genome-wide approaches that are, in principle, able to evaluate the entire system of a living cell without any preconceived notion or bias, thereby uncovering often unexpected or unprecedented biological processes.

The most important prerequisite for functional genomics is genome sequence information that allows the design of essential tools (e.g. microarrays) and/or final data analysis (e.g. mass spectroscopy-based proteomics, next generation sequencing). Hence, the completion of the first genomes of the malaria parasites species, *Plasmodium falciparum* (Gardner *et al.*, 2002) and *P. yoelii* (Carlton *et al.*, 2002) in October 2002 marked the onset of functional genomic studies of malaria parasites. Since then functional genomics has gained further momentum, fuelled by the completion of genomes of an additional five *Plasmodium* species, including another important human pathogen *P. vivax* (Carlton *et al.*, 2008), two rodent models *P chabaudi* and *P. berghei* (Hall *et al.*, 2005) and the primate pathogen *P. knowlesi* (Pain *et al.*, 2008) that is also capable of human infection (Singh *et al.*, 2004). In fact, many functional genomics studies of *Plasmodium* parasites were initiated concurrently with the final stages of the genome project, thanks to the public access to the wealth of genomics sequence information that was provided by the three sequencing centres (The Institute for Genomic Research (TIGR), Wellcome Trust Sanger Institute, and Stanford University) and hosted by the *Plasmodium* genomic resource database PlasmoDB (PlasmoDB.org), which currently is the most important and most comprehensive data source for all *Plasmodium* genomics (Kissinger *et al.*, 2002). With this valuable resource, functional genomics and proteomics* brought major advances in our understanding of malaria parasites, uncovering a wide spectrum of biological insights into the basic processes associated with the *Plasmodium* parasite life cycle as well as other aspect of its biology such as drug sensitivities, virulence, and pathogenicity. Many of these results provided the scientific basis for the new intervention strategies considered by different programmes around the world that are currently at various stages of development.

Transcriptional profiling of the *Plasmodium* life cycle

The *Plasmodium* intraerythrocytic developmental cycle (IDC)

The IDC transcriptome

Even the first version of the partially completed genome sequence of *P. falciparum* in the early 2000s was sufficient to initiate genomics and proteomic surveys. In the decade that followed, numerous studies brought profound understudying of transcriptional and posttranscriptional regulatory processes associated with the progression of the *Plasmodium* life cycle (Fig. 5.1). The main focus of these early studies was the parasite's blood stage asexual development, encompassing ring (i.e. young trophozoite), trophozoite, schizont, and merozoite stages. Two DNA microarray platforms were created: the photolithography-generated high density oligonucleotide microarray produced by AFFYMETRIX technologies (Le Roch *et al.*, 2002, 2003), and a long oligonucleotide DNA microarray based on DNA spotting technology (Bozdech *et al.*, 2003a,b). One of the original studies, in which the transcriptome of the 48-hour asexual blood stage development was analysed at single-hour resolution, showed that transcriptional regulation plays a major role in targeting different functionalities to specific stages in a 'just-in-time' manner (Bozdech *et al.*, 2003a). Over 60% of genes, representing the vast majority of biological functions, were found to be transcriptionally regulated, exhibiting a single-peak mRNA abundance profile. This transcriptional regulation can target the expression of particular transcripts to a developmental stage with razor-sharp precision. For example, in the case of parasite nucleotide metabolism, transcription of factors involved in synthesis and salvage of ribonucleotides (RNA precursors) are synchronized with expression of genes involved in RNA metabolism (peak abundance at 18–22 h post invasion, hpi). On the other hand, expression of mRNA transcripts for enzymes which convert ribonucleotides into deoxyribonucleotides (DNA precursors), including DHFR-TS, and both subunits of ribonucleotide reductase, parallels transcription of genes of DNA replication, peaking at approximately 32 hpi. The wide extent of the transcriptional changes during intraerythrocytic development significantly differs from transcriptional changes associated with the classical cell cycle programs of eukaryotes such as yeast (Spellman *et al.*, 1998) and mammalian cells (Whitfield *et al.*, 2002), during which no more than 15% of genes show a periodic expression pattern. Instead, the generated single-hour transcriptome of *P. falciparum* during asexual blood stages resembled transcriptomic results from developmental processes such as *Drosophila melanogaster* embryogenesis (Arbeitman *et al.*, 2002) that were emerging at that the time of this study. Hence the asexual intraerythrocytic section of the *Plasmodium* life cycle has been termed the

*Adhering to the broad definition of functional genomics (see above), in this chapter, DNA-based genome-wide studies and large-scale proteomics will be discussed in the context of each other. This also follows the trend of many recent studies using both approaches in parallel.

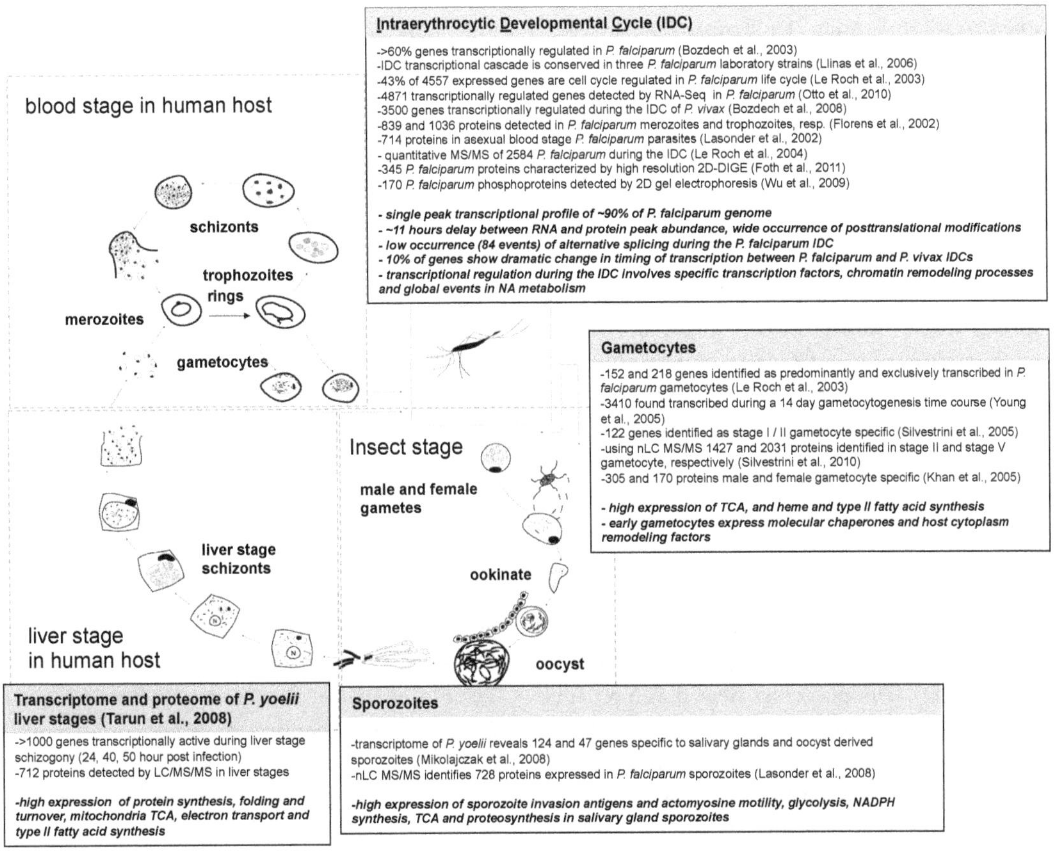

Figure 5.1 Overview of transcriptome and proteome surveys of individual developmental stages of the *Plasmodium* parasite life cycle. References to the individual studies and major observation are listed for four major sections of the *Plasmodium* life cycle. (The *Plasmodium* life cycle graphic was generated by Melody Munz, MPH.)

intraerythrocytic developmental cycle (IDC) (Bozdech *et al.*, 2003a).

A concurrent study focusing on different developmental stages of the parasite showed that at least 4557 *P. falciparum* genes (88% of the genome) are expressed in at least one stage and that their expression may vary by up to five orders of magnitude (Le Roch *et al.*, 2003). In particular, 43% of the detected genes were classified as 'cell cycle regulated', with their steady-state mRNA varying at least 1.5 fold ($P \leq 0.05$) between different stages. Moreover, many genes associated with basic metabolic and cellular functions – transcription, translation, and energy metabolism – were found to be expressed in the fast-growing ring and trophozoite stages, while genes encoding surface molecules involved in host–parasite interactions were mainly expressed in the invasive forms (merozoites and sporozoites) (Le Roch *et al.*, 2003), confirming the functional relationship between the gene expression pattern and the physiological needs of each developmental stage. The microarray-based IDC transcriptome data were recently validated using next-generation sequencing (NGS); mRNA abundance was analysed by quantitative sequencing via RNA-Seq using an Illumina Genome Analyser (Otto *et al.*, 2010). High correlations between the RNA-Seq and microarray data (Pearson correlation values ranging from 0.7 to 0.8) provided further confirmation of the just-in-time model of gene expression, extending it to 4871 genes (~90% of the genome). In addition to highly quantitative measurements of gene expression, the RNA-Seq results helped to correct existing models for 423 genes and identified 121 previously undetected open reading frames. Unfortunately, the current sequencing technology could not support assembly of the highly AT-rich untranslated regions (UTRs) but was able to confirm 75% of mRNA splicing sites originally predicted by the *in silico* gene models generated during the genome sequencing project (Gardner *et al.*, 2002). Interestingly, only 84 alternative splice events could be detected, in sharp contrast to higher eukaryotic species where over half of the genes are subject to alternative splicing (Zavolan and van Nimwegen, 2006). At the moment it is unclear whether this extremely low detection rate of alternative splicing is due to the limitations of the current NGS methodologies or if it is a genuine reflection of the mRNA expression pattern during the *P. falciparum* IDC. It could also mirror a previous suggestion that alternative splicing is highly underrepresented in the IDC compared to other developmental stages such as gametocytes (Iriko *et al.*, 2009). Undoubtedly, new developments in NGS and its wider application to functional genomics will bring new insights into the extent and character of alternative splicing as well as other features of RNA metabolism in *Plasmodium* parasites in the future.

One of the most striking features of the periodic transcriptional program of the IDC is its high level of conservation in different *P. falciparum* populations, including *P. falciparum* laboratory strains (Llinas *et al.*, 2006) and freshly adapted field isolates (Mackinnon *et al.*, 2009). In a study of three *P. falciparum* laboratory strains, including 3D7, HB3, and drug resistant DD2, Llinas and colleagues demonstrated that all genes facilitating basic molecular and cellular functions such as glycolysis, cytoplasmic transcription, translation, the tricarboxylic acid cycle, activity of the proteasome, and plastid biosynthesis, exhibit identical expression throughout the IDC in all three strains (Llinas *et al.*, 2006). A small group of genes show differences in temporal expression between the three strains, with only 69 genes (~ 1.3% of the genome) showing transcriptional shift by 12 or more hours. These strain-specific transcriptional differences are restricted to genes involved in parasite/host interactions, including cell adhesion, defence and immunity, and antigenic variation. Some of these differences represent individual members of gene families including PFC0110w, encoding the cytoadherence-linked asexual protein (CLAG 3.1) that is implicated in erythrocyte invasion; PF11_0512, encoding a ring-infected erythrocyte surface antigen 2 (RESA-2) potentially associated with host cell cytoskeleton; and PF08_0103 and PFB0010w, a pair of genes encoding erythrocyte membrane protein 1 (PfEMP1). Notwithstanding that some of the observed strain-specific differential expression may represent false positive microarray readings due to sequence polymorphism, these data suggest that in terms of timing of gene expression, *P. falciparum* has a limited degree of flexibility, suggesting that the IDC transcription cascade that characterizes the asexual erythrocytic stage is mediated by a highly conserved transcriptional program that does not fluctuate greatly amongst various strains or other parasite populations.

Hence, it was surprising that the transcription cascade of the second most pathogenic species, *P. vivax*, is not identical to *P. falciparum*, with a considerable proportion of syntenic genes being expressed at different times during the IDC (Bozdech *et al.*, 2008). The *P. vivax* transcription cascade consists of at least 3500 genes (70%) that exhibit >2-fold change in their mRNA abundance which, similarly to *P. falciparum*, follows the just-in-time gene expression model. The majority of genes facilitating basic cellular and metabolic functions are also highly regulated in terms of their expression, defining the distribution of these pathways along the IDC. However, while ~68% of the *P. falciparum-P. vivax* syntenic homologues exhibit the same expression timing between these two species, and an additional ~22% exhibit slight transcription timing shifts during the IDC. Moreover, about 10% of the genes show dramatic changes in their peak expression (Pearson correlation coefficient between –0.2 and –1.0). One of the striking examples is the knob-associated histidine-rich protein (KAHRP), which has two homologues (Pv081835 and Pv003520) in *P. vivax*, both of which are expressed in the late schizont stages. This is in sharp contrast with the single *kahrp* copy in *P. falciparum*, where KAHRP is expressed in the early ring stages coinciding with the formation of knobs on the surface of the infected red blood cells. This suggests that in *P.*

vivax, KAHRP is utilized for different purposes, likely associated with a schizont-specific function. This is in agreement with the fact that *P. vivax* does not form knobs on the surface of its host cells.

Two striking examples of dramatic transcriptional shifts that affect a large fraction of genes of a specific cellular pathway involve haemoglobin degradation and host cytoplasm remodelling. In *P. vivax*, most of the proteases involved in haemoglobin degradation are expressed in schizonts, as opposed to the early trophozoites expression observed in *P. falciparum*. This indicates that in *P. vivax*, IDC haemoglobin degradation progresses more gradually and is extended to the late schizont stage. This observation may explain the reduced sensitivity of *P. vivax* rings to chloroquine as being due to the low concentration of haem in these early stages (Sharrock *et al.*, 2008). Similarly, the *P. vivax* homologues of proteins transported to the parasitophorous vacuole (PV), which are essential for the establishment of the early *P. falciparum* rings in the host erythrocytes, are transcribed in the schizont stages. These include members of the ETRAMP family and two exported proteins, Exp1 and Exp2. Interestingly, Exp2 has been found recently to be an essential component of the translocon complex that has been implicated in the protein export across the *Plasmodium* plasma membrane (de Koning-Ward *et al.*, 2009). In addition to EXP2, another subunit of the translocon, thioredoxin-like protein TRX2 (MAL13P1.225), was shifted from its ring peak expression in *P. falciparum* to schizont expression in *P. vivax*; all other the members of the translocon show consistent peak expression in schizonts. The shift in the expression of these particular genes suggests major differences in host cell remodelling carried out by these two species.

In addition to haemoglobin degradation and host cell remodelling, essentially every biological process contains at least a small fraction of genes whose expression differs between *P. vivax* and *P. falciparum* (Bozdech *et al.*, 2008). These differences may either result in slight temporal shifts or extended activity of an otherwise conserved pathway. Taken together, these data suggest that evolution of *Plasmodium* species involves not only diversification of the primary protein sequences but also variations in gene expression regulation that further define the biological functions of each gene in the *Plasmodium* life cycle. Importantly, there is no evidence of specific clustering of the differentially expressed genes along the *P. falciparum-P. vivax* chromosomes, indicating that the transcriptional diversification between these two species is not associated with any chromosome domain but rather occurs independently at the level of individual genes and their *cis*- and *trans*-regulatory factors.

Transcriptional regulation during the IDC

We have only a limited understanding of molecular mechanisms that control gene expression during the IDC. Based on several recent genome-wide studies the emerging picture implicates three classes of contributing processes that are essential for the IDC transcriptome, including (i) chromatin remodelling and epigenetic regulatory mechanisms, (ii) a set of highly specialized transcription factors, and (iii) global processes of DNA and RNA metabolism such as nucleosomal occupancy at the *Plasmodium* genome, global modulators of transcriptional activity, and mechanisms of RNA stability (Fig. 5.2). Although an in-depth summary of gene expression regulation is provided in Chapter 7, here we focus solely on genome-wide functional genomics analyses of gene expression regulation during the *Plasmodium* IDC.

Epigenetics

There are at least 44 different modifications of the core histones of *Plasmodium* nucleosomes, including a number of canonical chromatin markers that have been associated with transcriptional regulation in other eukaryotes (Miao *et al.*, 2006; Trelle *et al.*, 2009). A substantial number of studies published in recent years show the involvement of histone modifications not just in silencing of surface variant antigens (reviewed in Deitsch *et al.*, 2007) but also affecting the core genes of metabolic and cellular pathways expressed during the IDC. Using chromatin immunoprecipitation linked with microarray detection (ChIP-on-chip) as well as a highly representative method of DNA amplification followed by mass parallel

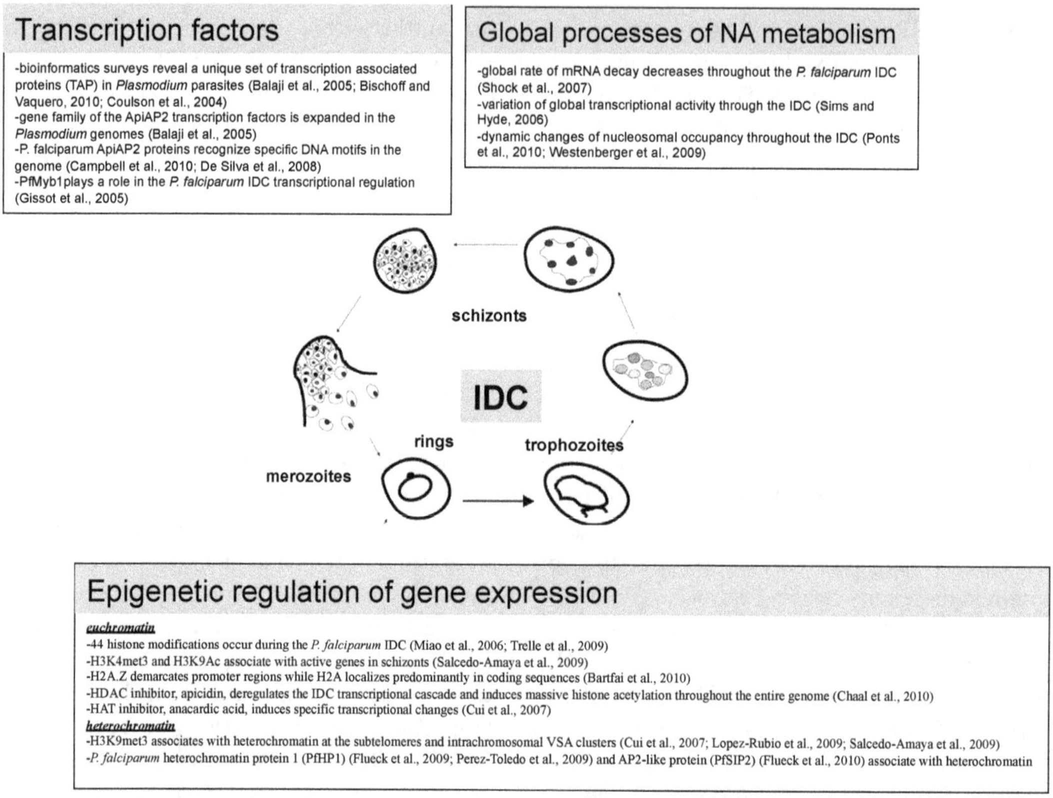

Figure 5.2 Overview of genome-wide analyses of transcriptional regulation associated with the *P. falciparum* IDC. References to the individual studies and major observation are listed for four major sections of the *Plasmodium* life cycle. (The *Plasmodium* life cycle graphic was generated by Melody Munz, MPH.)

sequencing of immunoprecipitated chromatin material (CHIP-Seq), reports over last 5 years have provided evidence that in *Plasmodium*, histone modifications can be broadly grouped into three categories including (i) modifications that show positional changes during the IDC, (ii) modifications that show variable abundance through the IDC, suggesting a role in the progression through these stages of the *Plasmodium* life cycle, and (iii) modifications that are associated with specific positions in the genome but do not change throughout the IDC. Salceo-Amaya and colleagues used CHIP-on-chip to demonstrate that three modifications of histone 3 – acetylation of lysine 9 (H3K9ac) and trimethylation at lysines 4 (H3K4me3) and 9 (H3K9me3) – exhibit a highly dynamic pattern through the IDC (Salcedo-Amaya *et al.*, 2009). While H3K9me3 is associated strictly with the subtelomeric regions – specifically, loci coding for variant surface antigen (VSA) gene families – H3K9ac and H3K4me3 are typical euchromatin markers that are widely distributed across the genome. Interestingly, in the ring stage, both of these euchromatin markers are associated with active as well as inactive genes equally. In the schizonts stage, however, H3K9ac and H3K4me3 are both enriched at the 5′ regions of actively transcribed genes, suggesting a function for these modifications in active transcription. In addition, CHIP-Seq has shown that two isoforms of histone 2 (H2A and H2A.Z) delineate distinct regions in the *P. falciparum* genome; H2A.Z 'demarcates' the intergenic regions while H2A is present predominantly in the coding sequences (Bartfai *et al.*, 2010). Moreover H2A.Z shows much higher accumulation in euchromatin and severe reduction in the heterochromatin domains (such as subtelomeres and intrachromosomal heterochromatin islands) that are characterized by high H3K9me3 content. Indeed, the euchromatin markers (H3K9ac and H3K4me3) co-localize well with H2A.Z. However, unlike the H3 modifications, which show steady abundance increases throughout the IDC, the levels of H2A.Z appears constant throughout all stages of the IDC. This suggests that H2A.Z by itself or in combination with other proteins, functions as a scaffold for recruiting chromatin modifying complexes that facilitate placement of more specific epigenetic markers. Overall the high enrichment of H2A.Z and H3K9ac at transcriptionally active genes suggests that both of these euchromatin markers play an active role in transcriptional activation.

In contrast to ChIP-on-chip results, a CHIP-Seq study failed to find any association between H3K4me3 and transcription (Bartfai *et al.*, 2010). It is possible that H3K4me3 may be involved in transcriptional regulation of only a small fraction of genes that appear statistically enriched in the ChIP-on-chip study, but the highly representative CHIP-Seq results mask this by revealing many additional H3K4me3 instances without link to transcription. More studies that involve rigorous statistics will be needed to resolve this apparent discrepancy.

The importance of epigenetic regulation of gene expression during the IDC was further demonstrated by analyses of the effect of specific inhibitors of chromatin remodelling enzymes on *Plasmodium* transcription. These include the histone deacetylase (HDAC) inhibitor apicidin (Chaal *et al.*, 2010) and the histone acetylase (HAT) (PfGCN5) inhibitor anacardic acid (Cui *et al.*, 2007, 2008), both of which can induce specific transcriptional changes linked to changes in euchromatin markers. In particular, H4K8ac and H4 tetra-acetylation at K5,8,12,16 (H4ac4) show a massive increase throughout the genome of *P. falciparum* cells treated with apicidin (Chaal *et al.*, 2010). Interestingly, there is a statistically significant overlap between the HDAC-induced H4K8ac and H3K4me3 and also between H3K9ac and H4ac4 chromosomal distribution patterns, which suggest the existence of two (possibly interconnected) pathways of histone acetylation. However, these rapid changes in the acetylation patterns correlate only partially with the apicidine-induced transcriptional changes, which suggests that epigenetics-based mechanisms of gene expression involves a wide spectrum of components, many of which remain unknown.

As for euchromatin markers, several recent functional genomics studies brought better understanding of heterochromatin domains in the *Plasmodium* genome and their role in gene silencing. One of the most important heterochromatin markers, H3K9me3, was initially shown to have a distribution pattern in the *P. falciparum* genome that differs dramatically from its euchromatic counterpart H3K9ac (Cui *et al.*, 2007). Using ChIP-on-chip, Lopez-Rubio *et al.* refined this initial finding further and demonstrated that H3K9me3 associates with the clonally variant surfaces antigen (VSA) gene families that are clustered in all subtelomeric regions and several internal regions of chromosomes 4, 6, 7, 8, 12 (Lopez-Rubio *et al.*, 2009). Although H3K9me3 was associated with the six repetitive DNA blocks (TARE 1–6) that are adjacent to the telomere repeats at the extremes of chromosomes, in *P. falciparum*, unlike other organisms, it is not associated with centromeres. Crucially, in the subtelomeric regions H3K9me3 is associated with seven previously defined gene families that include *var*, *rifin*, *stevor*, *Pfmc-2TM*, *clag*, *PfAcs*, and *PHIST*. Members of these families are characterized by highly variable expression that differs between *P. falciparum* strains and isolates, and their mRNA levels are sensitive to external stimuli (see below). Furthermore, many of these families encode surface proteins that contribute to phenotypic plasticity of the blood stage parasites and are essential for immune evasion (Scherf *et al.*, 2008). Based on the association with H3K9me3, five additional gene families could be identified and potentially linked with variable expression and thus phenotypic determination of *P. falciparum* parasites (Lopez-Rubio *et al.*, 2009). In all of these families, H3K9me3 accumulated only at the 5′ end of silenced genes, which strongly suggests a role in silencing. However, the matter was further complicated by the observations that the targeted disruption of PfSir, the main deacetylase of subtelomere-bound histones (Duraisingh *et al.*, 2005), resulted in an overall decrease of H3K9me3 occupancy on only two of the gene families, *var* and *rifin*. Moreover, H3K9me3 does not seem to be associated with three other gene families that are interspersed within the subtelomeric regions and whose members also exhibit differential expression, namely *ETRAMP, FIKK,* and *surfin* (Lopez-Rubio *et al.*, 2009).

In addition to elucidating histone modifications, genome-wide analyses provided many insights into several DNA-binding protein that play a role in chromatin remodelling-based regulation of gene expression. The first such study characterized the biological significance of heterochromatin protein-1 (PfHP1, PFL1005c), which was shown to specifically interact with the H3K9me3 epitope (Perez-Toledo *et al.*, 2009). PfHP1 is a homologue of highly conserved proteins that are implicated in heterochromatin formation in most eukaryotes (Grewal and Jia, 2007). High-resolution ChIP-on-chip analyses revealed an exclusive association of PfHP1 with all previously identified heterochromatin regions of the *P. falciparum* genome including the TARE1–6, and the VSA gene families (Flueck *et al.*, 2009). Interestingly, the immediate boundaries between PfHP1-occupied and PfHP1-free regions delineate syntenic break points in *Plasmodium* chromosomes as well as regions with high occurrence of species-specific indels. Of the 425 genes associated with PfHP1, 95% are *P. falciparum*-specific, including members of gene families involved in host–parasite interactions: *var, rifin, stevor, pfmc-2tm, surfin, pfac,* invasion factors *pfh, clag,* and liver-stage antigen (*lsa1*). Only a few genes conserved across *Plasmodium* species were found to be associated with PfHP1, among them genes involved in gametocytogenesis (*ccp1* and *pf230*), sporozoite functionalities (putative dynein heavy chains *crmp1* and *crmp4*), and PFL1085w (an AP2 transcription factor). The involvement of PfHP1 in heterochromatin formation and gene silencing is evident from its overlap with the chromosomal distribution of H3K9me3 and the fact that episomal overexpression of PfHP1 caused transcriptional down-regulation of at least 78 genes, and no gene up-regulation; 50 of these downregulated genes lie in PfHP1-enriched regions. Only 38 of the 425 PfHP1-bound genes were not associated with H3K9me3. Notwithstanding experimentally induced false positive results (ChIP-on-chip experiments were conducted with episomally overexpressed PfHP1–HA/GFP fusion proteins), this suggests strong link between these two heterochromatin markers; however, there is a distinct possibility that in some loci, PfHP1 is recruited via an H3K9me3-independent mechanism. One such example may be the FIKK protein kinase family that was found to be associated with PfHP1 (Flueck *et al.*, 2009) but not with H3K9met3 (Salcedo-Amaya *et al.*, 2009) (authors' note).

Another heterochromatin marker is PfSIP2, a member of the AP2 transcription factor family. PfSIP2 associated exclusively with the telomeric ends of *P. falciparum* chromosomes, recognizing specific SPE2 DNA motifs that are present in TARE3 and TARE4 upstream of subtelomeric *var* genes, both of which are also located within H3K9me3- and PfHP1-associated regions (Flueck *et al.*, 2009). Although overexpression of PfSIP2 in *P. falciparum* cells had no affect on gene expression, the presence of a substantial proportion of the SPE2 binding sites in the upstream regions of *uspB var* genes links this protein with silencing of this subset of surface antigens. Alternatively, PfSIP2 was proposed to play a role in maintenance and structural organization of telomeres that are closely linked with heterochromatin domains of the *P. falciparum* genome (Flueck *et al.*, 2009). In recent work, Volz *et al.* (2009) identified at least 24 novel *Plasmodium* nuclear proteins, many of which could be linked to chromatin remodelling. In the near future it will be interesting to investigate the significance of these proteins in transcriptional regulation of the *Plasmodium* life cycle.

Transcription factors

Several bioinformatics studies that emerged shortly after the release of the *P. falciparum* genome noted a paucity of transcription-associated proteins (TAPs) that could be discerned in the primary sequence (Balaji *et al.*, 2005; Bischoff and Vaquero, 2010; Coulson *et al.*, 2004). In particular, the lack of specific transcription factors led to the suggestion that during its evolution *Plasmodium* cells shifted the 'responsibility' for controlling gene expression onto other mechanisms such as epigenetic and posttranslational regulation. Nonetheless there is now compelling evidence that the mechanisms of transcriptional regulation in *Plasmodium* are more conventional than initially suggested and that many specific transcription factors may be still 'hidden' in the genome sequence (Bischoff and Vaquero, 2010). Much attention has been recently focused on Apicomplexan AP2 (ApiAP2), a class of transcription factors that are homologous to plant AP2 transcription factors predominantly involved in developmental processes and stress responses (Balaji *et al.*, 2005). The ApiAP2s are characterized by one or more DNA-binding motifs and a significant expansion to at least 27 genes in *P. falciparum* and similarly in other *Plasmodium* spp. (Balaji *et al.*, 2005). Several functional genomics efforts have now attempted to assess the role of ApiAP2 in the IDC transcriptional cascade. One such study utilized a protein binding microarray (PBM) containing all possible 10-mer DNA sequences spread across 44,000 double-stranded DNA 60-mer oligonucleotides, thereby providing a wide spectrum of specific targets as well as positive and negative controls (Berger *et al.*, 2006). This PBM was used to show that two member of the ApiAP2 family (PF14_0633 and PFF0200c) bound to palindromic motifs (TGCATGCA for PF14_0633 and GTGCAC for PFF0200c) similar to binding sites of other eukaryotic transcription factors (De Silva *et al.*, 2008). The binding site motifs identified for these two ApiAP2 can be located in multiple promoter regions of *P. falciparum* genes. Interestingly, the average expression profiles of genes containing the putative PFF0200c binding site correlates well with the transcription profile of *PFF0200c* itself, thereby further strengthening the original hypothesis that PfAP2s play a role in transcriptional regulation of the IDC. As of this writing, the 8-mer binding sites for 20 ApiAP2s have been defined (Campbell *et al.*, 2010). Remarkably, the PBM experiments showed that different AP2 domains from the same protein can have distinct DNA binding specificity and at the same time, an individual domain can bind anywhere from one to five different DNA sequences. This complexity in DNA binding was suggested to broaden the target specificity of the PfAP2 transcription factors in order to maximize their potential as master regulators of transcriptional regulation during the *P. falciparum* life cycle.

Comparing the ApiAP2 binding sites with previous bioinformatics analyses of *P. falciparum* *cis*-acting transcriptional elements revealed that the recognition site (A/GNGGGG(C/A) used by ApiAP2 PF13_0235 is analogous to a previously identified G-box that had been implicated in expression of *Plasmodium* heat shock proteins (Militello *et al.*, 2004). Electrophoretic mobility shift assay (EMSA) confirmed that PF13_0235 is indeed capable of specific binding to the G-box-containing DNA and that at the same time several G-box-containing *hsp* genes share a similar transcriptional profile with *PF13_0235*. Hence this particular PfAP2 is likely involved in the transcriptional regulation of heat shock genes during the IDC. In addition, these studies also established that the GTGCA motif found upstream of many genes involved in invasion, including *msp1* (*pfi1475w*), *msp10* (*pff0995*) and *rhopH3* (*pfi0265c*), is recognized by the DNA-binding domain of ApiAP2 PF10_0075, suggesting a role for that factor in the regulation of this important step in the *Plasmodium* life cycle (Young *et al.*, 2008; Zhou *et al.*, 2005).

In order to define a gene target set for all ApiAP2 factors and thus capture their functional involvement in the regulation of the IDC transcriptional cascade, Campbell and colleagues developed the so-called motif activity profile, which reflects the expression profile of genes regulated by each factor (Campbell *et al.*, 2010). Using this approach each AP2 transcription factor was assigned to a specific interval of the IDC in which its activity is required. Combining this with functional analysis to obtain the final result of this study, it was shown that each AP2 transcription factor regulates expression of genes involved in specific biological function(s) expressed at a particular stage of the IDC. Moreover, while some of the AP2 proteins function as transcriptional activators, others seem to have a suppressive role. The most intriguing example of the latter is presence of DNA recognition binding motifs for PFF0200c, PF14_0633 and PF11_0442 in the *var* gene clusters. It will be intriguing to investigate the role of these three factors in silencing of this important antigenic gene family.

In addition to the ApiAP2 family, several other transcription factors have been linked with transcriptional regulation of the *Plasmodium* life cycle. One study utilized a controversial gene knock-down method based on double-stranded (ds)RNA to analyse the biological function of PfMyb1, the homologue of Myb eukaryotic transcription factors (Gissot *et al.*, 2005). It was shown that dsRNA can mediate reductions of *PfMyb1* expression and by that means can affect mRNA levels of a considerable number of genes, some of whose promoters are associated with PfMyb1 protein. However, the relatively small numbers of genes analysed in this study need to be followed up with a full-scale genome-wide approach to assess the full extent of PfMyb1's role in *P. falciparum*.

Together, these studies provided a solid evidence of the role of specific transfection factors in gene expression during the *Plasmodium* life cycle and also help to derive new experimental approaches for their analyses in this challenging eukaryotic system. In future research, it will be interesting to expand such exploration to all predicted factors of transcriptional regulation (Bischoff and Vaquero, 2010) and potentially uncover new evolutionarily diverse proteins involved in these processes.

Global processes of nucleic acid metabolism

In addition to epigenetics and specific transcription factors, several global processes of nucleic acid metabolism have been shown to shape gene expression characteristics during the IDC. In particular, the global pattern of mRNA decay as well as overall levels of transcriptional activity that show great variation across the IDC both seem to be the major contributor to the steady-state mRNA levels in *Plasmodium* blood stage development. Shock *et al.* (2007) studied the global pattern of mRNA decay by monitoring the steady-state mRNA levels in *P. falciparum* cells treated with a global transcriptional inhibitor actinomycin D at four IDC stages (10, 20, 31 and 44 hpi). They showed that the global rate of mRNA decay decreases dramatically throughout the IDC, with an average mRNA half-life extending progressively from 9.5 min at the ring stage (10 hpi) to 20.5 min at 20 hpi, 49.9 min at 31 hpi, and finally reaching 65 min in the schizont stage (44 hpi). Interestingly, mRNA decay appears to impact all transcripts at each particular stage equally regardless of their functional assignment or phase of expression in the IDC. In other words, at each IDC stage, the vast majority of genes exhibited an mRNA decay pattern that is very similar to the average profile for that particular stage. These results suggest the existence of a global mechanism of mRNA stability throughout the progression of IDC. In particular, there is high mRNA stability in late schizonts that is likely essential for formation of invasive merozoites, in which a carefully regulated 'starter package' of mRNA transcripts is carried over through invasion, and rapidly activated after invasion. On the other hand, the high mRNA turnover at ring stages likely reflects vigorous remodelling of the cells during this fast-developing part of the IDC. K-means clustering of these mRNA decay rates allowed further sub-classification of *P. falciparum* genes, suggesting an added layer of complexity in this global mechanism of mRNA stability. Interestingly, each of the gene clusters exhibited significant functional enrichments that presumably reflects controlled mechanisms of functionally related genes. For example, genes involved in ribosomes, ribonucleoprotein complexes, cytoskeleton organization and biogenesis, and mitochondria exhibit lowest transcript half-life variability across the IDC, which agrees with their crucial function across all developmental stages. On the other hand, genes essential for pathogenesis and antigenic presentation exhibit short average half-lives in the ring and schizont stages but long average half-lives in the trophozoites. This profile may reflect a specific regulatory mechanism that facilitates parasites interactions with its host during the middle stage of IDC. In summary, the variable rate of mRNA stability observed during the IDC is unique amongst eukaryotic organisms and may represent another layer of gene expression regulation that governs the life cycle.

Nuclear run-on experiments provided strong evidence that, concurrent with variation in mRNA stability, there are also variations in the overall transcriptional activity throughout the IDC (Sims *et al.*, 2009). In these studies, it was shown that transcription remains relatively low in the first third of the IDC, rises gradually at 16–18 hpi, and reaches its peak in mid-schizonts (32–40hpi) before dropping towards invasion-stage levels. Interestingly, although the overall transcriptional activity correlated well with the number of nuclei in the parasite cell, this correlation was not linear, which is likely due to variable transcriptional activity in each individual nucleus. While the molecular basis of the global rate of transcription in *P. falciparum* IDC remains to be elucidated, it appears to be a function of the availability of the template DNA and the variable global transcriptional rates per nucleus. Interestingly, the transcriptional rates and mRNA steady-state levels of individual genes showed a strong positive correlation for only 7 out of 18 genes. The poor-to-no correlation of the remaining 11 transcripts was attributed to the opposing effect of their stability, measured by the abovementioned study (Shock *et al.*, 2007). The nuclear run-on experiments also detected anti-sense strain transcription for all the genes assayed in this study. While for many of these genes antisense transcription was very low, in two cases it exceeded the sense strand activity in at least one assayed time point. In these cases, the peak of antisense activity coincided with the lowest level of steady-state mRNA level, which opens an intriguing possibility for the importance of antisense RNA in gene expression control, namely, that it could act via mRNA stability (Sims *et al.*, 2009).

Nucleosome occupancy is another genome-wide factor in *Plasmodium* that contributes to gene expression regulation in the IDC (Ponts *et al.*, 2010; Westenberger *et al.*, 2009). Massively parallel sequencing of nucleosome-associated and nucleosome-free DNA regions revealed a dynamic pattern of nucleosome occupancy throughout the IDC (Ponts *et al.*, 2010). For this purpose micrococcal nuclease (MNase)-mediated purification of mononucleosomes to extract histone-bound DNA (MAINE) as well as formaldehyde-assisted isolation of regulatory elements to extract protein-free DNA (FAIRE) were used to isolate nucleosome-associated and nucleosome-free DNA, respectively. Quantitative analyses of DNA enrichment in both FAIRE and MAINE fractions of *P. falciparum* DNA isolated at 6hr intervals of the 48h IDC revealed that a disproportionately large fraction of the genome is protein-free, especially during the later developmental stages between 18 to 30h post invasion. It was suggested that this part of the genome represents an open form of chromatin that can be easily accessed by DNA-binding proteins and thus subjected to a high level of transcriptional activity. It was estimated that at the schizont stages, the 'accessible' chromatin accounts for up to 80% of the entire genome. On the other hand during the early ring stages (up to 12 hpi) as well as the late schizonts stage (past 36hpi), the *Plasmodium* chromatin seems to be more compact, with the nucleosome-bound fraction approaching 50%. Reconstruction of nucleosome positioning identified 45,000 (covering 26% of the genome) and 89,000 (50%) nucleosomes genome-wide in the late ring (18hpi) and the late schizont (36hpi) stage, respectively (Ponts *et al.*, 2010). Remarkably, there is a considerable preference of nucleosomes for coding regions, where the occupancy was up to 4-fold higher than in the promoter and terminator (3′) regions of *Plasmodium* genes. This supports the hypothesis that nucleosome-free DNA regions represent a more accessible substrate for *trans*-regulatory element of gene expression. Although nucleosome occupancy does not correlate precisely with transcriptional activity of individual genes, several parallels can be drawn. First the massive nucleosome depletion starting at 18 hpi and persisting until 30hpi correlates well with nuclear run-on experiments that show a gradual increase in transcriptional activity in the late ring stage and a peak at the mid-schizont stage (Sims *et al.*, 2009). Secondly, the interval of nucleosomal depletion between 18 to 24 hpi correlates with the start of DNA replication and a dramatic increase in the number of genes with maximum gains in mRNA levels, as measured by transcriptomic analyses (Bader and Hogue, 2003; Bozdech *et al.*, 2003a; Le Roch *et al.*, 2003). Finally, rapid repackaging of genomic DNA in the late schizont stage (36hpi) correlates with a dramatic decrease in transcriptional activity that is presumably compensated by the extended half-lives of the transcripts expressed at this stage (see above). The high-throughput sequencing results for nucleosomal occupancy were recently supported by ChIP-on-chip-based studies that confirmed the under-representation of nucleosomes in intergenic regions, and furthermore demonstrated a greater level of fluctuation in nucleosome occupancy in the subtelomeric gene families *var* and *rifin* (Westenberger *et al.*, 2009). Similarly, no overall correlation between nucleosome occupancy and steady-state mRNA levels was demonstrated in these ChIP-on-chip studies.

The IDC proteome

Similar to the transcriptome, much effort has been put into defining the proteome of the *Plasmodium* IDC. Two early studies published along with the *P. falciparum* genome in 2002 showed that the asexual blood stages express a wide array of proteins involved in many basic as well as specialized *Plasmodium* functions (Florens *et al.*, 2002; Lasonder *et al.*, 2002). In the first study, Florens *et al.* (2002) detected 839 and 1036 distinct proteins in the merozoite and trophozoite stages, respectively (Florens *et al.*, 2002), with proteins involved in host/parasite interactions, such as invasion and antigen presentation, being among the most abundant. This was confirmed by a concurrent study, in which 714 proteins were detected in the asexual blood stages, most of which were assigned to host/parasite interaction processes (Lasonder *et al.*, 2002). These studies provided the first insights into the proteome of the *Plasmodium* life cycle, showing that each developmental stage is characterized by a specific pattern of proteins. However, the outstanding question about the relationship between transcription and translation during the *Plasmodium* IDC remained unresolved at that time.

Transcription vs. translation

In a first attempt to address this issue Nirmalan *et al.* (2004a) developed a two-dimensional (2D) gel electrophoresis method using metabolically labelled parasite protein lysates in order quantify proteins throughout the IDC. Focusing on six proteins involved in glycolysis and other basic cellular functions, it was shown that there is a consistent delay between peak mRNA and protein abundance. In a subsequent study, Le Roch *et al.* (2004) correlated mRNA transcriptional profiles with semiquantitative measurements of protein abundance using the number of MS/MS spectra identified per protein (spectra count). This was successful for as many as 2584 proteins that were detected in at least one of the seven developmental stages of the life cycle in the human host (merozoites, rings, trophozoite, schizonts, gametocyte, gametes and sporozoites). Overall, it was shown that the mRNA profile in each developmental stage correlates better with the proteome profile of the subsequent stage (e.g. the ring mRNA profile correlates well with the trophozoite proteome and the trophozoite mRNA profile correlates well with the schizont proteome). This suggested a complex relationship between transcriptional and posttranscriptional regulation, with the majority of proteins reaching their peak abundance after their corresponding transcripts.

While these initial studies focused on the transcriptional and proteomic relationship of broad developmental stages, in order to gain a better understanding of these processes, the proteomic analyses had to be carried out for much narrower time windows. Quantitative 2D-differential gel electrophoresis (DIGE) that followed the *Plasmodium* proteome during a narrow window of schizont development between 34 and 46 hpi revealed that as many as 278 proteins exhibits a significant change in abundance, i.e. at least 1.4-fold (Fig. 5.3) (Foth *et al.*, 2008). For approximately 15% of the protein spots, there were significant discrepancies between mRNA and protein abundance. Although most of these cases simply represent a lag between mRNA and protein peak, there were some examples where transcripts and protein profiles were anticorrelated. To investigate this further, Foth B. *et al.* carried out experiments in which samples from every two hours of the 48 h IDC where analysed by both microarray and 2D-DIGE (Foth *et al.*, 2011). A total of 1183 protein/isoforms were found to change their abundance, with a median 5.4-fold change observed for the complete IDC set; a median 5.6-fold change was observed for all mRNA transcripts. Similar to mRNA, proteins also exhibit single peak abundance profiles, with each protein timed to a specific IDC stage. In contrast to mRNA, proteins display a more discontinuous cascade, with the majority of proteins peaking in either early ring stage or mid-schizont stage. Of the total 1183 protein spots, 363 were identified by MALDI-TOF/TOF and shown to correspond to 125 *P. falciparum* genes, representing several major functional groups. Given the fact that the mass spectroscopy-based identifications were prioritized by 2D-gel electrophoresis spot intensity, the identified proteins likely represent the most abundant functionalities in the parasites cytoplasm. Importantly, the majority of the proteins exhibit 6–16 h (median 11 h) delay in their peak abundance compared to their transcripts. Further analysis of this data showed good correlation between the steady-state mRNA levels and mathematical derivatives of the protein profiles that is a function of the slope and therefore the increase or decrease in abundance. This leads to the hypothesis that protein synthesis can be related directly to the steady-state mRNA level and thus, the translational rates may represent the fundamental component determining protein levels. Further mathematical modelling confirmed this hypothesis and suggested that the majority of proteins expressed during the IDC are not (necessarily) subjected to complex posttranslational regulation; instead their abundance is governed mainly by the rate of translation (directly proportional to the mRNA steady-state level) and protein turnover (Foth *et al.*, 2011).

Posttranslational modifications

In the initial small-scale 2-D gel analyses, 10 out of the 28 (Nirmalan *et al.*, 2004b) and 24 out of the 54 (Foth *et al.*, 2008) proteins identified were found to be present as two or more isoforms. Furthermore, the 363 protein spots identified in the large-scale 2-D gel proteomics analysis only represented 125 *P. falciparum* proteins, which correspond to an average of 2.9 discernible protein isoforms per gene (Foth *et al.*, 2011). Altogether, this suggests that extensive posttranslational modifications (PTMs) characterize the proteome of the IDC and possibly the entire life cycle. In many cases, the individual isoforms exhibit abundance profiles that are distinct from those of the full-length nascent polypeptides. Hence PTMs may help further diversify the *Plasmodium* proteome during the IDC and may also be essential for the cellular targeting of individual proteins at the different developmental stages (Bhowmick *et al.*, 2009). Over the last two decades a wide variety of PTMs have been demonstrated in *P. falciparum* including phosphorylation, acetylation, methylation, ubiquitination, glycosylation, lipidation, as well as proteolytic cleavage (reviewed in Chung *et al.*, 2009). Phosphorylation is one of the most abundant PTMs and has been implicated in various functions, including cell growth, differentiation, division, as well as regulation of metabolic pathways, enzyme activities, cytoskeleton organization, and chromatin remodelling. Using 2-D gel technology as well as LC/MS/MS analysis, it was shown that during the IDC, close to 170 proteins are phosphorylated at a serine or tyrosine residue (Wu *et al.*, 2009). While some phosphoproteins appear in specific developmental stages, others are present constitutively. Besides the many previously known phosphoproteins – including adenylate kinase 2, calcycline proteins and phosphoethanolamine-N-methyltransferase – this analyses revealed strong phosphorylation patterns for several protein families that have *Plasmodium*-specific functions, such as the CLAG family and a large group of heat shock proteins (HSPs). Another important PTM is proteolytic processing, which is known to play a major role in regulation of several crucial *Plasmodium* functions including haemoglobin degradation, merozoite egress, and invasion (Chung *et al.*, 2009). Of the 125 proteins identified by the large scale 2-D survey, 46 were detected as truncated (Foth *et al.*, 2011). Indeed, several haemoglobinases, including falcipains 2A and 2B and plasmepsins I, II, III (HAP) and IV, were detected at lower than expected molecular weights, which probably represent their proteolytically activated forms. In addition, several proteins known to be involved in invasion, glycolysis, and regulation of gene expression were found to be extensively processed. At this stage, hardly anything is known about the biological significance of these post-translationally modified polypetides; however, these results open a new door for functional analyses of genes and proteins and their role in the malaria parasite's physiology.

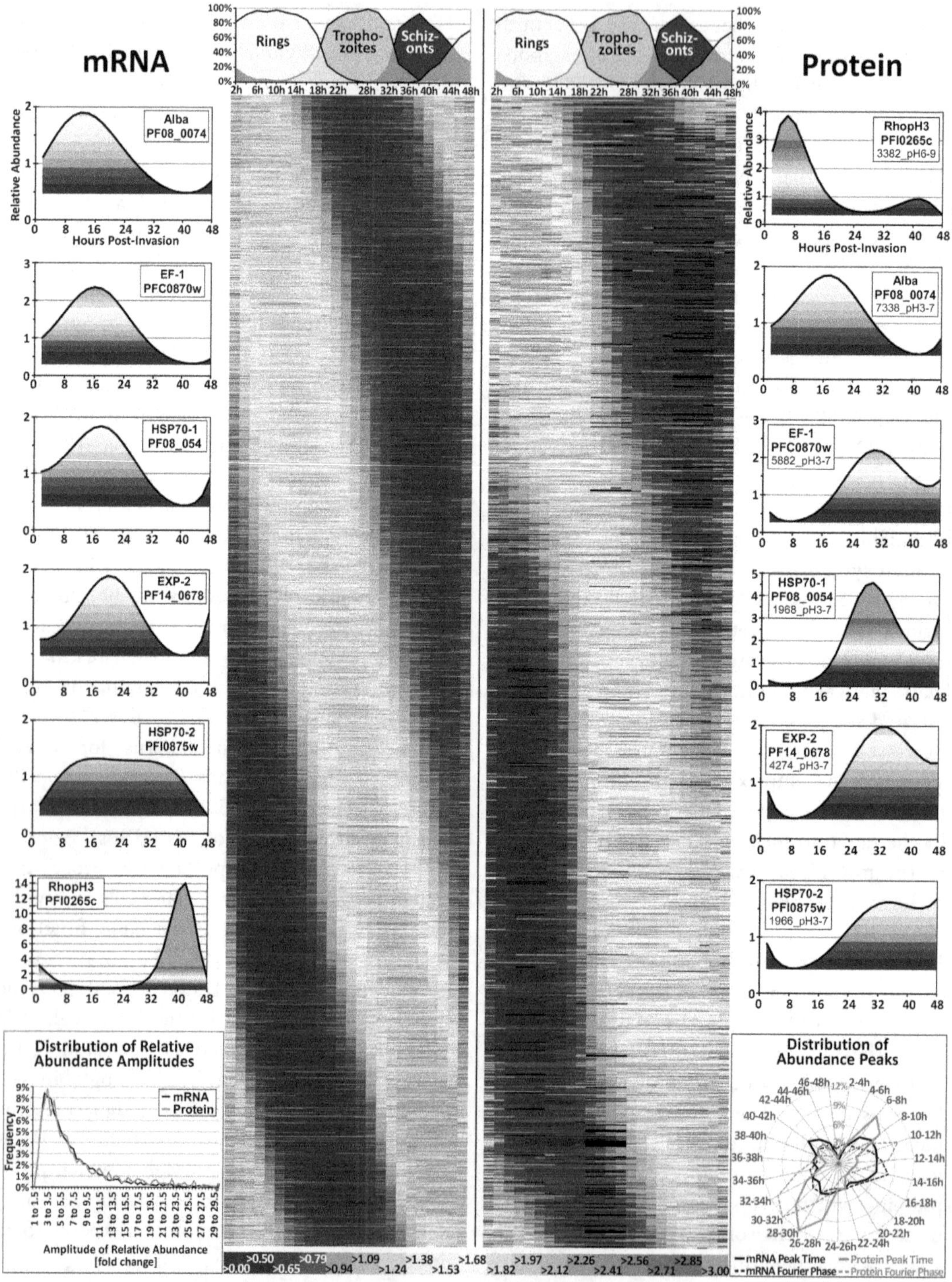

Figure 5.3 Overview of the transcript and protein abundance profiles. The left half of the central main figure depicts 4670 transcript abundance profiles whereas the right half shows 1183 protein abundance profiles. In these central panels each line represents one distinct gene (for the transcripts) on the left or one protein isoform on the right, while the 24 columns indicate the timepoint (in hpi = hours post-invasion) of the sample. Importantly, in this figure the lines representing transcripts (on the left) and proteins (on the right) do not correspond to one another, and the abundance profiles were sorted according to their Fourier phase. The profiles shown derive from log2-transformed and mean-centred quantitative data that were polynomial-fitted

and then converted back to a non-logarithmic scale. The colour scale indicating relative transcript or protein abundance is provided at the bottom of the figure, while the progression of the parasite population through the IDC is indicated at the top. Six examples of transcript abundance profiles and selected corresponding protein abundance profiles are shown at the sides. Lower left panel: the distribution of relative abundance amplitudes. Lower right panel: the distribution over time of abundance peaks as calculated by Fourier transformation and as evidenced by the major peak of the polynomial fit. [This figure is adapted from Foth *et al.* (2011) with permission.]

Transcriptome and proteome of extra-erythrocytic stages

Liver stages

The availability of new genomics and proteomics tools were also instrumental for in-depth analyses of the extra-erythrocytic stages that until recently had remained elusive to large scale molecular-based research. Due to its relevance to vaccine development (Alonso *et al.*, 2005; Hoffman and Doolan, 2000; Luke and Hoffman, 2003), much effort went into the establishment of the transcriptomes during the parasite liver stage development. The main motivation for understanding the basic biology of the hepatic stage is driven by the aim to develop genetically attenuated parasite lines that can develop liver stage infection but fail to proceed to the blood stages. Such attenuated parasites have the potential to provide complete protection to malaria infections, making this a powerful vaccination strategy (Hoffman *et al.*, 2009; Vaughan *et al.*, 2009b). At the same time, an understanding of liver stage development will be crucial for developing new drug therapies, particularly to target the recrudescence of *P. vivax* malaria from latent liver stage hypnozoites (Baird and Hoffman, 2004).

A combined transcriptome and proteome survey of three time points of the hepatocytic development of *P. yoelii* parasites including liver-stage schizont [24 hpi, i.e. small early schizonts after two to four rounds of nuclear division (LS24)]; LS40 (40 h post infection, i.e. large late schizont after approximately 13 rounds of nuclear division); and LS50 (50 h post infection, i.e. liver stage schizonts that have commenced merozoite formation) were studied using a genome-wide DNA microarray and LC MS/MS (Tarun *et al.*, 2008). Comparison of the transcriptome of the three liver stages with each other, as well as to two mosquito stages (midgut oocyst and salivary gland sporozoites) and to the erythrocytic stages, showed that the transcriptomes of the three liver stages are more similar to each other than to any other non-liver stage. Moreover, the transcriptomes of LS40 and LS50 were more similar to each other than to LS24. This likely reflects the fact that different host environments require a specific parasite adaption that is at least partly mediated by specific gene expression. More than 1000 genes were found to be transcriptionally active, with approximately 500 genes being expressed in at least two of the studied hepatic stages and at least 400 expressed consistently throughout intrahepatic development. A concurrent proteomic survey of the same hepatic stages identified 712 proteins, of which 422 show identical mRNA and protein abundance profiles and an additional 246 correspond to genes that are constitutively transcribed throughput the entire life cycle (Tarun *et al.*, 2008). This indicates that, similar to the IDC, gene expression in the liver stage is tightly regulated. Functional analyses of the transcriptome and proteome data revealed that all liver stages are characterized by strong expression of genes associated with protein synthesis (ribosomal proteins, translational initiation, elongation, protein translocation), transcription (including RNA polymerase components), protein turnover (proteosome subunits), and factors involved in protein folding (heat shock proteins). The metabolic pathways transcribed during the liver stages include those involved in redox metabolism, the mitochondrial TCA cycle and electron transport system, and the fatty acid synthesis II (FASII) pathway. Overall, liver stage development the *Plasmodium* parasites appears highly metabolically active and represents perhaps the fastest-growing developmental stage in the context of the entire life cycle, synthesizing large amounts of protein and fatty acids required to produce tens of thousands of exoerythrocytic merozoites.

In contrast to asexual (blood stage) development but similar to gametocytogenesis, liver stage development is characterized by high mitochondrial activity (see below), which suggests a specialized role for this organelle as a source of energy and metabolites during this rapidly growing parasite form. It has been suggested that the metabolites of primaquine, the only antimalarial effective against *P. vivax* relapses from liver hypnozoites (Baird and Hoffman, 2004), may involve inhibition of mitochondrial function (Wiesner *et al.*, 2003). In addition to mitochondrial functions, the liver stages exhibit high expression levels of the FASII pathway that is normally found only in plants, prokaryotes, and Archaea. This suggests that specific inhibitors of this pathway may serve as new drugs against the liver stage parasite. In addition, 93 and 26 out of the 712 identified proteins contain a signal peptide and a transmembrane domain, respectively, including two previously known secreted proteins Hep17 and UIS4. These proteins are likely transported to the surface of the cell and thus represent potential antigenic determinants of the liver stages. Microscopic analysis confirmed liver stage parasite surface expression of one of these newly identified proteins, PY04499 (Tarun *et al.*, 2008). Moreover, four of the identified proteins (PY0455, PY04544, PY03459 and PY10566) were also found to be expressed in sporozoites of *P. falciparum*, and to stimulate immune cells obtained from volunteers immunized by attenuated sporozoites. These proteins provide excellent new candidates for vaccine development against the pre-erythrocytic stages of malaria parasites. The FASII pathway that is highly expressed in the liver stages was suggested to be another suitable target for development of attenuated sporozoites that could be produced by gene inactivation of some of their factors (Vaughan *et al.*, 2009a; Yu *et al.*, 2008).

Sporozoites

Transcriptional analysis of *P. yoelii* oocyst sporozoites (ooSpz – collected 10 days after infection) and sporozoites isolated from salivary glands (sgSpz – collected 15 days after infection) were carried out in order to characterize molecular processes associated with mosquito stage development. These studies identified 124 and 47 unique genes up-regulated in sgSpz and ooSpz, respectively (Mikolajczak *et al.*, 2008). Overall these results were consistent with a previous microarray analysis of *P. yoelii* salivary gland sporozoites (Zhou *et al.*, 2008) which showed that 87% sgSpz genes and 77% ooSpz genes exhibit up-regulation in the equivalent developmental stage. Of the newly identified genes, 31 sgSpz- and 15 ooSpz-specific genes encode proteins that carry a signal peptide motif or a transmembrane domain and thus likely take part in liver cell and salivary gland invasion, respectively. The importance of one of the identified gene, UOS3 (PY04986), during sporozoite development was clearly shown by targeted gene deletion. While parasites lacking this gene were able to develop normally during the erythrocytic stage and also produced normal oocytes, in the mosquito the resulting midgut sporozoites failed to invade salivary glands. For that reason the infected mosquitoes were not able to transmit the infection into the mammalian host. Intracellular localization studies showed that UOS3 localizes at the apical end of both the oocyst and midgut-derived sporozoites migrating to the haemolymph. Overall, this study demonstrated the power of transcriptional profiling for function predictions of proteins with unknown function in the *Plasmodium* parasites, providing a set of new genes/proteins whose role in sporozoite formation and invasion could be studied further.

To complement the transcriptional study, Lasonder and colleagues completed a proteomics survey of *P. falciparum* oocysts, oocyst-derived sporozoites (ODS), and salivary glands sporozoites (SGS), using nano-liquid chromatography tandem mass spectroscopy (nLC MS/MS) (Lasonder *et al.*, 2008). A total of 728 non-redundant *P. falciparum* proteins were identified, distributed over the three stages as 127, 450 and 477 proteins, respectively. Comparison of this protein set with the previously identified proteome of the blood stages resulted in the identification of 250 *Plasmodium* proteins that are specific to the mosquito stages. Although there was no statistical enrichment of any particular biological function amongst the 250 mosquito proteins, the presence of 15 out of 23 previously verified mosquito stage proteins verified the quality of the analysis. In particular, proteins with multiple roles during sporozoite maturation, such as the circumsporozoite protein (CSP) and TRAP, were identified in all three stages, while two proteins previously linked with hepatocyte invasion, sporozoite-microneme protein essential for cell transversal 1 and 2 (SPECT1 and SPECT2), were identified exclusively in the SGS. In addition, a number of factors involved in actomyosin motor complex, including TRAP, myosinA, MyoA, tail domain interacting protein (MTIP), actin, and aldolase, were readily distinguished amongst the mosquito stage proteins. Moreover, a large number of proteins were found mainly in SGS and were absent or expressed at low levels in ODS. This proteome change most likely reflects the conversion of ODS, which have generally low levels of infectivity to the mammalian host, to SGS that are highly infective. In particular, SGS show increased expression of many metabolic pathways such as glycolysis, NADPH synthesis, and the TCA cycle. Along with this metabolic shift, SGS are characterized by enhanced protein synthesis, with a number of tRNA ligases, ribosomal proteins, translation elongation factors, and TCP chaperoning complex proteins that are either exclusively detected in SGS or are represented in the SGS proteome by substantially more peptides (determined by the emPAI peptide counting method). Further functional analysis of eight of the 250 mosquito stage-specific proteins using targeted gene disruption of their functional homologues in *P. berghei* showed that two of the proteins were essential for blood stage development, while a further three (PF11_0528, PF14_0074, and PFF1195c) had no detectable effect on the parasite growth. The remaining three proteins appear to play crucial roles in mosquito stage development; PF14_0435 is essential for oocyst development, PFD0425w is essential for egress of the sporozoites from the oocyst, and MAL8P1.66 is important for infection of the mammalian hepatocyte host cells. Hence, similarly to the transcriptome analyses (above), this proteomic survey uncovered several crucial factors in the pre-hepatocytic stages, and could be utilized for new malaria intervention strategies that are focusing on these parasite forms.

Gametocytes and sexual development

Until recently, gametocytogenesis and sexual development of *Plasmodium* parasites remained among the least characterized parts of the life cycle. However, due to the widely recognized potential for transmission-blocking vaccines (Kaslow, 1997; Stowers and Carter, 2001), these stages have now attracted the attention of functional genomics and proteomics studies. The initial transcriptome analysis of the *P. falciparum* life cycle included sexual stages and identified 152 genes as predominantly and 218 genes as exclusively expressed in gametocytes (Le Roch *et al.*, 2003). While some of these represented previously known genes, approximately 80% were classified as hypothetical. Similarly, the early proteomics surveys identified 1147 proteins in gametocytes, of which 376 were exclusive to this stage (Florens *et al.*, 2002; Lasonder *et al.*, 2002). Among these there is a high overrepresentation of genes involved in RNA metabolism, ribosomal structure, cell cycle/DNA processing, and energy metabolism. In addition, a wide spectrum of surface antigens, many of which are also expressed at other stages of the life cycle, were detected, including a narrow *rifin* and *var* gene repertoire.

To further explore the unique properties of the sexual stages, Young J.A. *et al.* (2005) characterized the genome-wide mRNA abundance profile of a fourteen-day time course of gametocytogenesis from stage I to stage V (Young *et al.*, 2005). On average, 3410 genes were expressed at any given time point, including 15 well known genes such as *pfg27/25*, *pfs16*, *alpha tubulin ii* and *pfs48/45* that were expressed in stage I-III, and *pfs25*,*pfs28*, *pf77*,and *dmc1-like protein* expressed in stage IV-V gametocytes. Using the ontology-based pattern identification algorithm (OPI) that utilizes the principle of guilt by association (GBA) (see Functional genomics in *Plasmodium* systems biology), 246 genes with the highest likelihood of being involved in the sexual development process were classified. Remarkably, 75% of these are categorized as hypothetical, which is in line with the unique character of *Plasmodium* sexual development. In addition, the expression of several metabolic enzymes, including factors of the type II fatty acid, haem biosynthesis, and mitochondrial

TCA pathways, indicates that metabolic requirements of the gametocytes are similar to the liver stage but distinct from the asexual blood stages (Young *et al.*, 2005). Analysis of the upstream regulatory sequences of the 246 genes predicted to be involved in sexual development identified a putative *cis*-regulatory element represented by a palindromic sequence TGTANT-NTACA in 65 genes. Although this DNA motif remains to be further validated, its presence in the promoter of *pfs25* was shown to be important for transcriptional activity (Dechering *et al.*, 1999). Comparison of the transcriptomes of a gametocyte-producing strain (3D7) and a strain defective in gametocyte production (F12) showed narrow but statistically significant differences in mRNA profiles during the very early stages of gametocyte development (Silvestrini *et al.*, 2005). As expected, two previously characterized gametocyte genes *pfg27* and *pfs16* exhibited marked decrease in their expression in the F12 clone. Together with these, 122 genes were classified using gap statistical analysis as early gametocyte-specific, of which only 32% were functionally annotated. This group of genes included diacylglycerol kinase (PFI1485c) and a putative cAMP-specific 3′,5′-cyclic phosphodiesterase, both of which are involved in cyclic AMP-dependent signalling that was previously implicated in gametocytogenesis commitment (Inselburg, 1983). In addition, three putative transcription factors (PF11_0477, MAL7P1.86, PFB02190c), and several factors of protein phosphorylation and actomyosin motility were present in this early gametocyte cluster, which links these functionalities with sexual commitment and differentiation. Based on this dataset, two novel proteins (PFL0795c and PF10_0164) were shown to be specifically expressed in stage I and II of gametocytogenesis and termed *P. falciparum* protein of early gametocytes *Pfpeg3* and *Pfpeg4*, respectively (Silvestrini *et al.*, 2005).

The gametocyte-deficient F12 strain also served as a negative control for analyses of the early gametocyte proteome. Here, transgenic parasites that expressed GFP under the control of the *pfg27* early gametocyte promoter were used to generate a high purity (≥95%) gametocyte population by fluorescence-activated cell sorting (FACS) (Silvestrini *et al.*, 2010). The proteomes of F12 trophozoites (devoid of sexual stages), 3D7-*pfg27gfp* purified stage I/early stage II gametocytes, and stage V gametocytes were analysed by nLC MS/MS. In total, this analysis identified 2400 non-redundant *P. falciparum* proteins across these three stages: 1345 in trophozoites, 1427 in early (stage I/II) gametocytes and 2031 in mature (stage V) gametocytes. More than half (1459) of these proteins were detected in more than one stage and 1090 were found in all three stages, indicating significant overlaps between the proteomes throughout asexual and sexual development. To fully understand the protein expression pattern during gametocytogenesis, quantitative differences in protein abundance (based on emPAI peptide counting) were utilized to classify 369, 251, and 1069 proteins with peak abundance in trophozoites, early, and late gametocytes, respectively. Using this classification it was shown that early gametocytes express a large number of exported proteins, molecular chaperones, and factors involved in host cytoplasm remodelling. Overall there were 63 putative exported proteins expressed in stage I/II gametocytes, of which 26 were either unique or highly enriched in this stage. This set of *P. falciparum* gametocyte exported proteins (PfGEXPs) included several, such as Pfg14.774, Pfg14.748, and PfRex-3, that had previously been shown to be exported to the host cell cytoplasm during gametocytogenesis. On the other hand, many protein factors responsible for export and processing of polymorphic parasite adhesin PfEMP1 were not detected amongst the PfGEXPs, which suggests significant mechanistic and functional differences between gametocyte and trophozoite exportomes. Overall, there was a 2.2-fold overrepresentation of exported proteins in the early gametocyte proteome compared to the total proteome, but intriguingly, a 3.2-fold depletion of these proteins in the proteome of the mature gametocytes. Overall this indicates that protein export is highly active in early gametocyte development but gradually decays with gametocyte maturation. Two possible explanations have been put forth (Silvestrini *et al.*, 2010): first, the exported proteins facilitate the modifications that are necessary for longevity of the PV during the uniquely long gametocytogenesis of *P. falciparum* compared to other *Plasmodium* species. This is supported by the fact that only four out of the 26 PfGEXPs were conserved in *P. berghei*, in which gametocytogenesis takes just slightly over 24 h. Alternatively, the exported proteins might be involved in a new mechanism of cytoadherence distinct from the knob-dependent cytoadherence that characterizes the asexual stages (Day *et al.*, 1998; Sinden, 1982). The decline in PfGEXPs throughout sexual development is in line with the lack of sequestration reported for stage IV and V gametocytes (Rogers *et al.*, 2000).

Stage V gametocytes circulate in the blood in a state of developmental arrest until they are triggered to transform into male or female gametes in the mosquito gut. While male gametes carry out multiple rounds of DNA replication, mitosis, and finally gamete production via exflagellation, female gametes prepare for meiosis and postzygote growth in the midgut. In order to understand the molecular bases for these two distinct sets of processes, and regulatory mechanisms that controls them, it was important to be able to isolate distinct populations of male and female gametocytes. For this purpose, pure populations of *P. berghei* gametocyte were generated using FACS in which GFP was expressed under the control of sex-specific promoters of *a-tubulin II* for male and elongation factor 1*a* for female gametocytes (Khan *et al.*, 2005). Proteomic comparison of the different stages using LC MS/MS allowed the identification of a gametocyte-specific set of proteins, of which 353 were found in both stages while 305 and 170 proteins were specific for male and female gametocytes, respectively (Khan *et al.*, 2005). As expected, there was a marked bias for ribosomal proteins and mitochondrial factors in the distribution of housekeeping genes in female gametocytes. In contrast, proteins expressed in the apicoplast, which is believed to be closely associated with mitochondria, did not show any enrichment in female gametocytes, suggesting a minor role for this organelle in their development. Male gametocytes were characterized by an overrepresentation of DNA replication factors and proteins associated with axoneme. The overall gametocyte proteome also contains 60 protein kinases and sixteen protein phosphatases, many of which are expressed exclusively during the sexual stages. Of these, two could be classified as female- and six as male-specific with a potential role in sex determination. This was confirmed for two of these protein candidates, a female-specific putative NIMA-related kinase NEK4 (PB 001094.00.0; *P. falciparum* orthologue of MAL7P1.100) and a male-specific putative mitogen-activated protein kinase (MAP2) (PB000659.00.0; *P. falciparum* orthologue of PF11_0147). Deletion of either gene resulted in severe disruption of sexual stage development, with the MAP2 deletion causing a defect in male gametocyte production while the NEK4-deletion results in the developmental arrest of zygotes at 16–18 h after fertilization (Khan *et al.*, 2005). In summary, this proteomic study not only provided a comprehensive list of gametocyte-specific proteins and the first list of sex-specific factors, but also set a precedent for follow-up functional analyses in which crucial regulatory proteins controlling the progress of the *Plasmodium* life cycle can be identified and characterized.

Transcriptomic and proteomic profiling of growth perturbations

Transcriptional responses of *P. falciparum* to growth perturbation

Transcriptional responses to growth perturbations have been shown to be highly informative about regulatory mechanisms of gene expression as well as gene function in model organisms (Hughes *et al.*, 2000; MacCarthy *et al.*, 2005). Functional analyses of such data are particularly useful for annotation of novel uncharacterized genes via the guilt by association (GBA) principle but also for understanding the mode of action of chemotherapeutic agents and other (lead) compounds in drug development (Hughes *et al.*, 2000). Perturbation analyses of *Plasmodium* parasites reported over the last 7 years, however, created a dual view of transcriptional regulation in this pathogenic organism. First, a number of growth perturbations, including treatments with several important antimalarial drugs, induced minimal transcriptional responses, which lead to the proposition that *Plasmodium* transcriptional regulation is 'hard-wired', i.e. without the ability to response to external stimuli. However, there were other studies that detected specific high amplitude transcriptional changes in *P. falciparum* exposed to other types of external factors (including artemisinin treatments), indicating that some transcription-linked signalling pathways were retained.

The hard-wired model of *Plasmodium* transcription

The hard-wired model of transcriptional regulations was first hinted at in IDC transcriptome analyses of three *P. falciparum* strains (HB3, DD2, and 3D7) showing identical expression profiles for essentially all genes despite the fact that they represented independent culturing experiments of phenotypically different parasites (Llinas *et al.*, 2006). Minimal differences in the genome-wide transcriptional profile were also observed between *P. falciparum* cells grown in culture media supplemented with human serum or lipid-rich bovine serum albumin fraction (Albumax I and II) (Singh *et al.*, 2007). Finally, functional genomic analyses of the effect of chloroquine on *P. falciparum* failed to identify any specific high amplitude changes in mRNA abundance (Gunasekera *et al.*, 2007a). In that study, parasite cultures were treated with 120 and 400 nM chloroquine for 30 min and 6 h and the transcriptional responses were analysed using a high resolution Affymetrix microarray. Approximately 600 genes showed statistically significant differences in transcription; however, these differences were low amplitude (<2-fold) and not reproducible between two replicate experiments, possibly due to different stage profiles of the starting cultures (*authors note*)Although 36 genes showed consistent changes, none of these could be directly linked to chloroquine's mode of action. Altogether it was concluded that no signature response to chloroquine exists in *Plasmodium* cells, which is highly surprising given the strong toxic effects of this antimalarial drug (Gunasekera *et al.*, 2007a). Nonetheless, and somewhat surprisingly, the detected chloroquine-induced (low amplitude) transcriptional differences are linked by a putative regulatory mechanism involving a specific DNA motif (GAGAGAA). This motif is enriched in the promoters of the genes that were affected by chloroquine, and it is able to drive transcription in transient transfection experiments (Gunasekera *et al.*, 2007b).

A genetically hard-wired metabolic transcriptome in *P. falciparum* was again suggested by Ganesan *et al.*, who reported unexpectedly low transcriptional responses to antifolate drugs (Ganesan *et al.*, 2008). In these studies, *P. falciparum* parasites were treated with 10 nM WR 99210, a potent DHFR inhibitor that renders all parasites in the culture nonviable after 12 h. Microarray analysis of parasites treated for up to 24 h did not detect a single gene that was reproducibly upregulated by more than 2-fold. In addition, the dihydrofolate reductase-thymidylate synthetase (*dhfr-ts*) gene, the direct target of WR99210, or other members of the folate and pyrimidine biosynthesis pathway, were virtually unresponsive to this treatment. Nonetheless, numerous death-related mRNA changes with very low amplitude (termed 'RNA whispers') were detected at various stages of the 24 h WR99210 treatment. These subtle changes were seen for mRNAs for ribonucleotide reductase, DNA repair factors, and one nuclease, and may represent relevant downstream responses following DHFR inhibition. In addition, several signalling molecules such as GDP binding proteins, calcium-independent protein kinases, proteases, and proteasome subunits were recognized amongst these RNA whispers. Due to the low amplitudes of these changes, which are only slightly above statistical significance, it is difficult to determine their impact on the metabolism of parasites under drug pressure. Low amplitude but reproducible transcriptional changes were also observed in *P. falciparum* parasites treated with purfalcamine, an inhibitor of *P. falciparum* calcium-dependent protein kinase (PfCDPK1) (Kato *et al.*, 2008). Several hundred genes that show less than two-fold changes in mRNA abundance were reproducibly detected in multiple treatment experiments, with a high statistical enrichment of genes involved in invasion ($\log_{10} P \sim -10.27$ to -19.55) as well as several trophozoite-specific functions ($\log_{10} P \sim -2.26$). Again, the subtle character of these transcriptional changes sheds doubt on their biological relevance.

Taken together, the low amplitude transcriptional responses detected in these studies are in sharp contrast to transcriptional responses observed in many model organism such as yeast that readily exhibit mRNA changes by several orders of magnitude (Oshiro *et al.*, 2002). Nonetheless, many of the small changes observed in *P. falciparum* are statistically significant and reproducible and may ultimately be biologically relevant. More studies will be needed to understand whether these subtle responses have an impact on parasite physiology or whether they represent non-functional 'whispers'.

Specific (and non-specific) transcriptional responses to external perturbations

However, not all growth perturbations go unnoticed by *Plasmodium* transcriptional regulation. The bisthiazolium compound T4 – a choline analogue with a strong potency against malaria parasites – elicited dramatic transcriptional changes, with 137 genes showing >2-fold differential expression after 24h treatments, and this response increased to 1140 and 1825 genes after 30 and 36h of treatment, respectively (Le Roch *et al.*, 2008). There was no particular overrepresentation of genes involved in lipid metabolism, the putative target of compound T4, but instead these transcriptional changes involved genes of sexual development, fatty acid biosynthesis in the apicoplast, microtubule cytoskeleton, and several generic cellular pathways including glycolysis, protein metabolism, RNA processing, cell cycle and DNA replication. These transcriptional profiles are consistent with a developmental arrest and induction of gametocytes and thus may represent a nonspecific stress response of *Plasmodium* (Le Roch *et al.*, 2008). Similarly, a lethal dose of artesunate, one of the important derivate of artemisinin, induced dramatic transcriptional responses involving 398 (8.5%) genes with 244 over-expressed and 154 under-expressed (Natalang *et al.*, 2008). Most of these genes were related to chaperones, transporters, cell cycle regulation, Zn finger proteins, transcription activity, and redox metabolism. In addition, these responses involved a considerable number of surface antigens and proteins exported to the host cell cytoplasm. Based on this the authors proposed that increased expression of exportome genes relates to an activation of host–parasite interaction which contributes to the parasite stress response (Natalang *et al.*, 2008).

Remarkably, activation of exported proteins was also observed in parasites exposed to febrile temperatures (Oakley *et al.*, 2007). Here it was shown that exposure of *P. falciparum* parasites to 41°C triggers an apoptosis-like cell death mechanism that ultimately leads to parasite death. However, prior to that the parasites exhibited a transcription-dependent stress response mechanism that includes at least 398 genes whose expression is altered as a result of these conditions. Interestingly, 47% of these genes are predicted to be related to transmembrane or secreted proteins. Hence, a large proportion of the temperature-induced response of *P. falciparum* is targeted to the cell surface and interactions with the host. In addition, increased temperature induced expression of a number of molecular chaperones, such as HSP90, HSP70, and the plant-related DnaJ proteins, and regulators of protein stability, including proteosomal subunits and ubiquitin metabolism factors. These results were indeed the first to suggest the existence of a direct feedback mechanism responding to generic protein damage inflicted by increased temperatures.

In agreement with this, a sizable number of studies have been reported over last 5 years in which growth perturbations affect expression of genes directly linked with the targeted functions, including chemical inhibition of sphingomyelin synthase that induces expression of proteins linked with the targeted vesicle compartment (Tamez *et al.*, 2008), co-inhibition of S-adenosylmethionine decarboxylase/ornithine decarboxylase that induces expression of compensatory enzymes lysine decarboxylase and ornithine aminotransferases (van Brummelen *et al.*, 2009), impairment of apicoplast gene expression by doxycycline (Dahl *et al.*, 2006), and the hyperoxia-induced up-regulation of antioxidant systems and down-regulation of vacuole metabolism and glycolysis in favour of mitochondrial respiration (Torrentino-Madamet *et al.*, 2010). In most of these cases, the growth perturbations affected additional off-target genes linked with biological processes that are directly or indirectly related to the targeted mechanisms, and which contribute at least in part to the total transcriptional responses to these external perturbations.

To explore this systematically, Hu *et al.* (2009) carried out extensive transcriptional profiling of a panel of small molecular inhibitors that affect *P. falciparum* growth and/or development. Using a tightly controlled microarray dataset that consists of 23 time-course experiments with 20 chemical perturbations (144 microarray data points) it was confirmed that indeed some perturbations, such as inhibitors of proteases or microtubule formation, induce minimal mRNA changes ('RNA whispers'?) affecting less than 1% of the genome. On the other hand, other compounds, such as specific inhibitors of calcium/calmodulin-dependent kinases and the calcineurin pathway, can cause specific high-amplitude transcriptional responses (>2-fold) that involves 5–10% of *Plasmodium* genes. In addition, some inhibitors, including staurosporine (PKs inhibitor), apicidin and trichostatin A (HDAC inhibitors), and EDTA (chelator of ambient Ca^{2+}) cause wide-range transcriptional changes that are consistent with developmental arrests. These results are in agreement with earlier speculations (Kato *et al.*, 2008) that while some response mechanisms were lost during evolution, *Plasmodium* parasites retain a level of transcriptional plasticity sufficient to respond to external perturbation or changing growth conditions. These specific responses consist of up- or down-regulation of the target or compensatory pathways as well as developmental arrest of the life cycle progression, possibly giving the parasite cell the time to repair the inflicted damage. An important observation is that transcriptional changes induced by growth perturbation are highly consistent with the functional assignment of the affected genes, which indicates common transcriptional regulation for functionally related genes (Hu *et al.*, 2009). Thus transcriptional profiling of extreme perturbations provide additional information that can be used for GBA-based functional gene predictions, and can be used by systems biology approaches for functional explorations of the malaria parasites (see below).

Proteomic analyses of growth perturbations

The notion that post-transcriptional (along with transcriptional) regulation plays a role in responses to perturbations was first suggested by the up-regulation of antifolate drug targets in *P. falciparum* cells treated with the DHFR inhibitor WR99210 (Nirmalan *et al.*, 2004b). This was also true for the related inhibitor pyrimethamine, that could in addition cause significant down-regulation of several key metabolic enzymes and cellular factors, unrelated to the folate biosynthesis pathway (Nirmalan *et al.*, 2004a). In the absence of a transcriptional response to antifolate inhibitors (see above and Ganesan *et al.*, 2008) this suggests that proteomics analyses may provide additional information about parasite-specific responses to particular treatments as well as shed additional light on the mode of action of a drug in *Plasmodium*. Investigating the mode of action of two of the most important antimalarial drugs, artemisinin and chloroquine, Prieto *et al.* (2008) conducted a quantitative proteomics survey of *P. falciparum* parasites treated with these two drugs at $2 \times IC_{50}$ concentration for 72 h. From the total ~800 proteins identified, 41 and 38 proteins were up-regulated and 14 and 8 proteins were down-regulated in the chloroquine and artemisinin treatments, respectively. Interestingly, the majority of the proteins up-regulated during chloroquine treatment were nuclear factors and proteins involved in translation, consistent with the mode of action of chloroquine, which involves a release of parasitotoxic haem that subsequently interacts with DNA and at the same time causes specific proteome oxidative damage (Lelievre *et al.*, 2007; Radfar *et al.*, 2008). Although the proteomic response to artemisinin was much less defined, two important observations were made. There was a significant down-regulation of the vacuolar ATPase subunits α and γ, which is consistent with artemisinin-mediated alkylation of PfATP6, the Ca^{2+} ATP-dependent pump of the endoplasmic reticulum. Second, the multidrug-resistant transporter PfMdr1, previously linked with drug-resistant phenotypes of *P. falciparum*, was consistently up-regulated in both artemisinin and chloroquine treated cells. These proteomic responses to artemisinin and chloroquine share common features with two results analysing the activity of doxycycline (Briolant *et al.*, 2010) and the two core compounds of the Coartem formulation arthemer and lumefantrine (Makanga *et al.*, 2005), where a relatively small number of proteins, including molecular chaperones as well as proteins involved in glycolysis and redox metabolism, are up-regulated. Although these results likely represent indirect responses to general stress, both studies identified specific protein expression differences that can be directly linked with drug activity. A significant proportion of proteins affected by doxycycline were linked with mitochondrial function, which is consistent with the previous model in which doxycycline-related tetracyclines were found to inhibit mitochondrial protein synthesis. Analogously, doxycycline caused up-regulation of a considerable number of proteins targeted to the apicoplast (19% of the overall response). This correlates well with the proposed mode of action of doxycycline, namely, inhibition of apicoplast proteosynthesis (Briolant *et al.*, 2010). In contrast to the responses to DHFR inhibitors, the doxycycline- and the Coartem-induced proteomic variations were also reflected at the mRNA level (Briolant *et al.*, 2010; Makanga *et al.*, 2005).

Similar to antimalarial drugs, proteomic profiling of the effect of specific small molecular inhibitors and other physiological perturbations could elicit specific response at the protein level. Following on the transcriptome profiling of the choline inhibitor compound T4 that identified only nonspecific transcriptional changes (see above), much more specific responses were detected at the protein level (Le Roch *et al.*, 2008). A significant increase of proteins associated with digestive vacuolar function is consistent with the accumulation of choline in this compartment, and its possible interaction with haem (Biagini *et al.*, 2003). Moreover, compound T4 suppressed expression of choline/ethanolamine phosphotransferase (PfCEPT, MAL6P1.145), which is involved in the final step of phosphoethanolamine synthesis, the direct target of this inhibitor.

While protein changes induced by compound T4 were independent of transcription, the combined transcriptomic and proteomic analysis of PfAdoMetDC/ODC co-inhibition revealed changes in polyamine and methionine metabolism (the presumed target of the co-inhibition) that were reflected at both mRNA and protein levels (van Brummelen *et al.*, 2009). Similarly, the proteomic changes induced by hyperoxia – mainly up-regulation of 'indirect targets' such as proteins involved in chaperone-mediated proteins folding, translation, redox metabolism, and glycolysis – were also reflected at the level of transcription (Torrentino-Madamet *et al.*, 2010).

From these data it is clear that transcriptional profiling of growth perturbation can enhance our understanding of *Plasmodium* biology and especially the mode of action of antimalarial drugs. However, these data also show that in the future, it will be crucial to include proteomic approaches into these studies, as many of the drug-induced changes are only reflected at the protein level, without variation in the corresponding transcripts. This was recently reiterated by a quantitative proteomic study of the invasion machinery, showing that invasion phenotype switches can be associated with large scale protein abundance changes that are not reflected in transcription (Kuss *et al.*, 2011).

Transcriptional profiling of *Plasmodium* strains and field isolates

Functional genomics of genetically distinct strains and lines of individual species captures another dimension of gene expression regulation in malaria parasites and is also extremely helpful in gene annotation efforts. In combination with genetic studies, transcriptional profiling of *Plasmodium* strains and isolates provides particularly useful insights into phenotypic variations that could be linked to pathogenesis, virulence, and other aspects of malaria infections. Although this research area is still in its initial stage, many recent advances have demonstrated its usefulness for many future explorations.

Transcriptional profiling of *in vivo* field isolates

Since the onset of *Plasmodium* transcriptomics in 2002, much attention has been focused on field isolates. The main aim has been to identify virulence factors that could be linked to various types of disease outcomes, ranging from asymptomatic or mild malaria to severe disease and death. An important challenge was to reliably utilize the relatively low amount of parasite material obtained from patient blood samples (as little as 2.5 ml of blood with 0.3% parasitaemia) for microarray hybridizations (Daily *et al.*, 2004). Initial studies analysing the genome-wide gene expression profile of a few samples from Senegal and Nigeria demonstrated that the global mRNA profile of *P. falciparum* parasites circulating in the peripheral blood is consistent with the ring stage transcriptome of *in vitro* cultures. However, in comparison to the *in vitro* 3D7 strain, the field isolates show marked differences in expression of genes encoding sexual stage antigens and surface proteins. The most significant observation was made regarding the three subtelomeric gene families *var*, *rifin* and *stevor*, with the field isolates having a much narrower expression repertoire of *var* genes (Daily *et al.*, 2005). The narrower set of expressed *var* genes may reflect their extensive polymorphism in various *in vivo* isolates that prevents their microarray detection. In contrast, neither *rifins* nor *stevors* exhibited these constraints, and in fact larger numbers of their members appear to be transcribed in the field isolates compared to lab strains. This suggests a distinct role of these two gene families in antigenic presentation where a natural immune environment requires their stronger expression. The field isolates showed increased expression levels of several other genes, including RESA-2 (PF11_0512), putative long-chain fatty acid ligase (PFC0050c), and one member of a hypothetical protein family (PF14_0752). Interestingly, this hypothetical gene family was later annotated as the *Plasmodium* helical interspersed subtelomeric family (PHIST) (Sergeant *et al.*, 2006) and its variable expression was further demonstrated in both *P. falciparum* and *P. vivax* field isolates (see below).

Expanding on their first two studies, Daily *et al.* conducted a large-scale *in vivo* transcriptional profiling survey that includes 43 patient samples from Senegal. This study identified major transcriptional differences for many metabolic enzymes and basic cellular processes, suggesting the existence of distinct physiological states of the *P. falciparum* parasites in infected patients (Daily *et al.*, 2007). Overall, the *P. falciparum* isolates could be assigned to one of three transcription profiles that are analogous to different metabolic states of *Saccharomyces cerevisiae*. The first state represents normal fermentative glycolytic growth associated with high expression levels for genes involved in glycolysis, amino acid and nitrogen metabolism, and general growth related processes such as nuclear transcription and cytoplasmic translation. The second state corresponds to energy starvation and is characterized by induction of genes for oxidative phosphorylation, respiration, mitochondrial biogenesis, and apicoplast fatty acid metabolism. Finally, the third physiological state, which is considerably more variable, is reminiscent of environmental stress responses, with strong induction of genes for heat shock proteins and oxidative and osmotic stress responses. Interestingly, the first physiological state is also analogous to the *in vitro* ring stage 3D7 reference transcriptome profile (Daily *et al.*, 2007).

The biological significance of these physiological states is unclear; however, some may be correlated with commitment to sexual development. This was proposed by Lemieux *et al.* using a statistical estimation algorithm by which *in vivo* transcriptome data can be correlated to the *in vitro* reference in order to estimate the age and the synchronicity of *in vivo* or *ex vivo* parasite samples (Lemieux *et al.*, 2009). Using this approach it was shown that a component of the global mRNA patterns generated in the Senegal study (Daily *et al.*, 2007) could be related to the transcriptional program of early gametocytes (Note: This profile also corresponds to trophozoites at 30 h post invasion). This was particularly evident for the transcription profiles that correspond to energy starvation and stress responses detected in the Senegal study (Daily *et al.*, 2007). At the present time it is difficult to evaluate the contribution of sexual development to these metabolic states, and Daily and coworkers argued retrospectively that the fact that no gametocytes have been observed in the corresponding blood samples suggests its impact on the overall transcriptome profile in the Senegal study is limited at best (Wirth *et al.*, 2009). In addition, the magnitude of the transcriptional differences – which in some cases could reach above 20-fold – could not be explained by gametocyte cross-contamination. Although this issue remains unresolved, Lemieux and colleagues addressed an important technical aspect of transcriptional (and other) analyses of *P. falciparum* field isolates, showing that the rate of maturation of *ex vivo* cultures varies significantly and that 48 h *ex vivo* culture generates parasite populations that fall between young trophozoites (20 hpi) and late schizonts (44 hpi) (Lemieux *et al.*, 2009).

The physiological states observed in the Senegal study could not be associated with any patient phenotype, with the exception of the stress response-like transcriptional signature that was predominantly found in patients with high fevers (Daily *et al.*, 2007). Hence, more work will be required to understand the underlying mechanisms of the parasite's adaptations to variable host environments. Transcriptional profiling of malaria infections during pregnancy was more successful in linking specific transcriptional profiles with a disease profile. Malaria during pregnancy, particularly first pregnancy, has a unique disease profile that is characterized by a lack of protective antibodies even in hyper-endemic regions, indicating the expression of novel surface molecules (Beeson *et al.*, 1999). Interestingly, all parasites obtained from pregnancy malaria exhibit highly uniform cytoadherence that is independent of their geographical origin (Fried *et al.*, 1998). Hence, identification of genes expressed in pregnancy malaria may provide important clues for understanding the basis of this disease phenotype. A study analysing the transcription profiles of *P. falciparum* parasites collected directly from infected placentas has identified six genes that are specific to pregnancy malaria infections (Francis *et al.*, 2007). One of these is *var2csa* (PFL0300c), a semi-conserved member of the *var* gene family whose protein product was previously implicated as a mediator of cytoadherence to the placental receptor CSA (Salanti *et al.*, 2003). High expression of this gene in placental and peripheral blood parasites obtained from pregnant women is consistent with this function and further validates the transcriptional data (Francis *et al.*, 2007). The additional five identified genes lack any functional annotation, but contain the PEXEL transport motif and thus might be involved in host–parasite interactions. Hence, these genes could represent additional candidates for vaccine development against pregnancy malaria. In a parallel study of pregnancy malaria, 84 genes (including *var2csa)* were found to be consistently overexpressed in parasites collected from infected placentas in patients in Senegal (Tuikue Ndam *et al.*, 2008). There was a significant enrichment of subtelomeric genes, as well as genes encoding exported proteins, amongst the 84 genes. Follow-up studies of the microarray data identified PFI1175w as being exclusively expressed in placenta-derived *P. falciparum*, with no expression being detected in parasites from peripheral blood of pregnant women or in *in vitro*-selected CSA-binding laboratory lines (Tuikue Ndam *et al.*, 2008). PFI1175w is a member of the subtelomeric PHIST gene family, whose other member PF14_0752 showed strong over-expression in the earlier field isolate studies (Daily *et al.*, 2005). Moreover, PFI1175w appears to be preferentially recognized by immune sera of pregnant infected women as compared to non-pregnant malaria patients (Tuikue Ndam *et al.*, 2008). This suggests that members of the PHIST protein family are involved directly in host–parasite interactions, with each member serving different roles during the disease profile.

Transcriptional profiling of *P. falciparum* field isolates was also shown to be useful in deciphering mechanisms of drug resistance. In a recent publication, Mok *et al.* analysed transcription profiles of artemisinin resistant *P. falciparum* parasites that were collected in the western Cambodian region, Pailin (Mok *et al.*, 2011). In this region an increasing number of malaria cases were reported that require a prolonged course of artemisinin combination therapy (ACT) treatment as the parasite clearance time is extended from the usual 48 h (interquantile range 36 to 66 h) to 84 h (60–96 h) (Dondorp *et al.*, 2009). This prolonged clearance time may signal the rise of resistance of the *P. falciparum* to artemisinin, which is the main line of treatment and prevention of malaria infection. Spread of artemisinin resistance throughout malaria endemic regions might have devastating affects on the malaria pandemic and hence much effort is being put into understanding, containing and eliminating these parasite populations in western Cambodia (Dondorp *et al.*, 2009). Using transcriptional profiling it was shown that the artemisinin-resistant parasites are characterized by specific shifts in the IDC transcriptional cascade. In particular, the majority of basic biochemical and metabolic activities of the ring and trophozoite stages were shown to be down-regulated, and subsequently, many functionalities associated with protein metabolism up-regulated in the schizont stages (*Mok et al.*, 2011). These results corroborated the current view of artemisinin mode of action and artemisinin resistance. First, the lowered metabolic activity of the earlier stages may result in lower level of drug activation and hence increase survival of the early ring forms. This is also in line with the observation that quiescent ring stage parasites can be induced by prolonged exposure to artemisinin (Teuscher *et al.*, 2010; Witkowski *et al.*, 2010). Second, the increased level of protein synthesis, trafficking and turnover that is represented by up-regulation of all involved transcripts in the schizont stages may provide the parasites with stronger survival potential by repairing the damage caused by an oxidative stress and alkylation effect of artemisinin (Meshnick *et al.*, 1996).

Similar to *P. falciparum*, *P. vivax* exhibits considerable transcriptional variation between individual isolates. The transcriptomes of three isolates from northern Thailand showed significant differences in transcript levels of genes associated with invasion (such as members of the merozoite surface protein families pv*msp3* and pv*msp8*, and reticulocyte binding proteins, pv*rbp*), antigenic variation (*vir* and *trag* genes), and other factors proposed to be involved in host–parasite interactions, including the *pvphist* genes (Bozdech *et al.*, 2008). In addition there was considerable variability in expression of genes involved in several metabolic functions, in particular glycolysis. This was further explored by *in vivo* profiling of eight *P. vivax* isolates collected in Peru (Westenberger *et al.*, 2010). Interestingly, while the first three enzymes of the glycolytic pathway were expressed at similar levels between isolates, the eight downstream proteins in the pathway exhibited up to 100-fold difference in expression between individual isolates.

Taken together, extensive *in vivo* transcriptional studies clearly show that *Plasmodium* parasites are able to modulate expression of their genes in order to adapt to their host environment and/or withstand external stresses and stimuli generated by this environment. The transcriptionally flexible genes can be loosely linked to four types of functionalities: (i) host–parasite interactions (subtelomeric gene families and PEXEL-containing proteins), (ii) biosynthesis and energy metabolism (glycolysis, amino acid metabolism, TCA cycle, and apicoplast fatty acid synthesis), (iii) stress responses (molecular chaperones and redox metabolism), and (iv) sexual development. Interestingly, these four types of functionalities were often found amongst genes induced (or suppressed) by external perturbations applied to the *P. falciparum* parasites *in vitro* (see above), which suggests their utility in parasite growth and adaptation to its environment.

Transcriptional profiling of *in vitro*-derived cell lines

Transcriptional profiling of culture-adapted isolates as well as laboratory strains has also proven to be an important tool to understand transcriptional differences in *Plasmodium*, particularly in relation to mechanisms of their regulation. Expanding on the original Tanzania study of pregnancy malaria (Francis *et al.*, 2007), the Duffy's group has analysed transcription profiles of *in vitro* culture-adapted parasites obtained from the placenta of six pregnant women and the peripheral blood of eight children (Vignali *et al.*, 2011). These *in vitro* cultures generated sufficient amount of material to allow RNA-Seq analysis combined with not-so-random (NSR) primers for selective *Plasmodium* RNA amplification. This approach expanded the originally identified group of six genes associated with pregnancy by an additional four members of *var*, three members of *phist*, S-antigen, one ribosomal protein, and one additional hypothetical protein. Most importantly, these studies showed that *in vivo* gene expression profiles correlate with 'freshly' *in vitro*-adapted parasites, and thus many of the *in vivo* phenotypes can be still studied.

To explore differences between freshly adapted parasites and long-term laboratory strains, Mackinnon *et al.* (2009) assembled the full IDC transcriptomes for five *P. falciparum* isolates (grown for less than 40 generations) collected in Kenya (Mackinnon *et al.*, 2009). These were compared to the IDC transcriptomes of three long-term *P. falciparum* laboratory strains: 3D7, Dd2, and T9/96. A total of 259 genes showed significant differential expression ($P < 0.01$) between the field isolates and the laboratory strains. Not surprisingly, there was significant enrichment of genes involved in host–parasite interactions, sexual differentiation, transcription and translation, protein folding, and biosynthesis. The majority (78%) of these genes were up-regulated in the field isolates, which indicates that during the long-term culture adaptation many of these functions are being diminished or lost. Interestingly, the differential expressed genes were localized to 30 chromosomal regions, with 19 of these being located in the subtelomeres. Overlaying these regions with copy number variations (CNV) detected by comparative genomic hybridization (CGH) revealed statistically significant positive or negative correlation. These data suggests that CNV may be one of the important factors regulating differential gene expression, with gene amplification being associated with both increased and decreased expression (Mackinnon *et al.*, 2009).

More insights into the transcriptional regulation of differentially expressed genes in *P. falciparum* came from gene expression quantitative trait loci (eQTL) analysis that evaluated expression level polymorphisms (ELPs) in 34 progeny from the HB3/DD2 genetic cross (Gonzales *et al.*, 2008). Approximately 18% (874) of all *P. falciparum* genes displayed a significant genetic component (*P*-value < 0.05) of transcriptional variation amongst the progeny measured at 18 hpi. Overall, 1063 significant eQTLs were identified, with 807 genes mapping to a single locus while another 74 map to multiple loci. The vast majority of the eQTLs-linked expression differences were 20 to 4-fold, with some being as low as 1.65-fold between the highest and lowest expression level amongst the progeny. The strong statistical significance of these traits further validates the idea that even subtle transcriptional changes can have considerable biological implications. Of the traits mapping to single eQTLs, 24% were local effects, suggesting that local structural determinants of gene expression such as chromatin organization and/or CNVs play a role in their regulation. On the other hand, 76% of eQTLs mapped to distant ELPs, which indicates their link with *trans*-regulatory elements. The global distribution of eQTLs subsequently identified 12 regulatory hot spots (chromosomal regions), each of which drove transcriptional changes of 14 to 182 genes and accounted for 63% of the detected traits. Of these, only a single hot spot on chromosome 12 corresponded to mostly local regulation, with the rest being linked to *trans*-regulation. The most predominant hotspot corresponded to a chromosome 5 amplicon (multiple eQTLs) that regulates a group of 228 transcripts, with a strong preference for genes involved in protein modifications. Three proteins with homologies to DNA-binding proteins were identified in the amplicon on chromosome 5, including PFE1145w, containing two tandem zinc-finger CCCH-type domains; PFE1095c, which has weak homology to YL1 nuclear protein; and PFE1110w, a putative transcription initiation factor TFIID. These proteins are primary candidates for regulatory modulators of gene expression levels between *P. falciparum* strains (Gonzales *et al.*, 2008). Taken together, both datasets indicate that CNV plus other structural polymorphisms may cause variable gene expression via either *cis*- or *trans*-regulatory mechanisms. Future studies of these properties open an exciting research area that will allow us to evaluate gene expression differences between *Plasmodium* strains and connect them to various phenotypic properties of the *P. falciparum* parasites on both individual as well as population levels.

Functional genomics in *Plasmodium* systems biology

Integration of functional genomics and proteomics data into systems biology approaches represents the most powerful utilization of such results. Systems biology, or integrative biology, combines various types of information, including many types of results from functional genomics, e.g. gene and protein expression, protein–protein interactions, and genetic interactions. For many model organisms such as yeast, these approaches represent the most comprehensive information for functional gene annotations as well as elucidating the biological relevance of many metabolic and cellular processes (Han *et al.*, 2004; Jansen *et al.*, 2003). In general, systems biology utilizes various probabilistic methods to integrate large-scale genome-wide dataset into a single gene/protein interaction network (interactome) and subsequently uses the existing gene annotations to predict the function of uncharacterized genes via their location within the network. The well-annotated yeast genome provided an excellent training ground for the development of integrative approaches that could then be further applied to more challenging biological systems such as the poorly characterized malaria parasite genomes. Thanks to the rising number of genome-wide studies that emerged since the completion of the first *Plasmodium* genome in 2002, the feasibility of systems biology methods for *Plasmodium* studies is steadily increasing (Winzeler, 2006).

In one such study, Zhou *et al.* (2008) combined gene expression data obtained from *P. falciparum* blood, sporozoite, zygote, and ookinete stages, as well as *P. yoelii* mosquito oocyst and salivary gland sporozoite, blood, and liver stages. Using ontology-based pattern identification (OPI) clustering, these expression data were stratified according to expression patterns throughout the life cycle. This demonstrated that together with genes involved in many well defined cellular and biochemical pathways, a large number of uncharacterized genes followed a coordinated transcription program throughout the *Plasmodium* life cycle; hence, their function could be predicted using the GBA principle. Integrating this OPI-generated dataset with unbiased literature mining made it possible to link many of the hypothetical genes with specific roles in many biological processes, such as gliding motility, hosts–parasite interaction, and rhoptry function. Finally, the expression-based functional predictions were overlaid with yeast two-hybrid generated protein–protein interaction data (LaCount *et al.*, 2005) in order to validate the OPI clustering and at the same time generate a probabilistic gene network. Two-hybrid analysis of the *Plasmodium* genome identified 2823 interactions between 1267 *P. falciparum* proteins, and in many cases captured essential *Plasmodium* function participants, such as the chromatin remodelling complex and a subnetwork of invasion proteins. Presently, this study remains the only protein–protein interaction analysis that has been used for a number of systems biology studies of *Plasmodium* parasites. Integrating two-hybrid system data with OPI clustering generated functional predictions for 926 previously uncharacterized *Plasmodium* genes. including several new invasion genes, genes interacting with glycolytic enzymes (such as MAL8P1.17, a protein disulphide isomerase), and ribosome interacting proteins, as well as a number of proteins-linked with a methyl transferase activity (Zhou *et al.*, 2008).

As an alternative approach, Date Stoeckert (2006) employed a Bayesian framework to integrate data from large-scale studies, including life cycle transcriptome analyses and phylogenetic profiling, as well as two-hybrid system- and Rosetta Stone database-predicted protein–protein interactions. The basis of this approach is measuring individual accuracies of the input data sets by comparing these with established 'gold standards' of functional annotations such the Kyoto Encyclopaedia Genes and Genomes (KEGG) database (Kanehisa *et al.*, 2004), the Gene Ontology (GO) Consortium database (Ashburner *et al.*, 2000) and, for *P. falciparum*, the Malaria Parasite Metabolic Pathways (MPMP) database (Ginsburg, 2008). This strategy has two main strengths compared with other types of integrations. First, it can integrate data from diverse large-scale studies that may vary considerably in their accuracy. Second, compared to more the specialized approaches of single-type genomics or proteomics studies, it can capture a wide spectrum of biological processes and reconstruct essentially the entire physiology of *Plasmodium* parasites (Date and Stoeckert, 2006). Hence, the Bayesian framework-based interactome established in these studies provided functional information for 4343 unique proteins (~81% of the proteome) from all life cycle stages. The interactome network based on 50% confidence thresholds contained 3667 proteins linked with 388,969 functional linkages with functional evidence from at least one of the inputs (e.g. transcriptional profile correlations, two-hybrid interaction, phylogenetic co-conservation, or Rosetta Stone-predicted domain–domain interactions). Taking advantage of the assembled interactome (termed PlasmoMAP), Date and colleagues explored several interesting protein functionalities in *P. falciparum*, including a subnetwork of molecular chaperones involving the TCP-1 ring complex, a set of HSP proteins, as well as a DnaJ-containing protein family (also termed ring-infected erythrocyte surface antigens, RESAs) that is highly expanded in *P. falciparum*.

A marked improvement to the overall confidence and coverage of the interactome network approach was achieved by integrating an additional large transcriptome dataset that included transcriptional responses of the *P. falciparum* cells to 20 diverse growth-inhibiting compounds (Hu *et al.*, 2009) [see Specific (and nonspecific) transcriptional responses to external perturbations]. Using an analogous Bayesian-based method, the newly assembled interactome network (PlasmoINT) captured functional information for a total of 5374 *P. falciparum* proteins (99.2% of the proteome), from which 4817 were linked with 339,721 linkages with 50% confidence and 3475 proteins with 72,748 linkages with 90% confidence. These major improvements, particularly for the 90% confidence network, which contains 2.5 times more proteins and six times more linkages than PlasmoMAP (Date and Stoeckert, 2006), allowed high confidence functional gene predictions for up to 95% of all predicted hypothetical *P. falciparum* proteins (,545). This was achieved using the weighted neighbour-counting (WNC) method that evaluates functional assignments of the 'environment' (up to two levels of neighbouring genes) of each gene in the 90% confidence network and scores it based on the confidence of each functional linkage. The WNC functional predictions enabled the assignment of novel genes into gene groups like *P. falciparum*-exclusive antigenic presentation and immune invasion mechanisms, apicomplexan specific invasion and host cytoplasm remodelling, and prokaryote-originating shikimate and isoprenoid synthesis, as well as apicoplast protein synthesis, that are linked by several classes of functionalities and have a high potential as intervention targets. Finally, focusing on the invasion process, the 90% confidence interactome network was used to assemble a sub-network of 418 proteins linked to 25 previously established invasion-associated proteins. This sub-network contains many proteins functionally linked with invasion, including surface antigens, factors of actomyosin motility, protein kinases, and proteases, as well as an additional 263 hypothetical proteins. Intracellular localization studies of 42 of these hypothetical proteins showed that 20 were localized to the apical region of the maturing merozoite, seven were associated with the inner membrane complex (IMC), and four were present at the surface of the freshly released merozoite. Although the remaining eleven proteins were localized in the parasite cytoplasm, their links with invasion could not be excluded. Taken together, experimental verifications of the probabilistic functional predictions gave good confidence that the PlasmoINT functional predictions are comparable to similar types of studies of well-defined model organisms such as yeast. This was further validated by a 'leave-one-out' analysis of the 90% confidence network that estimated the recall precision at 47%, similar to the precision seen in model systems (Hu *et al.*, 2009).

In addition to these comprehensive direct approaches, Wuchty and colleagues reported a series of studies in which protein interactome networks were inferred from various types of orthology-based searches where *Plasmodium* proteins were searched against complete sequence information from organisms such as *E. coli*, *S cerevisiae*, *C. elegans* and *H. sapiens*, deposited in the InParanoid database (Wuchty, 2006, 2007; Wuchty *et al.*, 2009; Wuchty *et al.*, 2006; Wuchty and Ipsaro, 2007). In these studies, *in silico* data were typically overlaid with yeast two-hybrid protein–protein interaction data (LaCount *et al.*, 2005) and the resulting functional predictions were scrutinized using various types of bioinformatics approaches. The goal was to understand the overall topology of protein interaction in malaria parasites. This work showed that the overall individual connectivity elements within the network followed a power law distribution, which indicates the existence of large number of small complexes with only a small number of highly connected proteins in the network (Wuchty *et al.*, 2009). (Similar observations were made from the transcriptome-based interactome approaches; Date and Stoeckert, 2006; Hu *et al.*, 2009.) However, these studies also uncovered a small number of highly connected nodes, termed 'rich club' nodes, that are represented by disproportionately high connectivity compared with the expected average of the underlying network as well as the randomized network. Many of these rich club nodes are, however, driven by the two-hybrid system results and so far have not been confirmed by any other genome-wide analyses. Although more work is necessary to verify the existence of these highly connected proteins in *Plasmodium*, they may represent essential factors that could be ideal targets for malaria intervention strategies.

In addition to these systematic approaches of integrative biology, several online searchable databases have been established that take advantage of the recent advances in functional genomics and proteomics, as well as comparative genomics data, and allow researchers to search for functional clues of *Plasmodium* proteins. One of these, PlasmoDraft, utilizes a GO-based GBA method named Gonna that takes into account protein sequence homology, functional domain predictions, and many functional genomics and proteomics results. Using this approach, PlasmoDraft provides GO-based functional predictions for up to 2434 uncharacterized *Plasmodium* genes (Brehelin *et al.*, 2008). In a similar fashion another database called PlasmoPredict allows the user to search the *Plasmodium* genome based on various types of parameters associated with the protein sequence as well as gene and protein expression. These two databases comprise a large volume of functional genomics and proteomics data that could be explored for various types of research.

Conclusion

The availability of completed genomes for a number of different *Plasmodium* species has provided us with a wide range of new information that is essential for understanding the complex biology of this parasite. Of equal importance is that it has created the foundation for using functional genomics approaches to study a wide range of questions about functional annotation of unknown genes, regulation of expression, parasite pathology, and host–parasite interaction. As highlighted in this chapter, in a relatively short time the amount of new understanding about the malaria parasites that we have gained through the use of different functional genomic approaches is impressive, and it is certain that the amount of new data will increase immensely over the next few years. This provides a great opportunity for identifying new targets for anti-malaria intervention, but also presents with new challenges on how to best analyse and evaluate this mountain of information. What is very clear, though, is that functional genomics provides data that, if made publicly available to the research community, can be the basis of many new scientific inquiries that had never been envisaged by the original study.

References

Alonso, P.L., Sacarlal, J., Aponte, J.J., Leach, A., Macete, E., Aide, P., Sigauque, B., Milman, J., Mandomando, I., Bassat, Q., *et al.* (2005). Duration of protection with RTS, S/AS02A malaria vaccine in prevention of *Plasmodium falciparum* disease in Mozambican children: single-blind extended follow-up of a randomised controlled trial. Lancet *366*, 2012–2018.

Arbeitman, M.N., Furlong, E.E., Imam, F., Johnson, E., Null, B.H., Baker, B.S., Krasnow, M.A., Scott, M.P., Davis, R.W., and White, K.P. (2002). Gene expression during the life cycle of *Drosophila melanogaster*. Science *297*, 2270–2275.

Ashburner, M., Ball, C.A., Blake, J.A., Botstein, D., Butler, H., Cherry, J.M., Davis, A.P., Dolinski, K., Dwight, S.S., Eppig, J.T., *et al.* (2000). Gene ontology: tool for the unification of biology. The Gene Ontology Consortium. Nat. Genet. *25*, 25–29.

Bader, G.D., and Hogue, C.W. (2003). An automated method for finding molecular complexes in large protein interaction networks. BMC Bioinformatics *4*, 2.

Baird, J.K., and Hoffman, S.L. (2004). Primaquine therapy for malaria. Clin. Infect. Dis. *39*, 1336–1345.

Balaji, S., Babu, M.M., Iyer, L.M., and Aravind, L. (2005). Discovery of the principal specific transcription factors of Apicomplexa and their implication for the evolution of the AP2-integrase DNA binding domains. Nucleic Acids Res. *33*, 3994–4006.

Bartfai, R., Hoeijmakers, W.A., Salcedo-Amaya, A.M., Smits, A.H., Janssen-Megens, E., Kaan, A., Treeck, M., Gilberger, T.W., Francoijs, K.J., and Stunnenberg, H.G. (2010). H2A.Z demarcates intergenic regions of the *Plasmodium falciparum* epigenome that are dynamically marked by H3K9ac and H3K4me3. PLoS Pathog. *6*, e1001223.

Beeson, J.G., Brown, G.V., Molyneux, M.E., Mhango, C., Dzinjalamala, F., and Rogerson, S.J. (1999). *Plasmodium falciparum* isolates from infected pregnant women and children are associated with distinct adhesive and antigenic properties. J. Infect. Dis. *180*, 464–472.

Berger, M.F., Philippakis, A.A., Qureshi, A.M., He, F.S., Estep, P.W., 3rd, and Bulyk, M.L. (2006). Compact, universal DNA microarrays to comprehensively determine transcription-factor binding site specificities. Nat. Biotechnol. *24*, 1429–1435.

Bhowmick, I.P., Kumar, N., Sharma, S., Coppens, I., and Jarori, G.K. (2009). *Plasmodium falciparum* enolase:

stage-specific expression and sub-cellular localization. Malar. J. *8*, 179.

Biagini, G.A., Richier, E., Bray, P.G., Calas, M., Vial, H., and Ward, S.A. (2003). Heme binding contributes to antimalarial activity of bis-quaternary ammoniums. Antimicrob. Agents Chemother. *47*, 2584–2589.

Bischoff, E., and Vaquero, C. (2010). In silico and biological survey of transcription-associated proteins implicated in the transcriptional machinery during the erythrocytic development of *Plasmodium falciparum*. BMC Genomics *11*, 34.

Bozdech, Z., Llinas, M., Pulliam, B.L., Wong, E.D., Zhu, J., and DeRisi, J.L. (2003a). The transcriptome of the intraerythrocytic developmental cycle of *Plasmodium falciparum*. PLoS Biol. *1*, E5.

Bozdech, Z., Zhu, J., Joachimiak, M.P., Cohen, F.E., Pulliam, B., and DeRisi, J.L. (2003b). Expression profiling of the schizont and trophozoite stages of *Plasmodium falciparum* with a long-oligonucleotide microarray. Genome Biol. *4*, R9.

Bozdech, Z., Mok, S., Hu, G., Imwong, M., Jaidee, A., Russell, B., Ginsburg, H., Nosten, F., Day, N.P., White, N.J., *et al.* (2008). The transcriptome of *Plasmodium* vivax reveals divergence and diversity of transcriptional regulation in malaria parasites. Proc. Natl. Acad. Sci. U.S.A. *105*, 16290–16295.

Brehelin, L., Dufayard, J.F., and Gascuel, O. (2008). PlasmoDraft: a database of *Plasmodium falciparum* gene function predictions based on postgenomic data. BMC Bioinformatics *9*, 440.

Briolant, S., Almeras, L., Belghazi, M., Boucomont-Chapeaublanc, E., Wurtz, N., Fontaine, A., Granjeaud, S., Fusai, T., Rogier, C., and Pradines, B. (2010). *Plasmodium falciparum* proteome changes in response to doxycycline treatment. Malar. J. *9*, 141.

Campbell, T.L., De Silva, E.K., Olszewski, K.L., Elemento, O., and Llinas, M. (2010). Identification and genome-wide prediction of DNA binding specificities for the ApiAP2 family of regulators from the malaria parasite. PLoS Pathog. *6*, e1001165.

Carlton, J.M., Adams, J.H., Silva, J.C., Bidwell, S.L., Lorenzi, H., Caler, E., Crabtree, J., Angiuoli, S.V., Merino, E.F., Amedeo, P., *et al.* (2008). Comparative genomics of the neglected human malaria parasite *Plasmodium* vivax. Nature *455*, 757–763.

Carlton, J.M., Angiuoli, S.V., Suh, B.B., Kooij, T.W., Pertea, M., Silva, J.C., Ermolaeva, M.D., Allen, J.E., Selengut, J.D., Koo, H.L., *et al.* (2002). Genome sequence and comparative analysis of the model rodent malaria parasite *Plasmodium yoelii yoelii*. Nature *419*, 512–519.

Chaal, B.K., Gupta, A.P., Wastuwidyaningtyas, B.D., Luah, Y.H., and Bozdech, Z. (2010). Histone deacetylases play a major role in the transcriptional regulation of the *Plasmodium falciparum* life cycle. PLoS Pathog. *6*, e1000737.

Chung, D.W., Ponts, N., Cervantes, S., and Le Roch, K.G. (2009). Post-translational modifications in *Plasmodium*: more than you think! Mol. Biochem. Parasitol. *168*, 123–134.

Coulson, R.M., Hall, N., and Ouzounis, C.A. (2004). Comparative genomics of transcriptional control in the human malaria parasite *Plasmodium falciparum*. Genome Res. *14*, 1548–1554.

Cui, L., Miao, J., Furuya, T., Fan, Q., Li, X., Rathod, P.K., and Su, X.Z. (2008). Histone acetyltransferase inhibitor anacardic acid causes changes in global gene expression during in vitro *Plasmodium falciparum* development. Eukaryot. Cell 7, 1200–1210.

Cui, L., Miao, J., Furuya, T., Li, X., and Su, X.Z. (2007). PfGCN5-mediated histone H3 acetylation plays a key role in gene expression in *Plasmodium falciparum*. Eukaryot Cell.

Dahl, E.L., Shock, J.L., Shenai, B.R., Gut, J., DeRisi, J.L., and Rosenthal, P.J. (2006). Tetracyclines specifically target the apicoplast of the malaria parasite *Plasmodium falciparum*. Antimicrob. Agents Chemother. *50*, 3124–3131.

Daily, J.P., Le Roch, K.G., Sarr, O., Fang, X., Zhou, Y., Ndir, O., Mboup, S., Sultan, A., Winzeler, E.A., and Wirth, D.F. (2004). In vivo transcriptional profiling of *Plasmodium falciparum*. Malar. J. *3*, 30.

Daily, J.P., Le Roch, K.G., Sarr, O., Ndiaye, D., Lukens, A., Zhou, Y., Ndir, O., Mboup, S., Sultan, A., Winzeler, E.A., *et al.* (2005). In vivo transcriptome of *Plasmodium falciparum* reveals overexpression of transcripts that encode surface proteins. J. Infect. Dis. *191*, 1196–1203.

Daily, J.P., Scanfeld, D., Pochet, N., Le Roch, K., Plouffe, D., Kamal, M., Sarr, O., Mboup, S., Ndir, O., Wypij, D., *et al.* (2007). Distinct physiological states of *Plasmodium falciparum* in malaria-infected patients. Nature *450*, 1091–1095.

Date, S.V., and Stoeckert, C.J., Jr. (2006). Computational modeling of the *Plasmodium falciparum* interactome reveals protein function on a genome-wide scale. Genome Res. *16*, 542–549.

Day, K.P., Hayward, R.E., Smith, D., and Culvenor, J.G. (1998). CD36-dependent adhesion and knob expression of the transmission stages of *Plasmodium falciparum* is stage specific. Mol. Biochem. Parasitol. *93*, 167–177.

de Koning-Ward, T.F., Gilson, P.R., Boddey, J.A., Rug, M., Smith, B.J., Papenfuss, A.T., Sanders, P.R., Lundie, R.J., Maier, A.G., Cowman, A.F., *et al.* (2009). A newly discovered protein export machine in malaria parasites. Nature *459*, 945–949.

De Silva, E.K., Gehrke, A.R., Olszewski, K., Leon, I., Chahal, J.S., Bulyk, M.L., and Llinas, M. (2008). Specific DNA-binding by apicomplexan AP2 transcription factors. Proc. Natl. Acad. Sci. U.S.A. *105*, 8393–8398.

Dechering, K.J., Kaan, A.M., Mbacham, W., Wirth, D.F., Eling, W., Konings, R.N., and Stunnenberg, H.G. (1999). Isolation and functional characterization of two distinct sexual-stage-specific promoters of the human malaria parasite *Plasmodium falciparum*. Mol. Cell Biol. *19*, 967–978.

Deitsch, K., Duraisingh, M., Dzikowski, R., Gunasekera, A., Khan, S., Le Roch, K., Llinas, M., Mair, G., McGovern, V., Roos, D., *et al.* (2007). Mechanisms of gene regulation in *Plasmodium*. Am. J. Trop. Med. Hyg. *77*, 201–208.

Duraisingh, M.T., Voss, T.S., Marty, A.J., Duffy, M.F., Good, R.T., Thompson, J.K., Freitas-Junior, L.H., Scherf, A., Crabb, B.S., and Cowman, A.F. (2005). Heterochromatin silencing and locus repositioning linked to regulation of virulence genes in *Plasmodium falciparum*. Cell *121*, 13–24.

Florens, L., Washburn, M.P., Raine, J.D., Anthony, R.M., Grainger, M., Haynes, J.D., Moch, J.K., Muster, N., Sacci, J.B., Tabb, D.L., *et al.* (2002). A proteomic view of the *Plasmodium falciparum* life cycle. Nature *419*, 520–526.

Flueck, C., Bartfai, R., Volz, J., Niederwieser, I., Salcedo-Amaya, A.M., Alako, B.T., Ehlgen, F., Ralph, S.A., Cowman, A.F., Bozdech, Z., *et al.* (2009). *Plasmodium falciparum* heterochromatin protein 1 marks genomic loci linked to phenotypic variation of exported virulence factors. PLoS Pathog. *5*, e1000569.

Foth, B.J., Zhang, N., Chaal, B.K., Sze, S.K., Preiser, P.R., and Bozdech, Z. (2011). Quantitative time-course profiling of parasite and host cell proteins in the human malaria parasite *Plasmodium falciparum*. Mol. Cell. Proteomics *10*, M110 006411.

Foth, B.J., Zhang, N., Mok, S., Preiser, P.R., and Bozdech, Z. (2008). Quantitative protein expression profiling reveals extensive post-transcriptional regulation and post-translational modifications in schizont-stage malaria parasites. Genome Biol. *9*, R177.

Francis, S.E., Malkov, V.A., Oleinikov, A.V., Rossnagle, E., Wendler, J.P., Mutabingwa, T.K., Fried, M., and Duffy, P.E. (2007). Six genes are preferentially transcribed by the circulating and sequestered forms of *Plasmodium falciparum* parasites that infect pregnant women. Infect. Immun. *75*, 4838–4850.

Fried, M., Nosten, F., Brockman, A., Brabin, B.J., and Duffy, P.E. (1998). Maternal antibodies block malaria. Nature *395*, 851–852.

Ganesan, K., Ponmee, N., Jiang, L., Fowble, J.W., White, J., Kamchonwongpaisan, S., Yuthavong, Y., Wilairat, P., and Rathod, P.K. (2008). A genetically hard-wired metabolic transcriptome in *Plasmodium falciparum* fails to mount protective responses to lethal antifolates. PLoS Pathog. *4*, e1000214.

Gardner, M.J., Hall, N., Fung, E., White, O., Berriman, M., Hyman, R.W., Carlton, J.M., Pain, A., Nelson, K.E., Bowman, S., *et al.* (2002). Genome sequence of the human malaria parasite *Plasmodium falciparum*. Nature *419*, 498–511.

Ginsburg, H. (2008). Caveat emptor: limitations of the automated reconstruction of metabolic pathways in *Plasmodium*. Trends Parasitol. *25*, 37–43.

Gissot, M., Briquet, S., Refour, P., Boschet, C., and Vaquero, C. (2005). PfMyb1, a *Plasmodium falciparum* transcription factor, is required for intra-erythrocytic growth and controls key genes for cell cycle regulation. J. Mol. Biol. *346*, 29–42.

Gonzales, J.M., Patel, J.J., Ponmee, N., Jiang, L., Tan, A., Maher, S.P., Wuchty, S., Rathod, P.K., and Ferdig, M.T. (2008). Regulatory hotspots in the malaria parasite genome dictate transcriptional variation. PLoS Biol. *6*, e238.

Grewal, S.I., and Jia, S. (2007). Heterochromatin revisited. Nat. Rev. Genet. *8*, 35–46.

Gunasekera, A.M., Myrick, A., Le Roch, K., Winzeler, E., and Wirth, D.F. (2007a). *Plasmodium falciparum*: genome-wide perturbations in transcript profiles among mixed stage cultures after chloroquine treatment. Exp. Parasitol. *117*, 87–92.

Gunasekera, A.M., Myrick, A., Militello, K.T., Sims, J.S., Dong, C.K., Gierahn, T., Le Roch, K., Winzeler, E., and Wirth, D.F. (2007b). Regulatory motifs uncovered among gene expression clusters in *Plasmodium falciparum*. Mol. Biochem. Parasitol. *153*, 19–30.

Hall, N., Karras, M., Raine, J.D., Carlton, J.M., Kooij, T.W., Berriman, M., Florens, L., Janssen, C.S., Pain, A., Christophides, G.K., *et al.* (2005). A comprehensive survey of the *Plasmodium* life cycle by genomic, transcriptomic, and proteomic analyses. Science *307*, 82–86.

Han, J.D., Bertin, N., Hao, T., Goldberg, D.S., Berriz, G.F., Zhang, L.V., Dupuy, D., Walhout, A.J., Cusick, M.E., Roth, F.P., *et al.* (2004). Evidence for dynamically organized modularity in the yeast protein–protein interaction network. Nature *430*, 88–93.

Hoffman, S.L., Billingsley, P.F., James, E., Richman, A., Loyevsky, M., Li, T., Chakravarty, S., Gunasekera, A., Chattopadhyay, R., Li, M., *et al.* (2009). Development of a metabolically active, non-replicating sporozoite vaccine to prevent *Plasmodium falciparum* malaria. Hum. Vaccin. *6*, 97–106.

Hoffman, S.L., and Doolan, D.L. (2000). Malaria vaccines-targeting infected hepatocytes. Nat. Med. *6*, 1218–1219.

Hu, G., Cabrera, A., Kono, M., Mok, S., Chaal, B.K., Haase, S., Engelberg, K., Cheemadan, S., Spielmann, T., Preiser, P.R., *et al.* (2009). Transcriptional profiling of growth perturbations of the human malaria parasite *Plasmodium falciparum*. Nat. Biotechnol. *28*, 91–98.

Hughes, T.R., Marton, M.J., Jones, A.R., Roberts, C.J., Stoughton, R., Armour, C.D., Bennett, H.A., Coffey, E., Dai, H., He, Y.D., *et al.* (2000). Functional discovery via a compendium of expression profiles. Cell *102*, 109–126.

Inselburg, J. (1983). Stage-specific inhibitory effect of cyclic AMP on asexual maturation and gametocyte formation of *Plasmodium falciparum*. J. Parasitol. *69*, 592–597.

Iriko, H., Jin, L., Kaneko, O., Takeo, S., Han, E.T., Tachibana, M., Otsuki, H., Torii, M., and Tsuboi, T. (2009). A small-scale systematic analysis of alternative splicing in *Plasmodium falciparum*. Parasitol. Int. *58*, 196–199.

Jansen, R., Yu, H., Greenbaum, D., Kluger, Y., Krogan, N.J., Chung, S., Emili, A., Snyder, M., Greenblatt, J.F., and Gerstein, M. (2003). A Bayesian networks approach for predicting protein–protein interactions from genomic data. Science *302*, 449–453.

Kanehisa, M., Goto, S., Kawashima, S., Okuno, Y., and Hattori, M. (2004). The KEGG resource for deciphering the genome. Nucleic Acids Res *32*, D277–280.

Kaslow, D.C. (1997). Transmission-blocking vaccines: uses and current status of development. Int. J. Parasitol. *27*, 183–189.

Kato, N., Sakata, T., Breton, G., Le Roch, K.G., Nagle, A., Andersen, C., Bursulaya, B., Henson, K., Johnson, J., Kumar, K.A., *et al.* (2008). Gene expression signatures and small-molecule compounds link a protein kinase to *Plasmodium falciparum* motility. Nat. Chem. Biol. *4*, 347–356.

Khan, S.M., Franke-Fayard, B., Mair, G.R., Lasonder, E., Janse, C.J., Mann, M., and Waters, A.P. (2005). Proteome analysis of separated male and female gametocytes reveals novel sex-specific *Plasmodium* biology. Cell *121*, 675–687.

Kissinger, J.C., Brunk, B.P., Crabtree, J., Fraunholz, M.J., Gajria, B., Milgram, A.J., Pearson, D.S., Schug, J., Bahl, A., Diskin, S.J., *et al.* (2002). The *Plasmodium* genome database. Nature *419*, 490–492.

LaCount, D.J., Vignali, M., Chettier, R., Phansalkar, A., Bell, R., Hesselberth, J.R., Schoenfeld, L.W., Ota, I., Sahasrabudhe, S., Kurschner, C., *et al.* (2005). A protein interaction network of the malaria parasite *Plasmodium falciparum*. Nature *438*, 103–107.

Lasonder, E., Ishihama, Y., Andersen, J.S., Vermunt, A.M., Pain, A., Sauerwein, R.W., Eling, W.M., Hall, N., Waters, A.P., Stunnenberg, H.G., *et al.* (2002). Analysis of the *Plasmodium falciparum* proteome by high-accuracy mass spectrometry. Nature *419*, 537–542.

Lasonder, E., Janse, C.J., van Gemert, G.J., Mair, G.R., Vermunt, A.M., Douradinha, B.G., van Noort, V., Huynen, M.A., Luty, A.J., Kroeze, H., *et al.* (2008). Proteomic profiling of *Plasmodium* sporozoite maturation identifies new proteins essential for parasite development and infectivity. PLoS Pathog. *4*, e1000195.

Le Roch, K.G., Johnson, J.R., Ahiboh, H., Chung, D.W., Prudhomme, J., Plouffe, D., Henson, K., Zhou, Y., Witola, W., Yates, J.R., *et al.* (2008). A systematic approach to understand the mechanism of action of the bisthiazolium compound T4 on the human malaria parasite, *Plasmodium falciparum*. BMC Genomics *9*, 513.

Le Roch, K.G., Johnson, J.R., Florens, L., Zhou, Y., Santrosyan, A., Grainger, M., Yan, S.F., Williamson, K.C., Holder, A.A., Carucci, D.J., *et al.* (2004). Global analysis of transcript and protein levels across the *Plasmodium falciparum* life cycle. Genome Res. *14*, 2308–2318.

Le Roch, K.G., Zhou, Y., Batalov, S., and Winzeler, E.A. (2002). Monitoring the chromosome 2 intraerythrocytic transcriptome of *Plasmodium falciparum* using oligonucleotide arrays. Am. J. Trop. Med. Hyg. *67*, 233–243.

Le Roch, K.G., Zhou, Y., Blair, P.L., Grainger, M., Moch, J.K., Haynes, J.D., De La Vega, P., Holder, A.A., Batalov, S., Carucci, D.J., *et al.* (2003). Discovery of gene function by expression profiling of the malaria parasite life cycle. Science *301*, 1503–1508.

Lelievre, J., Berry, A., and Benoit-Vical, F. (2007). Artemisinin and chloroquine: do mode of action and mechanism of resistance involve the same protagonists? Curr. Opin. Investig. Drugs *8*, 117–124.

Lemieux, J.E., Gomez-Escobar, N., Feller, A., Carret, C., Amambua-Ngwa, A., Pinches, R., Day, F., Kyes, S.A., Conway, D.J., Holmes, C.C., *et al.* (2009). Statistical estimation of cell-cycle progression and lineage commitment in *Plasmodium falciparum* reveals a homogeneous pattern of transcription in ex vivo culture. Proc. Natl. Acad. Sci. U.S.A. *106*, 7559–7564.

Llinas, M., Bozdech, Z., Wong, E.D., Adai, A.T., and DeRisi, J.L. (2006). Comparative whole genome transcriptome analysis of three *Plasmodium falciparum* strains. Nucleic Acids Res. *34*, 1166–1173.

Lopez-Rubio, J.J., Mancio-Silva, L., and Scherf, A. (2009). Genome-wide analysis of heterochromatin associates clonally variant gene regulation with perinuclear repressive centers in malaria parasites. Cell Host Microbe *5*, 179–190.

Luke, T.C., and Hoffman, S.L. (2003). Rationale and plans for developing a non-replicating, metabolically active, radiation-attenuated *Plasmodium falciparum* sporozoite vaccine. J. Exp. Biol. *206*, 3803–3808.

MacCarthy, T., Pomiankowski, A., and Seymour, R. (2005). Using large-scale perturbations in gene network reconstruction. BMC Bioinformatics *6*, 11.

Mackinnon, M.J., Li, J., Mok, S., Kortok, M.M., Marsh, K., Preiser, P.R., and Bozdech, Z. (2009). Comparative transcriptional and genomic analysis of *Plasmodium falciparum* field isolates. PLoS Pathog. *5*, e1000644.

Makanga, M., Bray, P.G., Horrocks, P., and Ward, S.A. (2005). Towards a proteomic definition of CoArtem action in *Plasmodium falciparum* malaria. Proteomics *5*, 1849–1858.

Marcotte, E.M., Pellegrini, M., Thompson, M.J., Yeates, T.O., and Eisenberg, D. (1999). A combined algorithm for genome-wide prediction of protein function. Nature *402*, 83–86.

Miao, J., Fan, Q., Cui, L., and Li, J. (2006). The malaria parasite *Plasmodium falciparum* histones: organization, expression, and acetylation. Gene *369*, 53–65.

Mikolajczak, S.A., Silva-Rivera, H., Peng, X., Tarun, A.S., Camargo, N., Jacobs-Lorena, V., Daly, T.M., Bergman, L.W., de la Vega, P., Williams, J., *et al.* (2008). Distinct malaria parasite sporozoites reveal transcriptional changes that cause differential tissue infection competence in the mosquito vector and mammalian host. Mol. Cell Biol. *28*, 6196–6207.

Militello, K.T., Dodge, M., Bethke, L., and Wirth, D.F. (2004). Identification of regulatory elements in the *Plasmodium falciparum* genome. Mol. Biochem. Parasitol. *134*, 75–88.

Natalang, O., Bischoff, E., Deplaine, G., Proux, C., Dillies, M.A., Sismeiro, O., Guigon, G., Bonnefoy, S., Patarapotikul, J., Mercereau-Puijalon, O., *et al.* (2008). Dynamic RNA profiling in *Plasmodium falciparum* synchronized blood stages exposed to lethal doses of artesunate. BMC Genomics *9*, 388.

Nirmalan, N., Sims, P.F., and Hyde, J.E. (2004a). Quantitative proteomics of the human malaria parasite *Plasmodium falciparum* and its application to studies of development and inhibition. Mol. Microbiol. *52*, 1187–1199.

Nirmalan, N., Sims, P.F., and Hyde, J.E. (2004b). Translational up-regulation of antifolate drug targets in the human malaria parasite *Plasmodium falciparum* upon challenge with inhibitors. Mol. Biochem. Parasitol. *136*, 63–70.

Oakley, M.S., Kumar, S., Anantharaman, V., Zheng, H., Mahajan, B., Haynes, J.D., Moch, J.K., Fairhurst, R., McCutchan, T.F., and Aravind, L. (2007). Molecular factors and biochemical pathways induced by febrile temperature in intraerythrocytic *Plasmodium falciparum* parasites. Infect. Immun. *75*, 2012–2025.

Oshiro, G., Wodicka, L.M., Washburn, M.P., Yates, J.R., 3rd, Lockhart, D.J., and Winzeler, E.A. (2002). Parallel identification of new genes in Saccharomyces cerevisiae. Genome Res. *12*, 1210–1220.

Otto, T.D., Wilinski, D., Assefa, S., Keane, T.M., Sarry, L.R., Bohme, U., Lemieux, J., Barrell, B., Pain, A., Berriman, M., *et al.* (2010). New insights into the blood-stage transcriptome of *Plasmodium falciparum* using RNA-Seq. Mol. Microbiol. *76*, 12–24.

Pain, A., Bohme, U., Berry, A.E., Mungall, K., Finn, R.D., Jackson, A.P., Mourier, T., Mistry, J., Pasini, E.M., Aslett, M.A., *et al.* (2008). The genome of the simian and human malaria parasite *Plasmodium knowlesi*. Nature *455*, 799–803.

Perez-Toledo, K., Rojas-Meza, A.P., Mancio-Silva, L., Hernandez-Cuevas, N.A., Delgadillo, D.M., Vargas, M., Martinez-Calvillo, S., Scherf, A., and Hernandez-Rivas, R. (2009). *Plasmodium falciparum* heterochromatin protein 1 binds to tri-methylated histone 3 lysine 9 and is linked to mutually exclusive expression of var genes. Nucleic Acids Res. *37*, 2596–2606.

Ponts, N., Harris, E.Y., Prudhomme, J., Wick, I., Eckhardt-Ludka, C., Hicks, G.R., Hardiman, G., Lonardi, S., and Le Roch, K.G. (2010). Nucleosome landscape and control of transcription in the human malaria parasite. Genome Res. *20*, 228–238.

Prieto, J.H., Koncarevic, S., Park, S.K., Yates, J., 3rd, and Becker, K. (2008). Large-scale differential proteome analysis in *Plasmodium falciparum* under drug treatment. PLoS One *3*, e4098.

Radfar, A., Diez, A., and Bautista, J.M. (2008). Chloroquine mediates specific proteome oxidative damage across the erythrocytic cycle of resistant *Plasmodium falciparum*. Free Radic. Biol. Med. *44*, 2034–2042.

Rogers, N.J., Hall, B.S., Obiero, J., Targett, G.A., and Sutherland, C.J. (2000). A model for sequestration of the transmission stages of *Plasmodium falciparum*: adhesion of gametocyte-infected erythrocytes to human bone marrow cells. Infect. Immun. *68*, 3455–3462.

Salanti, A., Staalsoe, T., Lavstsen, T., Jensen, A.T., Sowa, M.P., Arnot, D.E., Hviid, L., and Theander, T.G. (2003). Selective upregulation of a single distinctly structured var gene in chondroitin sulphate A-adhering *Plasmodium falciparum* involved in pregnancy-associated malaria. Mol. Microbiol. *49*, 179–191.

Salcedo-Amaya, A.M., van Driel, M.A., Alako, B.T., Trelle, M.B., van den Elzen, A.M., Cohen, A.M., Janssen-Megens, E.M., van de Vegte-Bolmer, M., Selzer, R.R., Iniguez, A.L., *et al.* (2009). Dynamic histone H3 epigenome marking during the intraerythrocytic cycle of *Plasmodium falciparum*. Proc. Natl. Acad. Sci. U.S.A. *106*, 9655–9660.

Sargeant, T.J., Marti, M., Caler, E., Carlton, J.M., Simpson, K., Speed, T.P., and Cowman, A.F. (2006). Lineage-specific expansion of proteins exported to erythrocytes in malaria parasites. Genome Biol. *7*, R12.

Scherf, A., Lopez-Rubio, J.J., and Riviere, L. (2008). Antigenic variation in *Plasmodium falciparum*. Annu. Rev. Microbiol. *62*, 445–470.

Sharrock, W.W., Suwanarusk, R., Lek-Uthai, U., Edstein, M.D., Kosaisavee, V., Travers, T., Jaidee, A., Sriprawat, K., Price, R.N., Nosten, F., *et al.* (2008). *Plasmodium* vivax trophozoites insensitive to chloroquine. Malar. J. *7*, 94.

Shock, J.L., Fischer, K.F., and DeRisi, J.L. (2007). Whole-genome analysis of mRNA decay in *Plasmodium falciparum* reveals a global lengthening of mRNA half-life during the intra-erythrocytic development cycle. Genome Biol. *8*, R134.

Silvestrini, F., Bozdech, Z., Lanfrancotti, A., Di Giulio, E., Bultrini, E., Picci, L., Derisi, J.L., Pizzi, E., and Alano, P. (2005). Genome-wide identification of genes upregulated at the onset of gametocytogenesis in *Plasmodium falciparum*. Mol. Biochem. Parasitol. *143*, 100–110.

Silvestrini, F., Lasonder, E., Olivieri, A., Camarda, G., van Schaijk, B., Sanchez, M., Younis Younis, S., Sauerwein, R., and Alano, P. (2010). Protein export marks the early phase of gametocytogenesis of the human malaria parasite *Plasmodium falciparum*. Mol. Cell Proteomics *9*, 1437–1448.

Sims, J.S., Militello, K.T., Sims, P.A., Patel, V.P., Kasper, J.M., and Wirth, D.F. (2009). Patterns of gene-specific and total transcriptional activity during the *Plasmodium falciparum* intraerythrocytic developmental cycle. Eukaryot Cell *8*, 327–338.

Sinden, R.E. (1982). Gametocytogenesis of *Plasmodium falciparum* in vitro: an electron microscopic study. Parasitology *84*, 1–11.

Singh, B., Kim Sung, L., Matusop, A., Radhakrishnan, A., Shamsul, S.S., Cox-Singh, J., Thomas, A., and Conway, D.J. (2004). A large focus of naturally acquired *Plasmodium knowlesi* infections in human beings. Lancet *363*, 1017–1024.

Singh, K., Agarwal, A., Khan, S.I., Walker, L.A., and Tekwani, B.L. (2007). Growth, drug susceptibility, and gene expression profiling of *Plasmodium falciparum* cultured in medium supplemented with human serum or lipid-rich bovine serum albumin [corrected]. J. Biomol. Screen. *12*, 1109–1114.

Spellman, P.T., Sherlock, G., Zhang, M.Q., Iyer, V.R., Anders, K., Eisen, M.B., Brown, P.O., Botstein, D., and Futcher, B. (1998). Comprehensive identification of cell cycle-regulated genes of the yeast *Saccharomyces cerevisiae* by microarray hybridization. Mol. Biol. Cell *9*, 3273–3297.

Stowers, A., and Carter, R. (2001). Current developments in malaria transmission-blocking vaccines. Expert Opin. Biol. Ther. *1*, 619–628.

Tamez, P.A., Bhattacharjee, S., van Ooij, C., Hiller, N.L., Llinas, M., Balu, B., Adams, J.H., and Haldar, K. (2008). An erythrocyte vesicle protein exported by the malaria parasite promotes tubovesicular lipid import from the host cell surface. PLoS Pathog. *4*, e1000118.

Tarun, A.S., Peng, X., Dumpit, R.F., Ogata, Y., Silva-Rivera, H., Camargo, N., Daly, T.M., Bergman, L.W., and Kappe, S.H. (2008). A combined transcriptome and proteome survey of malaria parasite liver stages. Proc. Natl. Acad. Sci. U.S.A. *105*, 305–310.

Torrentino-Madamet, M., Almeras, L., Desplans, J., Priol, Y.L., Belghazi, M., Pophillat, M., Fourquet, P., Jammes, Y., and Parzy, D. (2010). Global response of *Plasmodium falciparum* to hyperoxia: a combined transcriptomic and proteomic approach. Malar. J. *10*, 4.

Trelle, M.B., Salcedo-Amaya, A.M., Cohen, A.M., Stunnenberg, H.G., and Jensen, O.N. (2009). Global histone analysis by mass spectrometry reveals a high content of acetylated lysine residues in the malaria parasite *Plasmodium falciparum*. J. Proteome Res. *8*, 3439–3450.

Tuikue Ndam, N., Bischoff, E., Proux, C., Lavstsen, T., Salanti, A., Guitard, J., Nielsen, M.A., Coppee, J.Y., Gaye, A., Theander, T., *et al.* (2008). *Plasmodium falciparum* transcriptome analysis reveals pregnancy malaria associated gene expression. PLoS One *3*, e1855.

van Brummelen, A.C., Olszewski, K.L., Wilinski, D., Llinas, M., Louw, A.I., and Birkholtz, L.M. (2009). Co-inhibition of *Plasmodium falciparum* S-adenosylmethionine decarboxylase/ornithine decarboxylase reveals perturbation-specific compensatory mechanisms by transcriptome, proteome, and metabolome analyses. J. Biol. Chem. *284*, 4635–4646.

Vaughan, A.M., O'Neill, M.T., Tarun, A.S., Camargo, N., Phuong, T.M., Aly, A.S., Cowman, A.F., and Kappe, S.H. (2009a). Type II fatty acid synthesis is essential only for malaria parasite late liver stage development. Cell. Microbiol. *11*, 506–520.

Vaughan, A.M., Wang, R., and Kappe, S.H. (2009b). Genetically engineered, attenuated whole-cell vaccine approaches for malaria. Hum. Vaccin. *6*, 107–113.

Vignali, M., Armour, C.D., Chen, J., Morrison, R., Castle, J.C., Biery, M.C., Bouzek, H., Moon, W., Babak, T., Fried, M., *et al.* (2011). NSR-seq transcriptional profiling enables identification of a gene signature of *Plasmodium falciparum* parasites infecting children. J. Clin. Invest. *121*, 1119–1129.

Volz, J., Carvalho, T.G., Ralph, S.A., Gilson, P., Thompson, J., Tonkin, C.J., Langer, C., Crabb, B.S., and Cowman, A.F. (2009). Potential epigenetic regulatory proteins localise to distinct nuclear sub-compartments in *Plasmodium falciparum*. Int. J. Parasitol. *40*, 109–121.

Westenberger, S.J., Cui, L., Dharia, N., and Winzeler, E. (2009). Genome-wide nucleosome mapping of *Plasmodium falciparum* reveals histone-rich coding and histone-poor intergenic regions and chromatin remodeling of core and subtelomeric genes. BMC Genomics *10*, 610.

Westenberger, S.J., McClean, C.M., Chattopadhyay, R., Dharia, N.V., Carlton, J.M., Barnwell, J.W., Collins, W.E., Hoffman, S.L., Zhou, Y., Vinetz, J.M., *et al.* (2010). A systems-based analysis of *Plasmodium* vivax life cycle transcription from human to mosquito. PLoS Negl. Trop. Dis. *4*, e653.

Whitfield, M.L., Sherlock, G., Saldanha, A.J., Murray, J.I., Ball, C.A., Alexander, K.E., Matese, J.C., Perou, C.M., Hurt, M.M., Brown, P.O., *et al.* (2002). Identification of genes periodically expressed in the human cell cycle and their expression in tumors. Mol. Biol. Cell *13*, 1977–2000.

Wiesner, J., Ortmann, R., Jomaa, H., and Schlitzer, M. (2003). New antimalarial drugs. Angew Chem. Int. Ed. Engl. *42*, 5274–5293.

Winzeler, E.A. (2006). Applied systems biology and malaria. Nat. Rev. Microbiol. *4*, 145–151.

Wirth, D., Daily, J., Winzeler, E., Mesirov, J.P., and Regev, A. (2009). In vivo profiles in malaria are consistent with a novel physiological state. Proc. Natl. Acad. Sci. U.S.A. *106*, E70; author reply E71–72.

Wu, Y., Nelson, M.M., Quaile, A., Xia, D., Wastling, J.M., and Craig, A. (2009). Identification of phosphorylated proteins in erythrocytes infected by the human malaria parasite *Plasmodium falciparum*. Malar. J. *8*, 105.

Wuchty, S. (2006). Topology and weights in a protein domain interaction network – a novel way to predict protein interactions. BMC Genomics *7*, 122.

Wuchty, S. (2007). Rich-club phenomenon in the interactome of *P. falciparum* – artifact or signature of a parasitic life style? PLoS One *2*, e335.

Wuchty, S., Adams, J.H., and Ferdig, M.T. (2009). A comprehensive *Plasmodium falciparum* protein interaction map reveals a distinct architecture of a core interactome. Proteomics *9*, 1841–1849.

Wuchty, S., Barabasi, A.L., and Ferdig, M.T. (2006). Stable evolutionary signal in a yeast protein interaction network. BMC Evol. Biol. *6*, 8.

Wuchty, S., and Ipsaro, J.J. (2007). A draft of protein interactions in the malaria parasite *P. falciparum*. J. Proteome Res. *6*, 1461–1470.

Young, J.A., Fivelman, Q.L., Blair, P.L., de la Vega, P., Le Roch, K.G., Zhou, Y., Carucci, D.J., Baker, D.A., and Winzeler, E.A. (2005). The *Plasmodium falciparum* sexual development transcriptome: a microarray analysis using ontology-based pattern identification. Mol. Biochem. Parasitol. *143*, 67–79.

Young, J.A., Johnson, J.R., Benner, C., Yan, S.F., Chen, K., Le Roch, K.G., Zhou, Y., and Winzeler, E.A. (2008). In silico discovery of transcription regulatory elements in *Plasmodium falciparum*. BMC Genomics *9*, 70.

Yu, M., Kumar, T.R., Nkrumah, L.J., Coppi, A., Retzlaff, S., Li, C.D., Kelly, B.J., Moura, P.A., Lakshmanan, V., Freundlich, J.S., *et al.* (2008). The fatty acid biosynthesis enzyme FabI plays a key role in the development of liver-stage malarial parasites. Cell Host Microbe *4*, 567–578.

Zavolan, M., and van Nimwegen, E. (2006). The types and prevalence of alternative splice forms. Curr. Opin. Struct. Biol. *16*, 362–367.

Zhou, Y., Ramachandran, V., Kumar, K.A., Westenberger, S., Refour, P., Zhou, B., Li, F., Young, J.A., Chen, K., Plouffe, D., *et al.* (2008). Evidence-based annotation of the malaria parasite's genome using comparative expression profiling. PLoS One *3*, e1570.

Zhou, Y., Young, J.A., Santrosyan, A., Chen, K., Yan, S.F., and Winzeler, E.A. (2005). In silico gene function prediction using ontology-based pattern identification. Bioinformatics *21*, 1237–1245.

Plasmodium Experimental Genetic Crosses

6

Lisa C. Ranford-Cartwright, Karen L. Hayton and Michael T. Ferdig

Abstract

Experimental genetic crosses mimic the sexual reproduction process, and accompanying genetic recombination, that occurs between individuals of the same species during natural transmission. Experimental crosses performed using rodent and human species of *Plasmodium* have been used to link phenotype and genotype for a variety of traits, and have been particularly useful for understanding phenotypes for which no obvious candidate genes are known. In addition, analysis of experimental crosses has provided insights into the frequency and types of recombination that occur.

Some biological traits are explained by inheritance of single genes, whereas several loci, known as quantitative trait loci (qtl), contribute to a 'complex trait'. Linkage analyses of experimental genetic crosses of *Plasmodium falciparum* have identified parasite loci contributing to resistance to a number of antimalarial drugs such as chloroquine and quinine, as well as loci controlling the ability of parasites to invade erythrocytes of different primate species, the ability to infect mosquitoes, and intraerythrocytic growth rates. Genetic mapping can also identify genomic regions associated with the control of gene expression (expression qtl or eQTL). Genetic crosses and genetic mapping continue to play a significant role in our understanding of malaria parasite biology, transmission, and drug resistance.

Historical overview of genetic crossing experiments in *Plasmodium* parasites

Linkage analysis of experimental genetic crosses has been used to link phenotype and genotype in many eukaryotic organisms. Mating between two parental organisms results in progeny (offspring), some of which inherit a hybrid of the parental genomes through recombination during meiosis. Recombination includes both independent assortment of chromosomes, and crossing over or gene conversion events between homologous chromosomes. Analysis of genetic crosses relies on the processes of recombination to 'shuffle' the parental genomes and distribute them into the progeny. The extent of 'shuffling' varies between different species, but for *Plasmodium* appears to be extensive.

Malaria parasites are haploid throughout their life-cycle, with a brief diploid phase following fertilization of haploid gametes in the mosquito (Janse *et al.*, 1986). Meiosis begins immediately after zygote formation (Sinden and Hartley, 1985), resulting in the formation of haploid products of meiosis. Although it remains to be formally proven, it is likely that each zygote produces a maximum of four meiotic products (Fig. 6.1), each of which then replicates mitotically to produce an oocyst containing thousands of sporozoites of up to four different genotypes. A single mosquito can carry hundreds of oocysts, and thus can produce sporozoites with very many different genotypes, as a result of the random nature of the processes of recombination at meiosis, namely independent assortment of chromosomes and crossing over between homologous chromosomes (Fig. 6.1).

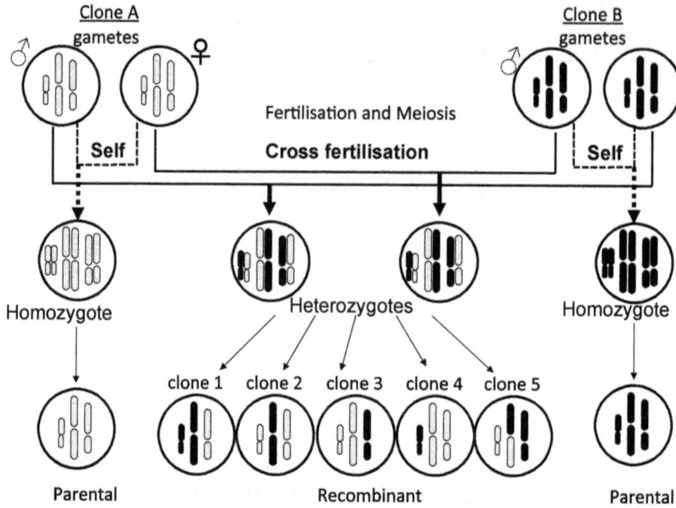

Figure 6.1 Genetics of a cross. Gametes from two parasite clones are shown with three chromosomes for representation purposes. Both self-fertilization events (dotted lines) and cross-fertilization events (solid lines) are equally likely to occur. Following meiosis, all of the progeny from self-fertilization events will be identical to one another and to the original parent clone. Progeny from cross-fertilization events will be different to either parent (recombinants) and probably to each other, as a consequence of independent assortment of chromosomes (e.g. clone 2) and crossing over events within a chromosome (e.g. clone 5, largest chromosome).

In the case of a trait controlled by variation in a single gene, in a haploid organism like *Plasmodium*, each progeny clone will inherit the allele from either one parent or the other, and so in theory the phenotype will fall into two parental groups, with no intermediate phenotypes. Variation in phenotype in this case would be due to environmental effects or experimental noise. The locus controlling the genetic variation in a phenotype can be investigated by examining the genomes of the progeny, and finding the region(s) of the genome that are shared among the parent and progeny that exhibit the phenotypic state. Analysis of genetic crosses is especially powerful for understanding phenotypes for which no obvious candidate genes are known, for example in cases of drug resistance where the mechanism of action of the drug is unknown. Here, the analysis of genetic crosses has a major advantage over population association studies, because variation within the progeny clones is limited to that existing between the two parents used to make the cross. Therefore, smaller numbers of parasites are needed to identify significant loci segregating in genetic crosses than for a genome-wide scan of unrelated isolates in populations, where there are multiple alleles at each locus reflecting a range of geographic and selective factors.

Analysis of experimental genetic crosses has allowed the identification of parasite loci controlling both simple and complex traits, including resistance to antimalarial drugs such as chloroquine and quinine, the ability to invade erythrocytes of different primate species, and the ability to infect mosquitoes.

The first experimental genetic cross performed with malaria parasites in the laboratory used the avian malaria parasite *Plasmodium gallinaceum* (Greenberg and Trembley, 1954). Two strains of parasite were transmitted simultaneously through *Aedes aegypti* mosquitoes: the non-lethal, pyrimethamine-sensitive M strain and the lethal, pyrimethamine –resistant BI strain. The resultant infections in chicks were found to be resistant to pyrimethamine but benign, suggesting that 'hybridization' had occurred between the strains, and drug resistance had been transferred from the more virulent strain to the other.

The field of malaria genetics became established in the early 1970s, when four species of rodent malaria parasites (*Plasmodium berghei, Plasmodium chabaudi, Plasmodium vinckei* and

Plasmodium yoelii) were isolated from African thicket-rats and adapted to laboratory mice (reviewed by Killick-Kendrick and Peters, 1978). The availability of different strains of parasites that could easily complete their life-cycle in the laboratory permitted the study of classical genetics in mammalian malaria parasites. Sexual recombination in *Plasmodium* was further demonstrated using the rodent malaria species *P. yoelii* and *P. chabaudi* (Walliker *et al.*, 1971; Walliker *et al.*, 1973; Walliker *et al.*, 1975; Knowles *et al.*, 1981). These studies utilized phenotypes that were readily distinguished (isoenzymes, drug resistance, virulence and, later, antigens) to show that cross-fertilization between different genotypes readily generated parasites with novel genotypes. Drug-resistant mutants of the rodent malaria species could readily be selected *in vivo*, and could then be crossed to unrelated drug-sensitive parasites, resulting in several genetic studies of drug resistance using *P. chabaudi* (reviewed by Carlton *et al.*, 2001).

Numerous obstacles had to be overcome before an experimental cross using human malaria parasites could be achieved. An *in vitro* culture system for *P. falciparum* was described in 1976 (Trager and Jensen, 1976; Haynes *et al.*, 1976) and it took a further 5 years for successful production of gametocytes (Ifediba and Vanderberg, 1981). It was possible to feed the culture-grown gametocytes to mosquitoes via membrane feeders, and obtain infections resulting in sporozoites (Ifediba and Vanderberg, 1981; Ponnudurai *et al.*, 1982), which were themselves infectious to non-human primates (Campbell *et al.*, 1982). However individual sporozoites could not be analysed to establish if they were recombinant, i.e. had inherited some characteristics from one parent in the cross, and some from the other parent. It was necessary to obtain blood-stage cultures, from which clones could be made and analysed by the limited typing methods available at that time. The sporozoites therefore needed to invade liver hepatocytes, and the merozoites resulting from exo-erythrocytic schizogony could then be used to infect erythrocytes *in vitro*. There was no suitable *in vitro* system for the liver stages, and *P. falciparum* sporozoites will only infect liver cells of certain primates (Coatney *et al.*, 1971; Bray, 1958). It was therefore necessary to use a non-human primate, entailing considerable ethical and logistical difficulties.

In May 1985, the first experimental cross was attempted using two parasite clones, denoted 3D7 and HB3 (Walliker *et al.*, 1987). 3D7 is a clone made for the experimental cross from isolate NF54, originating from a patient in The Netherlands who had never travelled overseas, and is believed to represent a case of airport malaria; the exact geographical origin of the parasite is therefore unknown (Delemarre and van der Kaay, 1979; Ponnudurai *et al.*, 1981). HB3 is a clone of isolate H1 from the Honduras (Trager *et al.*, 1981; Bhasin and Trager, 1984). Both NF54 and HB3 were known to produce gametocytes infectious to mosquitoes (Bhasin and Trager, 1984; Ponnudurai *et al.*, 1982). Gametocytes of each parent clone were grown *in vitro* (Ifediba and Vanderberg, 1981), and equal numbers of gametocytes from each parent were combined, and fed to *Anopheles* mosquitoes through a membrane feeder. Sixteen days later, approximately 500 infected mosquitoes were permitted to feed on a splenectomized chimpanzee, and a further 4×10^6 sporozoites were dissected from mosquito salivary glands and injected intravenously. Parasites appeared in the animal's bloodstream 10 days later, and blood was taken and cultures were quickly established *in vitro*. Individual progeny clones were isolated by limiting dilution for further genetic analysis. A small proportion of clones found to be 'mixtures' of two or more genotypes were discarded.

Analysis of the initial 12 progeny clones derived from the 3D7 × HB3 cross demonstrated that recombination occurred readily and frequently, with 8/12 clones exhibiting non–parental combinations among five markers. Subsequently 113 progeny clones have been isolated from the 3D7 × HB3 cross [L. Ranford-Cartwright, unpublished], of which 101 are recombinant, and 12 are identical to the parent clone 3D7. Amongst the recombinant clones there are at least 55 independent recombinants (based on at least nine markers).

A second experimental cross with *P. falciparum* was performed using HB3 and a chloroquine-resistant parasite, Dd2 (Wellems *et al.*, 1990). Dd2 is a clone of the W2-Mef line (Wellems *et al.*,

1988; Oduola *et al.*, 1988a), which was selected from the W2 clone of the Indochina III isolate originally derived from a Laotian patient who failed chloroquine therapy (Campbell *et al.*, 1982; Oduola *et al.*, 1988b). Seventy six progeny clones were isolated from the cross, of which 16 were found to be independent recombinants (Wellems *et al.*, 1990).

A third genetic cross was carried out using parasite 7G8, a clone of Brazilian isolate IMTM22 (Burkot *et al.*, 1984), and GB4, a clone of isolate Ghana III/CDC (Sullivan *et al.*, 2003). More than 200 progeny clones were obtained by limiting dilution and typed, of which 29 progeny clones were found to be parental. 33 unique independent recombinant clones were identified and analysed further (Hayton *et al.*, 2008; Sa *et al.*, 2009).

The methodology of the second and third *P. falciparum* crosses was identical to the original 3D7 × HB3 cross, except that the primate infections were established by natural mosquito feeds only –the labour-intensive step of dissecting sporozoites from mosquito salivary glands was not included.

Although the *P. falciparum* crosses continue to generate a wealth of information and provide an important resource to the malaria community, generating a cross remains challenging and it is uncertain whether additional crosses will be made in the future. While the complete life cycle has been described *in vitro* for the rodent malaria, *P. berghei* (Al-Olayan *et al.*, 2002), the insect and mammalian liver stages of *P. falciparum* remain frustratingly difficult to cultivate. The phenotypes of clinical interest are usually found in the asexual blood stages, so it is necessary to propagate recombinant sporozoites through the liver in order to obtain blood-stage parasites in tissue culture. In the absence of a major breakthrough in cell culture techniques, a human volunteer or chimpanzee is still needed to allow the cross progeny parasites to complete the liver stage. Crosses are possible in *Aotus* using a clone that is virulent in these monkeys, but the restriction is such that only one genotype can be obtained in a single animal (Hayton *et al.*, 2008). This requirement poses huge ethical and financial challenges to generating a cross, and such an effort will likely be justified in the future only by a pressing clinical need to identify a genetic locus, for example the investigation of the genetic basis of virulence or artemisinin resistance, should either a highly virulent or artemisinin-resistant parasite emerge.

Biased inheritance in the progeny of genetic crosses

All three of the crosses performed produced lower than expected numbers of parental-type clones. If mating between gametes of the parent clones in the mosquito is random (i.e. no favouring of cross-fertilization over self-fertilization), and equal numbers of gametocytes of the two parent clones are fed to mosquitoes, then half of the zygotes produced should be the result of self-fertilization events, from which only parental-type progeny can be produced (Fig. 6.1). Therefore half of the clones isolated were expected to be identical to one or the other parent. In all three crosses, parental-type progeny clones were in the vast minority of those isolated; for example, in the 3D7 × HB3 cross, 12/113 (11%) of progeny clones were 3D7 parental, and none were HB3-like. There is no evidence of favoured cross-fertilization in the oocysts (Ranford-Cartwright *et al.*, 1993), so the predominance of recombinants in the progeny is thought to be due to favoured growth of parasites with certain combinations of alleles in the chimpanzee or culture stages of the crossing experiment. Several of the progeny clones obtained in each cross were identical, indicating the presence of a particular genotype at a higher frequency in the original culture from which clones were made. For example, in the 3D7 × HB3 cross, 25 clones had the same genotype, including clones made from the very first isolation of parasitized blood from the chimpanzee, as well as later clones isolated from blood taken from the chimpanzee several days later (L. Ranford-Cartwright, unpublished). This strongly suggests a parasite genotype with a growth advantage under the experimental crossing and culture conditions. Biased inheritance was observed at certain loci: in the 3D7 × HB3 cross it was noted that 12/12 progeny clones inherited the subtelomeric *HRP-III* gene from the 3D7 parent (Wellems *et al.*, 1987). The HB3 parent does not possess *HRP-III*, and the authors suggested that inheritance of the locus provided a strong selective advantage, possibly related to faster asexual proliferation. Strongly favoured linkage groups were also observed in the 7G8 × GB4 cross in a subtelomeric locus on chromosome 7 that contains the *eba-175* gene (Hayton *et al.*, 2008). EBA-175 is known to interact with glycophorin A on the erythrocyte during invasion (Camus and Hadley, 1985), and different alleles may confer faster growth rates in different primates (Martin *et al.*, 2005).

Linkage analysis methods

There are three components to any linkage analysis study of experimental genetic crosses. The first is to define the inheritance of the phenotype of interest in the progeny of a genetic cross. The second is to identify within the progeny clones the regions of each chromosome inherited from each of the two parents in the cross (generation of the genetic map). The final step is to compare the two, using statistical methods such as maximum likelihood analysis and LOD scores (Fisher, 1922a,b; Haldane and Smith, 1947; Barnard, 1949).

Generation of genetic maps

The principal of genetic mapping is to compare the inheritance pattern of a phenotypic trait with the inheritance pattern of chromosomal regions. The basic method has been in use for almost one hundred years (Sturtevant, 1913). Initial analyses of genetic crosses used markers based on phenotypes, where there was no information about the genomic location of the marker. These analyses were able to prove the existence, and to measure the extent, of recombination. Positional cloning, whereby genes can be isolated on the basis of chromosomal location alone, developed with the advent of molecular biology techniques (Bender *et al.*, 1983; Spierer *et al.*, 1983; Botstein *et al.*, 1980).

Genetic maps define the relative order of loci along a chromosome, where the genetic distance between two loci on the same chromosome is defined as the number of crossover events between those two loci, and is measured in morgans. Crossing over would occur between two markers located one centimorgan (cM) apart in 1 in 100 meioses (0.01 or 1% = 1 cM). Crossover events are more likely to occur between markers further apart on a chromosome than between those close together. The generation of a genetic map requires the measurement of crossover events between multiple loci throughout the genome measured in a large number of meioses.

A genetic map is not the same as a physical map, in which sequence data define the exact position of loci within the chromosome; an increase in genetic distance between two loci usually correlates with an increased physical distance between them. However there is sometimes poor correlation between genetic and physical maps. For example in regions of high recombination activity ('hotspots'), the genetic distance between loci will appear larger than from the physical map.

Choice of markers to establish a genetic map for *Plasmodium* parasites

Early analysis of *Plasmodium* genetic crosses relied on phenotypic markers such as virulence (pathogenicity) and drug resistance (Greenberg and Trembley, 1954) to show that recombination had indeed occurred and that progeny clones had inherited characteristics from both parents. Later studies of rodent genetic crosses (Walliker *et al.*, 1971, 1973, 1975) made extensive use variants of housekeeping enzymes (isoenzymes) which have different mobility during electrophoresis as a result of differences in net electrostatic charge (Carter, 1970, 1972, 1973). Isoenzymes were also used as markers in the first human malaria cross (Walliker *et al.*, 1987; Sanderson *et al.*, 1981). Other markers used to analyse this cross included variants of parasite surface antigens detected by specific antisera or monoclonal antibodies (McBride *et al.*, 1982), size variation of chromosomes run on pulsed field gels (Kemp *et al.*, 1985), and resistance to antimalarial drugs (Thaithong and Beale, 1981).

The use of these markers to analyse experimental genetic crosses established that recombination occurred within the mosquito, and markers appeared to segregate in a Mendelian fashion. To identify genetic loci influencing or causing phenotypic traits of interest, it is necessary to link such phenotypic markers to positions in the genome, or to use DNA-based markers.

Table 6.1 Genetic maps for experimental *Plasmodium* genetic crosses

Species	Genetic cross	Marker type	Number of markers	References
P. falciparum	Dd2 × HB3	RFLP	85	Walker-Jonah *et al.* (1992), Wellems *et al.* (1991)
		Microsatellite	901	Su and Wellems (1996), Su *et al.* (1999)
	3D7 × HB3	SNP (chip)	2870	Ranford-Cartwright *et al.* (unpublished)
	7G8 × GB4	Microsatellite	285	Hayton *et al.* (2008)
P. chabaudi	AJ × AS (3CQ)	RFLP	46	Carlton *et al.* (1998)
	AJ × AS (50S/P)	RFLP	31	Hayton *et al.* (2002)
	AJ × AS	AFLP	672	Martinelli *et al.* (2005b)

The first DNA-based markers to be used for construction of genetic maps were restriction fragment length polymorphisms (RFLPs) (Grodzicker *et al.*, 1974). RFLPs are generally single nucleotide polymorphisms within recognition sites for restriction enzymes. They are used as markers, and thus are not expected to be related to the gene of interest in the trait, but are sufficiently close to that gene to identify its position by genetic linkage. The advent of the polymerase chain reaction (Saiki *et al.*, 1985, 1988) led to the use of other markers such as microsatellites (Weber and May, 1989) and AFLP-based markers (Vos *et al.*, 1995). In recent years, high throughput sequencing methodology has allowed the development of single nucleotide polymorphism (SNP) markers for linkage analysis (Kruglyak, 1997).

The first DNA-based genetic maps of *Plasmodium* (Table 6.1) were developed using RFLP markers for *P. falciparum* (Walker-Jonah *et al.*, 1992) and for *P. chabaudi* (Carlton *et al.*, 1998), and were followed by AFLP marker-based maps for *P. chabaudi* (Grech *et al.*, 2002; Martinelli *et al.*, 2005b). Microsatellites identified within the *P. falciparum* genome were used to establish denser genetic maps containing hundreds of polymorphic markers (Su *et al.*, 1999; Su and Wellems, 1996). High-throughput sequencing allowed the identification of thousands of SNPs within the genome (Mu *et al.*, 2007; Volkman *et al.*, 2007), which in turn led to the development of Affymetrix Molecular Inversion probes and chips (Mu *et al.*, 2010) and a Nimblegen genotyping platform that assays 45,000 SNPs and copy number variation genome-wide (Tan *et al.*, 2011). Although maps based on these SNPs have not yet been published, they are currently in preparation for all three of the *P. falciparum* crosses (Table 6.1) (Jiang *et al.*, 2011; L.C. Ranford-Cartwright and M.T. Ferdig, unpublished).

Statistical analysis of linkage

Linkage analysis examines the co-segregation of a chromosomal region and a trait of interest. It is not possible to describe fully the procedures necessary to carry out statistical analyses of linkage, but a detailed description of the principles can be found in Ott (1999).

Maximum likelihood approaches are widely used to define the linear arrangement of genes, and the recombination frequencies between them (Fisher, 1922a,b). Several methods have been described to detect linkage, but the most commonly used is the LOD (log of the odds) score (Morton, 1955). The LOD score compares the likelihood of obtaining the test data if two loci are actually linked to the likelihood of observing the same data purely by chance (no linkage). A LOD score of 3 or more is generally taken as evidence for linkage (1000 to 1 in favour of linkage observed not occurring by chance). A number of statistical packages were developed to examine linkage of Mendelian, or single locus, traits, including LINKAGE (Lathrop *et al.*, 1984; Lathrop and Lalouel, 1984), LIPED (Ott, 1974) and Mapmaker (Lander *et al.*, 1987).

However most biological traits do not follow such a simple model of inheritance; these are referred to as 'complex' traits. In some cases, several loci contribute to the trait of interest (a polygenic trait) and the loci contributing are known as QTL (Lynch and Walsh, 1997). Linkage analyses of complex traits is more problematic, but typically these involve interval mapping in which maximum likelihood tests are computed on a framework of QTL locations (Lander and Botstein, 1989). Tests to examine multiple QTL have also been developed (reviewed in Zou and Zeng, 2008). Both Bayesian and frequentist approaches have been taken to produce statistical packages such as Pseudomarker (Sen and Churchill, 2001), Mapmaker/qtl (Paterson *et al.*, 1988) and QTL Express (Seaton *et al.*, 2002).

The classical algorithms used in linkage analysis assume that the alleles at the markers being analysed are independent of one another. For markers sparsely distributed throughout the genome this is likely to be the case, but for densely distributed markers (such as SNPs), adjacent SNPs are likely to be associated with one another, a phenomenon known as linkage disequilibrium (LD). Several methods have been proposed to define SNP markers that are in linkage equilibrium, and are therefore suitable for linkage analyses (Purcell *et al.*, 2007; Allen-Brady *et al.*, 2007; Abecasis *et al.*, 2002; Thomas, 2007).

Plasmodium crosses have been analysed for single-locus as well as QTL, which will be discussed below. Analysis is made simpler because the parasite stages undergoing phenotyping are haploid; for a single-locus study there are no heterozygotes, and parasite clones are treated as being fully homozygous at all loci.

Biological insights from experimental genetic crosses

Recombination frequencies

Meiotic recombination (crossing over) is initiated by the formation of double strand breaks (DSB) by a specific endonuclease Spo11 (Keeney *et al.*, 1997). The chromosome breaks are then repaired using the homologous chromosome as a template. Repair can involve a reciprocal exchange between chromosomes (crossing over) or non-reciprocal events (gene conversion) where the DNA sequence of one chromosome is copied onto the homologous pair with no exchange (San Filippo *et al.*, 2008; Chen *et al.*, 2007). The rate of recombination within the genome varies widely between different organisms e.g. 370 cM/Mb for yeast, 1.13 cM/Mb for humans (San Filippo *et al.*, 2008).

The ability to identify candidate genes is dependent on the location of the informative recombination events and an adequate number of markers to capture these recombination events to define the locus. Therefore, the recombination frequency of the organism determines how many progeny need to be phenotyped and genotyped to narrow a locus to identify candidate genes, as well as offering insights into the recombination rate and production of new genotypes during natural transmission of malaria.

In *P. falciparum*, the recombination rate was estimated to be ~11 kb/cM in the 3D7×HB3 cross (L.C. Ranford-Cartwright, unpublished), ~15 kb/cM in the Dd2×HB3 cross (Su *et al.*, 1999) and ~36 kb/cM in the 7G8×GB4 cross (Hayton *et al.*, 2008). Recent analysis of the third cross using comparative genome hybridization suggests recombination was under-estimated previously and the frequency is much closer to that observed in the earlier crosses (H. Jiang and X. Su, unpublished). The recombination frequencies for the three *P. falciparum* crosses are comparable to the 14 kb/cM obtained for *P. chabaudi* (Martinelli *et al.*, 2005b). Extensive recombination provides a mechanism by which the parasite is able to generate genetic diversity. In controlled crossing experiments, it facilitates the identification of loci and candidate genes using surprisingly few individual recombinant progeny clones. For example, the *Pfmdv1* locus underlying the male gametocyte defect, and the chloroquine-resistance locus on chromosome 7, were initially mapped using just 11 or 14 progeny clones, respectively (Wellems *et al.*, 1990; Vaidya *et al.*, 1995).

An alternative approach is to infer recombination frequencies from population genetic data (Mu *et al.*, 2005). In natural infections of *Plasmodium*, recombination can only be observed if the host is infected with parasites of different genotypes, so the effective recombination rate is influenced by multiplicity of infection, inbreeding and transmission rates. The results suggest a population recombination rate from 400/Mb in

American *P. falciparum* populations, to 10^5/Mb in African populations (Mu *et al.*, 2005).

Hotspots of recombination

Crossing over/gene conversion events are not randomly distributed throughout the genome. In human and yeast genomes, 'hotspots' -highly localized short regions (70–250 bp) with high frequencies of crossing over and gene conversion events- have been identified (Jeffreys *et al.*, 2001; Baudat and Nicolas, 1997). Whereas bacterial recombination often occurs at a 'chi' sequence (cross-over hot-spot instigator; Smith *et al.*, 1995), there appears to be with no consensus sequence for eukaryotic hotspots of recombination. Some hotpots are associated with open chromatin (e.g. transcriptionally active, or areas with nucleosome-excluding sequences) (reviewed in Nishant and Rao, 2006), but open chromatin by itself does not necessarily result in a hotspot (Wu and Lichten, 1995).

There is very little information on recombination hotspots in *Plasmodium*. Genetic crosses offer the opportunity to identify the position of recombination events but require high-density markers or complete sequence information, since the accurate resolution of crossover breakpoints relies on the density of polymorphic SNPs/sequence coverage. Crossing over/gene conversion events have been mapped at high resolution in two progeny clones from the Dd2 × HB3 cross (Samarakoon *et al.*, 2011). Crossover events occurred within a 80–100kb window, but a wide range of crossover/conversion lengths were observed. It has not been possible to locate hotspots yet, as only two progeny clones from one cross have been analysed, but this information is likely to emerge as sequencing of whole genomes (progeny clones) becomes cheaper and faster.

Many of the recombination events inferred from population data occurred in the subtelomeric regions of the chromosome (Mu *et al.*, 2005), possibly due to the presence of multi-copy gene families such as *var*, *stevor* and *rifin*. Recombination is thought to be an important factor in the generation of diversity in *var* genes (Taylor *et al.*, 2000; Ward *et al.*, 1999), although recombination during mitosis may explain the high frequencies seen for the telomeric *var* genes (Freitas-Junior *et al.*, 2000).

Mapping of simple phenotypes using linkage analysis: identifying drug resistance genes

Phenotypes that are largely determined by a single gene are the easiest to map because they follow classical Mendelian inheritance, such that most inherited variation in the trait is due to a simple difference (e.g. a point mutation in a gene) at a single locus. The majority of mapping studies to date in *P. falciparum* have focused on drug resistance, mainly because there is an urgent need to understand the mechanism of resistance, and also because drug-resistance is often thought to be genetically simple (i.e. conferred by a single mutation event), lending itself to identification of genes by linkage mapping.

Experimental genetic crosses between pyrimethamine-resistant and pyrimethamine-sensitive strains of the rodent species *P. chabaudi* and *P. yoelii* showed that resistance to the drug segregated as a single gene in both species (Walliker *et al.*, 1975; Knowles *et al.*, 1981). Pyrimethamine resistance also segregated as a single locus the 3D7 × HB3 cross (Walliker *et al.*, 1987). The identification of the locus as the dihydrofolate resistance gene (*dhfr*) and the role of specific mutations within the gene in the resistance phenotype were discovered without the use of linkage analysis (Chen *et al.*, 1987; Cowman *et al.*, 1988), but subsequent analysis of crosses for which pyrimethamine resistance was a segregating trait provided definitive proof of the role of *dhfr*, prior to the development of transfection techniques (Peterson *et al.*, 1988; Wu *et al.*, 1996).

Chloroquine resistance was the first phenotype to be mapped in *P. falciparum* for which the drug target was completely unknown. In a landmark study, Wellems and colleagues (1990) crossed a chloroquine-resistant clone (Dd2) with a chloroquine-sensitive clone (HB3) to produce progeny clones that could be cleanly categorized as sensitive or resistant to chloroquine. The *P. falciparum* P-glycoprotein homologue, *Pfmdr1*, a leading candidate gene, was shown not to be the major determinant of chloroquine response in the Dd2 × HB3 genetic background, because polymorphisms in that gene did not segregate with resistance. The chloroquine resistance locus was instead mapped to chromosome 7 (Wellems *et al.*, 1991). However, it took almost another 10 years to isolate additional progeny (now totalling 35 identified independent recombinants), including several with recombination events that narrowed the chromosome 7 locus (Kirkman *et al.*, 1996), develop a high density microsatellite marker map (Su and Wellems, 1996) and develop techniques to clone and sequence large stretches of AT-rich DNA, in order to narrow the locus to a 36 kb segment. At first, the region was thought to contain nine genes, of which candidate gene 2 (*cg2*) showed the strongest correlation with chloroquine response (Su *et al.*, 1997). However, allelic replacement studies showed that mutations in *cg2* did not confer chloroquine resistance (Fidock *et al.*, 2000a) and further analysis identified an additional gene, *Pfcrt*, that had been overlooked previously in the locus because of a highly unusual 13 exon gene structure (Fidock *et al.*, 2000b). Several mutations in *Pfcrt* were identified in chloroquine-resistant *P. falciparum* parasites, of which a lysine to threonine change at position 76 (K76T) proved to be an absolute requirement for resistance (Fidock *et al.*, 2000b; Djimde *et al.*, 2001; Sidhu *et al.*, 2002). However, despite knowing the identity of the mutations underlying chloroquine resistance for a just over a decade, the molecular mechanism by which a resistant parasite evades the toxic effects of chloroquine remains elusive (reviewed in Hayton and Su, 2008).

In a recent twist, a role for *Pfmdr1* in chloroquine resistance has been confirmed using genetic crosses. In 7G8 × GB4, a cross between two chloroquine-resistant clones from Brazil and Ghana, the lower-level chloroquine resistance observed in progeny harbouring the South American *Pfcrt* allele was recently demonstrated to be a result of the contributions from both the *Pfcrt* and *Pfmdr1* alleles (Sa *et al.*, 2009). This interaction might reflect different drug use histories on the two continents from which the parents, 7G8 and GB4, were derived, and provides an explanation for many earlier observations of differences between chloroquine-resistant parasites from Africa compared to South America. Moreover, *Pfmdr1* can influence the degree of resistance in chloroquine-resistant progeny of the Dd2 × HB3 cross (Patel *et al.*, 2010), further explaining conflicting reports of a role of *Pfmdr1* as a possible modifier of chloroquine resistance. Together these studies highlight the genetic complexity that can underpin even simple phenotypes, such that combinations of alleles may play different roles in distinct genetic backgrounds as captured in the different *P. falciparum* crosses (see 'Complex phenotypes' below). This illustrates the value of mapping the same traits in different crosses and the need for future genetic crosses.

Chloroquine resistance has also been investigated in the rodent parasite *P. chabaudi*. Analysis of a cross between parasite clones sensitive and resistant to the drug mapped a chloroquine resistance locus to *P. chabaudi* chromosome 11 (Carlton *et al.*, 1998; Hunt *et al.*, 2004b). Further analyses of the cross showed no association, either of *Pccg10*, the *P. chabaudi* homologue of *Pfcrt*, or of *Pcmdr*, to chloroquine resistance (Hunt *et al.*, 2004a). Chloroquine resistance in *P. chabaudi* may provide insights into chloroquine-resistance in *P. vivax*, which also is not associated with mutations in *Pvcg10* (Nomura *et al.*, 2001) or in *Pvmdr1* (Sa *et al.*, 2005), or the general mechanism by which the digestive vacuole overcomes chloroquine.

Mefloquine resistance has been investigated in the *P. chabaudi* model. A cross between parasite clones sensitive and resistant to this drug showed that over-expression of *Pcmdr*, as a result of chromosomal duplication and translocation, conferred mefloquine resistance, although other loci must be involved as a few resistant progeny inherited wild-type *Pcmdr* alleles (Cravo *et al.*, 2003). The identification of both *Pcdhfr* and *Pcmdr* in drug resistance mapping studies demonstrates the usefulness of the rodent model system for some traits.

Mapping of simple phenotypes using linkage analysis: non-drug resistance phenotypes

Following the efforts to map chloroquine resistance, the resources developed to analyse the Dd2 × HB3 cross permitted the investigation of other traits that were segregating in the cross. These studies include mapping a male gamete defect in the Dd2 parent to chromosome 12, later narrowed to a single gene, *Pfmdv1* (Vaidya *et al.*, 1995; Furuya *et al.*, 2005), confirmation of mutations in the gene encoding dihydropteroate synthase (*Pfdhps*) in conferring sulfadoxine resistance (Wang *et al.*, 1997), and the identification of a role for *Pfmdr1* in food vacuole solute import (Rohrbach *et al.*, 2006).

The completion of the 7G8 × GB4 cross in 2002 resulted in the first study to investigate the genetic basis of virulence in *P. falciparum*. The GB4 parasite is highly virulent in *Aotus nancymaae* monkeys, an animal model used in drug and vaccine development, while the 7G8 clone is unable to produce an infection in these animals due to its inability to invade *A. nancymaae* erythrocytes (Hayton *et al.*, 2008). Using the microsatellite markers developed for the Dd2 × HB3 cross (Su and Wellems, 1996), the erythrocyte invasion trait was mapped to a 14 kb region in the sub-telomeric region of chromosome 4, which contained just two genes, both members of the reticulocyte binding protein homologue (RH) family, *Pfrh4* and *Pfrh5* (Hayton *et al.*, 2008). Allelic replacement studies showed that a single polymorphism in the *Pfrh5* gene, which resulted in an amino acid change from isoleucine to lysine at codon 204, conferred the ability to invade *A. nancymaae* erythrocytes in the 7G8 × GB4 genetic background. Examination of the *Pfrh5* alleles found in the handful of *A. nancymaae* virulent clones characterized to date suggested that additional polymorphisms in *Pfrh5*, in particular the presence of a positive charge at amino acid 407, also confer the ability to invade *A. nancymaae* erythrocytes, and this has recently been confirmed using allelic modification (K.L. Hayton, unpublished observations).

Mapping of complex phenotypes using QTL analysis

Most phenotypes, including growth rate, cell cycle control, and drug susceptibilities are complex, i.e. are determined by multiple loci indicating contributions from two to many genes. Such complex phenotypes present greater challenges for mapping candidate genes, but also hold crucial clues about gene interactions and secondary pathways that underpin key traits. Currently, the statistical power for detailed association mapping of quantitative traits in all three *P. falciparum* genetic crosses is limited by the number of progeny clones available. It is likely that thousands of independent recombinant progeny are present in chimpanzee blood frozen from the 3D7 × HB3 (Kerr *et al.*, 1994) and the 7G8 × GB4 crosses (Su *et al.*, 2007), although it may be that due to a number of dominant clones, the number of progeny available from frozen stocks of the Dd2 × HB3 cross has reached a maximum (Kirkman *et al.*, 1996).

The first quantitative trait analysis in *P. falciparum* investigated genes contributing to low level quinine resistance in the Dd2 × HB3 cross (Ferdig *et al.*, 2004). The analysis revealed loci on chromosomes 5, 7 and 13, corresponding to *Pfmdr1*, *Pfcrt* and a locus containing *Pfnhe1*, acting in an additive fashion, and interactive effects between two loci on chromosomes 9 and 6 and the two QTL on chromosomes 13 and 7 (Ferdig *et al.*, 2004).

A QTL mapping approach can provide evolutionary insights about both past and future drug resistance. For example, Sa and colleagues (2009) demonstrated the pivotal influence on chloroquine susceptibility of the genetic backgrounds in which *Pfcrt* mutations are inherited, indicating that different drug use policies in different geographic regions led to differentially tuned allele combinations. These authors proposed that *Pfmdr1* mediates a more stable, persistent form of chloroquine resistance in South America and raised the concern that this more stable *Pfmdr1* × *Pfcrt* allelic combination of drug-resistant parasites could be induced in Africa by increasing use of amodiaquine. That drug pressures can shape the genome to influence other drug susceptibilities has important implications for how drugs and drug combinations are developed and employed.

Another QTL analysis of the Dd2 × HB3 cross revealed that intraerythrocytic cycle time was linked to a major QTL on chromosome 4, and two additive loci on chromosome 14, and a locus on chromosome 4 that interacted epistatically with the major QTL on chromosome 12. Together these loci accounted for 75% of the variation seen in cycle time in the progeny clones (Reilly *et al.*, 2007; Reilly Ayala *et al.*, 2010).

QTL analysis has also been carried out on phenotypes of the mosquito-stages of the parasite (L.C. Ranford-Cartwright and J.M. Mwangi, unpublished). QTL analysis of progeny from the 3D7 × HB3 cross revealed a major locus on chromosome 12 which accounted for 94% of the variation in infection intensity (number of oocysts per mosquito midgut) and 44% of the variation in infection prevalence (proportion of mosquitoes infected).

Recently, high-throughput chemical screens have permitted studies to investigate drug response traits in *P. falciparum* on a much larger scale. To identify parasite growth inhibitors, Yuan *et al.* (2009) screened seven parasite clones, including all the parents of the crosses, with over 1200 bioactive chemicals. Hundreds of anti-parasitic compounds were identified, 149 of which produced just over 600 differential chemical response phenotypes between the seven parasite clones. Three of these differential response phenotypes were selected for mapping analysis in the 7G8 × GB4 cross; dihydroergotamine methanesulfonate (DHMS, a serotonin receptor antagonist), trimethoprim (a DHFR inhibitor used to treat urinary tract infections) and triamterene (a Na^+ channel blocker). The differential response phenotype for both trimethoprim and triamterene mapped to *pfdhfr*, while parasite response to DHMS mapped to *pfmdr1*. This study showed that mapping differential parasite responses to bioactive chemicals can be used to investigate mechanisms of drug action and resistance in malaria and other diseases, as well as identifying many potential antimalarial compounds. It is likely that this is the first of many studies that exploit the genetic crosses using high throughput techniques (Fig. 6.2).

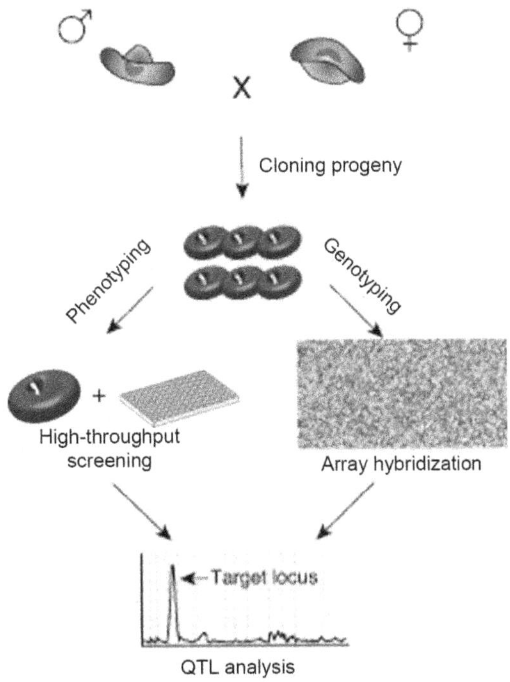

Figure 6.2 High-throughput mapping exploiting genetic crosses and new technologies (adapted from Yuan *et al.*, 2009).

Linkage group selection: an alternative method in rodent malaria parasites

An innovative method for mapping traits linked to particular phenotypes, known as linkage group selection (LGS), has recently been developed for rodent malaria parasite genetic studies (Culleton *et al.*, 2005). This powerful method eliminates the need for cloning individual recombinant progeny by exposing the entire uncloned progeny to selection and analysing the surviving progeny for regions of the genome under selection. Initial studies showed that selection on *Pcdhfr* could readily be detected following the pyrimethamine treatment of mice infected with progeny derived from a cross between a wild-type *P. chabaudi* clone and another clone with a mutant *Pcdhfr* (Culleton *et al.*, 2005). The LGS method has been used in the analysis of progeny from both *P. yoelii* and *P. chabaudi* crosses. The merozoite surface protein gene *msp1* mapped as the major parasite determinant of strain-specific immunity in two separate *P. chabaudi* crosses (Martinelli *et al.*, 2005a; Pattaradilokrat *et al.*, 2007). The apical membrane antigen gene *Pcama1* was also found to associate, albeit weakly, with strain-specific immunity (Pattaradilokrat *et al.*, 2007). The *P. yoelii* erythrocyte binding ligand (*Pyebl*) was identified as a major gene controlling growth rate in mice infected with *P. yoelii* (Pattaradilokrat *et al.*, 2009). LGS has also identified two different mutations in a gene encoding a deubiquitinating enzyme, selected by either chloroquine or artesunate pressure (Hunt *et al.*, 2007).

LGS has not yet been applied to human malaria, although in the Dd2 × HB3 cross, the absence of selection on *Pfmdr1* following chloroquine treatment of the uncloned progeny provided early evidence of the lack of an association between *Pfmdr1* and high-level chloroquine-resistance (Wellems *et al.*, 1990). Future crosses using human malaria parasites could use LGS before any attempt is made to clone and characterize individual recombinant progeny.

Systems genetics: mapping transcript levels (eQTL)

Polymorphisms in protein-coding sequence are often thought to be primary determinants of phenotypes, but regulatory variation is proving to be a powerful engine of phenotypic variation. Classical genetic linkage mapping can interface powerfully with large-scale genomic datasets. The path from DNA variation to downstream phenotypes must pass through several layers of a biological information, including transcripts, proteins, metabolites, and various physiological processes. It is becoming relatively cost-effective to capture these layers of information as 'phenotypes' using emerging high-throughput technologies. The relative abundance of each transcript in the genome can be easily measured as a quantitative trait in the progeny of a cross and mapped to the locus that controls its level.

Expression QTL (eQTL) mapping is part of the emerging field of 'systems genetics' and can provide a view of regulatory architecture of the genome's expression by revealing *cis* and *trans* regulatory loci (Gonzales *et al.*, 2008). Transcripts provide a comprehensive snapshot of biological activity generated by the genome; consequently, inherited transcript levels inherited in the progeny point to underlying genetic variation that regulates divergent strains, for example the shifts in pathways associated with drug susceptibilities or growth rates. These variable regulators could be promoters or transcription factors, but also any type of underlying DNA polymorphism that influences transcript pools, e.g. signalling cascades, epigenetic components, and non-coding RNAs. In this way, transcriptional variation contributes to strain-specific phenotype differences and gene regulation is an important target of evolution.

In *P. falciparum*, extensive heritable variation has been observed in transcript levels (Gonzales *et al.*, 2008), demonstrating that in spite of an apparently rigid transcriptional program through the developmental cycle of these parasites (Bozdech *et al.*, 2003), strain-to-strain differences in transcriptional control are prevalent. Transcript level variation at 18h post RBC invasion was strongly influenced by parasite genotype, such that nearly 18% of genes were regulated by a significant eQTL (Gonzales *et al.*, 2008). The controlling eQTLs were distributed throughout the genome with regulatory loci occurring both near to (implying a *cis* regulatory mechanism) and far from the genes they regulated. Several *trans* regulatory hotspots were identified, including one associated with the amplification event on chromosome 5 harbouring *Pfmdr1* and 13 other genes (inherited from Dd2) that influenced the levels of 269 transcripts around the genome. The 874 genes regulated by eQTL encode proteins are involved in numerous biochemical pathways and regulatory processes (Gonzales *et al.*, 2008; Huang *et al.*, 2009; Rider *et al.*, 2010) that contribute to phenotypic divergence between HB3 and Dd2. Mutation and copy number variation can effectively rewire the transcriptional programme in different malaria parasite strains with potentially profound effects on downstream traits.

Crosses in other human *Plasmodium* species

Experimental genetic crosses have so far only been reported for one of the *Plasmodium* species that infects humans. This is because the other four species could not be grown *in vitro*, which would be necessary to obtain clones suitable for genetic crossing work, and to obtain gametocytes which could be fed through membranes to mosquitoes. These problems have not yet been overcome, although some progress has been made (Golenda *et al.*, 1997). A genetic cross could also be performed using parasites taken from infected patients (*ex vivo*) but in this case it would be difficult to ensure that each patient carried a single genotype of a species, and that equal numbers of gametocytes of each genotype were present in the mixture fed to mosquitoes. However with the emergence and spread of resistance to chloroquine and primaquine in *P. vivax* (Baird, 2009), genetic analysis may offer a route to understanding the basis of resistance in that species.

Malaria genetics in the post-genome era

The power to detect loci under selection using genome-wide association studies (GWAS) is continuing to improve at an impressive rate (see Chapter 4). However, classical linkage mapping using progeny from genetic crosses will continue to offer unique opportunities for unbiased discovery of genetic mechanisms, because the variation in the progeny is derived from only two parents, and laboratory clones can be phenotyped repeatedly to eliminate both biological and measurement noise that can overwhelm studies using field isolates with diverse genetic backgrounds. The three existing *P. falciparum* crosses provide unique but complementary contrasts of various geographic and evolutionary histories, including independent origins of chloroquine resistance and different multi-drug resistance profiles. Moreover, these crosses have only begun to be tapped for the wealth of information they can provide regarding subtle divergence in strain-specific regulation of cellular pathways and molecular processes.

Ultimately, the resolving power of a genetic cross to narrow down the association with a phenotype to a small number of genes is limited by the extent of recombination within each progeny clone, and thus by the number of progeny clones available. For two of the existing crosses (3D7 × HB3 and GB4 × 7G8), it should be possible to obtain further independent recombinants (with crossover events in different places). High-density genetic maps are already available – even higher density maps are not more useful as they are affected by crossover interference between markers. Complete sequence information for the progeny clones can assist in identifying precise points of recombination (chromosome breakpoints) and the possible identification of hot spots of recombination.

In any genetic mapping study, identifying loci is often the easy part– the real challenge is mapping the precise polymorphism that determines the trait. This step, known as positional cloning, has been revolutionized by genome sequencing and high throughput technologies. A QTL peak in *P. falciparum* might contain from 10 to 100 genes. The genes serve as an entry point to query an ever-expanding gene database (PlasmoDB, http://www.plasmodb.org) to search for SNPs that distinguish the parents, identify the type of SNP (non-coding, nonsense, synonymous, non-synonymous), and to explore the available transcriptome data. Classical traits can be correlated with transcriptional profiles to find patterns of expression and functional enrichment in a co-expression network to look for pathways associated with, e.g. drug resistance or virulence phenotypes. The genetic linkage maps continue to be refined at increasing resolution, revealing new crossovers and facilitating positional mapping of candidate genes. This resolution will eventually include the complete genome sequence for all progeny as a community resource that will empower a broad effort to comprehensively track segregating variation and its influence on phenotypes.

It is increasingly evident that understanding the role of any gene in building a phenotype requires knowledge of the networks of gene products in which that gene functions. Such network views require large-scale datasets such as the complete set of transcripts, metabolites and interacting proteins in a cell. Moreover, chemical perturbations through small-molecule screens can also provide a high dimensional view of parasite responses, an approach described as 'chemical genetics', that was recently conducted in malaria parasite progeny by Yuan and colleagues (2009).

Web resources

PlasmoDB database for malaria (http://www.plasmodb.org)

Laboratory of Statistical Genetics, Rockefeller University (http://linkage.rockefeller.edu/)

R/qtl: Software for mapping qtl in experimental crosses (http://www.rqtl.org/)

References

Abecasis, G.R., Cherny, S.S., Cookson, W.O., and Cardon, L.R. (2002). Merlin-rapid analysis of dense genetic maps using sparse gene flow trees. Nat. Genet. *30*, 97–101.

Al-Olayan, E.M., Beetsma, A.L., Butcher, G.A., Sinden, R.E., and Hurd, H. (2002). Complete development of mosquito phases of the malaria parasite *in vitro*. Science *295*, 677–679.

Allen-Brady, K., Horne, B.D., Malhotra, A., Teerlink, C., Camp, N.J., and Thomas, A. (2007). Analysis of high-density single-nucleotide polymorphism data: three novel methods that control for linkage disequilibrium between markers in a linkage analysis. BMC. Proc. *1* (Suppl. 1), S160.

Baird, J.K. (2009). Resistance to therapies for infection by *Plasmodium vivax*. Clin. Microbiol. Rev. *22*, 508–534.

Barnard, G.A. (1949). Statistical inference. J. Roy. Stat. Soc. Series B (Methodological) *11*, 115–149.

Baudat, F., and Nicolas, A. (1997). Clustering of meiotic double-strand breaks on yeast chromosome III. Proc. Natl. Acad. Sci. U.S.A. *94*, 5213–5218.

Bender, W., Spierer, P., and Hogness, D.S. (1983). Chromosomal walking and jumping to isolate DNA from the Ace and rosy loci and the bithorax complex in *Drosophila melanogaster*. J. Mol. Biol. *168*, 17–33.

Bhasin, V.K., and Trager, W. (1984). Gametocyte-forming and non-gametocyte-forming clones of *Plasmodium falciparum*. Am. J. Trop. Med. Hyg. *33*, 534–537.

Botstein, D., White, R.L., Skolnick, M., and Davis, R.W. (1980). Construction of a genetic linkage map in man using restriction fragment length polymorphisms. Am. J. Hum. Genet. *32*, 314–331.

Bozdech, Z., Llinas, M., Pulliam, B.L., Wong, E.D., Zhu, J., and DeRisi, J.L. (2003). The transcriptome of the intraerythrocytic developmental cycle of *Plasmodium falciparum*. PLoS Biol. *1*, E5.

Bray, R.S. (1958). Studies on malaria in chimpanzees. VI. *Laverania falciparum*. Am. J. Trop. Med. Hyg. *7*, 20–24.

Burkot, T.R., Williams, J.L., and Schneider, I. (1984). Infectivity to mosquitoes of *Plasmodium falciparum* clones grown *in vitro* from the same isolate. Trans. R. Soc. Trop. Med. Hyg. *78*, 339–341.

Campbell, C.C., Collins, W.E., Nguyen-Dinh, P., Barber, A., and Broderson, J.R. (1982). *Plasmodium falciparum* gametocytes from culture *in vitro* develop to sporozoites that are infectious to primates. Science *217*, 1048–1050.

Camus, D., and Hadley, T.J. (1985). A *Plasmodium falciparum* antigen that binds to host erythrocytes and merozoites. Science *230*, 553–556.

Carlton, J., Mackinnon, M., and Walliker, D. (1998). A chloroquine resistance locus in the rodent malaria parasite *Plasmodium chabaudi*. Mol. Biochem. Parasitol. *93*, 57–72.

Carlton, J.M., Hayton, K., Cravo, P.V., and Walliker, D. (2001). Of mice and malaria mutants: unravelling the genetics of drug resistance using rodent malaria models. Trends Parasitol. *17*, 236–242.

Carter, R. (1970). Enzyme variation in *Plasmodium berghei*. Trans. R. Soc. Trop. Med. Hyg. *64*, 401–406.

Carter, R. (1972). Electrophoretic forms of glucose phosphate isomerase in oocysts and blood parasites of *Plasmodium berghei*. Trans. R. Soc. Trop. Med. Hyg. *66*, 542.

Carter, R. (1973). Enzyme variation in *Plasmodium berghei* and *Plasmodium vinckei*. Parasitology *66*, 297–307.

Chen, G.X., Mueller, C., Wendlinger, M., and Zolg, J.W. (1987). Kinetic and molecular properties of the dihydrofolate reductase from pyrimethamine-sensitive and pyrimethamine-resistant clones of the human malaria parasite *Plasmodium falciparum*. Mol. Pharmacol. *31*, 430–437.

Chen, J.M., Cooper, D.N., Chuzhanova, N., Ferec, C., and Patrinos, G.P. (2007). Gene conversion: mechanisms, evolution and human disease. Nat. Rev. Genet. *8*, 762–775.

Coatney, G.R., Collins, W.E., Warren, M., and Contacos, P.G. (1971). The Primate malarias. (Bethesda, MD: National Institute of Allergy and Infectious Diseases).

Cowman, A.F., Morry, M.J., Biggs, B.A., Cross, G.A., and Foote, S.J. (1988). Amino acid changes linked to pyrimethamine resistance in the dihydrofolate reductase-thymidylate synthase gene of *Plasmodium falciparum*. Proc. Natl. Acad. Sci. U.S.A. *85*, 9109–9113.

Cravo, P.V., Carlton, J.M., Hunt, P., Bisoni, L., Padua, R.A., and Walliker, D. (2003). Genetics of mefloquine resistance in the rodent malaria parasite *Plasmodium chabaudi*. Antimicrob. Agents Chemother. *47*, 709–718.

Culleton, R., Martinelli, A., Hunt, P., and Carter, R. (2005). Linkage group selection: rapid gene discovery in malaria parasites. Genome Res. *15*, 92–97.

Delemarre, B.J., and van der Kaay, H.J. (1979). [Tropical malaria contracted the natural way in the Netherlands]. Ned. Tijdschr. Geneeskd. *123*, 1981–1982.

Djimde, A., Doumbo, O.K., Cortese, J.F., Kayentao, K., Doumbo, S., Diourte, Y., Dicko, A., Su, X.Z., Nomura, T., Fidock, D.A., Wellems, T.E., Plowe, C.V., and Coulibaly, D. (2001). A molecular marker for chloroquine-resistant falciparum malaria. N. Engl. J. Med. *344*, 257–263.

Ferdig, M.T., Cooper, R.A., Mu, J., Deng, B., Joy, D.A., Su, X.Z., and Wellems, T.E. (2004). Dissecting the loci of low-level quinine resistance in malaria parasites. Mol. Microbiol. *52*, 985–997.

Fidock, D.A., Nomura, T., Cooper, R.A., Su, X., Talley, A.K., and Wellems, T.E. (2000a). Allelic modifications of the cg2 and cg1 genes do not alter the chloroquine response of drug-resistant *Plasmodium falciparum*. Mol. Biochem. Parasitol. *110*, 1–10.

Fidock, D.A., Nomura, T., Talley, A.K., Cooper, R.A., Dzekunov, S.M., Ferdig, M.T., Ursos, L.M., Sidhu, A.B., Naude, B., Deitsch, K.W., Su, X.Z., Wootton, J.C., Roepe, P.D., and Wellems, T.E. (2000b). Mutations in the *P. falciparum* digestive vacuole transmembrane protein PfCRT and evidence for their role in chloroquine resistance. Mol. Cell *6*, 861–871.

Fisher, R.A. (1922a). On the mathematical foundations of theoretical statistics. Phil. Trans. Roy. Soc. of Lond. A *222*, 309–368.

Fisher, R.A. (1922b). The systematic location of genes by means of crossover observations. Am. Nat. *56*, 406.

Freitas-Junior, L.H., Bottius, E., Pirrit, L.A., Deitsch, K.W., Scheidig, C., Guinet, F., Nehrbass, U., Wellems, T.E., and Scherf, A. (2000). Frequent ectopic recombination of virulence factor genes in telomeric chromosome clusters of *P. falciparum*. Nature *407*, 1018–1022.

Furuya, T., Mu, J., Hayton, K., Liu, A., Duan, J., Nkrumah, L., Joy, D.A., Fidock, D.A., Fujioka, H., Vaidya, A.B.,

Wellems, T.E., and Su, X.Z. (2005). Disruption of a *Plasmodium falciparum* gene linked to male sexual development causes early arrest in gametocytogenesis. Proc. Natl. Acad. Sci. U.S.A. *102*, 16813–16818.

Golenda, C.F., Li, J., and Rosenberg, R. (1997). Continuous *in vitro* propagation of the malaria parasite *Plasmodium vivax*. Proc. Natl. Acad. Sci. U.S.A. *94*, 6786–6791.

Gonzales, J.M., Patel, J.J., Ponmee, N., Jiang, L., Tan, A., Maher, S.P., Wuchty, S., Rathod, P.K., and Ferdig, M.T. (2008). Regulatory hotspots in the malaria parasite genome dictate transcriptional variation. PLoS Biol. *6*, e238.

Grech, K., Martinelli, A., Pathirana, S., Walliker, D., Hunt, P., and Carter, R. (2002). Numerous, robust genetic markers for *Plasmodium chabaudi* by the method of amplified fragment length polymorphism. Mol. Biochem. Parasitol. *123*, 95–104.

Greenberg, J., and Trembley, H.L. (1954). The apparent transfer of pyrimethamine-resistance from the BI strain of *Plasmodium gallinaceum* to the M strain. J. Parasitol. *40*, 667–672.

Grodzicker, T., Williams, J., Sharp, P., and Sambrook, J. (1974). Physical mapping of temperature-sensitive mutations of adenoviruses. Cold Spring Harbor Symposia on Quantitative Biology *39*, 439–446.

Haldane, J.B., and Smith, C.A. (1947). A new estimate of the linkage between the genes for colourblindness and haemophilia in man. Ann. Eugen. *14*, 10–31.

Haynes, J.D., Diggs, C.L., Hines, F.A., and Desjardins, R.E. (1976). Culture of human malaria parasites *Plasmodium falciparum*. Nature *263*, 767–769.

Hayton, K., Gaur, D., Liu, A., Takahashi, J., Henschen, B., Singh, S., Lambert, L., Furuya, T., Bouttenot, R., Doll, M., Nawaz, F., Mu, J., Jiang, L., Miller, L.H., and Wellems, T.E. (2008). Erythrocyte binding protein PfRH5 polymorphisms determine species-specific pathways of *Plasmodium falciparum* invasion. Cell Host Microbe *4*, 40–51.

Hayton, K., Ranford-Cartwright, L.C., and Walliker, D. (2002). Sulfadoxine-pyrimethamine resistance in the rodent malaria parasite *Plasmodium chabaudi*. Antimicrob. Agents Chemother. *46*, 2482–2489.

Hayton, K., and Su, X.Z. (2008). Drug resistance and genetic mapping in *Plasmodium falciparum*. Curr. Genet. *54*, 223–239.

Huang, Y., Wuchty, S., Ferdig, M.T., and Przytycka, T.M. (2009). Graph theoretical approach to study eQTL: a case study of *Plasmodium falciparum*. Bioinformatics *25*, i15-i20.

Hunt, P., Afonso, A., Creasey, A., Culleton, R., Sidhu, A.B., Logan, J., Valderramos, S.G., McNae, I., Cheesman, S., do, R., V., Carter, R., Fidock, D.A., and Cravo, P. (2007). Gene encoding a deubiquitinating enzyme is mutated in artesunate- and chloroquine-resistant rodent malaria parasites. Mol. Microbiol. *65*, 27–40.

Hunt, P., Cravo, P.V., Donleavy, P., Carlton, J.M., and Walliker, D. (2004a). Chloroquine resistance in *Plasmodium chabaudi*: are chloroquine-resistance transporter (crt) and multi-drug resistance (mdr1) orthologues involved? Mol. Biochem. Parasitol. *133*, 27–35.

Hunt, P., Martinelli, A., Fawcett, R., Carlton, J., Carter, R., and Walliker, D. (2004b). Gene synteny and chloroquine resistance in *Plasmodium chabaudi*. Mol. Biochem. Parasitol. *136*, 157–164.

Ifediba, T., and Vanderberg, J.P. (1981). Complete *in vitro* maturation of *Plasmodium falciparum* gametocytes. Nature *294*, 364–366.

Janse, C.J., van der Klooster, P.F., van der Kaay, H.J., van der Ploeg, M., and Overdulve, J.P. (1986). DNA synthesis in *Plasmodium berghei* during asexual and sexual development. Mol. Biochem. Parasitol. *20*, 173–182.

Jeffreys, A.J., Kauppi, L., and Neumann, R. (2001). Intensely punctate meiotic recombination in the class II region of the major histocompatibility complex. Nat. Genet. *29*, 217–222.

Jiang, H., Li, N., Gopalan, V., Zilversmit, M.M., Varma, S., Nagarajan, V., Li, J., Mu, J., Hayton, K., Henschen, B., Yi, M., Stephens, R., McVean, G., Awadalla, P., Wellems, T.E., and Su, X.Z. (2011). High recombination rates and hotspots in a *Plasmodium falciparum* genetic cross. Genome Biol. *12*, R33.

Keeney, S., Giroux, C.N., and Kleckner, N. (1997). Meiosis-specific DNA double-strand breaks are catalyzed by Spo11, a member of a widely conserved protein family. Cell *88*, 375–384.

Kemp, D.J., Corcoran, L.M., Coppel, R.L., Stahl, H.D., Bianco, A.E., Brown, G.V., and Anders, R.F. (1985). Size variation in chromosomes from independent cultured isolates of *Plasmodium falciparum*. Nature *315*, 347–350.

Kerr, P.J., Ranford-Cartwright, L.C., and Walliker, D. (1994). Proof of intragenic recombination in *Plasmodium falciparum*. Mol. Biochem. Parasitol. *66*, 241–248.

Killick-Kendrick, R., and Peters, W. (1978). Rodent Malaria. (London: Academic Press).

Kirkman, L.A., Su, X.Z., and Wellems, T.E. (1996). *Plasmodium falciparum*: isolation of large numbers of parasite clones from infected blood samples. Exp. Parasitol. *83*, 147–149.

Knowles, G., Sanderson, A., and Walliker, D. (1981). *Plasmodium yoelii*: genetic analysis of crosses between two rodent malaria subspecies. Exp. Parasitol. *52*, 243–247.

Kruglyak, L. (1997). The use of a genetic map of biallelic markers in linkage studies. Nat. Genet. *17*, 21–24.

Lander, E.S., and Botstein, D. (1989). Mapping Mendelian factors underlying quantitative traits using RFLP linkage maps. Genetics *121*, 185–199.

Lander, E.S., Green, P., Abrahamson, J., Barlow, A., Daly, M.J., Lincoln, S.E., and Newberg, L.A. (1987). MAPMAKER: an interactive computer package for constructing primary genetic linkage maps of experimental and natural populations. Genomics *1*, 174–181.

Lathrop, G.M., and Lalouel, J.M. (1984). Easy calculations of lod scores and genetic risks on small computers. Am. J. Hum. Genet. *36*, 460–465.

Lathrop, G.M., Lalouel, J.M., Julier, C., and Ott, J. (1984). Strategies for multilocus linkage analysis in humans. Proc. Natl. Acad. Sci. U.S.A. *81*, 3443–3446.

Lynch, M., and Walsh, B. (1997). Genetics and Analysis of Quantitative Traits. (Sunderland, MA: Sinauer Associates).

Martin, M.J., Rayner, J.C., Gagneux, P., Barnwell, J.W., and Varki, A. (2005). Evolution of human–chimpanzee differences in malaria susceptibility: relationship to human genetic loss of N-glycolylneuraminic acid. Proc. Natl. Acad. Sci. U.S.A. *102*, 12819–12824.

Martinelli, A., Cheesman, S., Hunt, P., Culleton, R., Raza, A., Mackinnon, M., and Carter, R. (2005a). A genetic approach to the *de novo* identification of targets of strain-specific immunity in malaria parasites. Proc. Natl. Acad. Sci. U.S.A. *102*, 814–819.

Martinelli, A., Hunt, P., Fawcett, R., Cravo, P.V., Walliker, D., and Carter, R. (2005b). An AFLP-based genetic linkage map of *Plasmodium chabaudi chabaudi*. Malaria J. *4*, 11.

McBride, J.S., Walliker, D., and Morgan, G. (1982). Antigenic diversity in the human malaria parasite *Plasmodium falciparum*. Science *217*, 254–257.

Morton, N.E. (1955). Sequential tests for the detection of linkage. Am. J. Hum. Genet. 7, 277–318.

Mu, J., Awadalla, P., Duan, J., McGee, K.M., Joy, D.A., McVean, G.A., and Su, X.Z. (2005). Recombination hotspots and population structure in *Plasmodium falciparum*. PLoS Biol. *3*, e335.

Mu, J., Awadalla, P., Duan, J., McGee, K.M., Keebler, J., Seydel, K., McVean, G.A., and Su, X.Z. (2007). Genome-wide variation and identification of vaccine targets in the *Plasmodium falciparum* genome. Nat. Genet. *39*, 126–130.

Mu, J., Myers, R.A., Jiang, H., Liu, S., Ricklefs, S., Waisberg, M., Chotivanich, K., Wilairatana, P., Krudsood, S., White, N.J., Udomsangpetch, R., Cui, L., Ho, M., Ou, F., Li, H., Song, J., Li, G., Wang, X., Seila, S., Sokunthea, S., Socheat, D., Sturdevant, D.E., Porcella, S.F., Fairhurst, R.M., Wellems, T.E., Awadalla, P., and Su, X.Z. (2010). *Plasmodium falciparum* genome-wide scans for positive selection, recombination hot spots and resistance to antimalarial drugs. Nat. Genet. *42*, 268–271.

Nishant, K.T., and Rao, M.R. (2006). Molecular features of meiotic recombination hot spots. Bioessays *28*, 45–56.

Nomura, T., Carlton, J.M., Baird, J.K., del Portillo, H.A., Fryauff, D.J., Rathore, D., Fidock, D.A., Su, X., Collins, W.E., McCutchan, T.F., Wootton, J.C., and Wellems, T.E. (2001). Evidence for different mechanisms of chloroquine resistance in 2 *Plasmodium* species that cause human malaria. J. Infect. Dis. *183*, 1653–1661.

Oduola, A.M., Milhous, W.K., Weatherly, N.F., Bowdre, J.H., and Desjardins, R.E. (1988a). *Plasmodium falciparum*: induction of resistance to mefloquine in cloned strains by continuous drug exposure *in vitro*. Exp. Parasitol. *67*, 354–360.

Oduola, A.M., Weatherly, N.F., Bowdre, J.H., and Desjardins, R.E. (1988b). *Plasmodium falciparum*: cloning by single-erythrocyte micromanipulation and heterogeneity *in vitro*. Exp. Parasitol. *66*, 86–95.

Ott, J. (1974). Estimation of the recombination fraction in human pedigrees: efficient computation of the likelihood for human linkage studies. Am. J. Hum. Genet. *26*, 588–597.

Ott, J. (1999). Analysis of Human Genetic Linkage. (Baltimore: The Johns Hopkins University Press).

Patel, J.J., Thacker, D., Tan, J.C., Pleeter, P., Checkley, L., Gonzales, J.M., Deng, B., Roepe, P.D., Cooper, R.A., and Ferdig, M.T. (2010). Chloroquine susceptibility and reversibility in a *Plasmodium falciparum* genetic cross. Mol. Microbiol. *78*, 770–787.

Paterson, A.H., Lander, E.S., Hewitt, J.D., Peterson, S., Lincoln, S.E., and Tanksley, S.D. (1988). Resolution of quantitative traits into Mendelian factors by using a complete linkage map of restriction fragment length polymorphisms. Nature *335*, 721–726.

Pattaradilokrat, S., Cheesman, S.J., and Carter, R. (2007). Linkage group selection: towards identifying genes controlling strain specific protective immunity in malaria. PLoS One *2*, e857.

Pattaradilokrat, S., Culleton, R.L., Cheesman, S.J., and Carter, R. (2009). Gene encoding erythrocyte binding ligand linked to blood stage multiplication rate phenotype in *Plasmodium yoelii yoelii*. Proc. Natl. Acad. Sci. U.S.A. *106*, 7161–7166.

Peterson, D.S., Walliker, D., and Wellems, T.E. (1988). Evidence that a point mutation in dihydrofolate reductase-thymidylate synthase confers resistance to pyrimethamine in falciparum malaria. Proc. Natl. Acad. Sci. U.S.A. *85*, 9114–9118.

Ponnudurai, T., Leeuwenberg, A.D., and Meuwissen, J.H. (1981). Chloroquine sensitivity of isolates of *Plasmodium falciparum* adapted to *in vitro* culture. Trop. Geogr. Med. 33, 50–54.

Ponnudurai, T., Meuwissen, J.H., Leeuwenberg, A.D., Verhave, J.P., and Lensen, A.H. (1982). The production of mature gametocytes of *Plasmodium falciparum* in continuous cultures of different isolates infective to mosquitoes. Trans. R. Soc. Trop. Med. Hyg. *76*, 242–250.

Purcell, S., Neale, B., Todd-Brown, K., Thomas, L., Ferreira, M.A., Bender, D., Maller, J., Sklar, P., de Bakker, P.I., Daly, M.J., and Sham, P.C. (2007). PLINK: a tool set for whole-genome association and population-based linkage analyses. Am. J. Hum. Genet. *81*, 559–575.

Ranford-Cartwright, L.C., Balfe, P., Carter, R., and Walliker, D. (1993). Frequency of cross-fertilization in the human malaria parasite *Plasmodium falciparum*. Parasitology *107*, 11–18.

Reilly Ayala, H.B., Wacker, M.A., Siwo, G., and Ferdig, M.T. (2010). Quantitative trait loci mapping reveals candidate pathways regulating cell cycle duration in *Plasmodium falciparum*. BMC. Genomics *11*, 577.

Reilly, H.B., Wang, H., Steuter, J.A., Marx, A.M., and Ferdig, M.T. (2007). Quantitative dissection of clone-specific growth rates in cultured malaria parasites. Int. J. Parasitol. 37, 1599–1607.

Rider, A.K., Siwo, G., Chawla, N.V., Ferdig, M., and Emrich, S.J. (2010). A statistical approach to finding overlooked genetic associations. BMC Bioinformatics *11*, 526.

Rohrbach, P., Sanchez, C.P., Hayton, K., Friedrich, O., Patel, J., Sidhu, A.B., Ferdig, M.T., Fidock, D.A., and

Lanzer, M. (2006). Genetic linkage of pfmdr1 with food vacuolar solute import in *Plasmodium falciparum*. EMBO J. *25*, 3000–3011.

Sa, J.M., Nomura, T., Neves, J., Baird, J.K., Wellems, T.E., and del Portillo, H.A. (2005). *Plasmodium vivax*: allele variants of the mdr1 gene do not associate with chloroquine resistance among isolates from Brazil, Papua, and monkey-adapted strains. Exp. Parasitol. *109*, 256–259.

Sa, J.M., Twu, O., Hayton, K., Reyes, S., Fay, M.P., Ringwald, P., and Wellems, T.E. (2009). Geographic patterns of *Plasmodium falciparum* drug resistance distinguished by differential responses to amodiaquine and chloroquine. Proc. Natl. Acad. Sci. U.S.A. *106*, 18883–18889.

Saiki, R.K., Gelfand, D.H., Stoffel, S., Scharf, S.J., Higuchi, R., Horn, G.T., Mullis, K.B., and Erlich, H.A. (1988). Primer-directed enzymatic amplification of DNA with a thermostable DNA polymerase. Science *239*, 487–491.

Saiki, R.K., Scharf, S., Faloona, F., Mullis, K.B., Horn, G.T., Erlich, H.A., and Arnheim, N. (1985). Enzymatic amplification of beta-globin genomic sequences and restriction site analysis for diagnosis of sickle cell anemia. Science *230*, 1350–1354.

Samarakoon, U., Regier, A., Tan, A., Desany, B.A., Collins, B., Tan, J.C., Emrich, S.J., and Ferdig, M.T. (2011). High-throughput 454 resequencing for allele discovery and recombination mapping in *Plasmodium falciparum*. BMC. Genomics *12*, 116.

San Filippo, J., Sung, P., and Klein, H. (2008). Mechanism of eukaryotic homologous recombination. Annu. Rev. Biochem. *77*, 229–257.

Sanderson, A., Walliker, D., and Molez, J.F. (1981). Enzyme typing of *Plasmodium falciparum* from African and some other Old World countries. Trans. R. Soc. Trop. Med. Hyg. *75*, 263–267.

Seaton, G., Haley, C.S., Knott, S.A., Kearsey, M., and Visscher, P.M. (2002). QTL Express: mapping quantitative trait loci in simple and complex pedigrees. Bioinformatics *18*, 339–340.

Sen, S., and Churchill, G.A. (2001). A Statistical Framework for Quantitative Trait Mapping. Genetics *159*, 371–387.

Sidhu, A.B., Verdier-Pinard, D., and Fidock, D.A. (2002). Chloroquine resistance in *Plasmodium falciparum* malaria parasites conferred by pfcrt mutations. Science *298*, 210–213.

Sinden, R.E., and Hartley, R.H. (1985). Identification of the meiotic division of malarial parasites. J. Protozool. *32*, 742–744.

Smith, G.R., Amundsen, S.K., Dabert, P., and Taylor, A.F. (1995). The initiation and control of homologous recombination in *Escherichia coli*. Philos. Trans. R. Soc. Lond. B. Biol. Sci. *347*, 13–20.

Spierer, P., Spierer, A., Bender, W., and Hogness, D.S. (1983). Molecular mapping of genetic and chromomeric units in *Drosophila melanogaster*. J. Mol. Biol. *168*, 35–50.

Sturtevant, A.H. (1913). The linear arrangement of six sex- linked factors in *Drosophila* as shown by their mode of association. J. Exp. Zool. *14*, 43–59.

Su, X., and Wellems, T.E. (1996). Toward a high-resolution *Plasmodium falciparum* linkage map: polymorphic markers from hundreds of simple sequence repeats. Genomics *33*, 430–444.

Su, X., Kirkman, L.A., Fujioka, H., and Wellems, T.E. (1997). Complex polymorphisms in an approximately 330 kDa protein are linked to chloroquine-resistant *P. falciparum* in Southeast Asia and Africa. Cell *91*, 593–603.

Su, X., Ferdig, M.T., Huang, Y., Huynh, C.Q., Liu, A., You, J., Wootton, J.C., and Wellems, T.E. (1999). A genetic map and recombination parameters of the human malaria parasite *Plasmodium falciparum*. Science *286*, 1351–1353.

Su, X., Hayton, K., and Wellems, T.E. (2007). Genetic linkage and association analyses for trait mapping in *Plasmodium falciparum*. Nat. Rev. Genet. *8*, 497–506.

Sullivan, J.S., Sullivan, J.J., Williams, A., Grady, K.K., Bounngaseng, A., Huber, C.S., Nace, D., Williams, T., Galland, G.G., Barnwell, J.W., and Collins, W.E. (2003). Adaptation of a strain of *Plasmodium falciparum* from Ghana to *Aotus lemurinus griseimembra*, *A. nancymai*, and *A. vociferans* monkeys. Am. J. Trop. Med. Hyg. *69*, 593–600.

Tan, J.C., Miller, B.A., Tan, A., Patel, J.J., Cheeseman, I.H., Anderson, T.J., Manske, M., Maslen, G., Kwiatkowski, D.P., and Ferdig, M.T. (2011). An optimized microarray platform for assaying genomic variation in *Plasmodium falciparum* field populations. Genome Biol. *12*, R35.

Taylor, H.M., Kyes, S.A., and Newbold, C.I. (2000). Var gene diversity in *Plasmodium falciparum* is generated by frequent recombination events. Mol. Biochem. Parasitol. *110*, 391–397.

Thaithong, S., and Beale, G.H. (1981). Resistance of ten Thai isolates of *Plasmodium falciparum* to chloroquine and pyrimethamine by *in vitro* tests. Trans. R. Soc. Trop. Med. Hyg. *75*, 271–273.

Thomas, A. (2007). Towards linkage analysis with markers in linkage disequilibrium by graphical modelling. Hum. Hered. *64*, 16–26.

Trager, W., and Jensen, J.B. (1976). Human malaria parasites in continuous culture. Science *193*, 673–675.

Trager, W., Tershakovec, M., Lyandvert, L., Stanley, H., Lanners, N., and Gubert, E. (1981). Clones of the malaria parasite *Plasmodium falciparum* obtained by microscopic selection: their characterization with regard to knobs, chloroquine sensitivity, and formation of gametocytes. Proc. Natl. Acad. Sci. U.S.A. *78*, 6527–6530.

Vaidya, A.B., Muratova, O., Guinet, F., Keister, D., Wellems, T.E., and Kaslow, D.C. (1995). A genetic locus on *Plasmodium falciparum* chromosome 12 linked to a defect in mosquito-infectivity and male gametogenesis. Mol. Biochem. Parasitol. *69*, 65–71.

Volkman, S.K., Sabeti, P.C., DeCaprio, D., Neafsey, D.E., Schaffner, S.F., Milner, D.A., Jr., Daily, J.P., Sarr, O., Ndiaye, D., Ndir, O., Mboup, S., Duraisingh, M.T., Lukens, A., Derr, A., Stange-Thomann, N., Waggoner, S., Onofrio, R., Ziaugra, L., Mauceli, E., Gnerre, S., Jaffe, D.B., Zainoun, J., Wiegand, R.C., Birren, B.W., Hartl, D.L., Galagan, J.E., Lander, E.S., and Wirth,

D.F. (2007). A genome-wide map of diversity in *Plasmodium falciparum*. Nat. Genet. *39*, 113–119.

Vos, P., Hogers, R., Bleeker, M., Reijans, M., van de Lee, T., Hornes, M., Frijters, A., Pot, J., Peleman, J., and Kuiper, M. (1995). AFLP: a new technique for DNA fingerprinting. Nucleic Acids Res. *23*, 4407–4414.

Walker-Jonah, A., Dolan, S.A., Gwadz, R.W., Panton, L.J., and Wellems, T.E. (1992). An RFLP map of the *Plasmodium falciparum* genome, recombination rates and favored linkage groups in a genetic cross. Mol. Biochem. Parasitol. *51*, 313–320.

Walliker, D., Carter, R., and Morgan, S. (1971). Genetic recombination in malaria parasites. Nature *232*, 561–562.

Walliker, D., Carter, R., and Morgan, S. (1973). Genetic recombination in *Plasmodium berghei*. Parasitology *66*, 309–320.

Walliker, D., Carter, R., and Sanderson, A. (1975). Genetic studies on *Plasmodium chabaudi*: recombination between enzyme markers. Parasitology *70*, 19–24.

Walliker, D., Quakyi, I.A., Wellems, T.E., McCutchan, T.F., Szarfman, A., London, W.T., Corcoran, L.M., Burkot, T.R., and Carter, R. (1987). Genetic analysis of the human malaria parasite *Plasmodium falciparum*. Science *236*, 1661–1666.

Wang, P., Read, M., Sims, P.F., and Hyde, J.E. (1997). Sulfadoxine resistance in the human malaria parasite *Plasmodium falciparum* is determined by mutations in dihydropteroate synthetase and an additional factor associated with folate utilization. Mol. Microbiol. *23*, 979–986.

Ward, C.P., Clottey, G.T., Dorris, M., Ji, D.D., and Arnot, D.E. (1999). Analysis of *Plasmodium falciparum* PfEMP-1/var genes suggests that recombination rearranges constrained sequences. Mol. Biochem. Parasitol. *102*, 167–177.

Weber, J.L., and May, P.E. (1989). Abundant class of human DNA polymorphisms which can be typed using the polymerase chain reaction. Am. J. Hum. Genet. *44*, 388–396.

Wellems, T.E., Oduola, A.M.J., Fenton, B., Desjardins, R., Panton, L.J., and do Rosario, V.E. (1988). Chromosome size variation occurs in cloned *Plasmodium falciparum* on *in vitro* cultivation. Revista Brasiliera de Genetica *11*, 813–825.

Wellems, T.E., Walker-Jonah, A., and Panton, L.J. (1991). Genetic Mapping of the Chloroquine-Resistance Locus on *Plasmodium falciparum* Chromosome 7. Proc. Nat. Acad. Sci. U.S.A. *88*, 3382–3386.

Wellems, T.E., Panton, L.J., Gluzman, I.Y., do Rosario, V.E., Gwadz, R.W., Walker-Jonah, A., and Krogstad, D.J. (1990). Chloroquine resistance not linked to mdr-like genes in a *Plasmodium falciparum* cross. Nature *345*, 253–255.

Wellems, T.E., Walliker, D., Smith, C.L., do Rosario, V.E., Maloy, W.L., Howard, R.J., Carter, R., and McCutchan, T.F. (1987). A histidine-rich protein gene marks a linkage group favored strongly in a genetic cross of *Plasmodium falciparum*. Cell *49*, 633–642.

Wu, T.C., and Lichten, M. (1995). Factors that affect the location and frequency of meiosis-induced double-strand breaks in *Saccharomyces cerevisiae*. Genetics *140*, 55–66.

Wu, Y., Kirkman, L.A., and Wellems, T.E. (1996). Transformation of *Plasmodium falciparum* malaria parasites by homologous integration of plasmids that confer resistance to pyrimethamine. Proc. Natl. Acad. Sci. U.S.A. *93*, 1130–1134.

Yuan, J., Johnson, R.L., Huang, R., Wichterman, J., Jiang, H., Hayton, K., Fidock, D.A., Wellems, T.E., Inglese, J., Austin, C.P., and Su, X.Z. (2009). Genetic mapping of targets mediating differential chemical phenotypes in *Plasmodium falciparum*. Nat. Chem. Biol. *5*, 765–771.

Zou, W., and Zeng, Z.B. (2008). Statistical Methods for Mapping Multiple QTL. Int. J. Plant Genomics *2008*, 286561.

Regulation of Gene Expression

7

Kirk W. Deitsch and Ron Dzikowski

Abstract

Malaria parasites represent an evolutionary lineage quite distant from the model organisms within the crown group of eukaryotes on which much of our current knowledge of basic biological mechanisms is based. With regard to regulation of gene expression, studies in *Plasmodium* have identified numerous aspects that are conserved in higher eukaryotic organisms; however, there are also several characteristics that appear to be quite different, and others that remain very poorly understood. In this chapter we review what is known about the many steps involved in regulating gene expression, from transcription initiation, through mRNA processing to protein synthesis. A better understanding of how specific gene expression patterns are controlled will shed light on such important features of parasite biology as antigenic variation, sexual differentiation, and cell cycle progression, to name a few.

Introduction

The genomes of eukaryotic organisms provide a relatively stable source of all the information necessary for building a cell and controlling its metabolism, differentiation and reproduction. However, translation of this information to the level of protein production must be tightly regulated to ensure proper cellular development. By closely controlling the various steps of gene expression, including gene transcription, mRNA processing and stability, as well as protein translation, cells can control the level and timing of protein synthesis, thus regulating virtually all aspects of cellular metabolism (Fig. 7.1). Because of its fundamental importance in cellular biology, the study of gene expression has become a focus of intense research in many fields of modern biology, and has yielded several groundbreaking discoveries in recent years, including a more sophisticated understanding of the role chromatin modification plays in regulating transcription and the function of noncoding RNAs in controlling mRNA stability.

The most detailed experimental work on various aspects of gene expression in eukaryotes has been conducted on model organisms ranging in complexity from yeasts to mammals. While these highly developed systems benefit from the existence of a great number of sophisticated molecular and genetic tools, all of these organisms are from the so called 'crown' group of eukaryotes and thus represent a fairly narrow view of the entire range of eukaryotic species. Owing to the significant evolutionary distance between the model organisms and malaria parasites, it is not yet clear which well-established paradigms of gene regulation will hold true for the various species of *Plasmodium*. A better understanding of the mechanisms controlling gene expression in malaria parasites therefore will provide significant insights into eukaryotic evolution as well as providing potential targets for novel disease intervention strategies.

mRNA production in *Plasmodium*

The complete life cycle of malaria parasites requires development through multiple morphological stages in several different tissues within two separate metazoan hosts. These rather stark

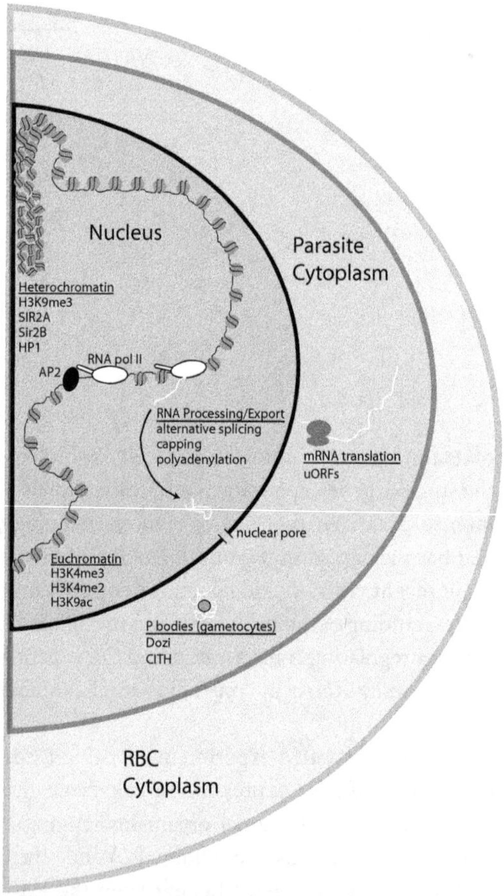

Figure 7.1 Depiction of many steps involved in regulating gene expression. Heterochromatin is associated with silent histone modifications, including trimethylation at lysine 9 of H3 (H3K9me3), which are put in place by such enzymes as Sir2A and B and HP1. Similarly, euchromatin is associated with the histone marks H3K4me2, H3K4me3 and H3K9ac. Members of the AP2 family of transcription factors are thought to represent the primary transcriptional activators in *Plasmodium*, serving to recruit RNA polymerase II to active genes. Several steps in RNA processing and export have been shown to be important, as has regulation of RNA translation through either P bodies or the presence of regulatory upstream open reading frames (uORFs).

changes in parasite morphology and metabolism require significant changes in gene expression that are tightly controlled. The completion of the *P. falciparum* genome project (and similar genome projects of additional *Plasmodium* species) (Carlton *et al.*, 2002, 2008a,b; Gardner *et al.*, 2002) provided the ability to analyse changes in gene expression at the level of the complete transcriptome and led to significant insights into how parasites alter their mRNA expression profiles as they transition through the different stages of development and differentiation.

Genome-wide studies of mRNA production

With the release of the nearly complete and annotated genome sequence of *P. falciparum*, two groups undertook a microarray-based analysis of gene expression patterns throughout the ~48 h asexual replicative cycle (Bozdech *et al.*, 2003; Le Roch *et al.*, 2003). These studies allowed the investigators to determine the steady-state level of each mRNA encoded in the parasite genome and thus visualize at which point in the cycle each individual gene is transcribed. They found that in general genes are transcribed at the point in the cycle when the encoded protein is required, leading them to propose a 'just in time' model for gene expression. By viewing expression of the genome as a whole, the researchers observed a tightly regulated cascade of mRNA production, suggesting that parasites have a 'hard-wired' gene expression program that results in highly reproducible progression through the cycle of schizogeny that is not easily perturbed. These studies also demonstrated that each gene is regulated as an individual transcription unit, similar to model eukaryotes like yeast but unlike the kinetoplastid parasites which utilize large, polycistronic transcription units and display little or no regulation at the level of transcription of individual genes (Palenchar *et al.*, 2006). These studies suggested that the elements controlling transcription should be found within the regulatory elements found at or near the transcription start site of each gene, and that regulatory proteins should exist that control the recruitment of transcriptional complexes to these start sites. However, at the time that these studies were published, the identity and nature of such transcription factors remained elusive.

Since the publication of these original genome-wide gene expression analyses, several additional studies have been completed that utilized similar broad-based approaches. Llinas and colleagues extended their original microarray analysis to two additional *P. falciparum* isolates and demonstrated that the basic expression pattern is very similar, suggesting that it should be possible to generalize the transcriptional profile defined in the original experiments to virtually all *P. falciparum* isolates (Llinas *et al.*, 2006). Bozdech *et al.* have more recently used a similar microarray approach to determine the genome-wide transcriptional profile of *P. vivax*, and compared it to that of *P. falciparum* (Bozdech *et al.*, 2008). While they documented several subtle differences between the two species, the overall conclusions were quite similar: there is a well-defined cascade of gene expression over the course of the asexual cycle and that there are very few differences between different parasite isolates. In a more focused study, Mikolajczak *et al.* used microarrays to identify transcriptional changes as sporozoites become competent to invade salivary glands within the mosquito (Mikolajczak *et al.*, 2008). They identified a number of genes that are differentially upregulated in the two different stages of sporozoite development, and using this knowledge, they were able to disrupt a gene specific for oocyst sporozoites and create parasites that are unable to successfully invade mosquito salivary glands. This shows the utility of the knowledge gained through this type of analysis. As technology has continued to advance, Otto *et al.* have now used massively parallel sequencing of cDNA (RNA-seq) to further refine the original expression studies, to identify transcription start sites and exon/intron boundaries, and to discover probable examples of alternative splicing (Otto *et al.*, 2010).

With the basic transcriptional profile of *P. falciparum* having been established through the genome-wide expression analyses described above, several groups investigated how treatment with various anti-malaria drugs might lead to alterations in gene expression. Messenger RNA profile changes resulting from treatment with antifolates (Ganesan *et al.*, 2008), chloroquine (Gunasekera *et al.*, 2003) and artemisinin (Natalang *et al.*, 2008) have all been determined using microarray analysis. In the studies of both chloroquine and antifolate treatment, the authors found remarkably few changes in the transcript profile of parasites under drug treatment, and in the artemisinin study, most changes were of low amplitude. These studies have been interpreted as indicating that malaria parasites posses a 'hard-wired' transcriptional cascade through the asexual cycle that does not respond to outside environmental influences. This has implications for understanding the mechanisms controlling gene expression and also suggests that the parasite's inability to react to outside pressures might make it uniquely vulnerable to intervention strategies in that it might not be able to substantially up or down regulate gene expression to overcome drug treatment.

In a unique study that linked genome-wide expression profiling by microarray with a classical genetic approach, Gonzales *et al.* utilized the parents and progeny from a previously described genetic cross (Gonzales *et al.*, 2008). By comparing the absolute expression levels of each gene in the genome and treating these expression levels as genetic traits, the authors were able to genetically map quantitative trait loci that serve as potential master regulators of gene expression. These 'eQTLs' might serve as ways to identify specific components of the transcriptional control mechanisms that regulate gene expression in malaria parasites. For example, one specific regulatory locus influenced the level of 269 transcripts and mapped to a chromosome 5 amplification event that encompasses 14 genes, including the *Pfmdr1* gene previously associated with drug resistance phenotypes. Which of the genes within this locus directly influence total transcript levels remains to be determined.

Transfection experiments

The ability to transfect and transform cultured malaria parasites has proven to be an extremely powerful tool to investigate various aspects of parasite biology. While this methodology is generally utilized for such studies as gene knockout experiments, over-expression analysis, and experiments designed to determine protein subcellular localization, the technique depends on efficient expression of genes included on the transfected plasmid constructs. This typically includes a gene encoding a drug selectable marker in addition to the gene being investigated. Therefore a significant amount of data regarding the DNA elements that control gene transcription has been gathered through these types of approaches.

Early transfection experiments using reporter gene constructs demonstrated that the DNA region found directly upstream of a gene was usually sufficient to accurately confer proper timing of expression, indicating that most regulatory elements exist within the region upstream of a gene's transcriptional start site (Dechering *et al.*, 1999; Koning-Ward *et al.*, 1999). In addition, promoter truncation studies indicated that these upstream regulatory regions likely conform to the typical bipartite structure common in most eukaryotic organisms, consisting of a basal promoter and associated upstream enhancer and silencer elements (Crabb *et al.*, 1996; Horrocks *et al.*, 1996; Mbacham *et al.*, 2001). In a couple of studies, promoter truncations were combined with electrophoretic mobility shift assays to map protein binding sites, thus more precisely mapping sites of activator or silencer binding (Dechering *et al.*, 1999; Flueck *et al.*, 2010; Horrocks *et al.*, 1999; Tham *et al.*, 2007; Voss *et al.*, 2003). The initial transfection experiments also determined that standard eukaryotic transcriptional terminator sequences (3′ UTRs) and polyadenylation signals (for example those from SV40) that are often provided with commercial reporter gene constructs were not recognized by malaria parasites and needed to be replaced by *Plasmodium* 3′ sequences (Wu *et al.*, 1995). However, promoters and terminators from one *Plasmodium* species generally do work in other species, suggesting that many transcriptional regulatory proteins are conserved within the *Plasmodium* lineage (van der Wel *et al.*, 1997).

In more recent years, some potential transcriptional regulatory proteins have been identified, in particular the ApiAP2 protein family (see below for details) (Balaji *et al.*, 2005). Transfection experiments have been used to help validate the function of these proteins, to identify their target genes, and to determine phenotypes of parasites in which they have been disrupted (Flueck *et al.*, 2010; Yuda *et al.*, 2009, 2010). As our understanding of transcriptional control mechanisms in *Plasmodium* continues to become more sophisticated, transfection will likely remain a central method in these investigations.

Genome annotations

The completion of the various *Plasmodium* genome projects and their annotations provided a great deal of information in addition to the ability to perform the genome-wide expression studies described above. The annotations also allowed the assembly of catalogues of genes likely involved in the regulation of gene expression (Aravind *et al.*, 2003; Bischoff *et al.*, 2010; Coulson *et al.*, 2004). These computational analyses were able to identify the orthologues of most of the proteins that bind to core promoters – the so-called basal transcription factors (Horrocks *et al.*, 2009; Meissner *et al.*, 2005). Investigators were also able to identify a full complement of histone modifiers, suggesting that mechanisms that control chromatin assembly and modification were conserved with those described in model organisms (Aravind *et al.*, 2003; Cui *et al.*, 2010). However, initial studies failed to identify many typical transcriptional activator or silencer proteins based on searches for motifs commonly found in eukaryotic transcriptional regulators. This led to the hypothesis that control of transcription might be distinctly different in *Plasmodium* than the standard model described for other organisms. This problem has been at least partly remedied by the discovery of the ApiAP2 family of transcription factors; however, the question remains whether additional families of transcription factors remain unidentified within the *Plasmodium* genome.

While there were initial reports of successful use of double stranded RNA to knock down gene expression in *P. falciparum* (Malhotra *et al.*, 2002; McRobert *et al.*, 2002), subsequent attempts to utilize this technique as a general tool for experiments with cultured parasites were unsuccessful. Close examination of the genome also failed to identify the major components of the RNAi pathway, and thus the current consensus is that this mechanism of regulating gene expression is not active in malaria parasites (Baum *et al.*, 2009). There are; however, examples of noncoding RNAs that have been implicated in structural and regulatory roles. Raabe *et al.* systematically categorized a large set of noncoding RNAs in *P. falciparum*, including several examples of antisense transcripts from annotated genes (Raabe *et al.*, 2010). More specifically, both sense and antisense noncoding

RNAs transcribed from the introns of *var* genes have been well characterized by several groups and have been implicated in the complex transcriptional regulation of this large gene family (Epp *et al.*, 2008; Kyes *et al.*, 2003; Su *et al.*, 1995). The predicted centromeres of *P. falciparum* have also been shown to be transcribed, leading to small transcripts that are retained in the nucleus (Li *et al.*, 2008). The exact mechanisms of action of these noncoding RNAs and how they influence gene expression and chromatin structure remain largely unknown.

Similar to model eukaryotes, malaria parasites utilize RNA polymerase II as the protein complex for the production of mRNAs. The largest subunit of RNA polymerase II (called Rpb1) contains a carboxy-terminal domain (CTD) that serves as an interface for recruitment of many factors necessary for efficient mRNA production, including initiation, elongation, splicing and termination (Chapman *et al.*, 2008; Egloff *et al.*, 2008). The CTD consists of a tandem array of the consensus heptapeptide YSPTSPS, varying in length from 52 tandem repeats in mammals to 26 in *S. cerevisiae*. Examination of this domain in malaria parasites has found that most species have a remarkably short CTD consisting of only eight repeats (Kishore *et al.*, 2009), a length that mutagenesis studies in yeast found to be the absolute minimum required for life (Liao *et al.*, 1991). Further, while the CTD length is highly conserved in most evolutionary lineages, it was found to be remarkably variable in *Plasmodium* (Kishore *et al.*, 2009). In particular, primate parasites display an expansion of repeat number up to 17 in some isolates of *P. falciparum*. Why the CTD length is variable in *Plasmodium* and why it appears to be specifically expanded in parasites that infect primates is not understood; however, given the key role this domain plays in transcriptional regulation in other eukaryotic organisms, these questions deserve exploration.

Post-transcriptional regulation

In addition to the genome-wide RNA analyses described above, several proteomic studies were also conducted to determine gene expression levels at the level of protein production in human (Florens *et al.*, 2002; Lasonder *et al.*, 2002), rodent (Hall *et al.*, 2005b; Khan *et al.*, 2005; Tarun *et al.*, 2008) and bird parasites (Patra *et al.*, 2008). In general, the researchers found a tight correlation between mRNA and protein levels, providing further support for the just-in-time paradigm for gene expression. However, more recent work examining progression through the schizont stage of the asexual cycle provided evidence for significant post-transcriptional regulation (Foth *et al.*, 2008). In addition, details regarding specific mechanisms controlling mRNA translation have now been provided for sexual differentiation (Mair *et al.*, 2006; Mair *et al.*, 2010) and for control of a specific variant antigen gene involved in pregnancy-associated malaria (Amulic *et al.*, 2009; Mok *et al.*, 2008).

Sexual differentiation: P granules

In metazoans, one highly conserved component of sexual development is the stabilization of a subclass of mRNAs through binding by protein complexes called P granules (Updike *et al.*, 2010). This binding prevents translation of the mRNAs into protein and serves to provide a store of 'pre-made' mRNAs for rapid protein production during early post-fertilization development. This system of post-transcriptional regulation has been well described in metazoan organisms; however, its evolutionary origins were not previously well understood. Recent work in *P. berghei* has demonstrated that P body-mediated translational repression appears to be tightly conserved in the sexual development of malaria parasites (Mair *et al.*, 2006, 2010). Considering the significant evolutionary distance between apicomplexan protozoans and metazoans, this molecular mechanism of controlling rates of protein production must have a very ancient evolutionary origin.

P body development occurs in female gametocytes through the formation of a multi-subunit messenger ribonucleoprotein (mRNP) that includes at least 16 major factors, including proteins with significant homology to those found in metazoans and archaea (Mair *et al.*, 2010). Two core components of the complex are the RNA helicase DOZI and the Sm-like factor CITH. Knockouts of either of the genes encoding these proteins disrupt the function of P bodies, preventing stabilization and storage of the relevant mRNAs. Female gametocytes that carry such mutations are fertilization competent, but the resulting zygotes fail to form ookinetes and thus cannot develop. With the recent identification of a significant number of the protein components found in P bodies, the mechanisms by which this mRNP controls translation efficiency and mRNA stability should begin to be elucidated, and this could provide novel targets for developing strategies to reduce or prevent malaria transmission.

Pregnancy-associated malaria and *var2csa*

One aspect of *P. falciparum* that is known to contribute to the severity of malaria caused by this species is its ability to induce infected RBCs to cytoadhere to endothelial cells within the deep tissue vasculature (Miller *et al.*, 2002). This phenomenon is mediated by expression of the parasite produced protein PfEMP1 on the infected cell surface (reviewed in Scherf *et al.*, 2008). Different forms of PfEMP1 are encoded by different *var* genes, and each parasite maintains a repertoire of around 60 of these hypervariable coding regions within its genome. Different forms of PfEMP1 bind to different host cell surface receptors, and there is a single, conserved *var* gene called *var2csa*, that when expressed results in cytoadherence of infected RBCs to chondroitin sulphate A, a molecule found on the vascular surface in placentas (Salanti *et al.*, 2003; Salanti *et al.*, 2004). Because PfEMP1 is prominently displayed on the parasitized RBC surface, infected individuals eventually make antibodies that will recognize PfEMP1, leading to destruction of the parasite. To avoid the antibody response, parasites employ antigenic variation by switching expression between different *var* genes, a process that will be discussed in detail in a subsequent section of this chapter. However, since *var2csa* appears to be the only *var* gene that mediates binding in the placenta, and because exposure to the encoded protein readily generates a protective immune response, it is important for the parasite that this gene not be expressed when no placenta is present within the infected host, i.e. in children or non-pregnant women. Otherwise women would likely be exposed to this form of PfEMP1 as children and would therefore have pre-existing antibodies that would render the protein useless for cytoadhesion when they became pregnant as adults.

Parasites utilize a system for translational repression to prevent expression of VAR2CSA in the absence of a placenta (Amulic *et al.*, 2009; Mok *et al.*, 2008). At the *var2csa* locus, between the transcriptional start site and the ATG codon of the open reading frame that encodes PfEMP1 exists an additional open reading frame that is absent in all other *var* genes (Lavstsen *et al.*, 2003). When ribosomes bind to the cap at the 5′ end of mRNAs transcribed from this gene and begin scanning for a translation start codon, they first encounter the additional, upstream open reading frame (uORF) and translate this sequence into protein, subsequently dissociating from the transcript at the stop codon of the uORF and prior to reaching the PfEMP1 encoding ORF. Thus the uORF serves as a translational repressor (Amulic *et al.*, 2009). However, when parasites infect a pregnant woman, the repressive effect of the uORF is somehow overcome, presumably either by inducing the ribosomes to skip the start codon of the uORF or by reinitiation of ribosomal scanning downstream of the uORF. The mechanism of activation of VAR2CSA translation has not been determined.

Alternative splicing

Alternative splicing is becoming widely recognized as a method by which an organism can greatly expand its total repertoire of expressed proteins well beyond the number of genes found in its genome. For example, it is estimated that greater than 90% of human genes display alternative splicing of their encoded mRNAs (Wang *et al.*, 2008). This type of post-transcriptional regulation can yield different 5′ or 3′ UTRs, thus changing the expression profile of the encoded protein, or result in alternative isoforms of a protein by directly changing the amino acid sequence. The process contributes substantially to the complexity of an organism's proteome, and can serve as yet another way to regulate various metabolic processes.

Alternative splicing in *Plasmodium* was first reported almost 20 years ago in the 41-3 antigen of *P. berghei* (Knapp *et al.*, 1991). Additional examples have since been reported for several genes in various metabolic pathways (Fonager *et al.*, 2007; Muhia *et al.*, 2003; Struck *et al.*, 2008; Wentzinger *et al.*, 2008), and in some cases specific isoforms of the protein are expressed during specific stages of development (Pace *et al.*, 1998; Saenz *et al.*, 2008). More systematic investigations into alternative splicing during the asexual cycle have found that this phenomenon is likely quite common (Iriko *et al.*, 2009). For example, a recent genome-wide investigation into the asexual transcriptome using RNA-seq identified 84 previously uncharacterized alternative splicing events (Otto *et al.*, 2010). These included examples of exon skipping, intron retention or deletion, and changes in the 5′ or 3′ end of an mRNA which can result in alternative start or stop codons. It is tantalizing to speculate that alternative splicing will play a significant role in many aspects of parasite development and differentiation; however, this type of post-transcriptional regulation remains underexplored.

Epigenetics

The term 'epigenetics' was originally coined to refer to changes in gene expression that were 'on top of' traditional genetic modifications. In other words, it refers to changes in the expression of any given gene that are not linked to either alterations in DNA sequence or the presence/absence of any specific transcription factor. In recent years this field of research has become largely focused on modifications to the structure of chromatin at or near genes that display variant expression, often mediated by histone modifications (acetylation, methylation, ubiquitinylation, etc.), DNA modifications (methylation) or the recruitment of specific chromatin binding proteins that alter the accessibility of portions of the genome to transcription factors. Malaria parasites appear to utilize this type of gene regulation widely, and it has become one of the most intensely studied aspects of gene expression in *Plasmodium*.

Mutually exclusive expression

Parasites of the genus *Plasmodium* usually become pathogenic when they multiply asexually within the circulating blood of their hosts, and many species have evolved the ability to maintain long-term chronic infections. This ensures successful transmission to mosquitoes and the completion of their life cycle even in regions where the mosquito's abundance is seasonal. In order to do so the parasite must evade the attack by the human immune system and particularly the antibody mediated response. Soon after invasion into an RBC, the parasite modifies the biology and morphology of the host cell in a variety of ways, including anchoring proteins on the surface of the RBC plasma membrane (Maier *et al.*, 2009). These proteins are believed to play a role in both pathogenicity and immune evasion. Advances in sequencing and annotation of the genome of several *Plasmodium* spp. have indicated that all malaria parasites have large, multi-copy gene families often located primarily in sub-telomeric regions (Hall *et al.*, 2005a). These variant super families were termed *pir* (Plasmodium interspersed repeats) and include *rif/stevor/Pfmc-2TM* in *P. falciparum*, *vir* in *P. vivax*, *kir* in *P. knowlesi* and the *cir/bir/yir* family in the three rodent malarias *P. chabaudi*, *P. berghei* and *P. yoelii*, respectively (Jemmely *et al.*, 2010). In addition to these families that encode rather small variant surface antigens (VSAs) that show high similarities across *Plasmodium* species, there are also larger VSAs found specifically in the simian malaria *P. knowlesi* known as SICA (schizont-infected cell agglutination) (Howard, 1984) and PfEMP1 (*Plasmodium falciparum* erythrocyte membrane protein 1) (Baruch *et al.*, 1995; Smith *et al.*, 1995; Su *et al.*, 1995) in the deadliest form of human malaria *P. falciparum*.

Among the VSAs found in the various species of *Plasmodium*, the protein that has been most extensively studied is PfEMP1. Different forms of PfEMP1 are encoded by different members of the *var* gene family, one of which is *var2csa* that was described above for its involvement in pregnancy-associated malaria (Salanti *et al.*, 2003, 2004). To avoid the exposure of their entire antigenic repertoire at once, each parasite expresses only a single *var* gene at a time, a phenomenon similar to what is known from other eukaryotes as 'allelic exclusion'. As the antibody response against the single expressed PfEMP1 variant starts to clear the infection, a small portion of the parasite population switches transcription to an alternative *var* gene and thereby expresses a different variant of

PfEMP1 in a mutually exclusive manner (Scherf *et al.*, 2008). The clinical manifestation of these antigenic switches is long-term chronic infections and relapses of parasitaemia in the blood of patients that are unable to clear the infection by the antibody mediated response (Miller *et al.*, 2002). Mutually exclusive expression of *var* genes is epigenetically regulated at the level of transcription so that only a single *var* gene is transcribed at a time while the rest of the family is maintained transcriptionally silent. The exact mechanism by which this process of gene expression is controlled is still elusive; however, extensive work in recent years has indicated that it involves several levels of regulation including DNA and RNA elements, chromatin reorganization and modifications as well dynamics in nuclear architecture (Dzikowski *et al.*, 2009; Scherf *et al.*, 2008).

DNA/RNA elements mediating mutually exclusive expression of *var* genes

One level of regulation of mutually exclusive expression lies within the *cis* elements surrounding each *var* gene. All 60 *var* genes found in the parasite's genome share a similar structure. Their coding regions are split into two exons: the longer first exon encodes the large extracellular portion of PfEMP1 while the shorter and more conserved second exon codes for the intracellular portion of PfEMP1 that anchors the protein to the RBC membrane (Su *et al.*, 1995). Each gene has also two promoters. The first promoter is found upstream of exon one and gives rise to the full-length mRNA encoding PfEMP1. Its transcription start site was mapped to about 1kb upstream of the start methionine (Deitsch *et al.*, 1999). A second promoter is found within the intron. It is bi-directional and drives both sense and anti-sense transcription of rather large (1.7–2.5kb) non-coding RNAs of unknown function (Epp *et al.*, 2008; Su *et al.*, 1995). Interestingly, the two promoters are not transcriptionally active at the same time during the parasite's development inside the erythrocyte (Kyes *et al.*, 2003). The *var* upstream promoter from the single expressed *var* gene is active within the first 12–18h after merozoite invasion, while the intron promoter begins to initiate transcription of the ncRNAs approximately 24h post invasion when the *var* upstream promoter is suppressed. The upstream promoter is also subject to mutually exclusive expression and therefore only a single gene is actively transcribed at a time. In contrast, the intron promoter appears to be active in all *var* genes that have been examined (Epp *et al.*, 2008; Su *et al.*, 1995), and thus late in the replicative cycle non-coding RNAs are likely being transcribed from the entire *var* gene family.

Once a *var* gene is 'chosen' to be the expressed member of the family, it remains imprinted as the single active gene for many subsequent cell cycles, even at points in the cycle when PfEMP1 is not being made (Lopez-Rubio *et al.*, 2007). The phenomenon that marks a single promoter to be active for so many generations is known as epigenetic memory and will be discussed in more detail below. Unlike in other eukaryotic system such as mammalian olfactory receptors and immunoglobulins, expression of the encoded proteins on the cell surface is not necessary for mutually exclusive expression of *var* genes. Two groups used similar approaches to show that mutually exclusive expression was maintained and properly regulated in parasites in which PfEMP1 expression was abolished (Dzikowski *et al.*, 2006; Voss *et al.*, 2006). For example, Dzikowski *et al.* (2006) created transgenic parasites in which the coding region of the first exon of an endogenous *var* gene (PFB1055c) was replaced with *bsd*, a drug-selectable marker that enables selection for activation of this particular *var* promoter. The resulting recombinant gene contained all the non-coding DNA elements of the chromosomal *var* gene, but instead of PfEMP1 it encoded blasticidin S deaminase. Under selection with blasticidin, the parasites were forced to express the selectable marker from the recombinant *var* gene; nonetheless, this gene was expressed in a mutually exclusive manner and all the rest of the *var* gene family remained transcriptionally silent, thus creating PfEMP1-null parasites. Without drug selection these parasites readily switch expression to other *var* genes in their genome, demonstrating that the proper regulation of mutually exclusive expression of the *var* gene family does not require PfEMP1 as long as the non-coding regulatory DNA elements are maintained. Voss and colleagues independently used a very similar approach, albeit with somewhat different constructs, and came to similar conclusions. (Voss *et al.*, 2006)

Bioinformatic comparison of the 5′ upstream region of all 60 *var* genes in the genome suggested that they can be classified into several subgroups based on the sequence similarity of the regions surrounding their promoters (Kraemer *et al.*, 2003; Lavstsen *et al.*, 2003). These subgroups were termed UpsA, UpsB, UpsC, and UpsE. Within these upstream regulatory regions DNA elements have been identified by deletion analysis, and their association with DNA-binding proteins was demonstrated by electron mobility shift assays (Voss *et al.*, 2003). Interestingly, the conserved motifs are promoter-type specific; SPE1 and SPE2 seem to be associated with subtelomeric UpsB promoters while CPE is specific for chromosome internal UpsC promoters. Recently, it was demonstrated that SPE2 specifically binds PfSIP2, a member of a newly discovered DNA-binding protein family named ApiAP2 (Flueck *et al.*, 2010). The possible role of these proteins in regulating gene expression is discussed below. Contrary to *var* promoters, the DNA sequence of all intron promoters is highly conserved and cannot be classified into subgroups.

It appears that promoter-pairing of the *var* upstream and intron promoters within each gene is necessary for the proper regulation of *var* gene expression (Frank *et al.*, 2006). A 'free' upstream promoter that is unpaired with an intron promoter cannot be silenced and therefore is constitutively active. In addition, this 'free' *var* promoter is not 'counted' as a *var* gene by the mechanism that ensures that only a single *var* gene is active at a time (Dzikowski *et al.*, 2007, 2008; Frank *et al.*, 2006). Specifically, a 'free' *var* upstream promoter is not expressed in mutually exclusive manner and therefore it can be actively transcribed simultaneously with another *var* gene in the nucleus of a single parasite. The nature of this promoter–promoter interaction and how it regulates *var* silencing and mutually exclusive expression is not fully understood; however, it is clear that the transcriptional activity of the intron plays a key role in this mechanism (Calderwood *et al.*, 2003; Dzikowski *et al.*, 2007). One piece of evidence supporting the importance of the transcriptional activity of the intron promoter for proper *var* gene regulation lies within a conserved *var* pseudogene named *var1csa*. This pseudogene is constitutively active and is not recognized as a *var* gene by the mechanism that controls mutually exclusive expression (Kyes *et al.*, 2003; Winter *et al.*, 2003), thus both the mechanisms for silencing and 'counting' seem not to recognize this particular pseudogene. Examination of the intron found within this gene revealed that it has a deletion exactly in the region which typically contains the promoter activity. Sequence alignment of *var* introns revealed a conserved sequence structure that can be divided into three regions, each differing in its base composition (Calderwood *et al.*, 2003). Transfection experiments demonstrated that the promoter activity of the intron is located in the middle region that is particularly A-T rich, and deletion of this region abolished the intron's ability to promote transcription as well as its ability to silence an adjacent *var* upstream promoter. These experiments are in agreement with the *var1csa* observation indicating that the promoter activity of *var* introns is required for both silencing and mutually exclusive expression.

Sequence analysis indicated that all three regions of *var* introns display a strong asymmetry in their nucleotide composition (Calderwood *et al.*, 2003). For example, region one contains a high proportion of G and T residues along the forward strand while region three consists of primarily Cs and As. The complementary nature of the asymmetry of regions one and three raised the hypothesis that the intron might possess a bi-directional promoter. This hypothesis was confirmed by transfection experiments showing that the intron promoter is indeed bi-directional and promotes active transcription of episomal reporter genes in both sense and antisense orientation (Epp *et al.*, 2008). Subsequent experiments demonstrated that native chromosomal *var* introns are indeed bi-directional and drive rather large 'sterile' transcripts: about 2–2.5 kb in the forward orientation and 1.7 kb in the reverse orientation (Epp *et al.*, 2008). Both transcripts are made late in the erythrocytic cycle, as expected knowing the temporal activity of the intron promoter. However, these transcripts do not have open reading frames that encode proteins and while they were found to be capped, they are not polyadenylated. This indicated that they might be retained and function in the nucleus similar to what was shown in other systems, such as *Xist* in mammals and Rox1 and 2 in *Drosophila* that mark specific loci for silencing by incorporating into chromatin structures (Chang *et al.*, 2006; Meller *et al.*, 2000). It is possible that the sterile transcripts of *var* genes have a similar function. RNA-FISH assays showed that both the sense and antisense ncRNAs transcribed from *var* introns appear in the nucleus in late stage parasites and accumulate in tightly localized spots (Epp *et al.*, 2008). In addition, chromatin immune-precipitation (ChIP) with antibodies for histone H3 indicated that *var* sterile transcripts are associated with chromatin. However, the nuclear role of these ncRNAs either in silencing or mutually exclusive expression is still elusive.

Histone modifications

Considering the paucity of known transcription factors identified in *Plasmodium* genomes it has been suggested that epigenetic processes might play an unusually large role in regulating gene expression (Aravind *et al.*, 2003). However, malaria parasites appear to lack typical DNA methylation of CpG islands, which is a major epigenetic mark in other eukaryotes. Nevertheless, the genomes of *Plasmodium* species contain many homologues to chromatin modifying enzymes including histone acetyl transferases (HATs, GCN5), deacetylases (HDAC), and lysine methyltransferases and demethylases (KMT, KDM) (Cui *et al.*, 2010). In recent years significant effort was put into research aiming to understand the epigenetic modifications (Miao *et al.*, 2006; Trelle *et al.*, 2009) found associated with the *Plasmodium* genome and the role they play in regulating gene expression. As a result the field has gained tremendous insights into the histone code of malaria parasites as well as their chromatin remodelling enzymes.

Similar to many other organisms, chromatin immune-precipitation (ChIP) showed that histone H4 at active genes is hyperacetylated while the H4 that is associated with silent genes is hypoacetylated (Freitas-Junior *et al.*, 2005). Further, ChIP analyses demonstrated that certain histone modifications are associated with the transcriptional regulation of virulence genes, specifically, enrichment of H3K9me3 was observed at the promoters of silent genes compared to active genes (Chookajorn *et al.*, 2007; Lopez-Rubio *et al.*, 2007). In contrast, the 5′ region of active *var* genes is enriched with H3K4me3 and H3K4me2 during the entire 48h of the parasite cycle, even at points in the cycle when the gene is no longer actively transcribed (Lopez-Rubio *et al.*, 2007). This is referred to as a poised state and results in the gene being re-activated in the next cycle after the progeny merozoites re-invade additional erythrocytes. In addition, while the H3K9me3 appears to be a mark for silent *var* promoters, acetylation of H3K9 is enriched at active *var* promoters. It was therefore suggested that these chromatin modifications contribute to the epigenetic regulation of *var* gene expression.

These pivotal studies have been followed by two comprehensive investigations that used a whole genome approach to characterize the entire epigenome of *P. falciparum* using ChIP-on-chip (Lopez-Rubio *et al.*, 2009; Salcedo-Amaya *et al.*, 2009). These studies indicated that the epigenome of *P. falciparum* exhibits a few unusual characteristics (Salcedo-Amaya *et al.*, 2009, 2010). First, H3K9me3 displays an atypical distribution, indicating that it is a mark for silencing that is specific for virulence gene families both at subtelomeric and internal chromosomal loci. Second, in asynchronous blood stage parasites the epigenome is mostly euchromatic with an unusual enrichment in H3K4me3 and H3K9ac, and the heterochromatin is restricted to loci containing mainly genes that encode for variant surface antigens. Surprisingly, there is a unique plasticity in the use of histone marks at different stages asexual cycle. In early rings H3K4me3 and H3K9ac are spread equally around active and inactive genes, whereas in schizonts, these marks are enriched at the promoter regions of active genes. It was suggested that this unusual epigenome could be a 'weak point' of the parasite that could be exploited to develop specific epi-drugs that selectively target *Plasmodium* enzymes without affecting the human host. It was recently shown that some HDAC inhibitors have anti-malarial activity both *in vivo* and *in vitro*; however, specific malarial HDAC inhibitors that will be potent and at the same time will not affect the human host still need to be developed (Andrews *et al.*, 2009). In addition to histone modifications, the histone variant H2AZ has also recently been shown to be play a role in marking active chromatin (Bartfai *et al.*, 2010; Petter *et al.*, 2011), adding an additional layer of complexity.

Histone modifiers

The first characterized histone modifying enzyme identified in malaria parasites was the *Plasmodium* homologue to the yeast silent information regulator 2 (SIR2), a member of the family of NAD$^+$-dependent histone deacetylases known as sirtuins (Duraisingh *et al.*, 2005; Freitas-Junior *et al.*, 2005). This enzyme modifies hyperacetylated euchromatin that is open for active transcription into hypoacetylated silent heterochromatin, thus repressing transcription. The *P. falciparum* genome contains two SIR2 homologues termed PfSIR2A and PfSIR2B (Gardner *et al.*, 2002; Tonkin *et al.*, 2009). Unlike yeast SIR2 that binds at telomeres, PfSIR2A was demonstrated to bind transcriptionally repressed regions extending from the telomeres to approximately 55 kb into the chromosome (Freitas-Junior *et al.*, 2005). These subtelomeric regions include many clonally variant gene families that are associated with parasite virulence such as *var*, *rif*, *stevor* and *Pfmc-2TM*. By selecting parasites for binding to either CD36 or chondroitin sulphate A (CSA), it is possible to isolate parasite populations that have activated or silenced specific *var* genes. This technique was utilized to show that PfSIR2A is associated with transcriptionally silent but not active genes (Freitas-Junior *et al.*, 2005). Genetic knockouts of *PfSir2A* or *PfSir2B* resulted in the abolishment of complete silencing of *var* genes and 'noisy' or 'leaky' transcription from numerous genes (Duraisingh *et al.*, 2005; Tonkin *et al.*, 2009). Interestingly, it seems that each of the SIR2 paralogues plays a role in silencing discrete subsets of *var* genes. PfSIR2A plays a suppressive role in expression of *var* genes with an UpsA or UpsC promoter type and PfSIR2B is involved in the suppression of most of the subtelomeric UpsB driven *var* genes. In addition to *var* genes, PfSIR2A has an apparent effect on the timing of transcription of *rif* genes. Transcription of several other gene families such as *stevors*, *Pfmc-2TM* and *PHIST* was also affected by the PfSIR2A k/o, supporting its effect on subtelomerically positioned genes. The two SIR2 paralogues are found in all *Plasmodium* species thus far analysed as well as in *Toxoplasma gondii*, which does not undergo antigenic variation. It is therefore intriguing that SIR2 has evolved to regulate virulence gene expression in *P. falciparum*.

Plasmodium spp. also possesses a homologue to the heterochromatin protein 1 (HP1) that is responsible for the propagation of silent heterochromatin along the chromosome. PfHP1 was shown to be able to create dimers *in vitro*, suggesting a mechanism by which it can propagate heterochromatin (Perez-Toledo *et al.*, 2009). In addition, it binds specifically to H3K9me3 but not to other histone marks that are associated with silencing (Flueck *et al.*, 2009; Perez-Toledo *et al.*, 2009). As mentioned previously, H3K9me3 is a mark for silencing of clonally variant gene families, and indeed whole genome ChIP-on-chip analysis indicated that PfHP1 occupancy is restricted to only 425 genes mostly associated with virulence but also genes implicated in invasion and cell cycle regulation. Interestingly however, some of the genes bound by PfHP1 are not marked with H3K9me3. PfHP1 is localized mainly at the nuclear periphery where its target genes are located (Flueck *et al.*, 2009). By incorporating the data regarding the different histone modifiers, a model was proposed by which a concerted interaction between PfSIR2, PfKMT1 and PfHP1 regulate gene silencing in subtelomeric regions. PfSIR2 deacetylates histone tails and is followed by PfKMT, which then methylates them and marks them for heterochromatin formation. Finally, PfHP1 binds H3K9me3 and propagates heterochromatin formation by dimerization. Similar whole-genome ChIP-on-chip analysis was used to map the global distribution of the histone acetyltransferase PfGCN5 that has been shown to acetylate histone H3 at K9 and K14 in *P. falciparum* (Cui *et al.*, 2007). This enzyme was associated with hyperacetylation (H3K9ac) at transcriptionally active genes.

Epigenetic memory

The maintenance of a chronic infection by many *Plasmodium* parasites relies on tightly regulated switches of expression between variable antigens. These switches must happen at a low frequency so that the parasite will not expose its entire antigenic repertoire to the immune system too quickly. At the beginning of the erythrocytic cycle, soon after merozoite invasion into the erythrocyte, the parasite starts to express these antigens and traffic them to the erythrocyte membrane. In *P. falciparum*, the single *var* gene that is being transcribed is epigenetically marked to be the 'on' gene for many generations over numerous replicative cycles. This process is termed epigenetic memory in which the specific gene is imprinted by an epigenetic mark to stay active cycle after cycle. Similarly, there is also epigenetic memory that ensures that transcriptionally silent genes will remain silent through many cell divisions. This epigenetic memory involves the specific histone marks mentioned above both for silent and active chromatin.

Recent work suggests that active transcription (the movement of the RNA polymerase complex through a gene) is required to maintain the epigenetic mark controlling antigenic switching (Dzikowski *et al.*, 2008). This conclusion was reached by transfecting parasites with a plasmid that carried a *var* promoter driving expression of the drug-selectable marker blasticidin S deaminase. This plasmid did not contain a *var* intron, therefore as described above, the *var* promoter on the plasmid was not recognized by the pathway that ensures mutually exclusive expression and was constitutively active. This resulted in the transfected parasites actively transcribing two *var* promoters simultaneously, the episomal copy and the endogenous chromosomal *var* gene. By increasing the dose of blasticidin that the parasites were exposed to, it was possible to select for parasite populations that carried increasing copy numbers of transcriptionally active episomal *var* promoters. The episomal *var* promoters effectively competed with the endogenous *var* gene for an essential, limited activation factor or factors, and high episome copy numbers resulted in complete silencing of all chromosomal *var* genes. This limited factor appears to be shared by several of the multicopy gene families since this technique also resulted in silencing of *rif*, *stevor* and *Pfmc-2TM* genes as well as *var* (Howitt *et al.*, 2009). Importantly, preventing the chromosomal *var* genes from being actively transcribed in this way erased all epigenetic memory. Specifically, once the parasites were removed from drug pressure and allowed to shed the competing episomes, they did not return to expressing the *var* gene that was previously active but rather displayed a heterogeneous transcription profile indicating that the epigenetic memory that was regulating *var* gene expression was completely lost. This led to the conclusion that active transcription is needed to maintain the epigenetic memory of the active gene.

It is possible that RNA polymerase II, in addition to its basic role in mRNA production, also has important functions in the maintenance of epigenetic marks during transcription. A model has been proposed suggesting that while moving along the DNA template during transcription, the C-terminal domain (CTD) of RNA polymerase II directly interacts with and recruits histone modifying enzymes that catalyse the histone marks needed for epigenetic memory (Cho, 2007). As described earlier in this chapter, the CTD is comprised of a tandem array of heptapeptide repeats that are essential for gene expression and coordinate transcriptional events by reversible phosphorylation. Analysis of the amino acid sequences from the various *Plasmodium* species showed that they all maintain the typical, conserved CTD structure. However, the CTDs in the rodent and bird parasites, *P. berghei*, *P. yoelii*, and *P. gallinaceum*, are remarkably short compared to the primate parasites *P. knowlesi*, *P. vivax*, *P. reichenowi*, and *P. falciparum*, leading to the suggestion that the expanded CTDs in primate parasites had evolved because it provided a strong functional advantage (Kishore *et al.*, 2009). One possibility for this advantage could be epigenetic memory. For example, since rodents tend to have short life spans, there might not be a need for the long-term, chronic infections that require the epigenetic memory described above. However, this could indeed be an advantage when infecting long-lived primates that have few offspring. This hypothesis is supported by a recent study showing that the rodent malaria parasite *P. yoelli* does not display epigenetic memory in controlling the expression of the *yir* multi gene family (Cunningham *et al.*, 2009). Further study will probably unveil the important role of RNA polymerase II in the unique transcriptional regulation of malaria parasite.

DNA assembly and nucleosome positioning

Recent advances in deep sequencing technology have allowed the first glimpse into the correlation between DNA assembly and nucleosome positioning and gene expression in *Plasmodium* parasites. The nucleosomal packaging of DNA influences its accessibility to the transcriptional machinery. Genome-wide maps of nucleosome occupancy over the *P. falciparum* erythrocytic cycle indicate that it is dynamic through the cycle (Bartfai *et al.*, 2010; Ponts *et al.*, 2010a,b; Westenberger *et al.*, 2009). Nucleosome assembly is much more compact at late schizont stages when total transcription decreases (Sims *et al.*, 2009), while in rings and trophozoite stages when the transcriptional machinery is highly active there is an increased relaxation of nucleosome assembly that allows transcription to happen. In addition, histones have been reported to be generally enriched in coding regions, and relatively poor in intergenic regions (Ponts *et al.*, 2010b; Westenberger *et al.*, 2009), although this view has been challenged recently as potentially resulting from a methodological error resulting from a PCR amplification step utilized in preparing samples for analysis (Bartfai *et al.*, 2010).

Nuclear sublocalization

An emerging aspect in epigenetic regulation of gene expression is the important role that nuclear architecture plays in this process. It is becoming clear in many eukaryotes that the nucleus as a whole organelle can be divided to different subnuclear compartments with different biological functions. There are regions where DNA is condensed into heterochromatin that is not accessible to the transcriptional machinery and therefore transcriptionally silent, as well as regions of loose euchromatin where transcriptional factors are centred (Sutherland *et al.*, 2009). Dynamics in nuclear architecture also play an important role in gene expression in *Plasmodium*. Fig. 7.2 shows several different methods used to visualize nuclear structure in *P. falciparum*. Special attention has

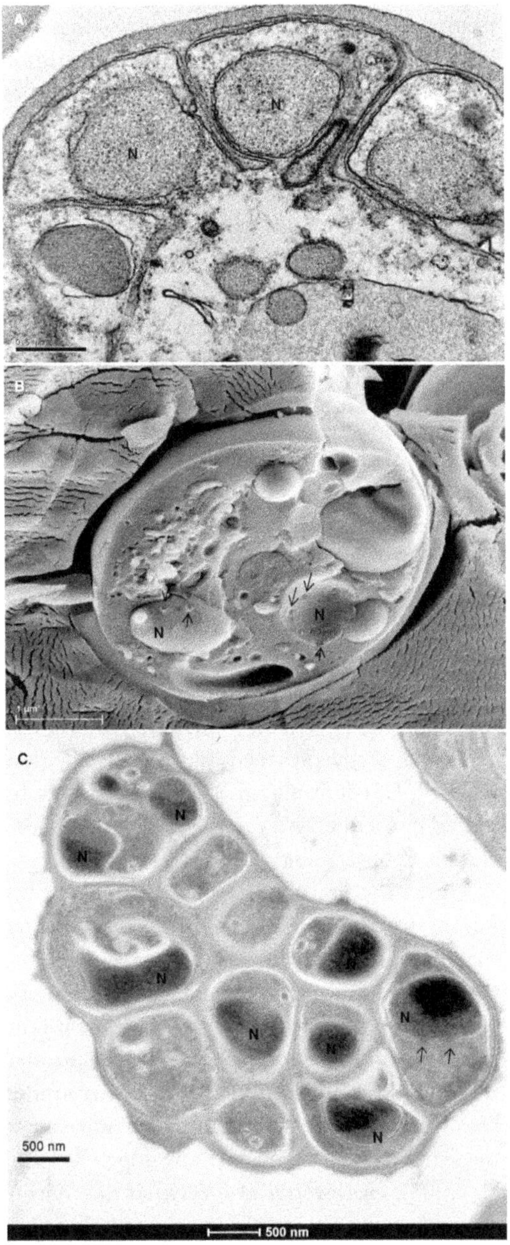

Figure 7.2 Different perspectives of *Plasmodium* nuclear architecture. A. TEM thin section through a red blood cell infected with schizont that was chemically fixed. B. Freeze fracture cryoSEM of an early schizont exposing nuclei and nuclear pores (arrowed). C. TEM thin section through a red blood cell infected with schizont that was cryo-fixed by high pressure freezing. It is possible to differentiate between electron dense and loose genetic matter. Nuclear pores are arrowed. N (nucleus); RBC (red blood cell). The EM was done at the Electron Microscopy Unit, Weizmann Institute of Science, Rehovot, Israel.

been given to the regulation of genes that have been implicated in the virulence of *P. falciparum*. Using DNA fluorescence *in situ* hybridization (FISH) with gene-specific probes, Ralph *et al.* showed for the first time dynamic changes in the subnuclear localization of genes in correlation to their transcriptional status (Duraisingh *et al.*, 2005; Ralph *et al.*, 2005). They used parasites selected by panning for their ability to bind chondroitin sulphate A (CSA). These parasites exclusively express *var2csa* that encodes a specific form of PfEMP1 that binds this placental receptor and underlies pregnancy associated malaria. Another parasite population was selected for binding to CD36 – an alternative endothelial receptor – in this parasite population *var2csa* is silent. They demonstrated that when *var2csa* is silent it co-localizes with the silent subtelomeric clusters at the nuclear periphery. However, upon activation it moves to a different location at the nuclear periphery and no longer co-localizes with the silent sub–telomeric clusters. Recently an additional study provided additional data from RNA-FISH, confirming that active *var* genes are indeed transcribed away from the silent telomeric clusters (Lopez-Rubio *et al.*, 2009).

Ultrastructural studies of the *P. falciparum* nucleus using simple EM thin sections indicated that in schizonts the nucleus can be divided into two distinct subcompartments. The inner region appears to be composed mainly of electron-faint genetic matter that corresponds with euchromatin (Fig. 7.2C). This region is usually surrounded by a thick layer of heterochromatin adjacent to the inner part of the nuclear envelope (Ralph *et al.*, 2005). However, it also seemed that within this perinuclear heterochromatic ring there is a small euchromatic region. A model was proposed suggesting that upon transcriptional activation, genes at the nuclear periphery move from the silent heterochromatic rim to the euchromatic region. Moreover, it was postulated that this can provide an explanation for the control of mutually exclusive expression of *var* genes by a mechanism similar to what was proposed for African trypanosomes (Navarro *et al.*, 2001). In *T. brucei*, antigenic variation is achieved by switches in the expression of *vsg* genes (Pays, 2005). These genes encode the hypervariable glycoprotein coat of the parasite and are also expressed in a mutually exclusive manner. The active *vsg* gene is located within a specific subnuclear expression site while silent *vsg* genes are found away from this expression site (Navarro *et al.*, 2001). In addition, in parasites in which mutually exclusive expression of *vsg* genes was interrupted and they express several *vsg* genes, there are also additional expression sites in the nucleus (Yang *et al.*, 2009). This implies that the expression site itself or something in it is limiting the parasite to express only a single gene at a time.

To determine whether regulation of antigenic variation in *Plasmodium* involves a subnuclear expression site, Voss and colleagues utilized transgenic parasite lines in which a transcriptionally active, episomal *var* promoter could be visualized using FISH (Dzikowski *et al.*, 2007; Voss *et al.*, 2006). In these studies, the active *var* promoter clearly moved to a specific subnuclear position. To specifically test the hypothesis that this subnuclear position possessed a capacity limited to only one active gene at a time, thus ensuring that the rest of the gene family is transcriptionally silent, Dzikowski *et al.* (2007) generated parasites were transfected with an episome that carried a 'free' *var* promoter that is not 'counted' by the mechanism that controls mutually exclusive expression. Two-colour DNA FISH was used to simultaneously visualize the intact, transcriptionally active, chromosomal *var* as well as the constitutively active *var* promoter on the plasmid. When both genes were active they co-localized within the same spot at the nuclear periphery; however, when the chromosomal gene was silent, it was located separately in a different spot than the transcriptionally active episomal *var* promoter, although it remained at the nuclear periphery. The rate that the two active *var* promoters co-localized was significantly higher than that found for an active *var* gene and an active housekeeping gene, implying that there might be a mechanism that is yet unknown that attracts active *var* genes to a particular subnuclear location. Thus the subnuclear expression site was clearly capable of accommodating more than one active *var* promoter at a time, and subsequent experiments demonstrated that it could include up to 20 active *var* promoters (Dzikowski *et al.*,

2008); therefore, limited accessibility to this subnuclear position appears not to be controlling mutually exclusive expression.

These observations raised the question of whether this expression site might also be occupied by transcriptionally active members of other clonally variant gene families that have been implicated in antigenic variation and pathogenicity, including the *rif*, *stevor* and *Pfmc-2TM* gene families. These clonally variant genes appear to share a common titratable factor that is required for their activation. However, the genes are transcribed at different times during the asexual cycle, with some overlap between *var* and *rif*. This overlap allowed an experiment to be performed asking whether active *var* and *rif* genes also localized to the same subnuclear expression site. DNA–FISH was performed with probes to active and silent *var* genes and to an active *rif* promoter on an episome that was selected to be transcriptionally active using a drug-selectable marker (Howitt *et al.*, 2009). These experiments demonstrated significant co-localization of active *rif* and *var* promoters compared with unrelated genes that are transcribed at the same time, providing additional evidence that virulence genes share a common subnuclear expression site.

Not only genes are located at distinct foci in the nucleus, but also the potential regulators that are involved in epigenetic control of gene expression. One of these enzymes is the *Plasmodium* homologue to the histone lysine 9 methyltransferase (PfKMT1) (Cui *et al.*, 2008) that is responsible for the methylation of H3K9. PfKMT1 is located largely at the periphery of *P. falciparum* nuclei in the same perinuclear space where H3K9me3 is enriched and at the same location where genes that are enriched with this histone mark are located (Lopez-Rubio *et al.*, 2009). Additional chromatin remodelling enzymes were also localized to distinct subnuclear compartments. Using gold labelling immune-EM it was possible to show that the sirtuin histone deacetylase (PfSIR2) and heterochromatin protein 1 (PfHP1) localized to the electron-dense heterochromatic region at the nuclear periphery exactly where one would expect it to be located (Flueck *et al.*, 2009; Freitas-Junior *et al.*, 2005). In addition, immunofluorescence assays have demonstrated subnuclear compartmentalization of additional epigenetic marks as well as possible epigenetic regulators (Volz *et al.*, 2010).

Altogether, nuclear architecture plays an important role in regulating gene expression in *Plasmodium*. However, the nature of the specific subnuclear compartments, such as the nuclear periphery and the nuclear envelope, as well as the dynamics in nuclear architecture associated with changes in gene expression, need to be further investigated.

New horizons-identification of the ApiAP2 gene family

As mentioned above, *Plasmodium* represents a phylogenic clade that is quite distant from the model organisms that have been the primary focus of research into the molecular mechanisms underlying transcription and gene expression. The concept that transcription might be fundamentally different in these protozoans was reinforced by the initial difficulty in identifying transcription factors encoded in the genomes of the various species of *Plasmodium* and their closely related relatives within the apicomplexa. However, this difficulty was at least in part alleviated with the identification of the ApiAP2 gene family, which now represents the primary class of transcriptional regulatory proteins found in this group of parasites.

Computational identification of the ApiAP2 gene family

Using sensitive sequence profile searches, investigators at the National Center for Biotechnology Information identified a lineage-specific expansion of a family of proteins that each contain one to four copies of a version of the Apetala 2-integrase DNA binding domain (called the AP2 domain) (Balaji *et al.*, 2005). This family of proteins was found in all of the apicomplexan parasites, including *Cryptosporidium*, *Theileria*, *Toxoplasma* and *Plasmodium*, and was therefore named the ApiAP2 family. While the AP2 domain itself has been identified as a DNA binding domain in mobile elements as well as transcription factors (Magnani *et al.*, 2004; Wessler, 2005; Wuitschick *et al.*, 2004), the authors proposed that in apicomplexan parasites it serves as the DNA binding domain in a family of transcriptional regulators, thereby filling the perceived hole in our understanding of the basic transcriptional machinery of malaria parasites and their close relatives among the protozoan parasites.

In *Plasmodium*, the ApiAP2 family consists of 27 members that each contain 1–4 AP2 domains (Balaji *et al.*, 2005). Using the genome-wide gene expression data, it was possible to identify the timing of expression of each member of the family, and thus construct a model cascade in which different members of the family are expressed in a specified order, with each serving as the principal transcriptional activator during the time point in development in which they are expressed. While the discovery of the ApiAP2 family of proteins was an entirely *in silico* endeavour, it represents one of the major intellectual and conceptual advances in the study of *Plasmodium* gene regulation to date.

Validation of the function of ApiAP2 family members as DNA-binding proteins

While the computational identification of the ApiAP2 domain allowed researchers to identify proteins with putative DNA binding properties, the rest of the encoded proteins displayed no homology to any known transcriptional activators or repressors. Therefore while it was attractive to hypothesize that the paucity of identifiable transcription factors had been at least partially filled, experimental data validating the putative function of these proteins needed to be obtained. This was first addressed by De Silva *et al.* who utilized protein binding microarrays and electrophoretic mobility shift assays (EMSA) to identify the DNA sequences bound by two members of the ApiAP2 family (De Silva *et al.*, 2008). This analysis was subsequently extended to all 27 members of the family (Campbell *et al.*, 2010). Once the consensus binding sequence had been identified for each protein, it was possible to search the genome for individual genes that contained this sequence within their upstream regulatory regions. Combined with gene expression data, the authors were able to provide additional evidence that these proteins serve as transcription factors.

Additional interesting information came from the same research group when they published the crystal structure of one ApiAP2 domain in complex with DNA (Lindner *et al.*, 2010). Interestingly, the AP2 domain dimerizes upon binding to DNA, a property that coincides with the identification of two putative binding sites in the DNA sequences of several putative target genes. This suggests that this protein (and potentially other members of the family) can serve to bring disparate DNA fragments together within the nucleus, thus potentially influencing the topology of the chromosomal regions to which they bind. This work provided significant insights into how members of the ApiAP2 family bind to DNA and assemble into a protein complex; however, experimental *in vivo* data were still lacking.

Genetic evidence for ApiAP2 proteins as transcription factors

The group of Yuda *et al.* (2009, 2010) chose to take advantage of the *P. berghei* system for investigating ApiAP2 proteins that might be involved in controlling transcription of genes expressed in the mosquito stages. Because genetic manipulations are performed in the asexual, erythrocytic stages of the parasite cycle, it is possible to obtain mutations that are lethal to the sexual/insect stages. In addition, *P. berghei* is the parasite of choice for investigating parasite development within the mosquito because of the ease with which transgenic parasites can be obtained and because infected mosquitoes do not pose a risk for transmitting parasites to humans and thus are logistically much easier to work with.

In the first set of experiments, Yuda and colleagues identified a member of the ApiAP2 family that is necessary for development of the ookinete stage of the parasite life cycle (Yuda *et al.*, 2009). This gene, called *AP2-O*, is transcribed in female gametocytes, but its translation into protein is repressed by the DOZI complex as described in a previous section of this chapter. Once fertilization occurs in the gut of the mosquito and an ookinete forms, the *AP2-O* mRNA is rapidly translated into protein, which then serves to activate a set of genes required for midgut invasion by binding to a six base pair element upstream of each of the target genes. Further evidence for the role of this protein was obtained by knocking out the gene, which prevented transcription of the target genes and disrupted the parasite's ability to complete the mosquito stage of development. Thus the authors concluded that AP2-O serves as a master regulator of a set of genes specifically required for ookinete development.

Using a similar approach, the same group identified a gene required for development of sporozoites and refer to it as *AP2-Sp* (Yuda *et al.*, 2010). This gene is expressed by parasites from the late ookinete stage through to salivary gland sporozoites. Parasites in which the gene had been knocked out failed to form sporozoites, demonstrating the importance of the protein for parasite development. The DNA binding sequence was identified using EMSA. Then, in an elegant experiment that provided definitive evidence that these two members of the ApiAP2 family act as specific transcriptional activators, the authors swapped the AP2 domains between the *AP2-O* and *AP2-Sp* genes in transgenic lines of parasites, thus attempting to reverse the timing of expression of the loci that are targeted by them. As predicted, several sporozoite-specific genes were expressed in ookinetes, thereby validating the model of ApiAP2 genes as major regulators of transcription.

Possible functions of ApiAP2 proteins other than transcription

Once the ApiAP2 family was identified and the DNA binding capacity of the proteins verified, the broad assumption in the field was that these proteins functioned as transcription factors. While this role has been demonstrated for the two members of the family described above (AP2-O and AP2-Sp), functions have not been described for the majority of the rest of the family. Since many proteins that do not function as transcription factors possess DNA binding activity, it is possible that some ApiAP2 proteins perform other roles within the cell. Recent work by Flueck *et al.* provides one such example of an alternative function (Flueck *et al.*, 2010).

In an effort to find enhancers and repressor sequences upstream of type B *var* genes, Voss and colleagues mapped the regions of DNA upstream of the protein coding region that were bound by proteins found in crude nuclear extracts (Voss *et al.*, 2003). One of these elements, called SPE2, was bound by a protein or protein complex found in erythrocyte stage parasites late in schizogeny at a point in the cycle when *var* genes are not actively transcribed, suggesting that it was not bound by a transcriptional activator. The protein complex that bound to SPE2 was then purified and found to be a member of the ApiAP2 family (PFF0200c) (Flueck *et al.*, 2010). Since binding of the protein to its DNA element did not coincide with activation of transcription, it was not clear exactly what function the protein had. A putative function was suggested by affinity purification and chromatin immunoprecipitation studies which found that that protein bound exclusively to subtelomeric DNA, and by the fact that this protein co-purified with proteins known to be involved in DNA repair and chromatin metabolism. Type B *var* genes reside immediately next to the telomere repeats of most chromosomes and they are oriented such that they are transcribed towards the centromere (Kraemer *et al.*, 2003; Lavstsen *et al.*, 2003). Therefore the upstream regions of these *var* genes are immediately adjacent to the telomeres, and thus the SPE2 element is as much as part of the subtelomeric DNA as it is the upstream region of the neighbouring *var* gene. It now appears that the primary function of this ApiAP2 protein is in fact in telomere maintenance rather than transcription, thereby demonstrating that this family of DNA binding proteins could have many functions outside of regulating transcription.

Possible targets for intervention strategies?

As the mechanisms regulating transcription in *Plasmodium* are becoming more clear, it is interesting to speculate about the possibilities for targeting this process for new therapies for the disease. While transcriptional repression during gamete development and the histone modifications responsible for epigenetic regulation appear to be largely conserved between parasites and their human hosts, several other aspects are unique and might serve as attractive targets for novel drugs. For example, the ApiAP2 family of transcription factors is not found in mammals; thus if pharmacological agents were discovered that specifically interacted with members of this family, they might very well have a parasite-specific effect. Similarly, the C-terminal domain of the large subunit of RNA polymerase II is unique in both its length and composition when compared to higher eukaryotes, suggesting that many of the core functions of the polymerase complex might be regulated somewhat differently. This could open up additional avenues for intervention strategies, and given how imperative transcription is for life, this could represent an effective target for disruption.

Conclusions

The study of mechanisms that control gene expression in malaria parasites has advanced significantly in recent years, and important insights have begun to provide us with a basic picture of the many points where gene expression is regulated (Fig. 7.1). Our understanding of epigenetic processes in particular has increased markedly with the application of the 'histone code' to *Plasmodium* genomics. The completion of the nearly fully assembled genomes of *P. falciparum*, followed by other species of *Plasmodium*, paved the way for the genome-wide analyses of gene expression that have become such a valuable resource to investigators working on many different aspects of parasite biology. Improving technology continues to expand our rapidly evolving database; for example new studies employing RNA-seq are providing unprecedented resolution to the catalogue of transcription start sites, polyadenylation sites, transcript abundance, alternatively spliced mRNAs, and nucleosome occupancy (Bartfai *et al.*, 2010; Otto *et al.*, 2010; Ponts *et al.*, 2010a,b; Westenberger *et al.*, 2009). The added knowledge should increase our understanding of parasite metabolism and development as well as improve the precision with which parasites can be genetically manipulated.

Studies into gene regulation in malaria parasites also continue to have broad implications for our understanding of the evolution of eukaryotic organisms. For example, malaria parasites and their close relatives among the apicomplexa display sharply divergent families of transcription factors, with the closest relatives of the ApiAP2 family being found in plants. In contrast, the post-transcriptional repression of sexual stage mRNA translation through P body formation is remarkably conserved even with the very distantly related metazoans. The significant evolutionary distance between these parasites and the more intensely studied 'crown' eukaryotes provides a powerful system for comparative studies and is likely to yield significant insights into the eukaryotic lineage. A better and more detailed understanding of the processes regulating gene expression will also provide a stronger basis for the development of novel strategies for disease intervention.

References

Amulic, B., Salanti, A., Lavstsen, T., Nielsen, M.A., and Deitsch, K.W. (2009). An upstream open reading frame controls translation of *var2csa*, a gene implicated in placental malaria. PLoS Pathog. *5*, e1000256.

Andrews, K.T., Tran, T.N., Wheatley, N.C., and Fairlie, D.P. (2009). Targeting histone deacetylase inhibitors for anti-malarial therapy. Curr. Top. Med. Chem. *9*, 292–308.

Aravind, L., Iyer, L.M., Wellems, T.E., and Miller, L.H. (2003). *Plasmodium* biology: genomic gleanings. Cell *115*, 771–785.

Balaji, S., Babu, M.M., Iyer, L.M., and Aravind, L. (2005). Discovery of the principal specific transcription factors of Apicomplexa and their implication for the evolution of the AP2-integrase DNA binding domains. Nucleic Acids Res. *33*, 3994–4006.

Bartfai, R., Hoeijmakers, W.A., Salcedo-Amaya, A.M., Smits, A.H., Janssen-Megens, E., Kaan, A., Treeck, M., Gilberger, T.W., Francoijs, K.J., and Stunnenberg, H.G. (2010). H2A.Z demarcates intergenic regions of the *Plasmodium falciparum* epigenome that are dynamically marked by H3K9ac and H3K4me3. PLoS Pathog. *6*, e1001223.

Baruch, D.I., Pasloske, B.L., Singh, H.B., Bi, X., Ma, X.C., Feldman, M., Taraschi, T.F., and Howard, R.J. (1995). Cloning the *P. falciparum* gene encoding PfEMP1, a malarial variant antigen and adherence receptor on the surface of parasitized human erythrocytes. Cell *82*, 77–87.

Baum, J., Papenfuss, A.T., Mair, G.R., Janse, C.J., Vlachou, D., Waters, A.P., Cowman, A.F., Crabb, B.S., and Koning-Ward, T.F. (2009). Molecular genetics and comparative genomics reveal RNAi is not functional in malaria parasites. Nucleic Acids Res *37*, 3788–3798.

Bischoff, E., and Vaquero, C. (2010). In silico and biological survey of transcription-associated proteins implicated in the transcriptional machinery during the erythrocytic development of *Plasmodium falciparum*. BMC. Genomics *11*, 34.

Bozdech, Z., Llinas, M., Pulliam, B.L., Wong, E.D., Zhu, J., and Derisi, J.L. (2003). The transcriptome of the intraerythrocytic developmental cycle of *Plasmodium falciparum*. PLoS Biol. *1*, E5.

Bozdech, Z., Mok, S., Hu, G., Imwong, M., Jaidee, A., Russell, B., Ginsburg, H., Nosten, F., Day, N.P., White,

N.J., Carlton, J.M., and Preiser, P.R. (2008). The transcriptome of *Plasmodium* vivax reveals divergence and diversity of transcriptional regulation in malaria parasites. Proc. Natl. Acad. Sci. U.S.A. *105*, 16290–16295.

Calderwood, M.S., Gannoun-Zaki, L., Wellems, T.E., and Deitsch, K.W. (2003). *Plasmodium falciparum var* genes are regulated by two regions with separate promoters, one upstream of the coding region and a second within the intron. J. Biol. Chem. *278*, 34125–34132.

Campbell, T.L., De Silva, E.K., Olszewski, K.L., Elemento, O., and Llinas, M. (2010). Identification and genome-wide prediction of DNA binding specificities for the ApiAP2 family of regulators from the malaria parasite. PLoS Pathog. *6*, e1001165.

Carlton, J.M., Adams, J.H., Silva, J.C., Bidwell, S.L., Lorenzi, H., Caler, E., Crabtree, J., Angiuoli, S.V., Merino, E.F., Amedeo, P., *et al.* (2008a). Comparative genomics of the neglected human malaria parasite *Plasmodium vivax*. Nature *455*, 757–763.

Carlton, J.M., Angiuoli, S.V., Suh, B.B., Kooij, T.W., Pertea, M., Silva, J.C., Ermolaeva, M.D., Allen, J.E., Selengut, J.D., Koo, H.L., *et al.* (2002). Genome sequence and comparative analysis of the model rodent malaria parasite *Plasmodium yoelii yoelii*. Nature *419*, 512–519.

Carlton, J.M., Escalante, A.A., Neafsey, D., and Volkman, S.K. (2008b). Comparative evolutionary genomics of human malaria parasites. Trends Parasitol. *24*, 545–550.

Chang, S.C., Tucker, T., Thorogood, N.P., and Brown, C.J. (2006). Mechanisms of X-chromosome inactivation. Front. Biosci. *11*, 852–866.

Chapman, R.D., Heidemann, M., Hintermair, C., and Eick, D. (2008). Molecular evolution of the RNA polymerase II CTD. Trends Genet. *24*, 289–296.

Cho, E.J. (2007). RNA polymerase II carboxy-terminal domain with multiple connections. Exp. Mol. Med. *39*, 247–254.

Chookajorn, T., Dzikowski, R., Frank, M., Li, F., Jiwani, A.Z., Hartl, D.L., and Deitsch, K.W. (2007). Epigenetic memory at malaria virulence genes. Proc. Natl. Acad. Sci. U.S.A. *104*, 899–902.

Coulson, R.M., Hall, N., and Ouzounis, C.A. (2004). Comparative genomics of transcriptional control in the human malaria parasite *Plasmodium falciparum*. Genome Res. *14*, 1548–1554.

Crabb, B.S., and Cowman, A.F. (1996). Characterization of promoters and stable transfection by homologous and nonhomologous recombination in *Plasmodium falciparum*. Proc. Natl. Acad. Sci. U.S.A. *93*, 7289–7294.

Cui, L., Fan, Q., Cui, L., and Miao, J. (2008). Histone lysine methyltransferases and demethylases in *Plasmodium falciparum*. Int. J. Parasitol. *38*, 1083–1097.

Cui, L., and Miao, J. (2010). Chromatin-mediated epigenetic regulation in the malaria parasite *Plasmodium falciparum*. Eukaryot. Cell *9*, 1138–1149.

Cui, L., Miao, J., Furuya, T., Li, X., Su, X.Z., and Cui, L. (2007). PfGCN5-mediated histone H3 acetylation plays a key role in gene expression in *Plasmodium falciparum*. Eukaryot. Cell *6*, 1219–1227.

Cunningham, D., Fonager, J., Jarra, W., Carret, C., Preiser, P., and Langhorne, J. (2009). Rapid changes in transcription profiles of the *Plasmodium yoelii* yir multigene family in clonal populations: lack of epigenetic memory? PLoS One *4*, e4285.

De Silva, E.K., Gehrke, A.R., Olszewski, K., Leon, I., Chahal, J.S., Bulyk, M.L., and Llinas, M. (2008). Specific DNA-binding by apicomplexan AP2 transcription factors. Proc. Natl. Acad. Sci. U.S.A. *105*, 8393–8398.

Dechering, K.J., Kaan, A.M., Mbacham, W., Wirth, D.F., Eling, W., Konings, R.N., and Stunnenberg, H.G. (1999). Isolation and functional characterization of two distinct sexual-stage- specific promoters of the human malaria parasite *Plasmodium falciparum*. Mol. Cell Biol. *19*, 967–978.

Deitsch, K.W., del Pinal, A., and Wellems, T.E. (1999). Intra-cluster recombination and var transcription switches in the antigenic variation of *Plasmodium falciparum*. Mol. Biochem. Parasitol. *101*, 107–116.

Duraisingh, M.T., Voss, T.S., Marty, A.J., Duffy, M.F., Good, R.T., Thompson, J.K., Freitas-Junior, L.H., Scherf, A., Crabb, B.S., and Cowman, A.F. (2005). Heterochromatin silencing and locus repositioning linked to regulation of virulence genes in *Plasmodium falciparum*. Cell *121*, 13–24.

Dzikowski, R., and Deitsch, K.W. (2008). Active transcription is required for maintenance of epigenetic memory in the malaria parasite *Plasmodium falciparum*. J. Mol. Biol. *382*, 288–297.

Dzikowski, R., and Deitsch, K.W. (2009). Genetics of antigenic variation in *Plasmodium falciparum*. Curr. Genet. *55*, 103–110.

Dzikowski, R., Frank, M., and Deitsch, K. (2006). Mutually exclusive expression of virulence genes by malaria parasites is regulated independently of antigen production. PLoS Pathog. *2*, e22.

Dzikowski, R., Li, F., Amulic, B., Eisberg, A., Frank, M., Patel, S., Wellems, T.E., and Deitsch, K.W. (2007). Mechanisms underlying mutually exclusive expression of virulence genes by malaria parasites. EMBO Rep. *8*, 959–965.

Egloff, S., and Murphy, S. (2008). Cracking the RNA polymerase II CTD code. Trends Genet. *24*, 280–288.

Epp, C., Li, F., Howitt, C.A., Chookajorn, T., and Deitsch, K.W. (2008). Chromatin associated sense and antisense noncoding RNAs are transcribed from the *var* gene family of virulence genes of the malaria parasite *Plasmodium falciparum*. RNA *15*, 116–127.

Florens, L., Washburn, M.P., Raine, J.D., Anthony, R.M., Grainger, M., Haynes, J.D., Moch, J.K., Muster, N., Sacci, J.B., Tabb, D.L., *et al.* (2002). A proteomic view of the *Plasmodium falciparum* life cycle. Nature *419*, 520–526.

Flueck, C., Bartfai, R., Niederwieser, I., Witmer, K., Alako, B.T., Moes, S., Bozdech, Z., Jenoe, P., Stunnenberg, H.G., and Voss, T.S. (2010). A major role for the *Plasmodium falciparum* ApiAP2 protein PfSIP2 in chromosome end biology. PLoS Pathog. *6*, e1000784.

Flueck, C., Bartfai, R., Volz, J., Niederwieser, I., Salcedo-Amaya, A.M., Alako, B.T., Ehlgen, F., Ralph, S.A.,

Cowman, A.F., Bozdech, Z., *et al.* (2009). *Plasmodium falciparum* heterochromatin protein 1 marks genomic loci linked to phenotypic variation of exported virulence factors. PLoS Pathog. *5*, e1000569.

Fonager, J., Cunningham, D., Jarra, W., Koernig, S., Henneman, A.A., Langhorne, J., and Preiser, P. (2007). Transcription and alternative splicing in the yir multigene family of the malaria parasite *Plasmodium* y. yoelii: identification of motifs suggesting epigenetic and post-transcriptional control of RNA expression. Mol. Biochem. Parasitol. *156*, 1–11.

Foth, B.J., Zhang, N., Mok, S., Preiser, P.R., and Bozdech, Z. (2008). Quantitative protein expression profiling reveals extensive post-transcriptional regulation and post-translational modifications in schizont-stage malaria parasites. Genome Biol. *9*, R177.

Frank, M., Dzikowski, R., Constantini, D., Amulic, B., Burdougo, E., and Deitsch, K. (2006). Strict pairing of var promoters and introns is required for *var* gene silencing in the malaria parasite *Plasmodium falciparum*. J. Biol. Chem. *281*, 9942–9952.

Freitas-Junior, L.H., Hernandez-Rivas, R., Ralph, S.A., Montiel-Condado, D., Ruvalcaba-Salazar, O.K., Rojas-Meza, A.P., Mancio-Silva, L., Leal-Silvestre, R.J., Gontijo, A.M., Shorte, S., and Scherf, A. (2005). Telomeric heterochromatin propagation and histone acetylation control mutually exclusive expression of antigenic variation genes in malaria parasites. Cell *121*, 25–36.

Ganesan, K., Ponmee, N., Jiang, L., Fowble, J.W., White, J., Kamchonwongpaisan, S., Yuthavong, Y., Wilairat, P., and Rathod, P.K. (2008). A genetically hard-wired metabolic transcriptome in *Plasmodium falciparum* fails to mount protective responses to lethal antifolates. PLoS Pathog. *4*, e1000214.

Gardner, M.J., Hall, N., Fung, E., White, O., Berriman, M., Hyman, R.W., Carlton, J.M., Pain, A., Nelson, K.E., Bowman, S., Paulsen, I.T., James, K., Eisen, J.A., Rutherford, K., Salzberg, S.L., Craig, A., Kyes, S., Chan, M.S., Nene, V., Shallom, S.J., Suh, B., Peterson, J., Angiuoli, S., Pertea, M., Allen, J., Selengut, J., Haft, D., Mather, M.W., Vaidya, A.B., Martin, D.M., Fairlamb, A.H., Fraunholz, M.J., Roos, D.S., Ralph, S.A., McFadden, G.I., Cummings, L.M., Subramanian, G.M., Mungall, C., Venter, J.C., Carucci, D.J., Hoffman, S.L., Newbold, C., Davis, R.W., Fraser, C.M., and Barrell, B. (2002). Genome sequence of the human malaria parasite *Plasmodium falciparum*. Nature *419*, 498–511.

Gonzales, J.M., Patel, J.J., Ponmee, N., Jiang, L., Tan, A., Maher, S.P., Wuchty, S., Rathod, P.K., and Ferdig, M.T. (2008). Regulatory hotspots in the malaria parasite genome dictate transcriptional variation. PLoS Biol. *6*, e238.

Gunasekera, A.M., Patankar, S., Schug, J., Eisen, G., and Wirth, D.F. (2003). Drug-induced alterations in gene expression of the asexual blood forms of *Plasmodium falciparum*. Mol. Microbiol. *50*, 1229–1239.

Hall, N., and Carlton, J. (2005a). Comparative genomics of malaria parasites. Curr. Opin. Genet. Dev. *15*, 609–613.

Hall, N., Karras, M., Raine, J.D., Carlton, J.M., Kooij, T.W., Berriman, M., Florens, L., Janssen, C.S., Pain, A., Christophides, G.K., James, K., Rutherford, K., Harris, B., Harris, D., Churcher, C., Quail, M.A., Ormond, D., Doggett, J., Trueman, H.E., Mendoza, J., Bidwell, S.L., Rajandream, M.A., Carucci, D.J., Yates, J.R., III, Kafatos, F.C., Janse, C.J., Barrell, B., Turner, C.M., Waters, A.P., and Sinden, R.E. (2005b). A comprehensive survey of the *Plasmodium* life cycle by genomic, transcriptomic, and proteomic analyses. Science *307*, 82–86.

Horrocks, P., and Kilbey, B.J. (1996). Physical and functional mapping of the transcriptional start sites of *Plasmodium falciparum* proliferating cell nuclear antigen. Mol. Biochem. Parasitol. *82*, 207–215.

Horrocks, P., and Lanzer, M. (1999). Mutational analysis identifies a five base pair cis-acting sequence essential for GBP130 promoter activity in *Plasmodium falciparum*. Mol. Biochem. Parasitol. *99*, 77–87.

Horrocks, P., Wong, E., Russell, K., and Emes, R.D. (2009). Control of gene expression in *Plasmodium falciparum* – ten years on. Mol. Biochem. Parasitol. *164*, 9–25.

Howard, R.J. (1984). Antigenic variation of bloodstage malaria parasites. Phil. Trans. R. Soc. Lond. B Biol. Sci. *307*, 141–158.

Howitt, C.A., Wilinski, D., Llinas, M., Templeton, T.J., Dzikowski, R., and Deitsch, K.W. (2009). Clonally variant gene families in *Plasmodium falciparum* share a common activation factor. Mol. Microbiol. *73*, 1171–1185.

Iriko, H., Jin, L., Kaneko, O., Takeo, S., Han, E.T., Tachibana, M., Otsuki, H., Torii, M., and Tsuboi, T. (2009). A small-scale systematic analysis of alternative splicing in *Plasmodium falciparum*. Parasitol. Int. *58*, 196–199.

Jemmely, N.Y., Niang, M., and Preiser, P.R. (2010). Small variant surface antigens and *Plasmodium* evasion of immunity. Future Microbiol. *5*, 663–682.

Khan, S.M., Franke-Fayard, B., Mair, G.R., Lasonder, E., Janse, C.J., Mann, M., and Waters, A.P. (2005). Proteome analysis of separated male and female gametocytes reveals novel sex-specific *Plasmodium* biology. Cell *121*, 675–687.

Kishore, S.P., Perkins, S.L., Templeton, T.J., and Deitsch, K.W. (2009). An unusual recent expansion of the C-terminal domain of RNA polymerase II in primate malaria parasites features a motif otherwise found only in mammalian polymerases. J. Mol. Evol. *68*, 706–714.

Knapp, B., Nau, U., Hundt, E., and Kupper, H.A. (1991). Demonstration of alternative splicing of a pre-mRNA expressed in the blood stage form of *Plasmodium falciparum*. J. Biol. Chem. *266*, 7148–7154.

Koning-Ward, T.F., Speranca, M.A., Waters, A.P., and Janse, C.J. (1999). Analysis of stage specificity of promoters in *Plasmodium berghei* using luciferase as a reporter. Mol. Biochem. Parasitol. *100*, 141–146.

Kraemer, S.M., and Smith, J.D. (2003). Evidence for the importance of genetic structuring to the structural and functional specialization of the *Plasmodium falciparum var* gene family. Mol. Microbiol. *50* 1527–1538.

Kyes, S.A., Christodoulou, Z., Raza, A., Horrocks, P., Pinches, R., Rowe, J.A., and Newbold, C.I. (2003). A well-conserved *Plasmodium falciparum var* gene shows an unusual stage-specific transcript pattern. Mol. Microbiol. *48* 1339–1348.

Lasonder, E., Ishihama, Y., Andersen, J.S., Vermunt, A.M.W., Pain, A., Sauerwein, R.W., Eling, W.M.C., Hall, N., Waters, A.P., Stunnenberg, H.G., and Mann, M. (2002). Analysis of the *Plasmodium falciparum* proteome by high-accuracy mass spectrometry. Nature *419*, 537–542.

Lavstsen, T., Salanti, A., Jensen, A.T.R., Arnot, D.E., and Theander, T.G. (2003). Sub-grouping of *Plasmodium falciparum* 3D7 *var* genes based on sequence analysis of coding and non-coding regions. Malar. J. *2*, 27.

Le Roch, K.G., Zhou, Y.Y., Blair, P.L., Grainger, M., Moch, J.K., Haynes, J.D., De la Vega, P., Holder, A.A., Batalov, S., Carucci, D.J., and Winzeler, E.A. (2003). Discovery of gene function by expression profiling of the malaria parasite life cycle. Science *301*, 1503–1508.

Li, F., Sonbuchner, L., Kyes, S.A., Epp, C., and Deitsch, K.W. (2008). Nuclear non-coding RNAs are transcribed from the centromeres of *Plasmodium falciparum* and are associated with centromeric chromatin. J. Biol. Chem. *283*, 5692–5698.

Liao, S.M., Taylor, I.C., Kingston, R.E., and Young, R.A. (1991). RNA polymerase II carboxy-terminal domain contributes to the response to multiple acidic activators in vitro. Genes Dev. *5*, 2431–2440.

Lindner, S.E., De Silva, E.K., Keck, J.L., and Llinas, M. (2010). Structural determinants of DNA binding by a *P. falciparum* ApiAP2 transcriptional regulator. J. Mol. Biol. *395*, 558–567.

Llinas, M., Bozdech, Z., Wong, E.D., Adai, A.T., and Derisi, J.L. (2006). Comparative whole genome transcriptome analysis of three *Plasmodium falciparum* strains. Nucleic Acids Res *34*, 1166–1173.

Lopez-Rubio, J.J., Gontijo, A.M., Nunes, M.C., Issar, N., Hernandez, R.R., and Scherf, A. (2007). 5′ flanking region of *var* genes nucleate histone modification patterns linked to phenotypic inheritance of virulence traits in malaria parasites. Mol. Microbiol. *66*, 1296–1305.

Lopez-Rubio, J.J., Mancio-Silva, L., and Scherf, A. (2009). Genome-wide localization of heterochromatin couples clonally variant gene regulation to perinuclear repressive centers in malaria parasites. Cell Host and Microbe *5*, 179–190.

Magnani, E., Sjolander, K., and Hake, S. (2004). From endonucleases to transcription factors: evolution of the AP2 DNA binding domain in plants. Plant Cell *16*, 2265–2277.

Maier, A.G., Cooke, B.M., Cowman, A.F., and Tilley, L. (2009). Malaria parasite proteins that remodel the host erythrocyte. Nat. Rev. Microbiol. *7*, 341–354.

Mair, G.R., Braks, J.A., Garver, L.S., Wiegant, J.C., Hall, N., Dirks, R.W., Khan, S.M., Dimopoulos, G., Janse, C.J., and Waters, A.P. (2006). Regulation of sexual development of *Plasmodium* by translational repression. Science *313*, 667–669.

Mair, G.R., Lasonder, E., Garver, L.S., Franke-Fayard, B.M., Carret, C.K., Wiegant, J.C., Dirks, R.W., Dimopoulos, G., Janse, C.J., and Waters, A.P. (2010). Universal features of post-transcriptional gene regulation are critical for *Plasmodium* zygote development. PLoS Pathog. *6*, e1000767.

Malhotra, P., Dasaradhi, P.V., Kumar, A., Mohmmed, A., Agrawal, N., Bhatnagar, R.K., and Chauhan, V.S. (2002). Double-stranded RNA-mediated gene silencing of cysteine proteases (falcipain-1 and -2) of *Plasmodium falciparum*. Mol. Microbiol. *45*, 1245–1254.

Mbacham, W.F., Chow, C.S., Daily, J., Golightly, L.M., and Wirth, D.F. (2001). Deletion analysis of the 5′ flanking sequence of the *Plasmodium* gallinaceum sexual stage specific gene pgs28 suggests a bipartite arrangement of cis-control elements. Mol. Biochem. Parasitol. *113*, 183–187.

McRobert, L., and McConkey, G.A. (2002). RNA interference (RNAi) inhibits growth of *Plasmodium falciparum*. Mol. Biochem. Parasitol. *119*, 273–278.

Meissner, M., and Soldati, D. (2005). The transcription machinery and the molecular toolbox to control gene expression in *Toxoplasma gondii* and other protozoan parasites. Microbes. Infect. *7*, 1376–1384.

Meller, V.H., Gordadze, P.R., Park, Y., Chu, X., Stuckenholz, C., Kelley, R.L., and Kuroda, M.I. (2000). Ordered assembly of roX RNAs into MSL complexes on the dosage-compensated X chromosome in *Drosophila*. Curr. Biol. *10*, 136–143.

Miao, J., Fan, Q., Cui, L., Li, J., Li, J., and Cui, L. (2006). The malaria parasite *Plasmodium falciparum* histones: organization, expression, and acetylation. Gene *369*, 53–65.

Mikolajczak, S.A., Silva-Rivera, H., Peng, X., Tarun, A.S., Camargo, N., Jacobs-Lorena, V., Daly, T.M., Bergman, L.W., de, l., V., Williams, J., Aly, A.S., and Kappe, S.H. (2008). Distinct malaria parasite sporozoites reveal transcriptional changes that cause differential tissue infection competence in the mosquito vector and mammalian host. Mol. Cell Biol. *28*, 6196–6207.

Miller, L.H., Baruch, D.I., Marsh, K., and Doumbo, O.K. (2002). The pathogenic basis of malaria. Nature *415*, 673–679.

Mok, B.W., Ribacke, U., Rasti, N., Kironde, F., Chen, Q., Nilsson, P., and Wahlgren, M. (2008). Default Pathway of var2csa switching and translational repression in *Plasmodium falciparum*. PLoS ONE 3, e1982.

Muhia, D.K., Swales, C.A., Eckstein-Ludwig, U., Saran, S., Polley, S.D., Kelly, J.M., Schaap, P., Krishna, S., and Baker, D.A. (2003). Multiple splice variants encode a novel adenylyl cyclase of possible plastid origin expressed in the sexual stage of the malaria parasite *Plasmodium falciparum*. J. Biol. Chem. *278*, 22014–22022.

Natalang, O., Bischoff, E., Deplaine, G., Proux, C., Dillies, M.A., Sismeiro, O., Guigon, G., Bonnefoy, S., Patarapotikul, J., Mercereau-Puijalon, O., Coppee, J.Y., and David, P.H. (2008). Dynamic RNA profiling in *Plasmodium falciparum* synchronized blood stages exposed to lethal doses of artesunate. BMC Genomics *9*, 388.

Navarro, M., and Gull, K. (2001). A pol I transcriptional body associated with VSG mono-allelic expression in *Trypanosoma brucei*. Nature *414*, 759–763.

Otto, T.D., Wilinski, D., Assefa, S., Keane, T.M., Sarry, L.R., Bohme, U., Lemieux, J., Barrell, B., Pain, A., Berriman, M., Newbold, C., and Llinas, M. (2010). New insights into the blood-stage transcriptome of *Plasmodium falciparum* using RNA-Seq. Mol. Microbiol. *76*, 12–24.

Pace, T., Birago, C., Janse, C.J., Picci, L., and Ponzi, M. (1998). Developmental regulation of a *Plasmodium* gene involves the generation of stage-specific 5′ untranslated sequences. Mol. Biochem. Parasitol. *97*, 45–53.

Palenchar, J.B., and Bellofatto, V. (2006). Gene transcription in trypanosomes. Mol. Biochem. Parasitol. *146*, 135–141.

Patra, K.P., Johnson, J.R., Cantin, G.T., Yates, J.R., III, and Vinetz, J.M. (2008). Proteomic analysis of zygote and ookinete stages of the avian malaria parasite *Plasmodium* gallinaceum delineates the homologous proteomes of the lethal human malaria parasite *Plasmodium falciparum*. Proteomics *8*, 2492–2499.

Pays, E. (2005). Regulation of antigen gene expression in Trypanosoma brucei. Trends Parasitol. *21*, 517–520.

Perez-Toledo, K., Rojas-Meza, A.P., Mancio-Silva, L., Hernandez-Cuevas, N.A., Delgadillo, D.M., Vargas, M., Martinez-Calvillo, S., Scherf, A., and Hernandez-Rivas, R. (2009). *Plasmodium falciparum* heterochromatin protein 1 binds to tri-methylated histone 3 lysine 9 and is linked to mutually exclusive expression of var genes. Nucleic Acids Res. *37*, 2596–2606.

Petter, M., Lee, C.C., Byrne, T.J., Boysen, K.E., Volz, J., Ralph, S.A., Cowman, A.F., Brown, G.V., and Duffy, M.F. (2011). Expression of *P. falciparum* var Genes Involves Exchange of the Histone Variant H2A.Z at the Promoter. PLoS Pathog. *7*, e1001292.

Ponts, N., Harris, E.Y., Lonardi, S., and Le Roch, K.G. (2010a). Nucleosome occupancy at transcription start sites in the human malaria parasite: A hard-wired evolution of virulence? Infect. Genet. Evol. *11*, 716–724.

Ponts, N., Harris, E.Y., Prudhomme, J., Wick, I., Eckhardt-Ludka, C., Hicks, G.R., Hardiman, G., Lonardi, S., and Le Roch, K.G. (2010b). Nucleosome landscape and control of transcription in the human malaria parasite. Genome Res. *20*, 228–238.

Raabe, C.A., Sanchez, C.P., Randau, G., Robeck, T., Skryabin, B.V., Chinni, S.V., Kube, M., Reinhardt, R., Ng, G.H., Manickam, R., Kuryshev, V.Y., Lanzer, M., Brosius, J., Tang, T.H., and Rozhdestvensky, T.S. (2010). A global view of the nonprotein-coding transcriptome in *Plasmodium falciparum*. Nucleic Acids Res *38*, 608–617.

Ralph, S.A., Scheidig-Benatar, C., and Scherf, A. (2005). Antigenic variation in *Plasmodium falciparum* is associated with movement of *var* loci between subnuclear locations. Proc. Natl. Acad. Sci. U.S.A. *102*, 5414–5419.

Saenz, F.E., Balu, B., Smith, J., Mendonca, S.R., and Adams, J.H. (2008). The transmembrane isoform of *Plasmodium falciparum* MAEBL is essential for the invasion of Anopheles salivary glands. PLoS One *3*, e2287.

Salanti, A., Dahlback, M., Turner, L., Nielsen, M.A., Barfod, L., Magistrado, P., Jensen, A.T., Lavstsen, T., Ofori, M.F., Marsh, K., Hviid, L., and Theander, T.G. (2004). Evidence for the involvement of VAR2CSA in pregnancy-associated malaria. J. Exp. Med. *200*, 1197–1203.

Salanti, A., Staalsoe, T., Lavstsen, T., Jensen, A.T.R., Sowa, M.P.K., Arnot, D.E., Hviid, L., and Theander, T.G. (2003). Selective upregulation of a single distinctly structured *var* gene in chondroitin sulphate A-adhering *Plasmodium falciparum* involved in pregnancy-associated malaria. Mol. Microbiol. *49*, 179–191.

Salcedo-Amaya, A.M., Hoeijmakers, W.A., Bartfai, R., and Stunnenberg, H.G. (2010). Malaria: could its unusual epigenome be the weak spot? Int. J. Biochem. Cell. Biol. *42*, 781–784.

Salcedo-Amaya, A.M., van Driel, M.A., Alako, B.T., Trelle, M.B., van den Elzen, A.M., Cohen, A.M., Janssen-Megens, E.M., van, d., V., Selzer, R.R., Iniguez, A.L., Green, R.D., Sauerwein, R.W., Jensen, O.N., and Stunnenberg, H.G. (2009). Dynamic histone H3 epigenome marking during the intraerythrocytic cycle of *Plasmodium falciparum*. Proc. Natl. Acad. Sci. U.S.A. *106*, 9655–9660.

Scherf, A., Lopez-Rubio, J.J., and Riviere, L. (2008). Antigenic variation in *Plasmodium falciparum*. Annu. Rev. Microbiol. *62*, 445–470.

Sims, J.S., Militello, K.T., Sims, P.A., Patel, V.P., Kasper, J.M., and Wirth, D.F. (2009). Patterns of gene-specific and total transcriptional activity during the *Plasmodium falciparum* intraerythrocytic developmental cycle. Eukaryot. Cell *8*, 327–338.

Smith, J.D., Chitnis, C.E., Craig, A.G., Roberts, D.J., Hudson-Taylor, D.E., Peterson, D.S., Pinches, R., Newbold, C.I., and Miller, L.H. (1995). Switches in expression of *Plasmodium falciparum var* genes correlate with changes in antigenic and cytoadherent phenotypes of infected erythrocytes. Cell *82*, 101–110.

Struck, N.S., Herrmann, S., Langer, C., Krueger, A., Foth, B.J., Engelberg, K., Cabrera, A.L., Haase, S., Treeck, M., Marti, M., Cowman, A.F., Spielmann, T., and Gilberger, T.W. (2008). *Plasmodium falciparum* possesses two GRASP proteins that are differentially targeted to the Golgi complex via a higher- and lower-eukaryote-like mechanism. J. Cell Sci. *121*, 2123–2129.

Su, X., Heatwole, V.M., Wertheimer, S.P., Guinet, F., Herrfeldt, J.V., Peterson, D.S., Ravetc.h, J.V., and Wellems, T.E. (1995). A large and diverse gene family (*var*) encodes 200–350 kD proteins implicated in the antigenic variation and cytoadherence of *Plasmodium falciparum*-infected erythrocytes. Cell *82*, 89–100.

Sutherland, H., and Bickmore, W.A. (2009). Transcription factories: gene expression in unions? Nat. Rev. Genet. *10*, 457–466.

Tarun, A.S., Peng, X., Dumpit, R.F., Ogata, Y., Silva-Rivera, H., Camargo, N., Daly, T.M., Bergman, L.W., and Kappe, S.H. (2008). A combined transcriptome and proteome survey of malaria parasite liver stages. Proc. Natl. Acad. Sci. U.S.A. *105*, 305–310.

Tham, W.H., Payne, P.D., Brown, G.V., and Rogerson, S.J. (2007). Identification of basic transcriptional elements required for rif gene transcription. Int. J. Parasitol. *37*, 605–615.

Tonkin, C.J., Carret, C.K., Duraisingh, M.T., Voss, T.S., Ralph, S.A., Hommel, M., Duffy, M.F., Silva, L.M., Scherf, A., Ivens, A., Speed, T.P., Beeson, J.G., and Cowman, A.F. (2009). Sir2 paralogues cooperate to regulate virulence genes and antigenic variation in *Plasmodium falciparum*. PLoS Biol. *7*, e84.

Trelle, M.B., Salcedo-Amaya, A.M., Cohen, A.M., Stunnenberg, H.G., and Jensen, O.N. (2009). Global histone analysis by mass spectrometry reveals a high content of acetylated lysine residues in the malaria parasite *Plasmodium falciparum*. J. Proteome. Res *8*, 3439–3450.

Updike, D., and Strome, S. (2010). P granule assembly and function in Caenorhabditis elegans germ cells. J. Androl. *31*, 53–60.

van der Wel, A.M., Tomas, A.M., Kocken, C.H., Malhotra, P., Janse, C.J., Waters, A.P., and Thomas, A.W. (1997). Transfection of the primate malaria parasite *Plasmodium knowlesi* using entirely heterologous constructs. J. Exp. Med. *185*, 1499–1503.

Volz, J., Carvalho, T.G., Ralph, S.A., Gilson, P., Thompson, J., Tonkin, C.J., Langer, C., Crabb, B.S., and Cowman, A.F. (2010). Potential epigenetic regulatory proteins localise to distinct nuclear sub-compartments in *Plasmodium falciparum*. Int. J. Parasitol. *40*, 109–121.

Voss, T.S., Healer, J., Marty, A.J., Duffy, M.F., Thompson, J.K., Beeson, J.G., Reeder, J.C., Crabb, B.S., and Cowman, A.F. (2006). A *var* gene promoter controls allelic exclusion of virulence genes in *Plasmodium falciparum* malaria. Nature *439*, 1004–1008.

Voss, T.S., Kaestli, M., Vogel, D., Bopp, S., and Beck, H.P. (2003). Identification of nuclear proteins that interact differentially with *Plasmodium falciparum var* gene promoters. Mol. Microbiol. *48*, 1593–1607.

Wang, E.T., Sandberg, R., Luo, S., Khrebtukova, I., Zhang, L., Mayr, C., Kingsmore, S.F., Schroth, G.P., and Burge, C.B. (2008). Alternative isoform regulation in human tissue transcriptomes. Nature *456*, 470–476.

Wentzinger, L., Bopp, S., Tenor, H., Klar, J., Brun, R., Beck, H.P., and Seebeck, T. (2008). Cyclic nucleotide-specific phosphodiesterases of *Plasmodium falciparum*: PfPDEalpha, a non-essential cGMP-specific PDE that is an integral membrane protein. Int. J. Parasitol. *38*, 1625–1637.

Wessler, S.R. (2005). Homing into the origin of the AP2 DNA binding domain. Trends Plant Sci. *10*, 54–56.

Westenberger, S.J., Cui, L., Dharia, N., Winzeler, E., and Cui, L. (2009). Genome-wide nucleosome mapping of *Plasmodium falciparum* reveals histone-rich coding and histone-poor intergenic regions and chromatin remodeling of core and subtelomeric genes. BMC Genomics *10*, 610.

Winter, G., Chen, Q.J., Flick, K., Kremsner, P., Fernandez, V., and Wahlgren, M. (2003). The 3D7var5.2 (var(COMMON)) type *var* gene family is commonly expressed in non-placental *Plasmodium falciparum* malaria. Mol. Biochem. Parasitol. *127*, 179–191.

Wu, Y., Sifri, C.D., Lei, H.-H., Su, X., and Wellems, T.E. (1995). Transfection of *Plasmodium falciparum* within human red blood cells. Proc. Natl. Acad. Sci. U.S.A. *92*, 973–977.

Wuitschick, J.D., Lindstrom, P.R., Meyer, A.E., and Karrer, K.M. (2004). Homing endonucleases encoded by germ line-limited genes in *Tetrahymena thermophila* have APETELA2 DNA binding domains. Eukaryot. Cell 3, 685–694.

Yang, X., Figueiredo, L.M., Espinal, A., Okubo, E., and Li, B. (2009). RAP1 is essential for silencing telomeric variant surface glycoprotein genes in *Trypanosoma brucei*. Cell *137*, 99–109.

Yuda, M., Iwanaga, S., Shigenobu, S., Kato, T., and Kaneko, I. (2010). Transcription factor AP2-Sp and its target genes in malarial sporozoites. Mol. Microbiol. *75*, 854–863.

Yuda, M., Iwanaga, S., Shigenobu, S., Mair, G.R., Janse, C.J., Waters, A.P., Kato, T., and Kaneko, I. (2009). Identification of a transcription factor in the mosquito-invasive stage of malaria parasites. Mol. Microbiol. *71*, 1402–1414.

Invasion of Host Red Blood Cells by Malaria Parasites

8

Amy K. Bei and Manoj T. Duraisingh

Abstract

The first challenge faced by the invasive merozoite form of the *Plasmodium* parasite when released into the bloodstream is to invade and establish itself within a host erythrocyte. Following hepatic schizont rupture, merozoites are released into the circulation and must successfully locate, bind to, and invade an erythrocyte lest they are cleared by the immune system. Subsequent cycles of erythrocytic egress, invasion and growth result in the escalation of parasitaemia, often associated with clinical symptoms. The parasite has developed molecular strategies that allow it to invade the host cell while avoiding the attack of the immune system and in response to diversity of the receptor repertoire of its human host. Depending on the species, there is a limited window of time in which the merozoite remains viable for invasion. The invasive potential of the merozoite is influenced by three levels of host cell selection: (1) selection between host species (species specific tropism), (2) selection among individuals within a host species (erythrocyte receptor diversity), and (3) selection of subpopulations within an individual (age-dependent invasion). The host–parasite interactions that allow for efficient host cell selection and invasion of erythrocytes will be the topic addressed by this chapter.

Malaria parasites and host tropism

Survey of natural tropism

Studies of malaria tropism began with the identification of the aetiological agent. Malaria parasites were first identified in human blood by Charles Louis Alphonse Laveran (Laveran, 1880), work for which he received a Nobel prize in 1907. Malaria parasites were later identified in the blood of wild birds by Danilewsky in 1884 (Garnham, 1966). Since those initial discoveries, malaria parasites have been identified in other primates, rodents, and reptiles. Interestingly, malaria has yet to be identified in domestic animals and livestock. While these animals are not infected by *Plasmodium*, erythrocytes of these species are susceptible to *Babesia*, a related haemolytic protozoan parasite.

While malaria parasites exist for many vertebrate hosts, a largely species-specific tropism exists which restricts each *Plasmodium* species to their respective host (Fig. 8.1). Such host-specific restriction of natural infections is due to many factors ranging from the distribution and habitat of the mosquito vector to compatibility between various stages of the host and parasite interaction: namely, the liver-stage, the blood-stage, factors influencing splenic clearance, and the sexual stage (gametocytes). At each stage in the parasite's life cycle, the ligand–receptor interactions must be compatible in order for infection to be established and propagate, including erythrocyte receptors and parasite ligands required for blood stage infection.

The delineation between species-tropism has historically been determined by what was observed in natural infections and what species could cause disease through experimental infection. With the exception of the recent zoonoses of *P. knowlesi*, which has been found to cause severe disease in humans in Southeast Asia (Cox-Singh

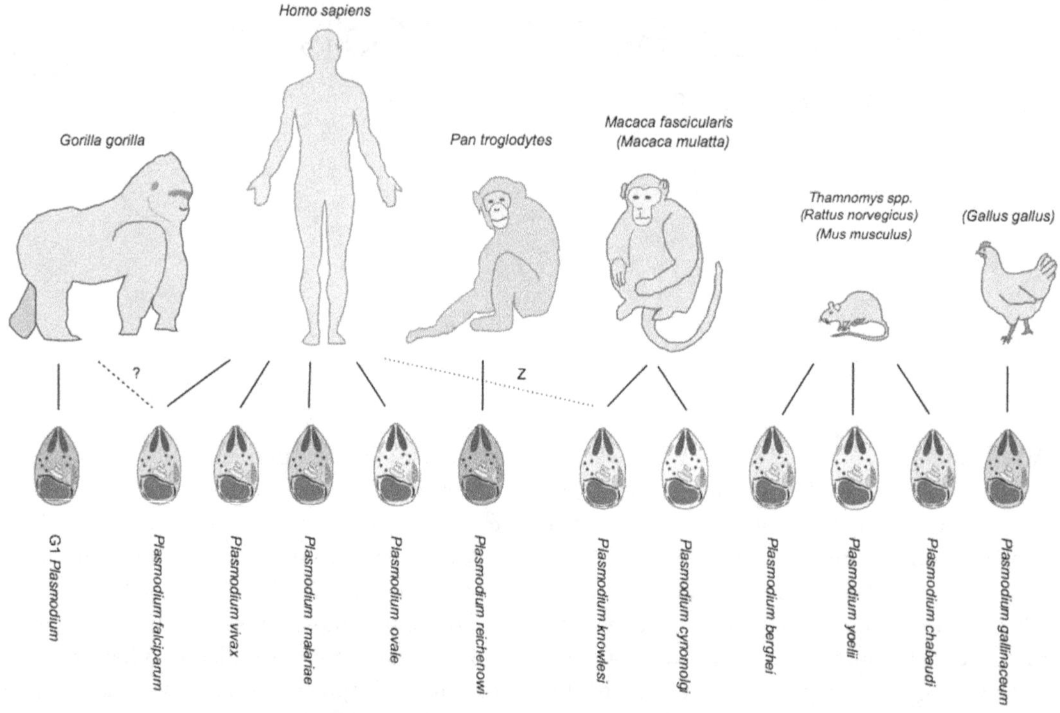

Figure 8.1 Host-specific tropism for *Plasmodium* species. Natural hosts of malaria parasites are shown. The natural hosts are written above the host species with experimental hosts denoted by parenthesis. The dashed line indicates a unique tropism with the recent description of *P. falciparum* like parasites identified in gorillas(Liu *et al.*, 2010). Further experimentation will clarify whether this represents a true zoonosis of *P. falciparum* or whether these parasites are merely phylogenetically similar. The dotted line (Z) refers to the zoonosis of *P. knowlesi* in humans in Southeast Asia (Cox-Singh *et al.*, 2008).

et al., 2008), and *P. cynomolgi* which was demonstrated to infect humans through accidental (and intentional) exposure to infected mosquitoes (Eyles *et al.*, 1960), no other primate malaria species has been documented to effectively infect humans (Coatney *et al.*, 1971), highlighting the specificity of malaria infection and the boundaries established by such tropism. While natural infections can be affected by host behaviour and vector habitat, experimental infections were viewed as fulfilment of the third of Koch's postulates – the ability to create infection in a healthy individual.

Two important distinctions must be made when discussing determinants of tropism – the ability to invade cells of a different species and the ability to establish infection within a host of another species. Most early experiments focused on the ability of parasites from one species to establish infection within another, whereas recent studies have focused on the ability of parasites to invade cells of the other species *in vitro*. These are important distinctions to make as *Plasmodium falciparum* can invade mouse erythrocytes, albeit inefficiently *in vitro*, however they are unable to grow and establish infection within the vertebrate host (Klotz *et al.*, 1987).

Experimental infections

The concept of tropism was classically defined as the inability of one species of parasite to establish infection in another non-natural host, through *in vivo* infection experiments. However, before experimental infectious were used to query the boundaries of species-specific tropism, experiments in the early 1900s were performed by Wagner-Jauregg in asylums to test malaria as a treatment for neurosyphilis (Wagner-Jauregg, 1922), studies which culminated in a Nobel Prize in 1927. In the 1920s, patients suffering from GPI (general paralysis of the insane) were injected with either *P. falciparum*, *P. vivax* or *P. malariae*, to stimulate high malarial fevers which some argued ameliorated the pathology by killing the syphilis spirochetes (although for such a mechanism to work, malarial fevers had to reach 41–42 degrees Celsius for 6 h to reach the thermal death point) (Chernin, 1984). While some patients were refractory to *P. vivax* (due to Duffy negativity in Africans and African Americans), it remained the treatment of choice due to its benign course of disease. The alternative treatments were associated with other problems as benign *P. malariae* sometimes caused renal problems, whereas malignant *P. falciparum* had to be monitored very carefully as infection was associated with mortality risks on its own. As avirulent infections were ideal, some researchers attempted to infect malaria therapy patients with primate malaria species, namely *Plasmodium reichenowi*, *Plasmodium schwetzi* and *Plasmodium rodhaini* (*P. malariae*) (Rodhain, 1939; Rodhain and Muylle, 1938) and these experiments introduced the concept of tropism boundaries between species. Some of the early pioneers in the field would go on to conduct experimental human and mosquito infections with both human and primate malaria species, experiments which were instrumental in determining the tropism boundaries as well as other critical findings in the biology of the parasite and malaria transmission.

Following these initial studies, many experimental infections were conducted in which infected human blood was injected into primates and primate blood was injected into humans (Coatney *et al.*, 1971). Often the results of these studies were difficult to interpret due to mixed species in the blood and the reliance on microscopy to detect blood stage parasitaemia and morphology. For many infections, it was noted that splenectomy of primates was essential to detect blood stages of various human malaria species. One of the curious aspects of such studies is that while *Plasmodium falciparum* was thought to have originated from *Plasmodium reichenowi* (Coatney *et al.*, 1971), *P. falciparum* cannot infect chimpanzees in the absence of splenectomy (Blacklock and Adler, 1922; Rodhain, 1939; Rodhain and Jadin, 1964). Additionally, human volunteers were unable to be infected with *P. reichenowi* (Blacklock and Adler, 1922; Rodhain, 1939). While *P. falciparum* has long been thought to have originated from *P. reichenowi*, recent evidence has challenged this dogma (Krief *et al.*, 2010; Liu *et al.*, 2010; Prugnolle *et al.*, 2010).

Experimentally, one of the challenges posed by species-specific tropism is that there is no ideal *in vivo* model for human malaria infection and pathology. While an *Aotus* model exists for *P. falciparum* and *P. vivax* infection, the model is not ideal in that is it permissive to only certain strains of parasites and the course of disease can often dramatically differ between vaccinated and control monkeys (Collins *et al.*, 1988; Collins *et al.*, 1987; Schmidt, 1978). In the absence of a robust and representative *in vivo* model that supports *Plasmodium falciparum* malaria, researchers are faced with the option of studying virulence processes in heterologous rodent malaria species, or more recently using partial humanized mouse models, which generally are able to capture only a single life cycle stage (liver, blood, etc.) and often inefficiently (Druilhe *et al.*, 2005; Moreno *et al.*, 2007; Morosan *et al.*, 2006).

Phylogeny

With advances in genetic and genomic methodologies, which are both more sensitive and more specific than microscopic evaluation, there has been a resurgence of interest in looking at *Plasmodium* speciation and host-restriction in light of phylogenic analyses. Recent reports have been published reporting the identification of *Plasmodium* species in African apes that are more closely related to *Plasmodium falciparum* than is *Plasmodium reichenowi*, by DNA sequence of both mitochondrial and nuclear encoded genes (Krief *et al.*, 2010; Prugnolle *et al.*, 2010). These reports have limitations as the sample sizes were small, the animals studied lived in close proximity to humans, and the methods used to sequence the marker genes were prone to errors. Whether these novel *Plasmodia* parasites represent *P. falciparum* infection of apes or a novel species (*Plasmodium gaboni*) is the topic of heated debate (Duval *et al.*, 2010).

Recently, a study using improved sequencing technology and a much larger sample set (Liu *et al.*, 2010), has proposed that *P. falciparum* originated from western gorillas and not from chimpanzees, the previously accepted dogma. Single-genome amplification methods were used to amplify cytochrome b from the faeces of non-habituated chimpanzees, bonobos, and gorillas. Phylogenic analyses reveal that *Plasmodium falciparum* clusters within a gorilla-specific clade, distinct from chimpanzee malaria species including *P. reichenowi*. The monophyletic nature of the relationship between *P. falciparum* and these gorilla *Plasmodium* species implies a potential single gorilla to human cross-species transmission event. The extent to which gorillas may serve as an animal reservoir for *P. falciparum* has yet to be addressed and will be critical to determine prior to malaria eradication efforts.

In addition to detection of new parasite species and low-level infections *in vivo*, genomic approaches are vastly improving our understanding of diversity and selective pressures acting on a given parasite species within a globally distributed population of hosts. Recently, the genomes of a globally diverse set of parasite isolates provided data as to the level of genome-wide diversity, and the phylogenic relationships between these strains (Volkman *et al.*, 2007). The depth that such genomic data provides can yield further insights into the dynamics of host cell selection operating between species and within global populations of the same species.

Host cell section – species, population, individual

In addition to considering the potential to infect members of another species, an important aspect of host-cell selection results from differential susceptibility of individuals within host populations. Such differential susceptibility can result from the selection of host genetic polymorphisms that confer resistance to infection or decreased disease severity. A striking example of resistant sub-populations are Duffy-negative humans in Africa which are resistant to *P. vivax* infection.

A further layer of host cell selection occurs within an *individual* (for characteristics such as the age and rigidity of erythrocytes) and can result in differences in the virulence of the parasite. The molecular basis for these three levels of host-cell selection during erythrocyte invasion (selection of species, selection of susceptible individuals, and selection of susceptible cells within an individual) will be discussed in this chapter.

Molecular and cellular biology of erythrocyte invasion

Mechanism of invasion

Invasion of the human erythrocyte by *Plasmodium* merozoites – the invasive blood stage of the malaria parasite – is a necessary step in its life cycle, and the mechanism of invasion is conserved among the different species. The molecular mechanism of invasion can be conceptually divided into five steps: (1) initial attachment, (2) apical reorientation, (3) tight junction formation, (4) apical organelle secretion followed by host cell entry, and (5) termination through resolution of ligand–receptor interactions by proteolytic cleavage (Fig. 8.2A). These steps of invasion were initially defined by elegant electron-microscopy studies using free *Plasmodium knowlesi* merozoites (Aikawa *et al.*, 1978). The tight junction formation is irreversible, after which the merozoite is committed to the process of invasion. Upon tight junction formation, the merozoite releases the contents of its secretory organelles (rhoptries and micronemes) (Fig. 8.2B) (Banister and Mitchell, 1989) and through the actin–myosin molecular motor (Baum *et al.*, 2006), powers its way into the erythrocyte. During the course of this process, the merozoite surrounds itself with a layer of membrane known as the parasitophorous vacuole. As the merozoite is surrounded by the erythrocyte, the ligand–receptor interactions which anchor it to the erythrocyte membrane are cleaved, presumably by merozoite surface and transmembrane proteases (O'Donnell and Blackman, 2005; O'Donnell *et al.*, 2006).

Molecules that mediate invasion

Prior to the assembly and annotation of the *Plasmodium falciparum*, *P. vivax*, and *P. knowlesi* genomes (Feng *et al.*, 2003; Gardner *et al.*, 2002; Pain *et al.*, 2008), sequence and structural data on invasion ligands had to be performed by cDNA cloning, sequencing, and alignment(Adams *et al.*, 1992). Initially, invasion ligands were defined

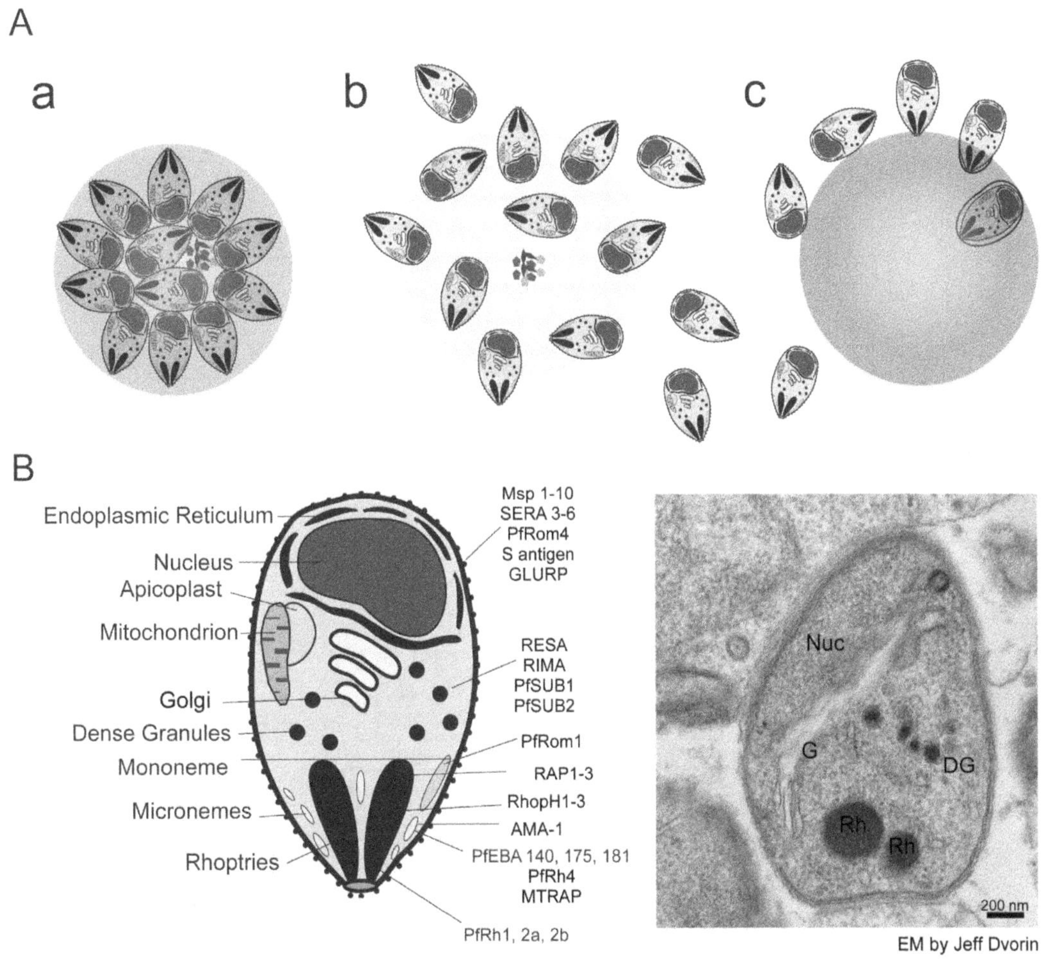

Figure 8.2 The *Plasmodium falciparum* merozoite (A) The process of invasion from the schizont stage involves schizont rupture and merozoite release followed by merozoite invasion represented by the following steps: (1) merozoite initial contact, (2) apical reorientation, (3) tight junction formation and commitment, (4) microneme/rhoptry release, (5) invasion, parasitophorous vacuole formation/sealing followed by establishment within the erythrocyte. (B) Left panel: Key proteins (right) and organelles (left) of the invasive merozoite are shown. Right panel: Electron microscopy image of a merozoite (courtesy of Dr Jeffrey Dvorin) showing subcellular organelles: Nuc, nucleus, G, Golgi; DG, dense granules; Rh, rhoptries.

based on the properties of native or recombinant protein binding to erythrocytes or reticulocytes. Later, genomics facilitated the identification of paralogues and orthologues. Even today, new molecules that mediate invasion are being identified and characterized.

The mechanism of invasion appears to be conserved among the *Plasmodium* species, as are the ligands that mediate invasion. Most of our knowledge results from studies on *P. falciparum* for which *in vitro* culture and genetics is available. Initial attachment involves members of the merozoite surface protein family (MSPs) (Fig. 8.3A), and apical reorientation is thought to be mediated by apical membrane antigen-1 (Mitchell *et al.*, 2004) (Fig. 8.3B), although recent data suggests that parasites are able to apically reorient but do not proceed further in invasion when AMA-1 is inhibited by a peptide which binds in the hydrophobic trough (Treeck *et al.*, 2009). This inhibition is thought to result from the inability of AMA-1 to interact with the RON complex, which is thought to be inserted into the erythrocyte membrane prior to rhoptry secretion and invasion (Richard

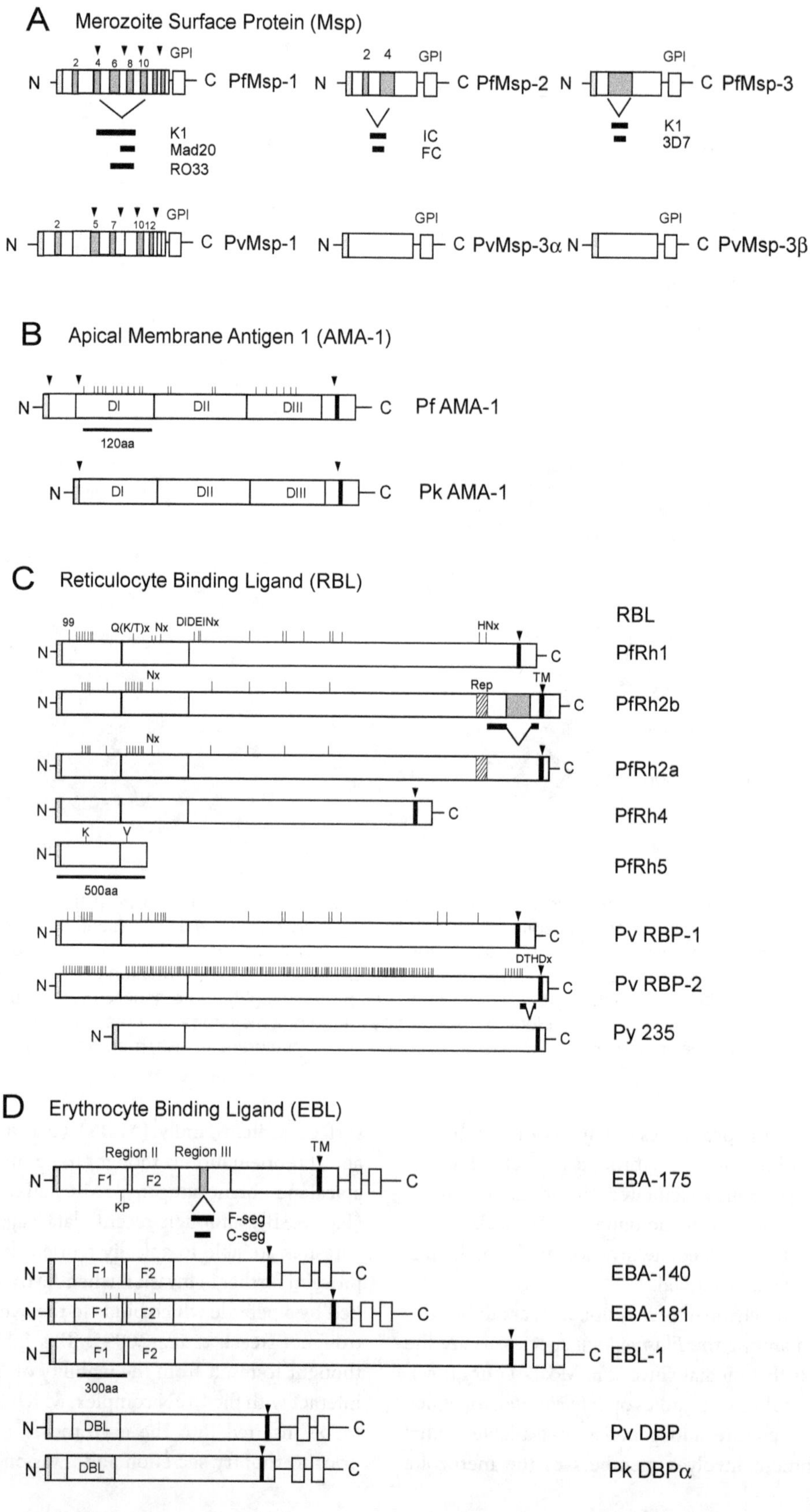
A Merozoite Surface Protein (Msp)
GPI
PfMsp-1
PfMsp-2
PfMsp-3
K1
Mad20
RO33
IC
FC
K1
3D7
PvMsp-1
PvMsp-3α
PvMsp-3β
B Apical Membrane Antigen 1 (AMA-1)
DI
DII
DIII
120aa
Pf AMA-1
Pk AMA-1
C Reticulocyte Binding Ligand (RBL)
RBL
99
Q(K/T)x
Nx
DIDEINx
HNx
PfRh1
Rep
TM
PfRh2b
PfRh2a
PfRh4
PfRh5
500aa
Pv RBP-1
DTHDx
Pv RBP-2
Py 235
D Erythrocyte Binding Ligand (EBL)
Region II
Region III
TM
F1
F2
KP
F-seg
C-seg
EBA-175
EBA-140
EBA-181
EBL-1
300aa
DBL
Pv DBP
Pk DBPα

Figure 8.3 *Plasmodium falciparum* invasion ligand domains and polymorphisms from selected proteins (A) The merozoite surface protein family (MSPs) contains 10 proteins, but the first 3 have been best characterized in terms of polymorphism and their role in invasion. Msp1 and 2 are so frequently polymorphic that they have been used in strain typing parasites based on polymorphic alleles in blocks 4–10 of Msp1 and blocks 2–4 of Msp2 (Tanabe *et al*., 1998). Msp-3, a current vaccine candidate, has a dimorphic domain consisting of 2 alleles. Orthologues PvMsp-1, PvMsp-3α, and PvMsp-3 in *P. vivax* are shown. (B) Apical membrane antigen-1 is a single copy gene which is a current vaccine candidate. Polymorphisms within R1, R2, and R3 have been described for AMA-1 and their functions in immune evasion have been genetically confirmed (Healer *et al*., 2004). The *P. knowlesi* orthologue PkAMA-1 is shown. (C) The RBL family of invasion ligands is composed of PfRh1, PfRh2a, PfRh2b, PfRh4, PfRh5 and the pseudogenes PfRh3 and PfRh6 (not shown). Orthologues PvRBP-1, PvRBP-2 in *P. vivax* and Py235 member from *P. yoelii* are shown. Polymorphisms and variant expression of these ligands have been identified which are associated with invasion pathway and disease severity. Important polymorphisms shown are from the following studies: PfRh1 (Chokejindachai and Conway, 2009), PfRh2a/2b (Rayner *et al*., 2005), PfRh2b (Jennings *et al*., 2007), PfRh5(Hayton *et al*., 2008). (D) The EBL family of invasion ligands is composed of EBA-140, EBA-175, EBA-181, EBL-1, and the pseudogene EBA-165 (not shown). EBA-175 contains a dimorphic domain within region III (Ware *et al*., 1993) and polymorphisms in region II of EBA-181 have been shown to alter erythrocyte receptor usage (Mayer *et al*., 2004). While polymorphisms in region II of EBA-175 have been identified, they have been predicted not to alter receptor binding (Tolia *et al*., 2005) whereas one polymorphism in EBA-175 has been associated with disease severity by unknown mechanisms (Chokejindachai and Conway, 2009).

et al., 2010). High-affinity interactions between the EBLs and RBLs (Fig. 8.3C and D) lead to the formation of the tight junction. These two invasion ligand multigene families have expanded to varying degrees in different *Plasmodium* species and will be discussed in detail in section IV.

Upon formation of the tight junction, the invading merozoite relies on an actin–myosin motor to power its way into the erythrocyte. The basic molecular components of this motor, namely a thrombospondin-related anonymous protein (TRAP) homologue, aldolase, actin, Myosin A, MyoA tail-interacting protein (MTIP) or myosin light chain (MLC), glideosome associated protein (Gap) 45 and Gap 50, are not only required for the gliding motility and invasion of the hepatocyte by the *Plasmodium* sporozoite, but they are also conserved among all apicomplexan parasites (Baum *et al.,* 2006). While apicomplexan parasites lack Arp2/3, *Plasmodium* formin 1 acts to nucleate actin and regulates actin polymerization at the moving junction during invasion (Baum *et al.,* 2008).

Dissolution of the tight junction through proteolytic cleavage is probably attributable to subtilisin-like protease 2 (SUB2) for MSPs and AMA-1, and Rhomboids (ROM1, 4) for a diverse range of invasion ligands such as AMA-1, EBLs, RBLs, and MTRAP (Baker *et al.,* 2006). There appears to be a degree of specificity in cleavage as the EBL, EBA-175, is cleaved by PfROM4, but not by other rhomboid proteases (O'Donnell *et al.,* 2006), while PfROM1, but no other rhomboids, can cleave AMA-1 (Baker *et al.,* 2006).

The functional conservation between species of erythrocyte ligands involved in the mechanism of invasion has been confirmed by genetic complementation experiments. Transgenic complementation experiments have demonstrated that both Msp-1_{19} and AMA-1 function are conserved when heterologous domains from rodent malaria species or *P. vivax* are introduced (O'Donnell *et al.,* 2000; Treeck *et al.,* 2009). The focus of the following section will be a discussion of the divergence in the ligands involved in tight junction formation – the EBL and RBL protein families.

Molecular evolution of invasion ligands

Host cell selection is thought to be critically dependent on the two families of invasion ligands that bind the erythrocyte with high affinity: the EBLs and RBLs. Current evidence suggests that a single parasite ligand binds a single erythrocyte receptor. In this paradigm, diversity in invasion can be generated through genomic expansion, genetic polymorphism, or variant expression of these invasion ligands. All three of these diversity-generating mechanisms have been described for the RBL and EBL proteins, resulting in a variety of invasion phenotypes.

Multiple invasion ligands: functional redundancy or orchestrated hierarchy?

While the protein players and the mechanism of invasion is largely conserved among *Plasmodium* species, some species contain expanded gene families such as the EBLs and RBLs, whereas others rely on a minimal set of invasion ligands. Such variable expansion prompts the question whether these alternative ligands serve redundant functions in the process of invasion. While some invasion ligands are clearly essential, such as AMA-1, MSP-1, and PfRh5 (Baum *et al.*, 2009; Cowman and Crabb, 2006; O'Donnell *et al.*, 2000; Triglia *et al.*, 2000), others appear to be dispensable through genetic knock-out experiments. However does non-essentiality necessarily imply redundancy? One possibility is that the EBL and RBL proteins act in a temporal manner in invasion, perhaps with an RBL initially 'sensing' the erythrocyte followed by the commitment to tight junction formation by the EBL (Duraisingh *et al.*, 2003b). Such a model is reminiscent of HIV receptor/co-receptor interactions – each interaction is necessary, but not independently sufficient, to promote pathogen binding and invasion. It is also possible that these EBL and RBL proteins have other functions in invasion besides their erythrocyte receptor binding functions, such as providing a scaffold for other invasion ligands or helping to organize the structure of the multiple proteins at the apical end of the invasive merozoite. Further studies using antibody/peptide inhibition, inducible knock-down, and high resolution microscopy could be applied to settle the debate whether these expanded multigene families contribute to functional redundancy or an orchestrated hierarchy of molecular interactions in invasion.

Erythrocyte-binding ligands (EBLs)

The EBL proteins of all *Plasmodium* species are remarkably similar in their structure. They all contain well-defined cysteine rich DBL domains – named from their initial characterization and similarity to the *P. vivax* 'Duffy binding like' domain of PvDBP – as well as conserved splice junctions and exon–intron boundaries. These DBL domains are present in other *Plasmodium* ligands as well, including the PfEMP-1 proteins that mediate cytoadherence (Smith *et al.*, 2000). The first EBL to be identified was EBA-175 from *Plasmodium falciparum.* EBA-175 was shown to bind human erythrocytes and in a sialic acid-dependent manner (Camus and Hadley, 1985). The *P. vivax* and *P. knowlesi* DBPs were characterized based on their interaction with Duffy-positive but not Duffy-negative erythrocytes (Haynes *et al.*, 1988; Wertheimer and Barnwell, 1989). DBP orthologues among malaria species play diverse roles: some are essential for invasion, others determine host tropism, and yet others play important roles in alternative invasion pathways.

The single EBL paralogue in *P. vivax*: essential for invasion?

The Duffy binding protein (DBP) of *Plasmodium vivax* is a unique member of the *Plasmodium* EBL family because it appears to be essential for the invasion of *P. vivax* into human erythrocytes. A seminal discovery was made when the Duffy receptor was shown to be necessary for *P. vivax* invasion of human cells (Miller *et al.*, 1976), in the first description of a human receptor polymorphism responsible for resistance to malaria infection. The identification of the parasite ligand responsible for the binding to the Duffy receptor (PvDBP) was discovered much later (Wertheimer and Barnwell, 1989). The crystal structure of PkDBPα, has been helpful in defining the structural features of PvDBPs interaction with the Duffy receptor by homology modelling (Singh *et al.*, 2006). It has been proposed that the high level of Duffy negativity in sub-Saharan Africa is the result of selective pressure by *Plasmodium vivax* (Livingstone, 1984; Miller *et al.*, 1976). This high frequency has resulted in the near disappearance of *P. vivax* in most regions of Africa. The essentiality of a single ligand–receptor interaction exposes an Achilles heel of the parasite and an represents an optimal target for intervention.

Sporadic reports of *P. vivax* in sub-Saharan Africa have often been thought to be due to infection within small populations of Duffy positive individuals, or mistaken diagnosis of *P. ovale* as *P. vivax*. However, there is recent evidence of *P. vivax* transmission among Duffy negative individuals in Western Kenya (Ryan *et al.*, 2006) and the Republic of Congo (Culleton *et al.*, 2009). Intriguingly, a recent study conducted in Madagascar confirms – both microscopically and molecularly – that *P. vivax* malaria occurs in Duffy negative individuals (Menard *et al.*, 2010). These observations have challenged the dogma that *P. vivax* is absolutely restricted to Duffy positive erythrocytes, and suggests that perhaps *P. vivax* is evolving to be less reliant on Duffy and is utilizing alternative ligand–receptor interactions, having important implications for the success of a PvDBP vaccine (Chitnis and Sharma, 2008)

An EBL determinant of tropism in the zoonotic parasite *P. knowlesi*

While *P. knowlesi* naturally infects *M. fascicularis* monkeys, most experiments have been conducted with the closely related Rhesus macaques, which are also very susceptible to infection. *P. knowlesi* can invade both human and rhesus erythrocytes, and while it relies on the Duffy receptor in human cells, it does not require the Duffy receptor for invasion into rhesus erythrocytes (Miller *et al.*, 1975). While *P. knowlesi* and *P. vivax* are very similar parasites, the DBP in *P. knowlesi* is expanded relative to *P. vivax*. *Plasmodium knowlesi* possesses three DBPs – DBPα. DBPβ, and DBPγ – each having a unique receptor binding specificity in invasion (Adams *et al.*, 1990). PkDBPα is similar to PvDBP in that it binds to (and requires) the Duffy receptor, on both human and rhesus erythrocytes. The DBL domain of DBPα has been shown biochemically and structurally to bind erythrocytes (McHenry and Adams, 2006; Ranjan and Chitnis, 1999; Singh *et al.*, 2006) and genetic disruption of this paralogue results in parasites which are unable to invade human erythrocytes.(Singh *et al.*, 2005). The other two DBPs (DBPβ and DBPγ) interact specifically, and alternatively, with two unidentified rhesus erythrocyte receptors. Thus, DBPs of *Plasmodium knowlesi* are true determinants of species-specific tropism.

Mediators of alternative invasion pathways in the lethal human parasite *P. falciparum*

P. falciparum is not restricted to the use of a single receptor for invasion as is *P. vivax*. This may reflect the fact that *P. falciparum* has evolved to utilize multiple receptor–ligand interactions known as invasion pathways (Baum *et al.*, 2003a). Evidence now suggests that members of the PfEBA family of invasion ligands are critical ligands responsible for invasion via sialic acid-dependent receptors, with each paralogue utilizing an alternative receptor.

There are five members of the *P. falciparum* EBA family of EBLs – EBA-175, EBA-140, EBA-181, EBL-1, and untranslated pseudogene EBA-165. Probably the best characterized of the *Pf*EBAs, EBA-175 was shown to bind erythrocytes in a manner dependent on the sialic acid residues on glycophorin A (Camus and Hadley, 1985). Co-crystallization studies of EBA-175 and its receptor glycophorin A have proposed that the interaction requires dimerization of EBA-175 Region II (RII) domains (Tolia *et al.*, 2005). EBA-175 knockout transgenic parasites have clearly genetically demonstrated that parasites without EBA-175 are no longer reliant on glycophorin A for invasion, and further that they can switch their invasion pathway utilization to invade in a sialic acid independent manner (Duraisingh *et al.*, 2003a; Reed *et al.*, 2000). Similar genetic and biochemical techniques were used to demonstrate that EBA-140 binds erythrocytes through glycophorin C (Lobo *et al.*, 2003; Maier *et al.*, 2003; Mayer *et al.*, 2001) and that EBA-181 binds to erythrocytes through an as yet unidentified receptor (Gilberger, 2003; Mayer *et al.*, 2004). By utilizing the genetic crosses between an EBL-1 containing parents (3D7, Dd2) and a parent that lacked the locus (HB3), EBL-1 was proposed as a determinant of efficient invasion, having a highly favoured inheritance among progeny (Peterson and Wellems, 2000). EBL-1 was later shown to bind to glycophorin B on the human erythrocyte surface (Mayer *et al.*, 2009).

Orthologues in *P. reichenowi* reveal strong selection between species

Most of the orthologues to the *Plasmodium falciparum* EBLs are present in the chimpanzee parasite *P. reichenowi*. Orthologues to EBA-140 (Baum *et al.*, 2003b), EBA-165 (Rayner *et al.*, 2004a), EBA-175 (Ozwara *et al.*, 2001), and EBA-181 (Rayner *et al.*, 2004a) exist in *P. reichenowi*, yet despite these similarities, *P. reichenowi* cannot invade human erythrocytes. Interestingly, unlike

PfEBA-140, there is no evidence of PrEBA-140 being under selection (Baum *et al.*, 2003b), implying that there are differential selective forces acting between the species. The *Plasmodium reichenowi* orthologue of *Plasmodium falciparum* EBA-175 is 83% identical at the amino acid level, and key amino acid differences occur in regions that define glycan binding sites (Chattopadhyay *et al.*, 2006). Recent evidence showing that PrEBA-175 preferentially binds chimpanzee erythrocytes whereas PfEBA-175 preferentially binds human erythrocytes corroborates this hypothesis (Martin *et al.*, 2005).

EBL influence on reticulocyte restriction of rodent malaria

P. yoelii invades mature murine erythrocytes in a Duffy dependent manner and contains a single EBL sharing similarity with *P. vivax*, *P. knowlesi*, and *P. falciparum* DBLs (Prasad *et al.*, 2003). The RII domain of PyDBP has been shown to bind erythrocytes, adding confidence that it is a *bona fide* DBL (Prasad *et al.*, 2003). Interestingly, when *P. yoelii* DBP is knocked out, the parasites are restricted to reticulocytes – implying that PyDBP is essential for normocyte invasion and that other ligands are responsible for invasion into reticulocytes. These reticulocyte binding ligands are discussed below.

Reticulocyte-binding ligands (RBLs)

The reticulocyte-binding ligands were named due to their preferential binding of reticulocytes in *P. vivax*. It is postulated that this family of proteins determines whether a given parasite line will preferentially invade reticulocytes and younger cells, and in some malaria species, this age preference has a dramatic impact on parasite virulence. While many of the receptors for the EBLs have been identified, with a few exceptions, the RBL receptors are largely unknown in all *Plasmodium* species.

Reticulocyte restriction by RBLs

P. vivax is restricted to reticulocytes that express DARC. While the DBP of *P. vivax* interacts with the DARC receptor, the RBLs which are involved in the recognition of reticulocytes in *P. vivax* are PvRBP-1 and PvRBP-2 (Galinski *et al.*, 1992). Since PvRBP-1 and PvRBP-2 were shown to bind human erythrocytes with enriched reticulocytes, they were termed as RBLs due to their functional binding preference.

In contrast to *P. vivax*, *P. knowlesi* is not reticulocyte restricted (in Rhesus macaques) and interestingly it has a limited repertoire of RBLs. With the recent completion of the *P. knowlesi* genome (Pain *et al.*, 2008), two RBP family members have been identified and characterized(Meyer *et al.*, 2009). Interestingly, PvRBP-1 appears to be a highly degenerated pseudogene in *P. knowlesi*, and no orthologue to PvRBP-2 exists. These two novel members of the RBL family (PkNBPXa and PkNBPXb) localized to the micronemes, have been described as alternatives to PvRBP-2 (as they are not direct orthologues), and have been shown to functionally bind rhesus macaque erythrocytes. In addition, a novel RBL was identified in *P. vivax* that is considered to be an orthologue of PkNBPXa; however, in *P. vivax* it appears to be a pseudogene (Meyer *et al.*, 2009).

RBLs as virulence determinants

The RBL family in *P. yoelii* is Py235, and like *P. vivax*, the YA strain of *P. yoelii* is restricted to reticulocytes. Although the level of sequence conservation is not high for the Py235 and PvRBP families, it is significant enough to hypothesize that they might serve homologous functions (Galinski *et al.*, 2000; Keen *et al.*, 1994). The Py235 virulence gene family was identified in *Plasmodium yoelii* by experimentally determining the differences in virulence of two experimental strains. While 17X YA results in controllable and avirulent infections in mice where parasite invasion is reticulocyte restricted, 17X YM is lethal and results in high parasitaemias marked by parasite invasion of erythrocytes of all ages. Passive transfer of monoclonal antibodies generated from mice resistant to the lethal YM strain resulted in survival of the susceptible mice and restricted invasion to reticulocytes (Freeman *et al.*, 1980). Further studies identified the target of these protective monoclonal antibodies as the Py235 family of invasion ligands, implying that this protein family might be involved in host cell selection and invasion of older erythrocytes (Holder and Freeman, 1984). These findings have been validated by vaccination of mice with purified Py235, resulting in protection of the mice against challenge by restricting parasite invasion to reticulocytes (Holder and Freeman, 1981). The Py235 protein was also shown to directly bind both erythrocytes and reticulocytes in a neuraminidase resistant, chymotrypsin sensitive manner; however, enhanced binding to erythrocytes was observed compared to reticulocytes, further implicating this protein family in age-dependent invasion (Ogun and Holder, 1996; Ogun *et al.*, 2000). In *P. yoelii*, it has been reported that the protein family is variantly expressed at the merozoite level – a single schizont will contain merozoites expressing different members of the Py235 family (Preiser *et al.*, 1999). This is in contrast to *P. falciparum*, which is able to variantly express members of the PfRh family, but variant expression is uniform within a schizont. Genomic and proteomic data indicates that there are between 14 and 50 members of the Py235 gene family (Carlton *et al.*, 2002; Iyer *et al.*, 2006; Khan *et al.*, 2001).

RBL mediators of alternative invasion pathways in *P. falciparum*

The *P. falciparum* PfRh proteins were originally identified as *P. falciparum* homologues of reticulocyte binding ligands of *Plasmodium yoelii* (Py235) (Iyer *et al.*, 2007b; Khan *et al.*, 2001; Preiser *et al.*, 2002) and *Plasmodium vivax* (PvRBP) (Galinski *et al.*, 1992). In *Plasmodium falciparum*, the PfRh family includes five members expressed in asexual stages (PfRh1, PfRh2a, PfRh2b, PfRh4, and PfRh5) and a pseudogene (PfRh3) that is transcribed but not translated (Baum *et al.*, 2009; Duraisingh *et al.*, 2003b; Hayton *et al.*, 2008; Rayner *et al.*, 2000, 2001; Rodriguez *et al.*, 2008; Stubbs *et al.*, 2005; Taylor *et al.*, 2002; Triglia *et al.*, 2001, 2005). Genetic knock-outs of these family members have helped define the invasion pathways that they utilize to enter the red cell (Duraisingh *et al.*, 2003b; Stubbs *et al.*, 2005; Triglia *et al.*, 2005).

The PfRh proteins are type-I transmembrane proteins with large ectodomains – ranging in size from approximately 200 to 383 kDa – which may directly engage receptors on the erythrocyte surface. While the structural details of erythrocyte binding by DBL domains have been delineated (Tolia *et al.*, 2005), the molecular details of comparable domains in the PfRh proteins have not yet been defined. However, both bioinformatic and genetic evidence are aiding in the definition of discrete domains within the PfRh proteins. While the unique domain of PfRh2b was previously thought to be critical determinant of invasion pathway (Duraisingh *et al.*, 2003b), increasing evidence implicates the N-terminal domains of PfRh1, PfRh4, and PfRh5 in erythrocyte binding (Baum *et al.*, 2009; Gao *et al.*, 2008; Gaur *et al.*, 2007; Hayton *et al.*, 2008; Rodriguez *et al.*, 2008; Tham *et al.*, 2009). Experiments conducted in our lab using chimeric parasite lines implicate the transmembrane domain and cytoplasmic tail in differential invasion pathway utilization observed for PfRh2a and PfRh2b (Dvorin *et al.*, 2010).

Orthologues in *P. reichenowi* reveal strong selection between species

Of the RBL proteins, the primary sequence conservation of the *P. reichenowi* and *P. falciparum* Rh proteins is the most similar. Both *P. reichenowi* and *P. falciparum* have 5 members; however, the paralogues differ in expression with *P. reichenowi* having PrRH1 as a pseudogene, yet a functional PrRH3, whereas in *P. falciparum*, PfRh1 is expressed and PfRh3 is a pseudogene. The dramatic degree of diversity in these proteins in very closely related *Plasmodium* species implies a strong selective pressure, perhaps through the event of speciation (Rayner *et al.*, 2004b).

Sequence and expression level polymorphism of invasion ligands

Evasive manoeuvres: variant expression and polymorphism in *Plasmodium* invasion ligands

The two main challenges facing a merozoite after egress into the bloodstream are the humoral immune system and the variability in erythrocyte receptors that it encounters, not only within the blood of a single individual, but within a population as well. The parasite is able to overcome these obstacles through variant expression and amino acid polymorphism of the invasion ligands.

Variant expression in field isolates

P. vivax relies on a single dominant receptor-ligand (DARC-DBP) interaction to invade erythrocytes. While other ligand–receptor interactions may also be necessary for *P. vivax* merozoite invasion to occur, they are not sufficient. This reliance on a single dominant ligand–receptor interaction has resulted in natural human polymorphisms in the DARC receptor that render the parasite unable to invade its human host. *P. falciparum* has a more sophisticated strategy for survival –the invasion ligands of *P. falciparum* exist in multi-gene families, and members of the family are variantly expressed in both laboratory and field isolates (Cortés, 2008). While some invasion ligands such as PfRh1 and PfRh2 are variantly expressed in a stable fashion (Duraisingh *et al.*, 2003b; Taylor *et al.*, 2002; Triglia *et al.*, 2005), others such as PfRh4 are dynamic and can be selected using erythrocytes bereft of receptors (Jiang *et al.*, 2010). It has been observed in the field that variant expression of PfRhs and PfEBAs can be associated with invasion pathway usage (Bei *et al.*, 2007; Lobo *et al.*, 2004; Nery *et al.*, 2006); however, such associations are not observed in all populations (Gomez-Escobar *et al.*, 2010; Jennings *et al.*, 2007).

Sequence polymorphisms in invasion ligands

Sequence polymorphisms in invasion ligands are likely to have two main roles – to alter receptor binding specificity and to evade the immune response. Polymorphisms of both of these types have been identified for invasion ligands (Fig. 8.3).

MSPs

The merozoite surface protein (MSP) family members are glycosylphosphatidylinositol (GPI) anchored proteins that have both conserved as well as dimorphic domains. The MSPs are so diverse and rapidly polymorphic that they have remained the gold standard for determining parasite clonality in the field (Snounou *et al.*, 1999; Tanabe *et al.*, 1998). MSP-1 is an essential gene and attempts to knock it out by targeted gene disruption have been unsuccessful (O'Donnell *et al.*, 2000); however, its extensive polymorphism generated the diversity needed to evade host humoral immune responses. Owing to its immunodominance, MSP-1 is also the classic marker of age-dependent acquisition of immunity (Egan *et al.*, 1996).

AMA-1

Like MSP-1, AMA-1 is an essential gene for parasite invasion and survival within the human host. While AMA-1 is a single copy gene (and as such, does not possess the potential for variant expression), it appears that polymorphisms, although more limited than that of MSP-1, play a significant role in differential immune mediated inhibition (Kocken *et al.*, 2000). Transgenic parasites expressing different alleles of AMA-1 (3D7 and W2mef) have shown differential inhibition when exposed to antibodies against the ectodomain (Healer *et al.*, 2004). Such differential inhibition has been observed in the Phase 1/2 vaccine trial as well, where inhibitory antibody responses were observed against the homologous vaccine strain, but not heterologous strains (Spring *et al.*, 2009).

EBL polymorphism

It has been observed that polymorphism in the binding regions of EBA-140 and EBA-181 alters the receptor specificity (Mayer *et al.*, 2002; Mayer *et al.*, 2004), however there are no known comparable polymorphisms in EBA-175. It has been proposed based on the structure of EBA-175 that mutations within the binding domain may not actually influence binding, but may affect immune recognition (Tolia *et al.*, 2005). While the mechanism of action is not yet clear, a case–control study in Thailand identified a polymorphism in EBA-175 associated with disease severity (Chokejindachai and Conway, 2009). Further genetic and population based studies are needed to determine the role of EBL polymorphism on invasion, immunity and virulence.

RBL polymorphism

Polymorphisms within the PfRh proteins have been recently identified that alter receptor tropism, and may play a role in immune evasion and virulence (Hayton *et al.*, 2008; Jennings *et al.*, 2007; Lantos *et al.*, 2009; Lobo *et al.*, 2004; Rayner *et al.*, 2004b). Recently, polymorphisms in the N-terminal domains of PfRh1 – predicted to be the region of the protein that binds the erythrocyte receptor – were reported to be associated with disease severity in Thailand (Chokejindachai and Conway, 2009). While associative studies have implicated a possible role for these polymorphisms in invasion pathway utilization (Jennings *et al.*, 2007; Lobo *et al.*, 2004), precise functional studies have yet to be conducted. A notable exception is in the case of PfRh5. Allelic differences in PfRh5, identified through progeny of the genetic cross of 7G8 × GB4 (Hayton *et al.*, 2008) were shown to alter the infectivity *in vivo* and invasion *in vitro* of *P. falciparum* strains into *Aotus* erythrocytes (Hayton *et al.*, 2008). The virulent allele from GB4 was genetically introduced into 7G8 and the invasion and virulence phenotype was restored. Further functional studies such as these are needed to precisely determine the role for RBL polymorphisms in tropism, invasion, and immune evasion.

Alternative invasion pathways, the use of different receptors for invasion

One strategic advantage to having multiple invasion ligands is the ability to invade erythrocytes with a diverse receptor repertoire. *Plasmodium falciparum* possesses a unique ability to adopt alternative invasion pathways for invasion into the erythrocyte (Dolan *et al.*, 1990) (Fig. 8.4). It is possible that both variant expression and sequence polymorphism contribute to alternative invasion pathway usage (Bei *et al.*, 2007; Duraisingh *et al.*, 2003a,b; Jennings *et al.*, 2007; Lantos *et al.*, 2009; Lobo *et al.*, 2004; Triglia *et al.*, 2005).

Recently, new evidence that *P. vivax* may not be completely restricted to invasion via the Duffy receptor has prompted the notion of alternative invasion pathways for this species as well (Menard *et al.*, 2010) (Fig. 8.4). While not strictly an 'alternative invasion pathway', the expanded EBL family of *P. knowlesi* allows it to invade erythrocytes of both humans and primates. While DBP-α

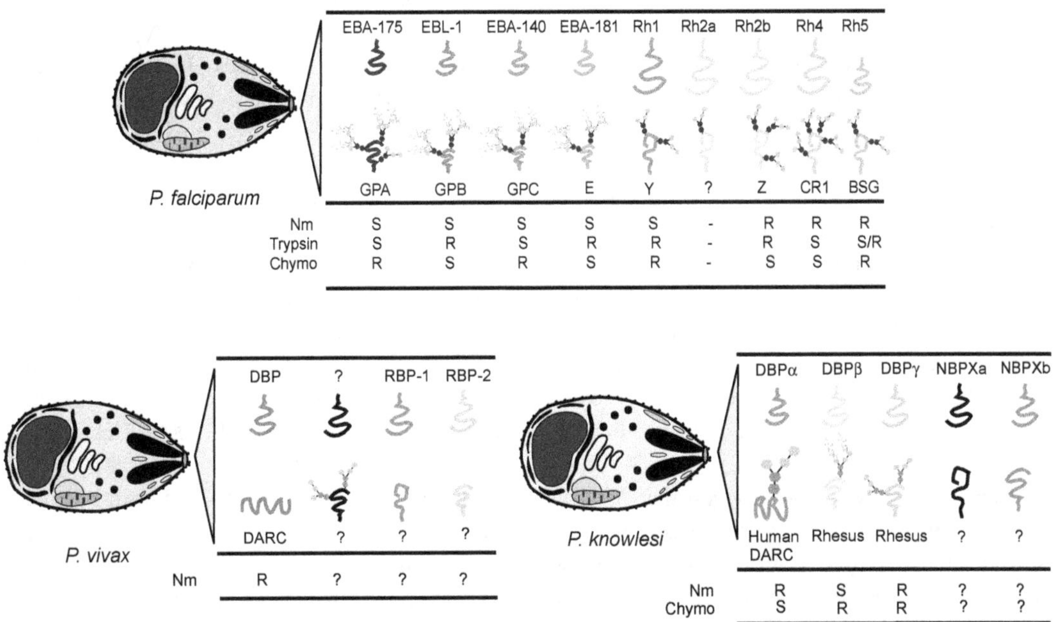

Figure 8.4 *Plasmodium* alternative erythrocyte invasion pathways Alternative invasion pathways for *P. falciparum*, *P. vivax*, and *P. knowlesi* are shown. Enzymatic sensitivities of known and proposed erythrocyte proteins and carbohydrates define erythrocyte invasion pathways; R – resistant, S – sensitive. In the case of the PfRh5 receptor, Rh5 has been shown to bind erythrocytes in a trypsin sensitive manner by some (Rodriguez *et al.*, 2008) and in a trypsin resistant manner by others (Baum *et al.*, 2009). The receptor for PfRh5 was recently identified as Basigin (Crosnier *et al.*, 2011) and Basigin is resistant to neuraminidase, trypsin, and chymotrypsin treatment.

binds the Duffy receptor on human cells (and blocking the interaction abolishes invasion into human cells) DBP-β and DBP-γ bind unidentified receptors on Rhesus erythrocytes experimentally and cynomolgous monkeys in the wild (Chitnis *et al.*, 1996; Ranjan and Chitnis, 1999).

Erythrocyte receptors involved in *Plasmodium* invasion

The majority of erythrocyte receptors that have been characterized have been identified using a combination of enzyme treated cells and cells that lack certain receptors (Fig. 8.4), and through naturally occurring polymorphisms. Rare erythrocyte mutants, such as En(a-) cells which lack glycophorin A, S-s-U erythrocytes which are glycophorin B null, and Gerbich negative erythrocytes which express a truncated version of glycophorin C through a deletion in exon 3, have been used to determine the receptors for *Plasmodium falciparum* invasion ligands EBA-175 (Camus and Hadley, 1985; Orlandi *et al.*, 1992), EBL-1 (Mayer *et al.*, 2009) and EBA-140 (Maier *et al.*, 2003) respectively. Enzymatic treatment of erythrocytes have demonstrated that these parasite ligands bind their cognate receptors in a sialic acid-dependent manner. Knock-outs of the parasite ligands confirmed these receptor specificities as parasites lacking EBA-175 and EBA-140 are able to invade via a sialic acid independent invasion pathway, not dependent on glycophorin A or glycophorin C respectively (Duraisingh *et al.*, 2003a; Maier *et al.*, 2003; Reed *et al.*, 2000). In the case of EBA-175, crystallographic studies of two DBL domains with a sialated glycan provide exquisite molecular detail of receptor binding (Tolia *et al.*, 2005).

Many associative studies suggest that Africans and African-Americans are less susceptible to *P. vivax* malaria than Caucasians (Bray, 1958; Young *et al.*, 1955); however, the resistance factor was definitively proven to be the Duffy receptor through *in vitro* experimentation with *P. knowlesi* merozoites and Duffy negative erythrocytes, and *in vivo* using *P. vivax* exposed Duffy-positive and Duffy-negative volunteers (Miller *et al.*, 1976). The region of the Duffy receptor which interacts with the DBP of *P. vivax* and *P. knowlesi* has been mapped to a 35-kDa extracellular portion of the N-terminal region of DARC (Chitnis *et al.*, 1996). The molecular details of the interaction of PvDBP interacting with DARC – the Duffy receptor – have now been well defined, the interaction is sialic acid independent, independent of N-linked oligosaccharides, and dependent on tyrosine sulphation of a single amino acid (Barnwell *et al.*, 1989; Chitnis *et al.*, 1996; Choe *et al.*, 2005; Hans *et al.*, 2005; Mason *et al.*, 1977; Singh *et al.*, 2006).

While many of the receptors have been described and even identified for the EBL invasion ligands, very few have been identified for the RBL proteins. Recently, CR1 – which was previously shown to influence erythrocyte rosetting – was proposed as a sialic acid independent ligand receptor (Spadafora *et al.*, 2010), and was later identified as the receptor for PfRh4, making it the first PfRh receptor definitively identified(Tham *et al.*, 2010). Even more recently, Basigin has been identified as the receptor for PfRh5 (Crosnier *et al.*, 2011). In contrast to other PfRh and PfEBA receptors, Basigin appears to be essential for *P. falciparum* invasion, as it is required regardless of alternative invasion pathway usage of the parasite(Crosnier *et al.*, 2011). The receptors for MSP-1 and AMA-1 have been postulated, but not definitively proven, to be Band 3 (Goel *et al.*, 2003) and Kell (Kx) respectively (Kato *et al.*, 2005). While *Plasmodium* parasites rely on host receptors for invasion, they are also able to modify the surface of the erythrocyte with their own proteins, as is the case for the RON proteins which are believed to be translocated from the merozoite to the erythrocyte membrane and whose interaction with AMA-1 facilitates rhoptry secretion and invasion (Besteiro *et al.*, 2009; Richard *et al.*, 2010; Straub *et al.*, 2009).

It has been well characterized both biochemically and genetically that invasion ligands can bind both protein and carbohydrate components of erythrocyte receptors; however, many of the receptors for these ligands remain unknown. Various erythrocyte receptors have been implicated in *Plasmodium* invasion, albeit by associative studies. One such association is the relationship between ABO blood group and *P. falciparum* invasion. Unfortunately, the literature thus far has been largely contradictory as to the role of specific blood group antigens in invasion (Uneke, 2007); however, this may be attributed to differences in *P. falciparum* field strains as well as other confounding host cell polymorphisms which cannot be controlled for in associative studies. One recent study implicated blood group O in protection from severe malaria by a mechanism of reduced rosetting (Rowe *et al.*, 2007). Precise genetic analysis in an isogenic host background is needed to truly determine the role of ABO blood group – and other putative host cell receptors – in *Plasmodium falciparum* invasion.

Selection by host receptor polymorphism: host and parasite co-evolution

Over the years of co-evolution between the parasite and its host, there has been a molecular co-evolution of both ligands and receptors. Malaria has had a detectable impact on the human genome (Sabeti *et al.*, 2002). From haemoglobin polymorphisms to G6PD deficiency to pyruvate kinase deficiency, the human host has developed to try to avoid invasion and growth of the malaria parasite. Some polymorphisms affect the oxidative state of the erythrocyte and the ability of the parasite to establish infection, as is the case for sickle cell disease and the thalassaemias (Ayi *et al.*, 2004), while others occur in erythrocyte surface proteins that serve as parasite receptors (Fig. 8.5). This section will explore some of the co-evolved mechanisms to promote parasite or host survival through promoting polymorphisms in ligands and receptors involved in invasion.

Polymorphism within individuals: erythrocyte age

Besides the species of the erythrocytes that can be invaded and the specific ligand–receptor interactions within hosts, *Plasmodium* species also display distinct preferences for erythrocytes of different ages (Garnham, 1966). As the erythrocyte ages, the erythrocyte loses surface area and volume, CD71 and sialic acid levels decrease, and there is an increase of phosphatidylserine (PS) on the cell surface and an emergence of senescent cell antigen, derived from the oxidation of Band3 (Durocher *et al.*, 1975; Kay *et al.*, 1983; Schroit *et al.*, 1985; Waugh *et al.*, 1992). These mechanisms make older cells more susceptible to splenic clearance. In addition to structural changes, differences in expression levels of certain blood groups have also been described in younger compared to older erythrocytes. CR1 (Moldenhauer *et al.*, 1988; Ripoche and Sim, 1986), DAF (Kanamaru *et al.*, 1988; Ueda *et al.*, 1990), and DARC (Woolley *et al.*, 2000) have been shown to be expressed more highly on reticulocytes than normocytes. It is thought that such differences in receptor levels might influence the age-preference of invasion for *Plasmodium* species.

Of the primate malarias, *Plasmodium knowlesi* can invade normocytes (Hegner, 1938) as can *Plasmodium cynomolgi*; however, like *P. falciparum*, *P. cynomolgi* has a preference for reticulocytes. Of the human malarias, while *Plasmodium vivax* (Simpson *et al.*, 1999) and *Plasmodium ovale* are restricted to reticulocytes, *Plasmodium falciparum* invades erythrocytes of all ages, yet exhibits a preference for reticulocytes (Pasvol *et al.*, 1980). *Plasmodium malariae* exhibits a preference for aged erythrocytes (Collins and Jeffery, 2007). Among the rodent malarias, *P. yoelii* avirulent strains are strongly reticulocyte restricted whereas virulent strains can infect reticulocytes and normocytes indiscriminately (Bungener, 1985; Walliker *et al.*, 1976). *P. chabaudi* is restricted to normocytes. *P. berghei* is similar to *P. falciparum* in that it exhibits a preference for reticulocytes, but can also invade normocytes, depending on the strain, and the rodent host species.

Quantitative measures of erythrocyte preference

A parasite's preference (or lack of preference) for different cell populations can be measured by the multiplicity of infection. Early observations noted that some *Plasmodium* parasites invade at lower multiplicities and higher parasitaemia whereas others seem restricted to certain erythrocyte subpopulations, resulting in more multiply infected cells (Wang, 1970). The number of multiply infected cells observed compared to those which would be predicted from random events (estimated by a Poisson distribution) can be calculated for different parasite strains and is referred to as the selectivity index (Simpson *et al.*, 1999). It

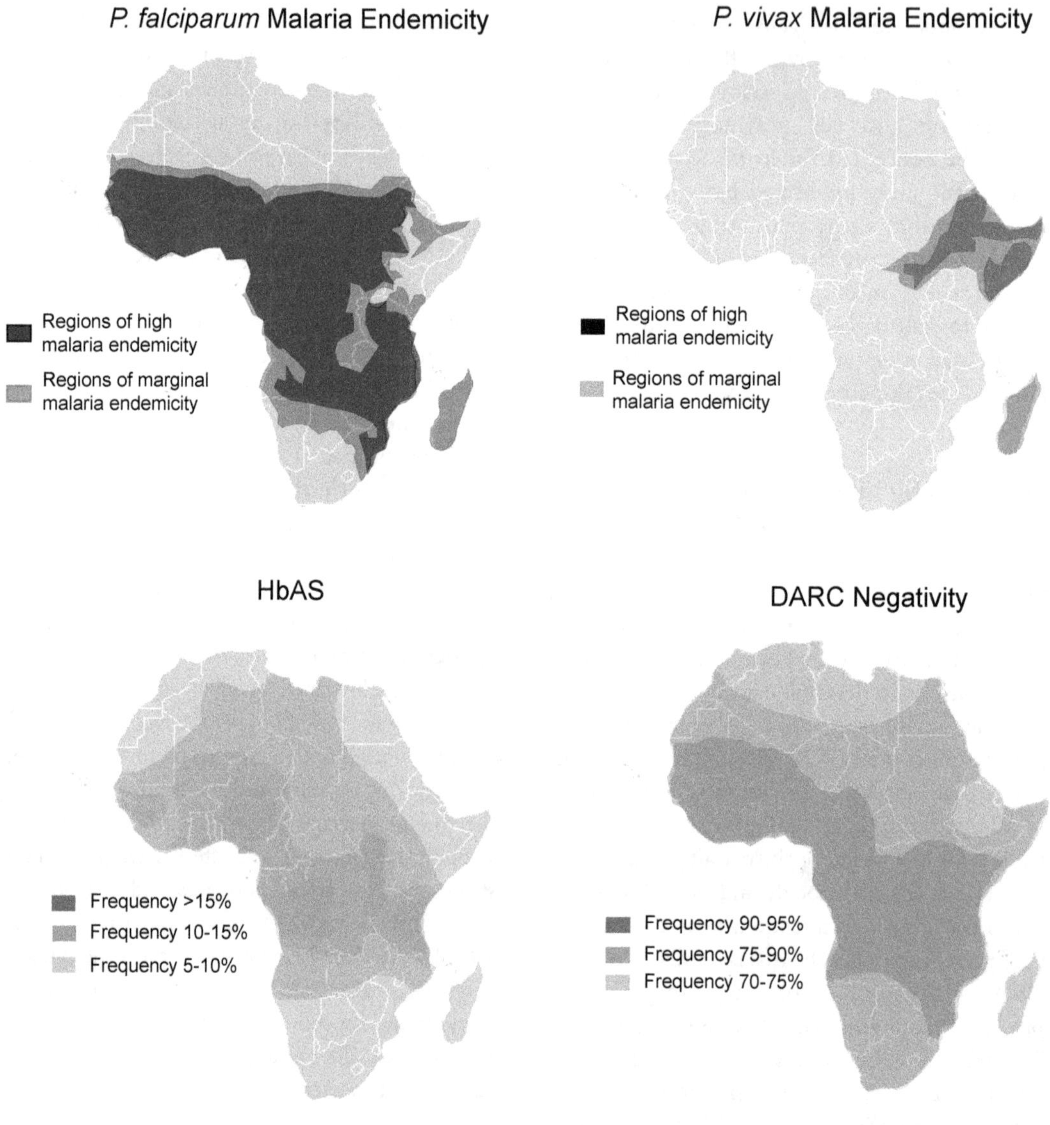

Figure 8.5 Overlapping distribution of erythrocyte polymorphisms with malaria endemicity. Erythrocyte polymorphisms have been proposed to be driving force in parasite evolution. Several polymorphisms that have been shown to provide resistance to malaria are spatially distributed in areas of malaria endemicity (Williams, 2006). HbAS (Allison, 2009) has been known to provide protection to malaria through mechanisms still under debate, ABO blood group (blood group O) (Mourant *et al.*, 1976) has been proposed to confer protection through reduced resetting (Rowe *et al.*, 2007), and DARC negativity (Cavalli-Sforza *et al.*, 1994; Guerra *et al.*, 2010; Rosenberg, 2007) prevents invasion through binding of *P. vivax* DBP.

has been well established with *P. yoelii* strains and clones that the virulence of the parasite is directly related to the selectivity of its invasion, which is in turn related to expression of key invasion ligands (Bungener, 1985; Iyer *et al.*, 2007a; Walliker *et al.*, 1976).

While *Plasmodium vivax* is restricted to reticulocytes, and relies on a single dominant and essential ligand–receptor interaction for invasion (DBP-DARC), *P. falciparum* has the capacity to engage diverse receptors on the erythrocyte by utilizing multiple members of two invasion protein families (EBL and RBL). These two families allow *P. falciparum* to invade all erythrocytes, normocytes as well as reticulocytes, and make it less reliant on any given erythrocyte receptor. Erythrocyte restriction is directly linked to the virulence and distribution of the disease since humans who lack the Duffy receptor (DARC) through naturally occurring polymorphisms, are resistant to *P. vivax* infection but remain susceptible to *P. falciparum*, as has been described in West Africa (Barnwell *et al.*, 1989; Miller *et al.*, 1976; Welch *et al.*, 1977). *P. falciparum* has evolved to variantly express multiple invasion ligands such that it can invade erythrocytes indiscriminately, which results in the manifestations of severe disease. The association between lack of reticulocyte restriction and virulence has also been observed for *P. knowlesi*, which can invade both normocytes and reticulocytes and has been found to result in hyperparasitaemia and fatal disease *in vivo* (Cox-Singh *et al.*, 2008).

It has long been hypothesized that the reticulocyte restriction of *Plasmodium vivax*, resulting in low parasitaemia and high selectivity, is what makes it a less virulent parasite than *Plasmodium falciparum*, similar to the differences in reticulocyte restriction between virulent and avirulent YM strains of *P. yoelii*. However, cases of severe malaria caused by *P. vivax* are increasingly being reported (Price *et al.*, 2009) and such observations may challenge this dogma. For *P. falciparum*, it was observed that a decrease in erythrocyte selectivity is associated with a corresponding increase in disease severity *in vivo* (Chotivanich *et al.*, 2000) in one study in Thailand, however this association was not observed in a study in Kenya (Deans *et al.*, 2007).

Polymorphisms between individuals: blood group polymorphism

Erythrocyte blood group polymorphisms occur in different individuals within a population, and can be under selection due to their ability to confer resistance to the malaria parasite (Miller, 1994; Rowe *et al.*, 2009; Williams, 2006). Duffy negativity is one of the best characterized receptor polymorphisms affecting malaria invasion which has been characterized in West Africa (Miller *et al.*, 1975) as well as recently in Papua New Guinea (PNG) (Kasehagen *et al.*, 2007). Duffy (DARC) negativity arises as the result of a mutation in the DARC promoter which affects the erythroid transcription factor GATA1, thus Duffy negativity occurs in erythrocytes but no other cell lineage (Tournamille *et al.*, 1995). Duffy negativity has reached high frequencies in West Africa, resulting in a relative restriction of *P. vivax* from the region (Miller *et al.*, 1976).

Gerbich negativity – the lack of glycophorin C – has a clearly documented impact on malaria parasite invasion and appears to be selected for in populations that experience frequent malaria exposure. The frequency of the Gerbich polymorphism is higher in Papua New Guinea in areas that experience high malaria exposure, Gerbich negative individuals were less likely to be infected with either *P. falciparum or P. vivax* (Serjeantson, 1989), and *P. falciparum* parasites invade Gerbich negative erythrocytes less efficiently (Maier *et al.*, 2003). Glycophorin Dantu (Unger *et al.*, 1987) – a hybrid of glycophorin A and B – has been shown to have a protective effect against *Plasmodium falciparum* invasion *in vitro* (Field *et al.*, 1994). Glycophorin B negative (S-s-U) individuals are common in Africans and African-Americans, and glycophorin B negativity is found at gene frequencies of 59% in the Ituri forest dwellers of PNG (Fraser *et al.*, 1966)

Another well-characterized erythrocyte polymorphism with clear influences on malaria severity is Southeast Asian ovalocytosis (SAO) – caused by deletions in the Band 3 gene. Lack of Band 3 results in a lack of anion transport resulting in erythrocytes that display reduced cellular deformability and increased membrane rigidity, in addition to oval shape (Mohandas *et al.*, 1984; Saul *et al.*, 1984) *In vitro* data suggests that most parasite lines show decreased invasion into SAO erythrocytes; however, there are strain specific differences in invasion (Cortes *et al.*, 2004).

The relationship between ABO blood group and malaria invasion/severity has been a contentious one for years (Uneke, 2007). Some studies claim a certain blood group is protective for malaria while others find associations to the contrary. It has been observed that blood group O is protective against severe *Plasmodium falciparum* malaria through a decrease in infected erythrocyte rosetting (Rowe *et al.*, 2007). Recently, a study was done using specific glycosidases to test the role of ABO blood group in cells from the same donor by converting A and B to O (Cohen *et al.*, 2009). Biochemical methods such as these, or genetic methods in haematopoietic stem cells (Bei *et al.*, 2010) could be employed to test the role of blood groups directly in isogenic backgrounds.

Recent studies have linked CR1 polymorphisms to malaria disease severity (Rout *et al.*, 2010; Teeranaipong *et al.*, 2008) as well as having implicated it as the receptor for PfRh4 (Tham *et al.*, 2010). Additionally, CR1 levels and polymorphisms seem to have been selected for in areas with malaria transmission (Sinha *et al.*, 2009). Other blood group polymorphisms have been observed at increased frequencies in malaria endemic areas (RhD, Jsa, Kell); however, detailed mechanistic or associative studies are lacking to establish a precise role in *Plasmodium* infection.

Polymorphism between host species

In terms of invasion, it has recently been proposed that the fundamental differences in sialic acid content on the erythrocytes of humans compared to that of great apes, mice, and other mammals may be responsible for species restriction (Varki, 2002). Sialic acids are expressed on the terminal end of glycoproteins in all cell types for all vertebrate animals and can serve as receptors for pathogens(Varki, 2007). While great apes (chimpanzees, bonobos, gorillas, orang-utans, etc.) predominantly display Neu5Gc on the surface of their erythrocytes, humans have a deletion in the CMAH gene responsible for the conversion of the Neu5Ac precursor to Neu5Gc, and as a result, express Neu5Ac only on the surface of cells (Chou *et al.*, 1998; Irie and Suzuki, 1998a,b). Recently is was shown that *Plasmodium reichenowi* and *P. falciparum* EBA-175, transgenically expressed in COS cells, showed specificity in binding erythrocytes from chimpanzees and humans respectively (Martin *et al.*, 2005). It has been proposed that the CMAH mutation could have arisen in humans to protect them from infection with *P. reichenowi*, but that eventually *P. falciparum* emerged to take advantage of the human-specific Neu5Ac displaying red cells (Rich *et al.*, 2009; Varki and Gagneux, 2009). Interestingly, *Aotus* cells, which can be invaded by *P. falciparum* have also express Neu5Ac on the surface of cells, similar to humans (Martin *et al.*, 2005). Such studies imply an important role for host cell surface glycosylation in species-specific tropism.

Selection of invasion ligand polymorphism by humoral immunity: prospects for a blood-stage vaccine?

Humoral immunity has been shown to play an important role in malaria infection through classical transfer experiments of hyperimmune sera into children suffering from acute clinical malaria (McGregor, 1964). Many of the targets of such protective humoral responses have been shown to be merozoite antigens which play key roles in invasion (Kyes *et al.*, 2001). As variant expression and sequence polymorphism of parasite proteins can affect immune recognition, clinically protective immune responses develop over years of exposure to different parasite variants, hence increasing with malaria exposure and patient age (Bull *et al.*, 2000; Dodoo *et al.*, 1999). Numerous methods have been employed to predict and determine the most effective vaccine candidate antigens, and such methods are discussed below.

Immune response to *Plasmodium* invasion proteins

Plasmodium invasion ligands are exposed to the humoral immune response during the process of invasion. Antibodies have been shown to play an important role in acquired immunity to *Plasmodium*, and in particular, invasion inhibitory antibodies are an important host defence against parasite replication and malaria-induced pathology (Brown *et al.*, 1982; Cohen *et al.*, 1969; Yazdani *et al.*, 2006). For this reason, many of the *P. falciparum* surface proteins involved in invasion – MSP-1, AMA-1, and EBA-175 – are current blood-stage vaccine candidates (Mahanty *et al.*, 2003; Wipasa *et al.*, 2002; Yazdani *et al.*, 2006).

Many studies characterizing the human immune response to invasion proteins have focused on antibody reactivity to whole schizont extracts and recombinantly expressed proteins.

Some antigens have been identified based on biochemical experiments in which endemic sera recognize immunodominant schizont-stage parasite proteins by western blot analysis. As genomic methods are becoming increasingly accessible, predictions of novel antigens or protein domains based on DNA sequence polymorphism has revealed signatures of balancing selection indicative of immune selection, which is increasing the number of vaccine candidate antigens deserving further evaluation (Conway *et al.*, 2000).

As candidates are identified, seroepidemiology is employed to determine the dynamics of acquisition, duration, titre, and subclass of the antibody response. While the association between high-titre antibody responses to *Plasmodium* antigens protection has been observed (Branch *et al.*, 1998; Egan *et al.*, 1996, 1999; Okenu *et al.*, 2000; Polley *et al.*, 2004), there is little mechanistic data regarding the acquisition of antibody and its role in conferring protection. One notable exception is MSP-1_{19}, in which detailed studies have been conducted and have mapped specific epitopes recognized by monoclonal antibodies that can either prevent invasion (inhibitory antibodies) or block productive antibodies (blocking antibodies) (Guevara Patino *et al.*, 1997; O'Donnell *et al.*, 2001; Okech *et al.*, 2004; Uthaipibull *et al.*, 2001). Interestingly, while the EBL family members *P. vivax* Duffy binding protein and *P. falciparum* EBA-175 are current vaccine candidates (Beeson and Crabb, 2007; Devi *et al.*, 2007; Fraser *et al.*, 1997; Grimberg *et al.*, 2007) the immune response against the RBL family of proteins has received little attention until recently (Gao *et al.*, 2008; Persson *et al.*, 2008; Tham *et al.*, 2009). Some groups have characterized the functional inhibitory capacity of naturally acquired antibodies in invasion inhibition experiments (Corran *et al.*, 2004; Spring *et al.*, 2009). These assays can either measure the level of invasion inhibition (Boyle *et al.*, 2010) or growth inhibition (Spring *et al.*, 2009), in the presence of antibodies. New advances are being made with transgenic parasites making such approaches higher throughput (Healer *et al.*, 2004; Persson *et al.*, 2008; Wilson *et al.*, 2010) and adding genetic specificity to results obtained.

Making progress towards a blood-stage vaccine

While the ultimate goal of a vaccine should be to prevent infection, even the best current vaccines are far from this goal. However, vaccines targeting blood stage antigens do aid in decreasing parasite burden and decreasing clinical severity, which may help the immune system in controlling parasitaemia. There is recent renewed interest in a blood stage vaccine (Beeson and Crabb, 2007; Crabb and Beeson, 2005; Lopaticki *et al.*, 2011).

P. vivax – a single-antigen PvDBP

P. vivax PvDBP is considered by some to be the most promising blood stage vaccine candidate in development (Beeson and Crabb, 2007; Chitnis and Sharma, 2008; Grimberg *et al.*, 2007). Due to the parasite's restricted invasion using a single pathway, targeting PvDBP is an excellent strategy for eliminating *P. vivax*. The experiment has been done through human evolution with the high levels of Duffy negativity essentially eliminating *P. vivax* from most of sub-Saharan Africa. While progress has been made towards validating PvDBP as a vaccine candidate antigen (Grimberg *et al.*, 2007) concerns have been raised regarding the allelic specificity and duration of inhibitory responses (Ceravolo *et al.*, 2009). Additionally, there is a concern that *P. vivax* is now adapting to invade Duffy negative cells, and thus might constitute invasion by alternative pathways (Menard *et al.*, 2010) (Fig. 8.3). Such claims must be carefully monitored if PvDBP is to continue as a vaccine candidate. However, despite these concerns, PvDBP remains the most promising blood stage malaria vaccine candidate to date and efforts are being made to pursue its development and move into clinical trials.

P. falciparum – the ever-elusive blood-stage vaccine

In the case of *P. falciparum*, its multiple invasion ligand repertoire and greater degree of polymorphism make a blood stage vaccine more of a challenge. Owing to these factors, targeting any single ligand might select for parasites that utilize alternative pathways. Even for single copy genes such as AMA-1, allele specific immunity has been observed following vaccination, which is largely ineffective at preventing heterologous challenge (Spring *et al.*, 2009). To avoid such vaccine-driven immune evasion, efforts have been made to include multiple alleles of the same antigen in blood stage vaccines (Barclay *et al.*, 2008). Additional evidence has shown that including members of both EBL and RBL invasion ligand families might increase invasion inhibition (Lopaticki *et al.*, 2011).

A few *P. falciparum* blood stage vaccine candidates have made it through pre-clinical development and even Phase I trials. These candidates – AMA-1 (Thera *et al.*, 2008), Msp-3/GLURP (Esen *et al.*, 2009; Mordmuller *et al.*, 2010), and EBA-175 (El Sahly *et al.*, 2010) – have reached the phase I clinical trial phase and have been shown to be safe and immunogenic. AMA-1 and Msp3/GLURP have tested safety and immunogenicity in disease-endemic countries (Mali and Gabon, respectively), whereas the phase 1 clinical trial of the EBA-175 RII vaccine was recently completed in the United States (El Sahly *et al.*, 2010) and a similar phase 1 trial is under way in Ghana to determine safety and immunogenicity in semi-immune adults. The challenge now remains whether these vaccines will provide allele specific immunity, as has been observed with AMA-1 (Spring *et al.*, 2009), or whether there is a candidate or combination of antigens that can overcome the diversity of *P. falciparum* isolates in the field.

Prospects for a malaria vaccine

The major challenge facing vaccine development is that the most important targets of inhibitory antibodies are also the most antigenically diverse and polymorphisms rapidly arise. Determining the minimal number of epitopes or alleles needed to produce a robust immune response while covering enough antigenic variants to protect a diverse population of parasites has been the holy grail of malaria vaccinology. Thus far, this approach is being developed for the AMA-1 vaccine to include multiple allelic forms which would hopefully prevent the allele-specific immunity overserved with the 3D7 allele AMA-1 vaccine (Spring *et al.*, 2009). A similar multiallelic version of the sporozoite stage RTS,S vaccine is being developed (Heppner *et al.*, 2005). Because allelic frequencies vary depending on geographic region, and can be selected for over time as has been observed for serotype replacement resulting from vaccination with multivalent pneumococcal vaccines (Obaro *et al.*, 1996; Weinberger *et al.*, 2011), the number of alleles and specific combination can be critical to global long-term vaccine efficacy. In addition to multiple alleles of the same protein, vaccine approaches are being suggested to include multiple proteins alternatives, such as alternative invasion ligands (Lopaticki *et al.*, 2011). In addition to evaluating vaccine efficacy, monitoring parasite diversity pre-and post-vaccination is critically important so as to prevent the emergence of non-vaccine strains.

Current vaccine efforts have focused on known and well characterized antigen targets. While population genetic approaches have yielded insights into antigens under selection (Binks *et al.*, 2001; Conway *et al.*, 2000), the genomics era allows for this methodology to be applied to the entire genome (Duraisingh *et al.*, 2006). Novel approaches including multiple antigens to account for alternative invasion pathways (Lopaticki *et al.*, 2011) as well as transgenic parasites for genetic specificity and high throughput assays (Healer *et al.*, 2004; O'Donnell *et al.*, 2001; O'Donnell *et al.*, 2000; Persson *et al.*, 2008; Wilson *et al.*, 2010) are being developed and employed. Additionally, new advances in structure–function relationships will aid the vaccine effort. Currently, few structures of merozoite invasion ligands have been solved (Garman *et al.*, 2003; Hodder *et al.*, 1996; McHenry and Adams, 2006; Singh *et al.*, 2006), but as these critical pieces of the puzzle are assembled, a clearer picture of the specificity of antibodies for invasion ligands can be visualized.

Future research directions

As malaria research has moved from the descriptive, yet elegant, microscopy studies of the 1800s, to *in vitro* culture of some malaria parasite species, to genetic identification of strains, to finally genomic level analyses, the possibilities are limitless for scientific inquiry. The dawn of the genomic area brings the promise of genome-wide association studies (GWAS) to identify the genetic determinants of parasite virulence traits like drug resistance and invasion (Mu *et al.*, 2010). Improvements in both forward and reverse genetic methodologies make analyses of such virulence determinants possible. Yet such genomic advances should not overshadow the need for *in vivo* experiments, particularly model systems and culture methods that will allow the elucidation of virulence phenotypes for some of the more intractable malaria species – such as *P. vivax*. As the genetic and genomic tools in the arsenal increase, questions of malaria host cell selection, invasion, and virulence become increasingly more tractable.

References

Adams, J.H., Hudson, D.E., Torii, M., Ward, G.E., Wellems, T.E., Aikawa, M., and Miller, L.H. (1990). The Duffy receptor family of *Plasmodium knowlesi* is located within the micronemes of invasive malaria merozoites. Cell *63*, 141–153.

Adams, J.H., Sim, B.K., Dolan, S.A., Fang, X., Kaslow, D.C., and Miller, L.H. (1992). A family of erythrocyte binding proteins of malaria parasites. Proc. Natl. Acad. Sci. U.S.A. *89*, 7085–7089.

Aikawa, M., Miller, L.H., Johnson, J., and Rabbege, J. (1978). Erythrocyte entry by malarial parasites. A moving junction between erythrocyte and parasite. J. Cell Biol. 77, 72–82.

Allison, A.C. (2009). Genetic control of resistance to human malaria. Curr. Opin. Immunol. *21*, 499–505.

Ayi, K., Turrini, F., Piga, A., and Arese, P. (2004). Enhanced phagocytosis of ring-parasitized mutant erythrocytes: a common mechanism that may explain protection against falciparum malaria in sickle trait and beta-thalassemia trait. Blood *104*, 3364–3371.

Baker, R.P., Wijetilaka, R., and Urban, S. (2006). Two *Plasmodium* rhomboid proteases preferentially cleave different adhesins implicated in all invasive stages of malaria. PLoS Pathog. *2*, e113.

Bannister, L.H., and Mitchell, G.H. (1989). The fine structure of secretion by *Plasmodium knowlesi* merozoites during red cell invasion. J. Protozool. *36*, 362–367.

Barclay, V.C., Chan, B.H., Anders, R.F., and Read, A.F. (2008). Mixed allele malaria vaccines: host protection and within-host selection. Vaccine *26*, 6099–6107.

Barnwell, J.W., Nichols, M.E., and Rubinstein, P. (1989). In vitro evaluation of the role of the Duffy blood group in erythrocyte invasion by *Plasmodium vivax*. J. Exp. Med. *169*, 1795–1802.

Baum, J., Chen, L., Healer, J., Lopaticki, S., Boyle, M., Triglia, T., Ehlgen, F., Ralph, S.A., Beeson, J.G., and Cowman, A.F. (2009). Reticulocyte-binding protein homologue 5 – an essential adhesin involved in invasion of human erythrocytes by *Plasmodium falciparum*. Int. J. Parasitol. 39, 371–380.

Baum, J., Pinder, M., and Conway, D.J. (2003a). Erythrocyte invasion phenotypes of *Plasmodium falciparum* in The Gambia. Infect. Immun. *71*, 1856–1863.

Baum, J., Richard, D., Healer, J., Rug, M., Krnajski, Z., Gilberger, T.-W., Green, J.L., Holder, A.A., and Cowman, A.F. (2006). A conserved molecular motor drives cell invasion and gliding motility across malaria life cycle stages and other apicomplexan parasites. J. Biol. Chem. *281*, 5197–5208.

Baum, J., Thomas, A.W., and Conway, D.J. (2003b). Evidence for diversifying selection on erythrocyte-binding antigens of *Plasmodium falciparum* and P. vivax. Genetics *163*, 1327–1336.

Baum, J., Tonkin, C.J., Paul, A.S., Rug, M., Smith, B.J., Gould, S.B., Richard, D., Pollard, T.D., and Cowman, A.F. (2008). A malaria parasite formin regulates actin polymerization and localizes to the parasite-erythrocyte moving junction during invasion. Cell Host Microbe 3, 188–198.

Beeson, J.G., and Crabb, B.S. (2007). Towards a vaccine against *Plasmodium vivax* malaria. PLoS Med. *4*, e350.

Bei, A.K., Brugnara, C., and Duraisingh, M.T. (2010). In vitro genetic analysis of an erythrocyte determinant of malaria infection. J. Infect. Dis. *202*, 1722–1727.

Bei, A.K., Membi, C.D., Rayner, J.C., Mubi, M., Ngasala, B., Sultan, A.A., Premji, Z., and Duraisingh, M.T. (2007). Variant merozoite protein expression is associated with erythrocyte invasion phenotypes in *Plasmodium falciparum* isolates from Tanzania. Mol. Biochem. Parasitol. *153*, 66–71.

Besteiro, S., Michelin, A., Poncet, J., Dubremetz, J.F., and Lebrun, M. (2009). Export of a Toxoplasma gondii rhoptry neck protein complex at the host cell membrane to form the moving junction during invasion. PLoS Pathog. *5*, e1000309.

Binks, R.H., Baum, J., Oduola, A.M., Arnot, D.E., Babiker, H.A., Kremsner, P.G., Roper, C., Greenwood, B.M., and Conway, D.J. (2001). Population genetic analysis of the *Plasmodium falciparum* erythrocyte binding antigen-175 (eba-175) gene. Mol. Biochem. Parasitol. *114*, 63–70.

Blacklock, B., and Adler, S. (1922). A parasite resembling *Plasmodium falciparum* in a Chimpanzee. Ann. Trop. Med. Parasitol. *160*, 99–106.

Boyle, M.J., Wilson, D.W., Richards, J.S., Riglar, D.T., Tetteh, K.K., Conway, D.J., Ralph, S.A., Baum, J., and Beeson, J.G. (2010). Isolation of viable *Plasmodium falciparum* merozoites to define erythrocyte invasion events and advance vaccine and drug development. Proc. Natl. Acad. Sci. U.S.A. *107*, 14378–14383.

Branch, O.H., Udhayakumar, V., Hightower, A.W., Oloo, A.J., Hawley, W.A., Nahlen, B.L., Bloland, P.B., Kaslow, D.C., and Lal, A.A. (1998). A longitudinal investigation of IgG and IgM antibody responses to the merozoite surface protein-1 19-kiloDalton domain of *Plasmodium falciparum* in pregnant women and infants: associations with febrile illness, parasitemia, and anemia. Am. J. Trop. Med. Hyg. *58*, 211–219.

Bray, R.S. (1958). The susceptibility of Liberians to the Madagascar strain of *Plasmodium vivax*. J. Parasitol. *44*, 371–373.

Brown, G.V., Anders, R.F., Mitchell, G.F., and Heywood, P.F. (1982). Target antigens of purified human immunoglobulins which inhibit growth of *Plasmodium falciparum* in vitro. Nature *297*, 591–593.

Bull, P.C., Kortok, M., Kai, O., Ndungu, F., Ross, A., Lowe, B.S., Newbold, C.I., and Marsh, K. (2000). *Plasmodium falciparum*-infected erythrocytes: agglutination by diverse Kenyan plasma is associated with severe disease and young host age. J. Infect. Dis. *182*, 252–259.

Bungener, W. (1985). Virulence of *Plasmodium yoelii* line YM is caused by altered preference for parasitized and unparasitized immature and mature erythrocytes. Trop. Med. Parasitol. *36*, 43–45.

Camus, D., and Hadley, T.J. (1985). A *Plasmodium falciparum* antigen that binds to host erythrocytes and merozoites. Science *230*, 553–556.

Carlton, J.M., Angiuoli, S.V., Suh, B.B., Kooij, T.W., Pertea, M., Silva, J.C., Ermolaeva, M.D., Allen, J.E., Selengut, J.D., Koo, H.L., *et al.* (2002). Genome sequence and comparative analysis of the model rodent malaria parasite *Plasmodium yoelii yoelii*. Nature *419*, 512–519.

Cavalli-Sforza, L., Menozzi, P., and Piazza, A. (1994). The History and Geography of Human Genes (New Jersey, Princeton University Press).

Ceravolo, I.P., Sanchez, B.A., Sousa, T.N., Guerra, B.M., Soares, I.S., Braga, E.M., McHenry, A.M., Adams, J.H., Brito, C.F., and Carvalho, L.H. (2009). Naturally acquired inhibitory antibodies to *Plasmodium vivax* Duffy binding protein are short-lived and allele-specific following a single malaria infection. Clin. Exp. Immunol. *156*, 502–510.

Chattopadhyay, D., Rayner, J., McHenry, A.M., and Adams, J.H. (2006). The structure of the *Plasmodium falciparum* EBA175 ligand domain and the molecular basis of host specificity. Trends Parasitol. *22*, 143–145.

Chernin, E. (1984). The malariatherapy of neurosyphilis. J. Parasitol. *70*, 611–617.

Chitnis, C.E., Chaudhuri, A., Horuk, R., Pogo, A.O., and Miller, L.H. (1996). The domain on the Duffy blood group antigen for binding *Plasmodium vivax* and P. knowlesi malarial parasites to erythrocytes. J. Exp. Med. *184*, 1531–1536.

Chitnis, C.E., and Sharma, A. (2008). Targeting the *Plasmodium vivax* Duffy-binding protein. Trends Parasitol. *24*, 29–34.

Choe, H., Moore, M.J., Owens, C.M., Wright, P.L., Vasilieva, N., Li, W., Singh, A.P., Shakri, R., Chitnis, C.E., and Farzan, M. (2005). Sulphated tyrosines mediate association of chemokines and *Plasmodium vivax* Duffy binding protein with the Duffy antigen/receptor for chemokines (DARC). Mol. Microbiol. *55*, 1413–1422.

Chokejindachai, W., and Conway, D.J. (2009). Case–control approach to identify *Plasmodium falciparum* polymorphisms associated with severe malaria. PLoS ONE *4*, e5454.

Chotivanich, K., Udomsangpetch, R., Simpson, J.A., Newton, P., Pukrittayakamee, S., Looareesuwan, S., and White, N.J. (2000). Parasite multiplication potential and the severity of Falciparum malaria. J. Infect. Dis. *181*, 1206–1209.

Chou, H.H., Takematsu, H., Diaz, S., Iber, J., Nickerson, E., Wright, K.L., Muchmore, E.A., Nelson, D.L., Warren, S.T., and Varki, A. (1998). A mutation in human CMP-sialic acid hydroxylase occurred after the Homo-Pan divergence. Proc. Natl. Acad. Sci. U.S.A. *95*, 11751–11756.

Coatney G.R., C.W., Warren, M., Contacos, P.G. (1971). The primate malarias (Washington D.C., U.S. Government Printing Office).

Cohen, M., Hurtado-Ziola, N., and Varki, A. (2009). ABO blood group glycans modulate sialic acid recognition on erythrocytes. Blood *114*, 3668–3676.

Cohen, S., Butcher, G.A., and Crandall, R.B. (1969). Action of malarial antibody in vitro. Nature *223*, 368–371.

Collins, W.E., and Jeffery, G.M. (2007). Plasmodium malariae: parasite and disease. Clin. Microbiol. Rev. *20*, 579–592.

Collins, W.E., Skinner, J.C., Pappaioanou, M., Broderson, J.R., Ma, N.S., Filipski, V., Stanfill, P.S., and Rogers, L. (1988). Infection of Peruvian *Aotus nancymai* monkeys with different strains of *Plasmodium falciparum, P. vivax*, and *P. malariae*. J. Parasitol. *74*, 392–398.

Collins, W.E., Skinner, J.C., Pappaioanou, M., Ma, N.S., Broderson, J.R., Sutton, B.B., and Stanfill, P.S. (1987). Infection of *Aotus vociferans* (karyotype V) monkeys with different strains of *Plasmodium vivax*. J. Parasitol. *73*, 536–540.

Conway, D.J., Cavanagh, D.R., Tanabe, K., Roper, C., Mikes, Z.S., Sakihama, N., Bojang, K.A., Oduola, A.M., Kremsner, P.G., Arnot, D.E., *et al.* (2000). A principal target of human immunity to malaria identified by molecular population genetic and immunological analyses. Nat. Med. *6*, 689–692.

Corran, P.H., O'Donnell, R.A., Todd, J., Uthaipibull, C., Holder, A.A., Crabb, B.S., and Riley, E.M. (2004). The fine specificity, but not the invasion inhibitory activity, of 19-kilodalton merozoite surface protein 1-specific antibodies is associated with resistance to malarial parasitemia in a cross-sectional survey in The Gambia. Infect. Immun. *72*, 6185–6189.

Cortés, A. (2008). Switching *Plasmodium falciparum* genes on and off for erythrocyte invasion. Trends Parasitol. *24*, 517–524.

Cortes, A., Benet, A., Cooke, B.M., Barnwell, J.W., and Reeder, J.C. (2004). Ability of *Plasmodium falciparum* to invade Southeast Asian ovalocytes varies between parasite lines. Blood *104*, 2961–2966.

Cowman, A.F., and Crabb, B.S. (2006). Invasion of red blood cells by malaria parasites. Cell *124*, 755–766.

Cox-Singh, J., Davis, T.M., Lee, K.S., Shamsul, S.S., Matusop, A., Ratnam, S., Rahman, H.A., Conway, D.J., and Singh, B. (2008). *Plasmodium knowlesi* malaria in humans is widely distributed and potentially life threatening. Clin. Infect. Dis. *46*, 165–171.

Crabb, B.S., and Beeson, J.G. (2005). Promising functional readouts of immunity in a blood-stage malaria vaccine trial. PLoS Med. *2*, e380.

Crosnier, C., Bustamante, L.Y., Bartholdson, S.J., Bei, A.K., Theron, M., Uchikawa, M., Mboup, S., Ndir, O., Kwiatkowski, D.P., Duraisingh, M.T., *et al.* (2011).

Basigin is a receptor essential for erythrocyte invasion by *Plasmodium falciparum*. Nature *480*, 534–537.

Culleton, R., Ndounga, M., Zeyrek, F.Y., Coban, C., Casimiro, P.N., Takeo, S., Tsuboi, T., Yadava, A., Carter, R., and Tanabe, K. (2009). Evidence for the transmission of *Plasmodium vivax* in the Republic of the Congo, West Central Africa. J. Infect. Dis. *200*, 1465–1469.

Deans, A.-M., Nery, S., Conway, D.J., Kai, O., Marsh, K., and Rowe, J.A. (2007). Invasion pathways and malaria severity in Kenyan *Plasmodium falciparum* clinical isolates. Infect. Immun. *75*, 3014–3020.

Devi, Y.S., Mukherjee, P., Yazdani, S.S., Shakri, A.R., Mazumdar, S., Pandey, S., Chitnis, C.E., and Chauhan, V.S. (2007). Immunogenicity of *Plasmodium vivax* combination subunit vaccine formulated with human compatible adjuvants in mice. Vaccine *25*, 5166–5174.

Dodoo, D., Theander, T.G., Kurtzhals, J.A., Koram, K., Riley, E., Akanmori, B.D., Nkrumah, F.K., and Hviid, L. (1999). Levels of antibody to conserved parts of *Plasmodium falciparum* merozoite surface protein 1 in Ghanaian children are not associated with protection from clinical malaria. Infect. Immun. *67*, 2131–2137.

Dolan, S.A., Miller, L.H., and Wellems, T.E. (1990). Evidence for a switching mechanism in the invasion of erythrocytes by *Plasmodium falciparum*. J. Clin. Invest. *86*, 618–624.

Druilhe, P., Spertini, F., Soesoe, D., Corradin, G., Mejia, P., Singh, S., Audran, R., Bouzidi, A., Oeuvray, C., and Roussilhon, C. (2005). A malaria vaccine that elicits in humans antibodies able to kill *Plasmodium falciparum*. PLoS Med. *2*, e344.

Duraisingh, M., Ferdig, M.T., Stoeckert, C.J., Volkman, S.K., and McGovern, V.P. (2006). *Plasmodium* research in the postgenomic era. Trends Parasitol. *22*, 1–4.

Duraisingh, M.T., Maier, A.G., Triglia, T., and Cowman, A.F. (2003a). Erythrocyte-binding antigen 175 mediates invasion in *Plasmodium falciparum* utilizing sialic acid-dependent and -independent pathways. Proc. Natl. Acad. Sci. U.S.A. *100*, 4796–4801.

Duraisingh, M.T., Triglia, T., Ralph, S.A., Rayner, J.C., Barnwell, J.W., McFadden, G.I., and Cowman, A.F. (2003b). Phenotypic variation of *Plasmodium falciparum* merozoite proteins directs receptor targeting for invasion of human erythrocytes. EMBO J. *22*, 1047–1057.

Durocher, J.R., Payne, R.C., and Conrad, M.E. (1975). Role of sialic acid in erythrocyte survival. Blood *45*, 11–20.

Duval, L., Fourment, M., Nerrienet, E., Rousset, D., Sadeuh, S.A., Goodman, S.M., Andriaholinirina, N.V., Randrianarivelojosia, M., Paul, R.E., Robert, V., *et al.* (2010). African apes as reservoirs of *Plasmodium falciparum* and the origin and diversification of the *Laverania* subgenus. Proc. Natl. Acad. Sci. U.S.A. *107*, 10561–10566.

Dvorin, J.D., Bei, A.K., Coleman, B.I., and Duraisingh, M.T. (2010). Functional diversification between two related *Plasmodium falciparum* merozoite invasion ligands is determined by changes in the cytoplasmic domain. Mol. Microbiol. *75*, 990–1006.

Egan, A.F., Burghaus, P., Druilhe, P., Holder, A.A., and Riley, E.M. (1999). Human antibodies to the 19kDa C-terminal fragment of *Plasmodium falciparum* merozoite surface protein 1 inhibit parasite growth in vitro. Parasite Immunol. *21*, 133–139.

Egan, A.F., Morris, J., Barnish, G., Allen, S., Greenwood, B.M., Kaslow, D.C., Holder, A.A., and Riley, E.M. (1996). Clinical immunity to *Plasmodium falciparum* malaria is associated with serum antibodies to the 19-kDa C-terminal fragment of the merozoite surface antigen, PfMSP-1. J. Infect. Dis. *173*, 765–769.

El Sahly, H.M., Patel, S.M., Atmar, R.L., Lanford, T.A., Dube, T., Thompson, D., Sim, B.K., Long, C., and Keitel, W.A. (2010). Safety and immunogenicity of a recombinant nonglycosylated erythrocyte binding antigen 175 Region II malaria vaccine in healthy adults living in an area where malaria is not endemic. Clin. Vaccine Immunol. *17*, 1552–1559.

Esen, M., Kremsner, P.G., Schleucher, R., Gassler, M., Imoukhuede, E.B., Imbault, N., Leroy, O., Jepsen, S., Knudsen, B.W., Schumm, M., *et al.* (2009). Safety and immunogenicity of GMZ2 – a MSP3-GLURP fusion protein malaria vaccine candidate. Vaccine *27*, 6862–6868.

Eyles, D.E., Coatney, G.R., and Getz, M.E. (1960). Vivax-type malaria parasite of macaques transmissible to man. Science *131*, 1812–1813.

Feng, X., Carlton, J.M., Joy, D.A., Mu, J., Furuya, T., Suh, B.B., Wang, Y., Barnwell, J.W., and Su, X.Z. (2003). Single-nucleotide polymorphisms and genome diversity in *Plasmodium vivax*. Proc. Natl. Acad. Sci. U.S.A. *100*, 8502–8507.

Field, S.P., Hempelmann, E., Mendelow, B.V., and Fleming, A.F. (1994). Glycophorin variants and *Plasmodium falciparum*: protective effect of the Dantu phenotype in vitro. Hum. Genet. *93*, 148–150.

Fraser, G.R., Giblett, E.R., and Motulsky, A.G. (1966). Population genetic studies in the Congo. 3. Blood groups (ABO, MNSs, Rh, Jsa). Am. J. Hum. Genet. *18*, 546–552.

Fraser, T., Michon, P., Barnwell, J.W., Noe, A.R., Al-Yaman, F., Kaslow, D.C., and Adams, J.H. (1997). Expression and serologic activity of a soluble recombinant *Plasmodium vivax* Duffy binding protein. Infect. Immun. *65*, 2772–2777.

Freeman, R.R., Trejdosiewicz, A.J., and Cross, G.A. (1980). Protective monoclonal antibodies recognising stage-specific merozoite antigens of a rodent malaria parasite. Nature *284*, 366–368.

Galinski, M.R., Medina, C.C., Ingravallo, P., and Barnwell, J.W. (1992). A reticulocyte-binding protein complex of *Plasmodium vivax* merozoites. Cell *69*, 1213–1226.

Galinski, M.R., Xu, M., and Barnwell, J.W. (2000). *Plasmodium vivax* reticulocyte binding protein-2 (PvRBP-2) shares structural features with PvRBP-1 and the *Plasmodium yoelii* 235 kDa rhoptry protein family. Mol. Biochem. Parasitol. *108*, 257–262.

Gao, X., Yeo, K.P., Aw, S.S., Kuss, C., Iyer, J.K., Genesan, S., Rajamanonmani, R., Lescar, J., Bozdech, Z., and Preiser, P.R. (2008). Antibodies targeting the PfRH1 binding domain inhibit invasion of *Plasmodium falciparum* merozoites. PLoS Pathog. *4*, e1000104.

Gardner, M.J., Hall, N., Fung, E., White, O., Berriman, M., Hyman, R.W., Carlton, J.M., Pain, A., Nelson, K.E., Bowman, S., *et al.* (2002). Genome sequence of the human malaria parasite *Plasmodium falciparum*. Nature *419*, 498–511.

Garman, S.C., Simcoke, W.N., Stowers, A.W., and Garboczi, D.N. (2003). Structure of the C-terminal domains of merozoite surface protein-1 from *Plasmodium knowlesi* reveals a novel histidine binding site. J. Biol. Chem. *278*, 7264–7269.

Garnham, P.C.C. (1966). Malaria Parasites and Other Haemosporidia (Oxford, Blackwell Scientific Publications).

Gaur, D., Singh, S., Singh, S., Jiang, L., Diouf, A., and Miller, L.H. (2007). Recombinant *Plasmodium falciparum* reticulocyte homology protein 4 binds to erythrocytes and blocks invasion. Proc. Natl. Acad. Sci. U.S.A. *104*, 17789–17794.

Gilberger, T.-W. (2003). A novel erythrocyte binding antigen-175 Paralogue from *Plasmodium falciparum* defines a new trypsin-resistant receptor on human erythrocytes. J. Biol. Chem. *278*, 14480–14486.

Goel, V.K., Li, X., Chen, H., Liu, S.C., Chishti, A.H., and Oh, S.S. (2003). Band 3 is a host receptor binding merozoite surface protein 1 during the *Plasmodium falciparum* invasion of erythrocytes. Proc. Natl. Acad. Sci. U.S.A. *100*, 5164–5169.

Gomez-Escobar, N., Amambua-Ngwa, A., Walther, M., Okebe, J., Ebonyi, A., and Conway, D.J. (2010). Erythrocyte invasion and merozoite ligand gene expression in severe and mild *Plasmodium falciparum* malaria. J. Infect. Dis. *201*, 444–452.

Grimberg, B.T., Udomsangpetch, R., Xainli, J., McHenry, A., Panichakul, T., Sattabongkot, J., Cui, L., Bockarie, M., Chitnis, C., Adams, J., *et al.* (2007). *Plasmodium vivax* invasion of human erythrocytes inhibited by antibodies directed against the Duffy binding protein. PLoS Med. *4*, e337.

Guerra, C.A., Howes, R.E., Patil, A.P., Gething, P.W., Van Boeckel, T.P., Temperley, W.H., Kabaria, C.W., Tatem, A.J., Manh, B.H., Elyazar, I.R., *et al.* (2010). The international limits and population at risk of *Plasmodium vivax* transmission in 2009. PLoS Negl. Trop. Dis. 4, e774.

Guevara Patino, J.A., Holder, A.A., McBride, J.S., and Blackman, M.J. (1997). Antibodies that inhibit malaria merozoite surface protein-1 processing and erythrocyte invasion are blocked by naturally acquired human antibodies. J. Exp. Med. *186*, 1689–1699.

Hans, D., Pattnaik, P., Bhattacharyya, A., Shakri, A.R., Yazdani, S.S., Sharma, M., Choe, H., Farzan, M., and Chitnis, C.E. (2005). Mapping binding residues in the *Plasmodium vivax* domain that binds Duffy antigen during red cell invasion. Mol. Microbiol. *55*, 1423–1434.

Haynes, J.D., Dalton, J.P., Klotz, F.W., McGinniss, M.H., Hadley, T.J., Hudson, D.E., and Miller, L.H. (1988). Receptor-like specificity of a *Plasmodium knowlesi* malarial protein that binds to Duffy antigen ligands on erythrocytes. J. Exp. Med. *167*, 1873–1881.

Hayton, K., Gaur, D., Liu, A., Takahashi, J., Henschen, B., Singh, S., Lambert, L., Furuya, T., Bouttenot, R., Doll, M., *et al.* (2008). Erythrocyte binding protein PfRH5 polymorphisms determine species-specific pathways of *Plasmodium falciparum* invasion. Cell Host Microbe *4*, 40–51.

Healer, J., Murphy, V., Hodder, A.N., Masciantonio, R., Gemmill, A.W., Anders, R.F., Cowman, A.F., and Batchelor, A. (2004). Allelic polymorphisms in apical membrane antigen-1 are responsible for evasion of antibody-mediated inhibition in *Plasmodium falciparum*. Mol. Microbiol. *52*, 159–168.

Hegner, R. (1938). Relative frequency of ring-stage plasmodia in reticulocytes and mature erythrocytes in man and monkey. Am. J. Epidemiol. *27*, 690–718.

Heppner, D.G., Jr., Kester, K.E., Ockenhouse, C.F., Tornieporth, N., Ofori, O., Lyon, J.A., Stewart, V.A., Dubois, P., Lanar, D.E., Krzych, U., *et al.* (2005). Towards an RTS, S-based, multi-stage, multi-antigen vaccine against *falciparum* malaria: progress at the Walter Reed Army Institute of Research. Vaccine *23*, 2243–2250.

Hodder, A.N., Crewther, P.E., Matthew, M.L., Reid, G.E., Moritz, R.L., Simpson, R.J., and Anders, R.F. (1996). The disulfide bond structure of Plasmodium apical membrane antigen-1. J. Biol. Chem. *271*, 29446–29452.

Holder, A.A., and Freeman, R.R. (1981). Immunization against blood-stage rodent malaria using purified parasite antigens. Nature *294*, 361–364.

Holder, A.A., and Freeman, R.R. (1984). Characterization of a high molecular weight protective antigen of *Plasmodium yoelii*. Parasitology *88*, 211–219.

Irie, A., and Suzuki, A. (1998a). CMP-N-Acetylneuraminic acid hydroxylase is exclusively inactive in humans. Biochem. Biophys. Res. Commun. *248*, 330–333.

Irie, A., and Suzuki, A. (1998b). [The molecular basis for the absence of N-glycolylneuraminic acid in humans]. Tanpakushitsu Kakusan Koso *43*, 2404–2409.

Iyer, J.K., Amaladoss, A., Ganesan, S., and Preiser, P.R. (2007a). Variable expression of the 235 kDa rhoptry protein of *Plasmodium yoelii* mediate host cell adaptation and immune evasion. Mol. Microbiol. *65*, 333–346.

Iyer, J.K., Amaladoss, A., Genesan, S., and Preiser, P.R. (2007b). Variable expression of the 235 kDa rhoptry protein of *Plasmodium yoelii* mediate host cell adaptation and immune evasion. Mol. Microbiol. *65*, 333–346.

Iyer, J.K., Fuller, K., and Preiser, P.R. (2006). Differences in the copy number of the py235 gene family in virulent and avirulent lines of *Plasmodium yoelii*. Mol. Biochem. Parasitol. *150*, 186–191.

Jennings, C.V., Ahouidi, A.D., Zilversmit, M., Bei, A.K., Rayner, J., Sarr, O., Ndir, O., Wirth, D.F., Mboup, S., and Duraisingh, M.T. (2007). Molecular analysis of erythrocyte invasion in *Plasmodium falciparum* isolates from Senegal. Infect. Immun. *75*, 3531–3538.

Jiang, L., López-Barragán, M.J., Jiang, H., Mu, J., Gaur, D., Zhao, K., Felsenfeld, G., and Miller, L.H. (2010). Epigenetic control of the variable expression of a *Plasmodium falciparum* receptor protein for erythrocyte invasion. Proc. Natl. Acad. Sci. U.S.A. *107*, 2224–2229.

Kanamaru, A., Okuda, K., Ueda, E., Kitani, T., Kinoshita, T., and Nagai, K. (1988). Different distribution of decay-accelerating factor on hematopoietic progenitors from normal individuals and patients with paroxysmal nocturnal hemoglobinuria. Blood 72, 507–511.

Kasehagen, L.J., Mueller, I., Kiniboro, B., Bockarie, M.J., Reeder, J.C., Kazura, J.W., Kastens, W., McNamara, D.T., King, C.H., Whalen, C.C., *et al.* (2007). Reduced *Plasmodium vivax* erythrocyte infection in PNG Duffy-negative heterozygotes. PLoS One 2, e336.

Kato, K., Mayer, D.C., Singh, S., Reid, M., and Miller, L.H. (2005). Domain III of *Plasmodium falciparum* apical membrane antigen 1 binds to the erythrocyte membrane protein Kx. Proc. Natl. Acad. Sci. U.S.A. *102*, 5552–5557.

Kay, M.M., Goodman, S.R., Sorensen, K., Whitfield, C.F., Wong, P., Zaki, L., and Rudloff, V. (1983). Senescent cell antigen is immunologically related to band 3. Proc. Natl. Acad. Sci. U.S.A. *80*, 1631–1635.

Keen, J.K., Sinha, K.A., Brown, K.N., and Holder, A.A. (1994). A gene coding for a high-molecular mass rhoptry protein of *Plasmodium yoelii*. Mol. Biochem. Parasitol. *65*, 171–177.

Khan, S.M., Jarra, W., Bayele, H., and Preiser, P.R. (2001). Distribution and characterisation of the 235 kDa rhoptry multigene family within the genomes of virulent and avirulent lines of *Plasmodium yoelii*. Mol. Biochem. Parasitol. *114*, 197–208.

Klotz, F.W., Chulay, J.D., Daniel, W., and Miller, L.H. (1987). Invasion of mouse erythrocytes by the human malaria parasite, *Plasmodium falciparum*. J. Exp. Med. *165*, 1713–1718.

Kocken, C.H., Narum, D.L., Massougbodji, A., Ayivi, B., Dubbeld, M.A., van der Wel, A., Conway, D.J., Sanni, A., and Thomas, A.W. (2000). Molecular characterisation of *Plasmodium reichenowi* apical membrane antigen-1 (AMA-1), comparison with *P. falciparum* AMA-1, and antibody-mediated inhibition of red cell invasion. Mol. Biochem. Parasitol. *109*, 147–156.

Krief, S., Escalante, A.A., Pacheco, M.A., Mugisha, L., André, C., Halbwax, M., Fischer, A., Krief, J.-M., Kasenene, J.M., Crandfield, M., *et al.* (2010). On the diversity of malaria parasites in African apes and the origin of *Plasmodium falciparum* from Bonobos. PLoS Pathog. *6*, e1000765.

Kyes, S., Horrocks, P., and Newbold, C. (2001). Antigenic variation at the infected red cell surface in malaria. Annu. Rev. Microbiol. *55*, 673–707.

Lantos, P.M., Ahouidi, A.D., Bei, A.K., Jennings, C.V., Sarr, O., Ndir, O., Wirth, D.F., Mboup, S., and Duraisingh, M.T. (2009). Erythrocyte invasion profiles are associated with a common invasion ligand polymorphism in Senegalese isolates of *Plasmodium falciparum*. Parasitology *136*, 1–9.

Laveran, C.L.A. (1880). [A newly discovered parasite in the blood of patients suffering from malaria. Parasitic etiology of attacks of malaria.]. Bull. Memoris Soc. Med. Hop. Paris *17*, 158–164.

Liu, W., Li, Y., Learn, G.H., Rudicell, R.S., Robertson, J.D., Keele, B.F., Ndjango, J.B., Sanz, C.M., Morgan, D.B., Locatelli, S., *et al.* (2010). Origin of the human malaria parasite *Plasmodium falciparum* in gorillas. Nature *467*, 420–425.

Livingstone, F.B. (1984). The Duffy blood groups, vivax malaria, and malaria selection in human populations: a review. Hum. Biol. *56*, 413–425.

Lobo, C.-A., de Frazao, K., Rodriguez, M., Reid, M., Zalis, M., and Lustigman, S. (2004). Invasion profiles of Brazilian field isolates of *Plasmodium falciparum*: phenotypic and genotypic analyses. Infect. Immun. 72, 5886–5891.

Lobo, C.A., Rodriguez, M., Reid, M., and Lustigman, S. (2003). Glycophorin C is the receptor for the *Plasmodium falciparum* erythrocyte binding ligand PfEBP-2 (baebl). Blood *101*, 4628–4631.

Lopaticki, S., Maier, A.G., Thompson, J., Wilson, D.W., Tham, W.H., Triglia, T., Gout, A., Speed, T.P., Beeson, J.G., Healer, J., *et al.* (2011). Reticulocyte and erythrocyte binding-like proteins function cooperatively in invasion of human erythrocytes by malaria parasites. Infect Immun. *79*, 1107–1117.

Mahanty, S., Saul, A., and Miller, L.H. (2003). Progress in the development of recombinant and synthetic blood-stage malaria vaccines. J. Exp. Biol. *206*, 3781–3788.

Maier, A.G., Duraisingh, M.T., Reeder, J.C., Patel, S.S., Kazura, J.W., Zimmerman, P.A., and Cowman, A.F. (2003). *Plasmodium falciparum* erythrocyte invasion through glycophorin C and selection for Gerbich negativity in human populations. Nat. Med. *9*, 87–92.

Martin, M.J., Rayner, J.C., Gagneux, P., Barnwell, J.W., and Varki, A. (2005). Evolution of human–chimpanzee differences in malaria susceptibility: relationship to human genetic loss of N-glycolylneuraminic acid. Proc. Natl. Acad. Sci. U.S.A. *102*, 12819–12824.

Mason, S.J., Miller, L.H., Shiroishi, T., Dvorak, J.A., and McGinniss, M.H. (1977). The Duffy blood group determinants: their role in the susceptibility of human and animal erythrocytes to *Plasmodium knowlesi* malaria. Br. J. Haematol. *36*, 327–335.

Mayer, D.C., Kaneko, O., Hudson-Taylor, D.E., Reid, M.E., and Miller, L.H. (2001). Characterization of a *Plasmodium falciparum* erythrocyte-binding protein paralogous to EBA-175. Proc. Natl. Acad. Sci. U.S.A. *98*, 5222–5227.

Mayer, D.C., Mu, J.B., Feng, X., Su, X.Z., and Miller, L.H. (2002). Polymorphism in a *Plasmodium falciparum* erythrocyte-binding ligand changes its receptor specificity. J. Exp. Med. *196*, 1523–1528.

Mayer, D.C., Mu, J.B., Kaneko, O., Duan, J., Su, X.Z., and Miller, L.H. (2004). Polymorphism in the *Plasmodium falciparum* erythrocyte-binding ligand JESEBL/EBA-181 alters its receptor specificity. Proc. Natl. Acad. Sci. U.S.A. *101*, 2518–2523.

Mayer, D.C.G., Cofie, J., Jiang, L., Hartl, D.L., Tracy, E., Kabat, J., Mendoza, L.H., and Miller, L.H. (2009). Glycophorin B is the erythrocyte receptor of *Plasmodium falciparum* erythrocyte-binding ligand, EBL-1. Proc. Natl. Acad. Sci. U.S.A. *106*, 5348–5352.

McGregor, I.A. (1964). The Passive Transfer of Human Malarial Immunity. Am. J. Trop. Med. Hyg. *13*, Suppl. 237–239.

McHenry, A.M., and Adams, J.H. (2006). The crystal structure of P. knowlesi DBPalpha DBL domain and its implications for immune evasion. Trends Biochem. Sci. *31*, 487–491.

Menard, D., Barnadas, C., Bouchier, C., Henry-Halldin, C., Gray, L.R., Ratsimbasoa, A., Thonier, V., Carod, J.F., Domarle, O., Colin, Y., *et al.* (2010). *Plasmodium vivax* clinical malaria is commonly observed in Duffy-negative Malagasy people. Proc. Natl. Acad. Sci. U.S.A. *107*, 5967–5971.

Meyer, E.V., Semenya, A.A., Okenu, D.M., Dluzewski, A.R., Bannister, L.H., Barnwell, J.W., and Galinski, M.R. (2009). The reticulocyte binding-like proteins of P. knowlesi locate to the micronemes of merozoites and define two new members of this invasion ligand family. Mol. Biochem. Parasitol. *165*, 111–121.

Miller, L.H. (1994). Impact of malaria on genetic polymorphism and genetic diseases in Africans and African Americans. Proc. Natl. Acad. Sci. U.S.A. *91*, 2415–2419.

Miller, L.H., Mason, S.J., Clyde, D.F., and McGinniss, M.H. (1976). The resistance factor to *Plasmodium vivax* in blacks. The Duffy-blood-group genotype, FyFy. N. Engl. J. Med. *295*, 302–304.

Miller, L.H., Mason, S.J., Dvorak, J.A., McGinniss, M.H., and Rothman, I.K. (1975). Erythrocyte receptors for (*Plasmodium knowlesi*) malaria: Duffy blood group determinants. Science *189*, 561–563.

Mitchell, G.H., Thomas, A.W., Margos, G., Dluzewski, A.R., and Bannister, L.H. (2004). Apical membrane antigen 1, a major malaria vaccine candidate, mediates the close attachment of invasive merozoites to host red blood cells. Infect. Immun. *72*, 154–158.

Mohandas, N., Lie-Injo, L.E., Friedman, M., and Mak, J.W. (1984). Rigid membranes of Malayan ovalocytes: a likely genetic barrier against malaria. Blood *63*, 1385–1392.

Moldenhauer, F., Botto, M., and Walport, M.J. (1988). The rate of loss of CR1 from ageing erythrocytes in vivo in normal subjects and SLE patients: no correlation with structural or numerical polymorphisms. Clin. Exp. Immunol. *72*, 74–78.

Mordmuller, B., Szywon, K., Greutelaers, B., Esen, M., Mewono, L., Treut, C., Murbeth, R.E., Chilengi, R., Noor, R., Kilama, W.L., *et al.* (2010). Safety and immunogenicity of the malaria vaccine candidate GMZ2 in malaria-exposed, adult individuals from Lambarene, Gabon. Vaccine *28*, 6698–6703.

Moreno, A., Perignon, J.L., Morosan, S., Mazier, D., and Benito, A. (2007). *Plasmodium falciparum*-infected mice: more than a tour de force. Trends Parasitol. *23*, 254–259.

Morosan, S., Hez-Deroubaix, S., Lunel, F., Renia, L., Giannini, C., Van Rooijen, N., Battaglia, S., Blanc, C., Eling, W., Sauerwein, R., *et al.* (2006). Liver-stage development of *Plasmodium falciparum*, in a humanized mouse model. J. Infect. Dis. *193*, 996–1004.

Mourant, A., Kopec, A., and Domaniewska-Sobczak, K. (1976). The Distribution of the Human Blood Groups (London: Oxford University Press).

Mu, J., Myers, R.A., Jiang, H., Liu, S., Ricklefs, S., Waisberg, M., Chotivanich, K., Wilairatana, P., Krudsood, S., White, N.J., *et al.* (2010). *Plasmodium falciparum* genome-wide scans for positive selection, recombination hot spots and resistance to antimalarial drugs. Nat. Genet. *42*, 268–271.

Nery, S., Deans, A.M., Mosobo, M., Marsh, K., Rowe, J.A., and Conway, D.J. (2006). Expression of *Plasmodium falciparum* genes involved in erythrocyte invasion varies among isolates cultured directly from patients. Mol. Biochem. Parasitol. *149*, 208–215.

O'Donnell, R.A., and Blackman, M.J. (2005). The role of malaria merozoite proteases in red blood cell invasion. Curr. Opin. Microbiol. *8*, 422–427.

O'Donnell, R.A., Saul, A., Cowman, A.F., and Crabb, B.S. (2000). Functional conservation of the malaria vaccine antigen MSP-119across distantly related *Plasmodium* species. Nat. Med. *6*, 91–95.

O'Donnell, R.A., de Koning-Ward, T.F., Burt, R.A., Bockarie, M., Reeder, J.C., Cowman, A.F., and Crabb, B.S. (2001). Antibodies against merozoite surface protein (MSP)–1(19) are a major component of the invasion-inhibitory response in individuals immune to malaria. J. Exp. Med. *193*, 1403–1412.

O'Donnell, R.A., Hackett, F., Howell, S.A., Treeck, M., Struck, N., Krnajski, Z., Withers-Martinez, C., Gilberger, T.W., and Blackman, M.J. (2006). Intramembrane proteolysis mediates shedding of a key adhesin during erythrocyte invasion by the malaria parasite. J. Cell Biol. *174*, 1023–1033.

Obaro, S.K., Adegbola, R.A., Banya, W.A., and Greenwood, B.M. (1996). Carriage of pneumococci after pneumococcal vaccination. Lancet *348*, 271–272.

Ogun, S.A., and Holder, A.A. (1996). A high molecular mass *Plasmodium yoelii* rhoptry protein binds to erythrocytes. Mol. Biochem. Parasitol. *76*, 321–324.

Ogun, S.A., Scott-Finnigan, T.J., Narum, D.L., and Holder, A.A. (2000). *Plasmodium yoelii*: effects of red blood cell modification and antibodies on the binding characteristics of the 235-kDa rhoptry protein. Exp. Parasitol. *95*, 187–195.

Okech, B.A., Corran, P.H., Todd, J., Joynson-Hicks, A., Uthaipibull, C., Egwang, T.G., Holder, A.A., and Riley, E.M. (2004). Fine specificity of serum antibodies to *Plasmodium falciparum* merozoite surface protein, PfMSP-1(19), predicts protection from malaria infection and high-density parasitemia. Infect. Immun. *72*, 1557–1567.

Okenu, D.M., Riley, E.M., Bickle, Q.D., Agomo, P.U., Barbosa, A., Daugherty, J.R., Lanar, D.E., and Conway, D.J. (2000). Analysis of human antibodies to erythrocyte binding antigen 175 of *Plasmodium falciparum*. Infect. Immun. *68*, 5559–5566.

Orlandi, P.A., Klotz, F.W., and Haynes, J.D. (1992). A malaria invasion receptor, the 175-kilodalton erythrocyte binding antigen of *Plasmodium falciparum* recognizes the terminal Neu5Ac(alpha 2–3)Gal-sequences of glycophorin A. J. Cell Biol. *116*, 901–909.

Ozwara, H., Kocken, C.H., Conway, D.J., Mwenda, J.M., and Thomas, A.W. (2001). Comparative analysis of *Plasmodium reichenowi* and *P. falciparum* erythrocyte-binding proteins reveals selection to maintain polymorphism in the erythrocyte-binding region of EBA-175. Mol. Biochem. Parasitol. *116*, 81–84.

Pain, A., Bohme, U., Berry, A.E., Mungall, K., Finn, R.D., Jackson, A.P., Mourier, T., Mistry, J., Pasini, E.M., Aslett, M.A., *et al.* (2008). The genome of the simian and human malaria parasite *Plasmodium knowlesi.* Nature *455*, 799–803.

Pasvol, G., Weatherall, D.J., and Wilson, R.J. (1980). The increased susceptibility of young red cells to invasion by the malarial parasite *Plasmodium falciparum*. Br. J. Haematol. *45*, 285–295.

Persson, K.E.M., McCallum, F.J., Reiling, L., Lister, N.A., Stubbs, J., Cowman, A.F., Marsh, K., and Beeson, J.G. (2008). Variation in use of erythrocyte invasion pathways by *Plasmodium falciparum* mediates evasion of human inhibitory antibodies. J. Clin. Invest. *118*, 342–351.

Peterson, D.S., and Wellems, T.E. (2000). EBL-1, a putative erythrocyte binding protein of *Plasmodium falciparum*, maps within a favored linkage group in two genetic crosses. Mol. Biochem. Parasitol. *105*, 105–113.

Polley, S.D., Mwangi, T., Kocken, C.H., Thomas, A.W., Dutta, S., Lanar, D.E., Remarque, E., Ross, A., Williams, T.N., Mwambingu, G., *et al.* (2004). Human antibodies to recombinant protein constructs of *Plasmodium falciparum* Apical Membrane Antigen 1 (AMA1) and their associations with protection from malaria. Vaccine *23*, 718–728.

Prasad, C.D., Prasad Singh, A., Chitnis, C.E., and Sharma, A. (2003). A *Plasmodium yoelii yoelii* erythrocyte binding protein that uses Duffy binding-like domain for invasion: a rodent model for studying erythrocyte invasion. Mol. Biochem. Parasitol. *128*, 101–105.

Preiser, P.R., Jarra, W., Capiod, T., and Snounou, G. (1999). A rhoptry-protein-associated mechanism of clonal phenotypic variation in rodent malaria. Nature *398*, 618–622.

Preiser, P.R., Khan, S., Costa, F.T., Jarra, W., Belnoue, E., Ogun, S., Holder, A.A., Voza, T., Landau, I., Snounou, G., *et al.* (2002). Stage-specific transcription of distinct repertoires of a multigene family during Plasmodium life cycle. Science *295*, 342–345.

Price, R.N., Douglas, N.M., and Anstey, N.M. (2009). New developments in *Plasmodium vivax* malaria: severe disease and the rise of chloroquine resistance. Curr Opin Infect Dis *22*, 430–435.

Prugnolle, F., Durand, P., Neel, C., Ollomo, B., Ayala, F.J., Arnathau, C., Etienne, L., Mpoudi-Ngole, E., Nkoghe, D., Leroy, E., *et al.* (2010). African great apes are natural hosts of multiple related malaria species, including *Plasmodium falciparum*. Proc. Natl. Acad. Sci. U.S.A. *107*, 1458–1463.

Ranjan, A., and Chitnis, C.E. (1999). Mapping regions containing binding residues within functional domains of *Plasmodium vivax* and *Plasmodium knowlesi* erythrocyte-binding proteins. Proc. Natl. Acad. Sci. U.S.A. *96*, 14067–14072.

Rayner, J.C., Galinski, M.R., Ingravallo, P., and Barnwell, J.W. (2000). Two *Plasmodium falciparum* genes express merozoite proteins that are related to *Plasmodium vivax* and *Plasmodium yoelii* adhesive proteins involved in host cell selection and invasion. Proc. Natl. Acad. Sci. U.S.A. 97, 9648–9653.

Rayner, J.C., Huber, C.S., and Barnwell, J.W. (2004a). Conservation and divergence in erythrocyte invasion ligands: *Plasmodium reichenowi* EBL genes. Mol. Biochem. Parasitol. *138*, 243–247.

Rayner, J.C., Huber, C.S., Galinski, M.R., and Barnwell, J.W. (2004b). Rapid evolution of an erythrocyte invasion gene family: the *Plasmodium reichenowi* Reticulocyte Binding Like (RBL) genes. Mol. Biochem. Parasitol. *133*, 287–296.

Rayner, J.C., Tran, T.M., Corredor, V., Huber, C.S., Barnwell, J.W., and Galinski, M.R. (2005). Dramatic difference in diversity between *Plasmodium falciparum* and *Plasmodium vivax* reticulocyte binding-like genes. Am. J. Trop. Med. Hyg. *72*, 666–674.

Rayner, J.C., Vargas-Serrato, E., Huber, C.S., Galinski, M.R., and Barnwell, J.W. (2001). A *Plasmodium falciparum* homologue of *Plasmodium vivax* reticulocyte binding protein (PvRBP1) defines a trypsin-resistant erythrocyte invasion pathway. J. Exp. Med. *194*, 1571–1581.

Reed, M.B., Caruana, S.R., Batchelor, A.H., Thompson, J.K., Crabb, B.S., and Cowman, A.F. (2000). Targeted disruption of an erythrocyte binding antigen in *Plasmodium falciparum* is associated with a switch toward a sialic acid-independent pathway of invasion. Proc. Natl. Acad. Sci. U.S.A. *97*, 7509–7514.

Rich, S.M., Leendertz, F.H., Xu, G., LeBreton, M., Djoko, C.F., Aminake, M.N., Takang, E.E., Diffo, J.L.D., Pike, B.L., Rosenthal, B.M., *et al.* (2009). The origin of malignant malaria. Proc. Natl. Acad. Sci. U.S.A. *106*, 14902–14907.

Richard, D., MacRaild, C.A., Riglar, D.T., Chan, J.A., Foley, M., Baum, J., Ralph, S.A., Norton, R.S., and Cowman, A.F. (2010). Interaction between *Plasmodium falciparum* apical membrane antigen 1 and the rhoptry neck protein complex defines a key step in the erythrocyte invasion process of malaria parasites. J. Biol. Chem. *285*, 14815–14822.

Ripoche, J., and Sim, R.B. (1986). Loss of complement receptor type 1 (CR1) on ageing of erythrocytes. Studies of proteolytic release of the receptor. Biochem. J. *235*, 815–821.

Rodhain, J. (1939). Les plasmodiums des anthropoides de l'Afrique centrale et leurs relations avec les plasmodiums humains. Ann. Soc. Belg. Med. Trop. *19*, 563–572.

Rodhain, J., and Jadin, J. (1964). [Transmission of *plasmodium falciparum* in the splenectomized chimpanzee.]. Ann. Soc. Belges Med. Trop. Parasitol. Mycol. *44*, 531–535.

Rodhain, J., and Muylle, G. (1938). Sur la speciflcite des *Plasmodium* des anthropoides de l'Afrique centrale. C. R. Soc. Biol. *127*, 1467–1468.

Rodriguez, M., Lustigman, S., Montero, E., Oksov, Y., and Lobo, C.A. (2008). PfRH5: a novel reticulocyte-binding family homolog of *plasmodium falciparum* that binds to the erythrocyte, and an investigation of its receptor. PLoS ONE 3, e3300.

Rosenberg, R. (2007). *Plasmodium vivax* in Africa: hidden in plain sight? Trends Parasitol. 23, 193–196.

Rout, R., Dhangadamajhi, G., Mohapatra, B.N., Kar, S.K., and Ranjit, M. (2010). High CR1 level and related polymorphic variants are associated with cerebral malaria in eastern-India. Infect. Genet. Evol. *11*, 139–144.

Rowe, J.A., Handel, I.G., Thera, M.A., Deans, A.-M., Lyke, K.E., Koné, A., Diallo, D.A., Raza, A., Kai, O., Marsh, K., *et al.* (2007). Blood group O protects against severe *Plasmodium falciparum* malaria through the mechanism of reduced rosetting. Proc. Natl. Acad. Sci. U.S.A. *104*, 17471–17476.

Rowe, J.A., Opi, D.H., and Williams, T.N. (2009). Blood groups and malaria: fresh insights into pathogenesis and identification of targets for intervention. Curr. Opin. Hematol. *16*, 480–487.

Ryan, J.R., Stoute, J.A., Amon, J., Dunton, R.F., Mtalib, R., Koros, J., Owour, B., Luckhart, S., Wirtz, R.A., Barnwell, J.W., *et al.* (2006). Evidence for transmission of *Plasmodium vivax* among a Duffy antigen negative population in Western Kenya. Am. J. Trop. Med. Hyg. *75*, 575–581.

Sabeti, P.C., Reich, D.E., Higgins, J.M., Levine, H.Z.P., Richter, D.J., Schaffner, S.F., Gabriel, S.B., Platko, J.V., Patterson, N.J., McDonald, G.J., *et al.* (2002). Detecting recent positive selection in the human genome from haplotype structure. Nature *419*, 832–837.

Saul, A., Lamont, G., Sawyer, W.H., and Kidson, C. (1984). Decreased membrane deformability in Melanesian ovalocytes from Papua New Guinea. J. Cell Biol. *98*, 1348–1354.

Schmidt, L.H. (1978). *Plasmodium falciparum* and *Plasmodium vivax* infections in the owl monkey (*Aotus trivirgatus*). I. The courses of untreated infections. Am. J. Trop. Med. Hyg. *27*, 671–702.

Schroit, A.J., Madsen, J.W., and Tanaka, Y. (1985). In vivo recognition and clearance of red blood cells containing phosphatidylserine in their plasma membranes. J. Biol. Chem. *260*, 5131–5138.

Serjeantson, S.W. (1989). A selective advantage for the Gerbich-negative phenotype in malarious areas of Papua New Guinea. PNG Med. J. *32*, 5–9.

Simpson, J.A., Silamut, K., Chotivanich, K., Pukrittayakamee, S., and White, N.J. (1999). Red cell selectivity in malaria: a study of multiple-infected erythrocytes. Trans. R. Soc. Trop. Med. Hyg. *93*, 165–168.

Singh, A.P., Ozwara, H., Kocken, C.H., Puri, S.K., Thomas, A.W., and Chitnis, C.E. (2005). Targeted deletion of *Plasmodium knowlesi* Duffy binding protein confirms its role in junction formation during invasion. Mol. Microbiol. *55*, 1925–1934.

Singh, S.K., Hora, R., Belrhali, H., Chitnis, C.E., and Sharma, A. (2006). Structural basis for Duffy recognition by the malaria parasite Duffy-binding-like domain. Nature *439*, 741–744.

Sinha, S., Jha, G.N., Anand, P., Qidwai, T., Pati, S.S., Mohanty, S., Mishra, S.K., Tyagi, P.K., Sharma, S.K., Venkatesh, V., *et al.* (2009). CR1 levels and gene polymorphisms exhibit differential association with falciparum malaria in regions of varying disease endemicity. Hum. Immunol. *70*, 244–250.

Smith, J.D., Subramanian, G., Gamain, B., Baruch, D.I., and Miller, L.H. (2000). Classification of adhesive domains in the *Plasmodium falciparum* erythrocyte membrane protein 1 family. Mol. Biochem. Parasitol. *110*, 293–310.

Snounou, G., Zhu, X., Siripoon, N., Jarra, W., Thaithong, S., Brown, K.N., and Viriyakosol, S. (1999). Biased distribution of msp1 and msp2 allelic variants in *Plasmodium falciparum* populations in Thailand. Trans. R. Soc. Trop. Med. Hyg. *93*, 369–374.

Spadafora, C., Awandare, G.A., Kopydlowski, K.M., Czege, J., Moch, J.K., Finberg, R.W., Tsokos, G.C., and Stoute, J.A. (2010). Complement receptor 1 is a sialic acid-independent erythrocyte receptor of *Plasmodium falciparum*. PLoS Pathog. *6*, e1000968.

Spring, M.D., Cummings, J.F., Ockenhouse, C.F., Dutta, S., Reidler, R., Angov, E., Bergmann-Leitner, E., Stewart, V.A., Bittner, S., Juompan, L., *et al.* (2009). Phase 1/2a study of the malaria vaccine candidate apical membrane antigen-1 (AMA-1) administered in adjuvant system AS01B or AS02A. PLoS ONE *4*, e5254.

Straub, K.W., Cheng, S.J., Sohn, C.S., and Bradley, P.J. (2009). Novel components of the Apicomplexan moving junction reveal conserved and coccidia-restricted elements. Cell. Microbiol. *11*, 590–603.

Stubbs, J., Simpson, K.M., Triglia, T., Plouffe, D., Tonkin, C.J., Duraisingh, M.T., Maier, A.G., Winzeler, E.A., and Cowman, A.F. (2005). Molecular mechanism for switching of *P. falciparum* invasion pathways into human erythrocytes. Science *309*, 1384–1387.

Tanabe, K., Sakihama, N., Kaneko, O., Saito-Ito, A., and Kimura, M. (1998). A PCR method for molecular epidemiology of *Plasmodium falciparum* Msp-1. Tokai J. Exp. Clin. Med. *23*, 375–381.

Taylor, H.M., Grainger, M., and Holder, A.A. (2002). Variation in the expression of a *Plasmodium falciparum* protein family implicated in erythrocyte invasion. Infect. Immun. *70*, 5779–5789.

Teeranaipong, P., Ohashi, J., Patarapotikul, J., Kimura, R., Nuchnoi, P., Hananantachai, H., Naka, I., Putaporntip, C., Jongwutiwes, S., and Tokunaga, K. (2008). A functional single-nucleotide polymorphism in the CR1 promoter region contributes to protection against cerebral malaria. J. Infect. Dis. *198*, 1880–1891.

Tham, W.-H., Wilson, D.W., Reiling, L., Chen, L., Beeson, J.G., and Cowman, A.F. (2009). Antibodies to reticulocyte binding protein-like homologue 4 inhibit invasion of *Plasmodium falciparum* into human erythrocytes. Infect. Immun. *77*, 2427–2435.

Tham, W.H., Wilson, D.W., Lopaticki, S., Schmidt, C.Q., Tetteh-Quarcoo, P.B., Barlow, P.N., Richard, D., Corbin, J.E., Beeson, J.G., and Cowman, A.F. (2010). Complement receptor 1 is the host erythrocyte receptor for *Plasmodium falciparum* PfRh4 invasion ligand. Proc. Natl. Acad. Sci. U.S.A. *107*, 17327–17332.

Thera, M.A., Doumbo, O.K., Coulibaly, D., Diallo, D.A., Kone, A.K., Guindo, A.B., Traore, K., Dicko, A., Sagara, I., Sissoko, M.S., *et al.* (2008). Safety and immunogenicity of an AMA-1 malaria vaccine in Malian adults: results of a phase 1 randomized controlled trial. PLoS One 3, e1465.

Tolia, N.H., Enemark, E.J., Sim, B.K.L., and Joshua-Tor, L. (2005). Structural basis for the EBA-175 erythrocyte invasion pathway of the malaria parasite *Plasmodium falciparum*. Cell *122*, 183–193.

Tournamille, C., Colin, Y., Cartron, J.P., and Le Van Kim, C. (1995). Disruption of a GATA motif in the Duffy gene promoter abolishes erythroid gene expression in Duffy-negative individuals. Nat. Genet. *10*, 224–228.

Treeck, M., Zacherl, S., Herrmann, S., Cabrera, A., Kono, M., Struck, N.S., Engelberg, K., Haase, S., Frischknecht, F., Miura, K., *et al.* (2009). Functional analysis of the leading malaria vaccine candidate AMA-1 reveals an essential role for the cytoplasmic domain in the invasion process. PLoS Pathog. *5*, e1000322.

Triglia, T., Duraisingh, M.T., Good, R.T., and Cowman, A.F. (2005). Reticulocyte-binding protein homologue 1 is required for sialic acid-dependent invasion into human erythrocytes by *Plasmodium falciparum*. Mol. Microbiol. *55*, 162–174.

Triglia, T., Healer, J., Caruana, S.R., Hodder, A.N., Anders, R.F., Crabb, B.S., and Cowman, A.F. (2000). Apical membrane antigen 1 plays a central role in erythrocyte invasion by Plasmodium species. Mol. Microbiol. *38*, 706–718.

Triglia, T., Thompson, J., Caruana, S.R., Delorenzi, M., Speed, T., and Cowman, A.F. (2001). Identification of proteins from *Plasmodium falciparum* that are homologous to reticulocyte binding proteins in *Plasmodium vivax*. Infect. Immun. *69*, 1084–1092.

Ueda, E., Kinoshita, T., Terasawa, T., Shichishima, T., Yawata, Y., Inoue, K., and Kitani, T. (1990). Acetylcholinesterase and lymphocyte function-associated antigen 3 found on decay-accelerating factor-negative erythrocytes from some patients with paroxysmal nocturnal hemoglobinuria are lost during erythrocyte aging. Blood *75*, 762–769.

Uneke, C.J. (2007). *Plasmodium falciparum* malaria and ABO blood group: is there any relationship? Parasitol. Res. *100*, 759–765.

Unger, P., Procter, J.L., Moulds, J.J., Moulds, M., Blanchard, D., Guizzo, M.L., McCall, L.A., Cartron, J.P., and Dahr, W. (1987). The Dantu erythrocyte phenotype of the NE variety. II. Serology, immunochemistry, genetics, and frequency. Blut *55*, 33–43.

Uthaipibull, C., Aufiero, B., Syed, S.E., Hansen, B., Guevara Patiño, J.A., Angov, E., Ling, I.T., Fegeding, K., Morgan, W.D., Ockenhouse, C., *et al.* (2001). Inhibitory and blocking monoclonal antibody epitopes on merozoite surface protein 1 of the malaria parasite *Plasmodium falciparum*. J. Mol. Biol. *307*, 1381–1394.

Varki, A. (2002). Loss of N-glycolylneuraminic acid in humans: mechanisms, consequences, and implications for hominid evolution. Am. J. Phys. Anthropol. *116*, 54–69.

Varki, A. (2007). Glycan-based interactions involving vertebrate sialic-acid-recognizing proteins. Nature *446*, 1023–1029.

Varki, A., and Gagneux, P. (2009). Human-specific evolution of sialic acid targets: explaining the malignant malaria mystery? Proc. Natl. Acad. Sci. U.S.A. *106*, 14739–14740.

Volkman, S.K., Sabeti, P.C., DeCaprio, D., Neafsey, D.E., Schaffner, S.F., Milner, D.A., Jr., Daily, J.P., Sarr, O., Ndiaye, D., Ndir, O., *et al.* (2007). A genome-wide map of diversity in *Plasmodium falciparum*. Nat. Genet. *39*, 113–119.

Wagner-Jauregg, J. (1922). The treatment of general paresis by inoculation of malaria.. J. Nerv. Ment. Dis. *55*, 369–375.

Walliker, D., Sanderson, A., Yoeli, M., and Hargreaves, B.J. (1976). A genetic investigation of virulence in a rodent malaria parasite. Parasitology *72*, 183–194.

Wang, C.C. (1970). Multiple invasion of erythrocyte by malaria parasites. Trans. R. Soc. Trop. Med. Hyg. *64*, 268–270.

Ware, L.A., Kain, K.C., Lee Sim, B.K., Haynes, J.D., Baird, J.K., and Lanar, D.E. (1993). Two alleles of the 175-kilodalton *Plasmodium falciparum* erythrocyte binding antigen. Mol. Biochem. Parasitol. *60*, 105–109.

Waugh, R.E., Narla, M., Jackson, C.W., Mueller, T.J., Suzuki, T., and Dale, G.L. (1992). Rheologic properties of senescent erythrocytes: loss of surface area and volume with red blood cell age. Blood *79*, 1351–1358.

Weinberger, D.M., Malley, R., and Lipsitch, M. (2011). Serotype replacement in disease after pneumococcal vaccination. Lancet. *378*, 1962–1973.

Welch, S.G., McGregor, I.A., and Williams, K. (1977). The Duffy blood group and malaria prevalence in Gambian West Africans. Trans. R. Soc. Trop. Med. Hyg. *71*, 295–296.

Wertheimer, S.P., and Barnwell, J.W. (1989). *Plasmodium vivax* interaction with the human Duffy blood group glycoprotein: identification of a parasite receptor-like protein. Exp. Parasitol. *69*, 340–350.

Williams, T.N. (2006). Human red blood cell polymorphisms and malaria. Curr. Opin. Microbiol. *9*, 388–394.

Wilson, D.W., Crabb, B.S., and Beeson, J.G. (2010). Development of fluorescent *Plasmodium falciparum* for in vitro growth inhibition assays. Malar. J. *9*, 152.

Wipasa, J., Elliott, S., Xu, H., and Good, M.F. (2002). Immunity to asexual blood stage malaria and vaccine approaches. Immunol. Cell Biol. *80*, 401–414.

Woolley, I.J., Hotmire, K.A., Sramkoski, R.M., Zimmerman, P.A., and Kazura, J.W. (2000). Differential expression of the Duffy antigen receptor for chemokines according to RBC age and FY genotype. Transfusion *40*, 949–953.

Yazdani, S.S., Mukherjee, P., Chauhan, V.S., and Chitnis, C.E. (2006). Immune responses to asexual blood-stages of malaria parasites. Curr. Mol. Med. *6*, 187–203.

Young, M.D., Eyles, D.E., Burgess, R.W., and Jeffery, G.M. (1955). Experimental testing of the immunity of Negroes to *Plasmodium vivax*. J. Parasitol. *41*, 315–318.

Host Cell Remodelling and Protein Trafficking

9

Silvia Haase, Hayley E. Bullen, Sarah C. Charnaud, Brendan S. Crabb, Paul R. Gilson and Tania F. de Koning-Ward

Abstract

Inside their respective vertebrate hosts, *Plasmodium* spp. spend most of their life residing within hepatocytes and erythrocytes, with large-scale infection of the latter responsible for the clinical symptoms associated with malaria. These parasites extensively remodel these host cells for a variety of purposes relating to both pathogenesis and maintaining growth. Remodelling of the erythrocytic stage has been most intensively studied in *P. falciparum* and is the subject of this chapter. To help remodel their hosts these parasites export hundreds of proteins into the erythrocytic compartment. This principally alters the architecture of the erythrocyte, rendering the host membrane more permeable to solutes and nutrients, and also increases the rigidity and adhesiveness of the infected erythrocyte. Moreover, because erythrocytes lack a secretory apparatus, the parasite must also export many additional proteins to help traffic other proteins to their correct destination within the host cell. The functions of some of these exported proteins will be discussed as will recent progress that has been made in unravelling how exported proteins gain access to the host compartment.

Plasmodium falciparum invasion into erythrocytes

The Apicomplexan parasite, *Plasmodium falciparum,* is the main causative agent of malaria in humans and results in approximately 800,000 deaths and a further 500 million infections annually (WHO, 2010). To cause such devastating outcomes, *P. falciparum* must infect human erythrocytes, which in their normal uninfected state are highly elastic (An and Mohandas, 2010; Safeukui *et al.*, 2008). The main component of the red blood cell (RBC) membrane skeleton responsible for this elasticity is spectrin, a protein consisting of both α and β chains, which interact to form hetero-dimers and in turn self-assemble to form hetero-tetramers. These hetero-tetramers subsequently form a latticework, held together at its nodes by actin filaments, and bound to the membrane via the integral membrane protein band 3 as well as glycophorins C and D (Dhermy *et al.*, 2007). During infection of these highly elastic cells, the invasive merozoite secretes multiple proteins from its apical organelles (rhoptries, micronemes and dense granules) onto the parasite's anterior surface (Harris *et al.*, 2005), resulting in the formation of a tight junction between parasite and host. This junction is associated with the merozoite-derived actin–myosin motor, causing it to be translocated from the parasite's anterior to posterior as invasion proceeds (Baum *et al.*, 2006).

During entry of the parasite into the erythrocyte, a specialized parasitophorous vacuole (PV) forms, protecting the parasite from the host immune system and thus enabling it to undergo its intracellular growth and development phase unabated. The membrane surrounding this vacuole, termed the parasitophorous vacuole membrane (PVM), is believed to be formed from a patch of erythrocyte membrane and is relatively free of host-cell proteins including the cytoskeletal protein spectrin (Dluzewski *et al.*, 1989, 1992), although several lipid-raft associated host membrane proteins are found in the PVM (Murphy *et al.*, 2004). The PVM also contains a number of parasite proteins embedded in a blend of both erythrocyte and parasite phospholipids (Johnson *et al.*, 1994).

Parasite-associated induction of new permeability pathways for transport of small molecules

To increase the membrane conductance of the infected erythrocyte whilst also permitting nutrient acquisition and waste disposal, the intracellular parasite induces the formation of new permeability pathways (NPPs) (Decherf *et al.*, 2004; Staines *et al.*, 2000, 2005, 2006). These pathways appear ~12–15 h after parasite invasion when the parasite is at ring-stages and act to increase the permeability of the erythrocyte plasma membrane to small molecules such as sugars, amino acids, nucleosides, anions and cations (Deitsch and Wellems, 1996) enabling the parasite to grow. Additionally, they act as pathways through which toxic parasite metabolites such as lactic acid can be removed from the cell (Merckx *et al.*, 2009), promoting longevity of the parasite. There are two main types of NPPs, the first of which only displays a low copy number per cell and is both charge and size selective, discriminating against cations. The second main type of NPP is far more abundant (~100 fold) and permits the passage of anions and nucleosides between the infected erythrocyte and the external medium (Bouyer *et al.*, 2006). Permeation through the parasite-induced NPPs appears to be dependent upon both hydrophobicity and size however the rate of permeation of solutes through these pathways is not known (Staines *et al.*, 2004). It is also unknown whether these channels are derived from the erythrocyte or are installed by the parasite for its specific intracellular requirements (Bouyer *et al.*, 2006; Staines *et al.*, 2005). Irrespective of this, NPPs are of great interest to the scientific community as they provide parasite-associated permeation pathways, which may form targets for novel antimalarial compounds. Initial work based on finding NPP-inhibitor compounds has revealed that furosemide based compounds are most potent at inhibiting these pathways (Staines *et al.*, 2005). Additionally, work is also under way to investigate whether these parasite-infected erythrocyte NPPs can act as routes for delivery of antimalarial compounds directly into the host erythrocyte (Staines *et al.*, 2005).

Parasite-induced structural modifications to the erythrocyte: Knobs and Maurer's clefts

During the first 20h after invasion the parasite also initiates a number of morphological alterations to its host cell. Membrane-bound extensions and whorls of the PVM known as the tubulovesicular network (TVN) (Elmendorf and Haldar, 1994) appear, as do heterogeneous membranous structures in the erythrocyte cytoplasm known as Maurer's clefts, which have an electron-dense coat and electron-lucent lumen (Fig. 9.1) (reviewed in Przyborski, 2008). Small electron-dense knob-like structures protrude from the RBC surface (Kilejian, 1979) and these act as platforms for presenting the major virulence determinant and cytoadherence factor, *P. falciparum* erythrocyte membrane protein 1 (PfEMP1, Fig. 9.1) (Baruch *et al.*, 1995; Su *et al.*, 1995). All of these parasite-induced structures as well as the erythrocyte cytoskeleton and surface are composed of proteins of parasite origin, for example the knobs are comprised of knob-associated histidine-rich protein (KAHRP) (Culvenor *et al.*, 1987; Pologe *et al.*, 1987), and thus the parasite has evolved an elaborate trafficking system to export its proteins to these various destinations within the host cell. Both the TVN and Maurer's cleft structures are implicated in the sorting and trafficking of parasite proteins, particularly those destined for the erythrocyte skeleton and surface [see Lanzer *et al.* (2006) and Sam-Yellowe (2009) for review]. However, the recent identification of two novel TVN specific markers, the TVN junction protein 1 (TVN-JP1) and the erythrocyte vacuolar protein 1 (EVP-1), the latter being required for maintenance of the TVN, suggests that the TVN is distinct and independent from the Maurer's clefts and therefore constitutes a separate membrane network for the transport of macromolecules (Fig. 9.1) (Tamez *et al.*, 2008).

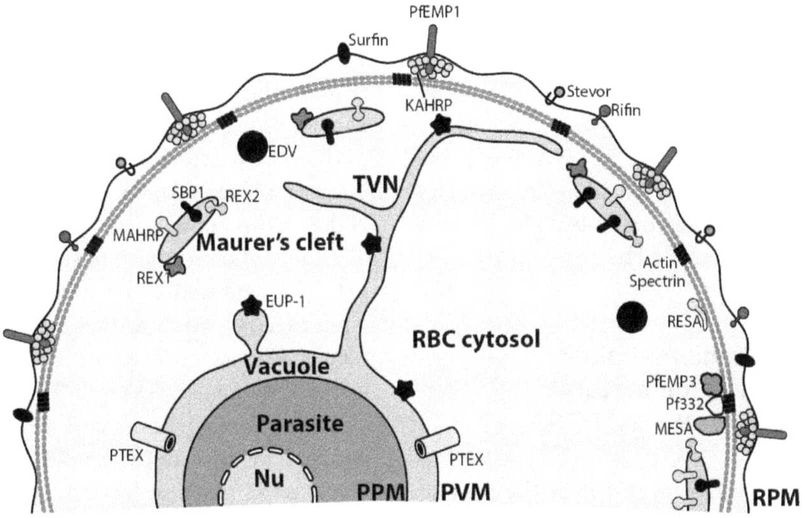

Figure 9.1 Schematic of a *Plasmodium falciparum* infected red blood cell showing the final locations of functionally characterized exported proteins. Abbreviations: EDV; electron dense vesicle; MC, Maurer's Cleft; Nu, nucleus of parasite; PPM, parasite plasma membrane; PV, parasitophorous vacuole; PVM, parasitophorous vacuole membrane; RPM, Red blood cell plasma membrane; and TVN, tubulovesicular network.

Maurer's clefts are highly polymorphic and convoluted disk-like compartments with slender membrane extensions connecting neighbouring clefts and stalk-like 'tethers' attaching the Maurer's clefts to the erythrocyte membrane and PVM. However, no continuum exists to allow molecules to freely flow between these three compartments (Hanssen *et al.*, 2008b; Spycher *et al.*, 2006; Tilley *et al.*, 2008). Little is known about Maurer's cleft biogenesis, although a number of parasite proteins localize to these compartments, including the resident proteins membrane associated histidine-rich protein 1 (MAHRP1) (Spycher *et al.*, 2003), skeleton binding protein 1 (SBP1) (Blisnick *et al.*, 2000), the ring exported proteins REX1 and REX2 (Hawthorne *et al.*, 2004; Spielmann *et al.*, 2006) and the PfMC-2TM proteins (Fig. 9.1) (Sam-Yellowe *et al.*, 2004; Tsarukyanova *et al.*, 2009). Whilst the function of REX2 and the PfMC-2TM proteins remain to be elucidated, SBP1, MAHRP1 and REX1 have been shown to be involved in the formation and architecture of the Maurer's clefts and/or PfEMP1 trafficking to the erythrocyte surface (Cooke *et al.*, 2006; Hanssen *et al.*, 2008a,b; Maier *et al.*, 2007; Spycher *et al.*, 2008). Other proteins transiently associate with the Maurer's clefts, including the soluble proteins KAHRP (Waterkeyn *et al.*, 2000; Wickham *et al.*, 2001) and *P. falciparum* erythrocyte membrane protein 3 (PfEMP3) (Pasloske *et al.*, 1993) that interact with the erythrocyte skeleton (Pei *et al.*, 2007b; Waller *et al.*, 2007; Wickham *et al.*, 2001). There is also a wide range of integral membrane proteins that associate with the Maurer's clefts. These include *P. falciparum* antigen 332 (Pf332 antigen) (Hinterberg *et al.*, 1994), which associates with the erythrocyte membrane skeleton and binds to receptors for adherence and invasion (Hodder *et al.*, 2009; Moll *et al.*, 2007), the variant antigen PfEMP1 that localizes to the erythrocyte membrane and some members of the STEVOR (Kaviratne *et al.*, 2002; Przyborski *et al.*, 2005) and RIFIN family of variant proteins (Khattab and Klinkert, 2006).

All of these parasite-induced modifications serve this deadly pathogen generally in one of three main ways: (1) they alter the erythrocyte architecture to provide a robust cell inside of which the parasite can safely grow and replicate (Maier *et al.*, 2008; Nash *et al.*, 1989; Parker *et al.*, 2004); (2) they allow the expression of parasite-derived adhesion proteins to facilitate adherence to the vascular endothelium, thus helping avoid phagocytic clearance by the spleen and hence securing the survival of the parasite (Luse and Miller, 1971; Miller *et al.*, 2002); and (3) they increase the membrane conductance of the erythrocyte membrane to facilitate nutrient and solute acquisition (Decherf *et al.*, 2004; Deitsch and Wellems, 1996; Staines *et al.*, 2006). A summary of some of the exported proteins that play a role in these processes are outlined below.

Exported proteins that alter the erythrocyte architecture: KAHRP, PfEMP3, MESA, RESA and Pf332

A variety of proteins exported from the parasite have been shown to interact and modify the architecture of the erythrocyte, including 11 novel proteins that affect the formation of knobs and the rigidity of infected host cells (Maier *et al.*, 2008). However, the best characterized proteins are KAHRP (Culvenor *et al.*, 1987; Pologe *et al.*, 1987), PfEMP3 (Pasloske *et al.*, 1993), the mature parasite-infected erythrocyte surface antigen (MESA, also known as PfEMP2) (Coppel *et al.*, 1988; Howard *et al.*, 1988), the ring-infected erythrocyte surface antigen (RESA) (Cowman *et al.*, 1984) and Pf332 (Fig. 9.1) (Hinterberg *et al.*, 1994; Mattei and Scherf, 1992b).

KAHRP is an 85–110kDa protein (Culvenor *et al.*, 1987; Pologe *et al.*, 1987) associated with parasite-induced membrane rigidity (Glenister *et al.*, 2002) and is essential for the formation of knobs (Crabb *et al.*, 1997). The N- and C-terminal regions of KAHRP contain binding sites to interact with the cytoplasmic ATS (acidic terminal segment) domain of PfEMP1 (Waller *et al.*, 1999; Waller *et al.*, 2002), whilst the His-rich region is also believed to interact with the Maurer's clefts (Lopez-Estrano *et al.*, 2003; Wickham *et al.*, 2001). KAHRP has been shown to self-associate in order to bind to spectrin protein complexes (Oh *et al.*, 2000) and latter studies mapped a specific region of KAHRP (residues 370–441) binding to a single repeat of α spectrin (Pei *et al.*, 2005; Rug *et al.*, 2006). Cytoadherence under flow conditions is dramatically reduced in KAHRP-deficient parasites (Crabb *et al.*, 1997; Rug *et al.*, 2006) and mutant versions of KAHRP reveal changes in the elastic properties of the infected erythrocyte membrane (Rug *et al.*, 2006). Whilst the formation of functional knobs requires the C-terminal repeat region, trafficking and translocation of PfEMP1 onto the surface of the erythrocyte was not found to be impaired in the absence of knobs or KAHRP (Rug *et al.*, 2006). However, an earlier study by Horrocks *et al.* (2005) showed that whilst the distribution of PfEMP1 appeared to be unaltered in knobless infected red blood cells, there was an approximate 50% reduction of PfEMP1 surface expression (Horrocks *et al.*, 2005).

PfEMP3 is a parasite-derived protein deposited directly beneath the erythrocyte membrane. The *pfemp3* gene is located on the telomeric side of *kahrp* on chromosome 2 and encodes a large, highly charged protein of about 315kDa in size (Pasloske *et al.*, 1993, 1994). PfEMP3 is transiently associated with the Maurer's clefts before it relocates to the cytoplasmic surface of the erythrocyte membrane. Its precise function is unknown, but it is not essential for knob formation or PfEMP1 surface expression; instead PfEMP3 is believed to play a role in transferring PfEMP1 from the Maurer's clefts to the erythrocyte surface (Waterkeyn *et al.*, 2000). PfEMP3 also appears to contribute to the reduced deformability of infected RBCs via a specific 60 amino acid region that interacts with the C-terminus of alpha spectrin near the actin-protein 4.1R junction, leading to a disruption of the actin–spectrin-4.1R interaction (Pei *et al.*, 2007b).

Another *P. falciparum* protein that associates with the host erythrocyte membrane skeletal protein 4.1 is MESA (Coppel *et al.*, 1988; Lustigman *et al.*, 1990). MESA is a 250–300kDa phosphoprotein expressed in trophozoites and schizonts and contains an unusually high amount of charged residues and repeat regions (Coppel, 1992; Coppel *et al.*, 1988; Howard *et al.*, 1988). MESA is exported from the parasite to the erythrocyte cytoskeleton and interacts via a 19-residue amphipathic helix near the N-terminus with a 51-residue region of the erythrocyte cytoskeletal protein 4.1 (Bennett *et al.*, 1997; Kun *et al.*, 1999; Waller *et al.*, 2003). The 19-residue 4.1 binding site was also found to be essential for the peripheral trafficking of MESA to the erythrocyte skeleton (Black *et al.*, 2008). Whilst its precise function remains unknown, its binding to protein 4.1 appears to be crucial for parasite survival (Magowan *et al.*, 1995).

RESA (Cowman *et al.*, 1984), which is also known as Pf155 (Perlmann *et al.*, 1984), is synthesized in mature schizonts where it is stored in the dense granules (Culvenor *et al.*, 1991). After merozoite invasion, RESA is released from these organelles into the nascent PV before being exported into the host cell cytosol (Culvenor *et al.*, 1991). Upon its association with the host cell membrane it becomes phosphorylated (Foley *et al.*, 1990). The 155-kDa protein contains a 5′ and 3′ repetitive sequence rich in acidic residues (Favaloro *et al.*, 1986) and a KLQKTY peptide with similarity to the cytoadherence-related motif in the erythrocyte band 3 protein (Iqbal *et al.*, 1995). It also contains a segment of 70 residues with similarity to J domains of *E. coli* and the human DnaJ domain, which suggests that RESA might have some chaperone-like properties (Da Silva *et al.*, 1994). Early studies identified spectrin as the primary attachment site for RESA (Foley *et al.*, 1991), which has been shown to modify the thermal stability of the erythrocyte (Da Silva *et al.*, 1994). Indeed, latter studies identified a 108 amino acid fragment of RESA that binds to a specific repeat unit in spectrin (the repeat 16 of beta-spectrin), thus stabilizing the spectrin tetramer against its dissociation into its dimers and enhancing the resistance of the erythrocyte membrane to mechanical and thermal degradation (Pei *et al.*, 2007a). Absence of RESA results in an increased susceptibility to heat shock, but does not influence merozoite invasion or the physical properties of the erythrocyte membrane (Silva *et al.*, 2005).

Pf332 is the largest exported protein of approximately 1MDa, localizing to the PVM and Maurer's clefts during the early stages of parasite development and later associating with the infected erythrocyte membrane skeleton (Hinterberg *et al.*, 1994; Mattei and Scherf, 1992a, b). It is characterized by a Duffy binding like (DBL) domain at the N-terminus, followed by a putative transmembrane domain and a very large cytoplasmic domain with a large number of degenerated Glu-rich repeats, a Val-rich region and an EB200 region comprising numerous XXXEEXXEEXX motifs (X = hydrophobic amino acid) (Moll *et al.*, 2007). Deletion or truncation of the *Pf332* gene has no influence on merozoite invasion, but leads to an altered rigidity of the infected erythrocyte, suggesting the involvement of Pf332 (and the DBL domain in particular) in modulating the overall deformability of the host cell (Glenister *et al.*, 2009; Hodder *et al.*, 2009). Interestingly, Pf332-deficient 3D7 parasites are less adhesive to CD36 and show decreased expression of PfEMP1 on the infected erythrocyte surface. These findings are believed to correlate with dramatic changes in the morphology of Maurer's clefts, suggesting that the highly negatively charged cytoplasmic domain may influence the formation of Maurer's clefts lamellae (Glenister *et al.*, 2009). However, CS2 parasites lacking Pf332 show no alterations in PfEMP1 levels (Hodder *et al.*, 2009), which may be a reflection in the differences and redundancy within the different parasite strains.

Expression of exported variant molecules: PfEMP1, RIFINs, STEVORs, PfMC-2TMs and SURFINs.

Antigenic variation is a key immune evasion strategy used by the parasite to escape immune destruction. The major virulence protein involved in this is PfEMP1, which is expressed on the surface of infected erythrocytes (Baruch *et al.*, 1995; Su *et al.*, 1995). There are approximately 60 different PfEMP1 molecules, ranging in size from 200–360kDa, which are encoded by members of the multicopy *var* gene family. Each parasite expresses only a single *var* gene at a time, maintaining all other members of the family in a transcriptionally silent state (reviewed in Scherf *et al.*, 2008). As an antibody response mounts against a single expressed PfEMP1, the few parasites that express other *var* genes are able to evade the immune system and begin to proliferate. Each PfEMP1 shares certain structural features, including an N-terminal segment, 2–7 Duffy binding-like (DBL) domains, either one or two cysteine-rich interdomain regions (CIDR), a transmembrane domain, a C2 domain and the conserved cytoplasmic ATS (Kraemer and Smith, 2006). The transmembrane domain is thought to be integrated into the erythrocyte membrane, thereby presenting the ectodomain with its adhesive DBL domains and cysteine-rich interdomain regions at the external surface. The extracellular domain is extremely variable in both sequence and length (Smith *et al.*, 2000), resulting in distinct adhesion properties towards a number of different host cell receptors, such as CD36, ICAM1, VCAM1 and CSA (Kraemer and Smith, 2006). These different binding phenotypes facilitate sequestration of *P. falciparum* infected erythrocytes within the microvasculature, enabling the parasite to escape clearance via the spleen and hence securing its survival. Sequestration in various tissues also leads to the different pathological outcomes of malaria, including cerebral or placental malaria. In contrast to the extracellular region, the PfEMP1 cytoplasmic tail is relatively conserved and interacts with KAHRP (Fig. 9.1) (Waller *et al.*, 1999, 2002). PfEMP1 is also expressed in early stage gametocytes (Sharp *et al.*, 2006) implying a similar cytoadherence function.

Apart from the *var* genes, several other multigene families encoding for proteins with putative antigenic properties have been described in *P. falciparum*, including the repetitive interspersed family (*rif*), the subtelomeric variable open reading frame (*stevor*) and *Pfmc-2TM* genes, which are frequently found in subtelomeric chromosome positions (reviewed in Jemmely *et al.*, 2010). These genes encode for proteins which share a conserved architecture, consisting of a highly conserved N-terminus (within the respective families), a hypervariable region between two conserved transmembrane domains (with each family differing in the number of amino acids between these domains) and a highly conserved charged C-terminal domain.

The *rif* multigene family encodes for membrane proteins of 27–45 kDa with approximately 150 members, thus being the largest multigene family in *P. falciparum* (Cheng *et al.*, 1998; Kyes *et al.*, 1999). RIFINs were first described as rosetting proteins and trypsin treatment of ^{125}I-labelled infected erythrocytes revealed they are exposed on the surface of the infected erythrocyte (Kyes *et al.*, 1999). Two types of RIFINs have subsequently been identified and are proposed to fulfil two different functions: A-type RIFINs have a 25 amino acid insertion, are exported and are trafficked via the Maurer's clefts towards the surface of infected erythrocytes, whereas B-type RIFINs are retained inside the parasite (Joannin *et al.*, 2008; Petter *et al.*, 2007). Differential expression patterns of A- and B-type RIFINs are also found in gametocytes, with A-type RIFIN variants accumulating in punctuate structures at the RBC membrane and B-type RIFINs associating with the parasite (Petter *et al.*, 2008). Interestingly, a latter study has proposed different structural organizations for the members of the two subfamilies (Bultrini *et al.*, 2009). So far, the biological role of the RIFINs remains to be unravelled, although their number and hypervariability support a role in antigenic variation (Fernandez *et al.*, 1999; Kyes *et al.*, 1999).

The *stevor* multigene family is the third largest in *Plasmodium falciparum*, with 30–40 genes encoding for proteins around 30–40kDa in size (Cheng *et al.*, 1998). STEVORS have a similar structure to RIFINs, but are expressed much later (22–32 h post invasion) than PfEMP1 and RIFIN (Kaviratne *et al.*, 2002), suggesting a distinct role in the later stages of the parasite development. Furthermore, the expression of STEVORs in the different stages of *P. falciparum* indicates a possible role in parasite survival (Blythe *et al.*, 2004). Transport of STEVOR occurs via the Maurer's clefts (Kaviratne *et al.*, 2002) before associating with the erythrocyte membrane in late stage schizonts (Lavazec *et al.*, 2006). It has been subsequently shown that expression of different STEVOR variants at the erythrocyte surface (as demonstrated by trypsin treatment or FACS analysis of infected erythrocytes using antibodies recognizing STEVORs) leads to changes in the immunogenic properties of the infected cell (Niang *et al.*, 2009). Furthermore, given the STEVORs two transmembrane domain topology and hence the location of the hypervariable loop, together with observations that the expression of *stevors* is not only clonally variant but that they also undergo switching at rates that vary amongst different variants and isogenic clones, provides evidence that the *stevor* multigene family is indeed involved in antigenic variation (Lavazec *et al.*, 2006, 2007). Similar observations have also been found for the *Pfmc-2TM* multigene family that comprises 13 members (Lavazec *et al.*, 2006,

2007; Sam-Yellowe *et al.*, 2004). These proteins are around 20–25 kDa and are expressed from late rings to early trophozoites. They localize to the PVM and the Maurer's clefts (Tsarukyanova *et al.*, 2009) before associating with the erythrocyte membrane (Lavazec *et al.*, 2006).

The smallest multigene family comprises the surface-associated interspersed genes (*surf* genes), which encode a novel polymorphic protein family of 10 members with a variable N-terminus and a semi-conserved C-terminus. SURFINs appear to be structurally related to other surface proteins of human, simian and rodent malaria parasites. These proteins comprise a transmembrane domain followed by four conserved tryptophan rich domains in the C-terminus. The *surf* variant $SURFIN_{4.2}$ is a 286-kDa protein that could be cleaved from the surface of infected erythrocytes by trypsin treatment and along with PfEMP1 and RIFIN, is trafficked via the Maurer's clefts (Fig. 9.1) (Winter *et al.*, 2005). It has also been found in the PV and at the apical tip of free merozoites, suggesting a possible role in invasion (Winter *et al.*, 2005). In contrast, $SURFIN_{4.1}$ (258 kDa) localizes to the PV and around free merozoites as 'merozoite-associated material', but not at the erythrocyte surface (Mphande *et al.*, 2008). Whilst the *surf* multigene family is differentially transcribed and expressed amongst different parasite stages, as well as in lab and clinical isolates, it remains to be established whether differential expression of the *surf* genes could suggest different functions (Mphande *et al.*, 2008; Winter *et al.*, 2005).

Exported proteins that localize to the Maurer's clefts: SBP1, MAHRP1, REX1 and REX2

Several exported proteins reside at the Maurer's clefts, either as integral membrane proteins or peripherally associated with these structures. SBP1 is a 48 kDa integral membrane Maurer's cleft resident protein, with its C-terminal domain exposed to the host cell cytosol and a suggested role in linking the Maurer's clefts to the erythrocyte membrane skeleton (Fig. 9.1) (Blisnick *et al.*, 2000). A later study revealed the host Lantibiotic synthetase component C-like protein (LANCL1) being recruited to the surface of Maurer's clefts and interacting with SBP1 in mature parasites, indicating a role in stabilizing the binding of clefts to the erythrocyte membrane (Blisnick *et al.*, 2005). Recently, SBP1 has been found to be associated with tether-like structures attaching the Maurer's clefts to the erythrocyte membrane and PVM (Hanssen *et al.*, 2008b). SBP1 knock-out studies have demonstrated that whilst this protein does play an important role in PfEMP1 trafficking it is not necessary for general Maurer's clefts formation nor for the trafficking of other Maurer's clefts resident proteins such as MAHRP1 and REX1 (Cooke *et al.*, 2006; Maier *et al.*, 2007). Despite this, SBP1 knock-out lines show altered Maurer's cleft morphology, with clefts appearing further from the erythrocyte plasma membrane and this may be the cause of the poor PfEMP1 surface expression seen in SBP1-deficient 3D7 parasites (Cooke *et al.*, 2006).

Like SBP1, MAHRP1 is an integral membrane Maurer's clefts resident protein that has its His-rich region facing towards the erythrocyte cytosol (Fig. 9.1) (Spycher *et al.*, 2003). This protein is believed to play a role in the organization of the clefts (Spycher *et al.*, 2008). In MAHRP1-deficient parasites, PfEMP1 accumulates within the PVM, indicating that this protein is involved in loading PfEMP1 from the PVM into nascent Maurer's clefts (Spycher *et al.*, 2008). MAHRP1 is expressed during ring stages but is present throughout the intraerythrocytic life cycle. It binds the toxic haemoglobin degradation product ferriprotoporphyrin and may, therefore, also play a protective role against oxidative stress (Spycher *et al.*, 2003).

The ring exported proteins REX1 and REX2 are peripheral membrane and integral membrane Maurer's clefts proteins, respectively (Fig. 9.1) (Hawthorne *et al.*, 2004; Spielmann *et al.*, 2006). They are encoded by a gene cluster (*rex1–rex4*) located within the subtelomeric region of the right arm on chromosome 9, a region which has been linked to cytoadherence (Spielmann and Beck, 2000). In contrast to REX1 and REX2, GFP-chimeras of REX3 and REX4 do not localize to the Maurer's clefts, and instead are exported into the erythrocyte cytoplasm, although no endogenous REX4 could be detected with REX4 specific antisera (Spielmann *et al.*, 2006). REX1 is the largest of the REX proteins and is characterized by a recessed N-terminal hydrophobic stretch, a predicted coiled-coil region and C-terminal repeats (Hawthorne *et al.*, 2004). REX1 associates with the cytoplasmic surface of the Maurer's clefts (Hawthorne *et al.*, 2004) where it accumulates at the edges of these structures, playing a crucial role in their overall morphology (Hanssen *et al.*, 2008a). The biological function of the REX2 (and the soluble proteins REX3 and REX4) still remains to be investigated.

Other exported proteins

Other parasite proteins that have been reported to be exported include the glycophorin binding protein GBP130, the histidine-rich protein II (HRPII) and the membrane associated histidine-rich protein 2 (MAHRP2). GBP130 is expressed in trophozoite and schizont stages, being released during schizont rupture (Bonnefoy *et al.*, 1988) and contains a charged 225 residue N-terminal region and 11 highly conserved repeats (Perkins, 1984). Deletion of this gene increases erythrocyte rigidity (Maier *et al.*, 2008).

HRPII is a soluble protein found in the digestive vacuole, but also in the erythrocyte cytosol where it binds to actin (Benedetti *et al.*, 2003; Howard *et al.*, 1986; Papalexis *et al.*, 2001). HRPII is synthesized during ring and trophozoite stages (Howard *et al.*, 1986) and has been implicated in the detoxification and the modulation of the redox activity of ferriprotoporphyrin IX (FP), a by-product of the digestion of haemoglobin (Mashima *et al.*, 2002; Papalexis *et al.*, 2001). These findings, and the location of HRPII near the erythrocyte membrane, suggest a function in protecting the erythrocyte membrane from oxidative stress to prevent premature host cell lysis.

The *mahrp2* gene was initially identified in a transcription analysis of early-transcribed genes (Spielmann and Beck, 2000). MAHRP2 is a 15.8-kDa protein and is maximally expressed in trophozoite stages. It localizes to the Maurer's clefts tethers and solubility assays indicates MAHRP2 to be peripherally associated with the membranes of these structures (Hanssen *et al.*, 2008b; Pachlatko *et al.*, 2010). The protein is characterized by a histidine-rich N-terminal domain and a central region with hydrophobic amino acids and efforts in disrupting the *mahrp2* gene have been unsuccessful, suggesting that the protein plays an important role in parasite development (Pachlatko *et al.*, 2010).

Exported proteins are trafficked through a novel secretory pathway

Since the host erythrocyte is devoid of any trafficking machinery, the malaria parasite has had to evolve its own unique system to export proteins beyond the confines of its own membranes to the various sites within the erythrocyte. Many exported parasite proteins such as PfEXP1 (Adisa *et al.*, 2003), PfHRPII, PfEMP2 (Chang *et al.*, 2008) and PfEMP1 (Knuepfer *et al.*, 2005b; Wickham *et al.*, 2001) are sensitive to brefeldin A (BFA), an inhibitor of transport from the ER to the Golgi and inducer of retrograde protein transport from the Golgi back to the ER. Peptides of PfHRPII and PfEMP2 that accumulate in the ER after brefeldin A treatment are proteolytically cleaved and acetylated at the N-terminus (Chang *et al.*, 2008) suggesting exported parasite proteins are initially secreted via the classic ER/Golgi secretory pathway, with co-translational processing on import into the ER (Fig. 9.2). By fusing signal peptides of various *Plasmodium* proteins to GFP, it could be demonstrated that the default trafficking pathway of proteins entering this secretory pathway is the PV (Waller *et al.*, 2000) and that a sequence additional to the signal peptide is required to target exported proteins to their ultimate destination (Fig. 9.2). Similarly, targeting to other locations requires additional sequences. For example, targeting to the apicoplast requires a bipartite N terminal sequence, comprising a signal peptide and N-terminal transit peptide (Waller *et al.*, 1998).

Export signal sequence discovery

In 2004, two groups independently identified a core pentameric motif that, together with a signal sequence, is necessary for entry into the ER and subsequent export into the erythrocyte cytosol

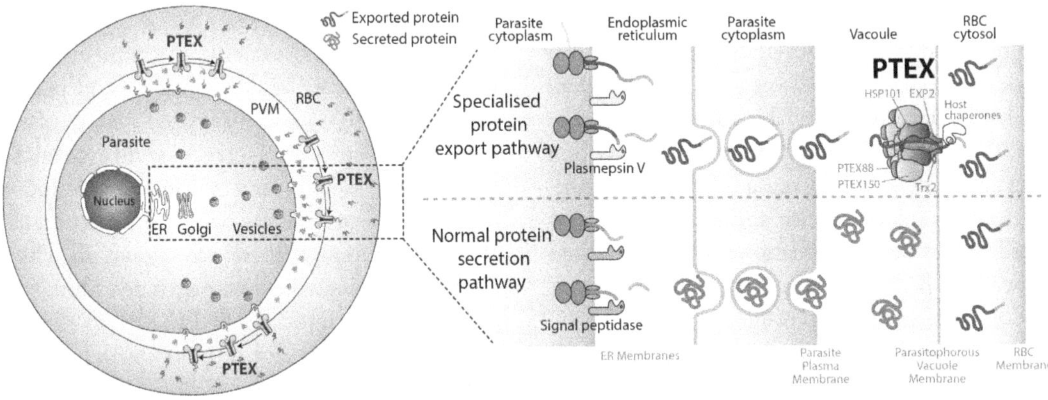

Figure 9.2 Illustration of the synthesis and export of *Plasmodium* proteins into their red blood cell hosts. Secreted and exported PEXEL proteins enter the endoplasmic reticulum by co- or post-translational insertion through the Sec61/SecY translocon. The secretion signal of secreted proteins is cleaved by signal peptidase and unless retained within the host these proteins traffic to the parasitophorous vacuole (PV) but no further. In contrast PEXEL proteins are cleaved within the PEXEL motif by Plasmepsin V, whereupon they traffic to the PV and are recognized by the PTEX complex that putatively translocates them into the host cytosol. Just how PTEX discriminates normal secreted proteins from PEXEL proteins that have had most of the PEXEL motif removed is not fully understood. The fact that proteolytic processing by Plasmepsin V as opposed to signal peptidase is required, even if identical termini are exposed, suggests Plasmepsin V may also help shunt exported proteins into a specialized secretory pathway that may deliver the proteins to PTEX.

(Hiller *et al.*, 2004; Marti *et al.*, 2004). This motif, termed *Plasmodium* EXport ELement (PEXEL) or vacuolar transport signal (VTS), consists of RxLxE/Q/D, where X is any uncharged amino acid. Hiller *et al.* (2004) identified an 11 amino acid primary pattern motif containing this pentameric motif. This motif is located within 60 amino acids downstream of the ER signal sequence, and N terminal fusions of the soluble exported proteins KAHRP, GBP130 and HRPII that contain these sequences to GFP facilitate export of GFP into the erythrocyte cytosol (Hiller *et al.*, 2004; Marti *et al.*, 2004). Mutations of any of the conserved residues in the PEXEL motif to alanine block the export of GFP, often resulting in accumulation in the PV (Hiller *et al.*, 2004; Marti *et al.*, 2004). Interestingly, a similar motif is found in secretory proteins of the potato blight pathogen *Phytophthora infestans*, and is necessary for export (Bhattacharjee *et al.*, 2006).

Many PEXEL-containing proteins have a transcription profile that peaks in late rings to early trophozoite stages of parasite development. They are also often two exon proteins, with the ER signal sequence found in exon one and the PEXEL signal found near the beginning of exon two. Using prediction algorithms followed by manual curation, approximately 300 *P. falciparum* proteins are predicted to be exported. Moreover, since only *Plasmodium* spp. reside in a membrane-bound compartment across which they need to export proteins, as expected no proteins are predicted to be exported in other Apicomplexa using this algorithm (Sergeant *et al.*, 2006). However, some proteins known to reside in the erythrocyte are absent from these predictions as they do not contain a classic PEXEL motif. Such PEXEL-negative exported proteins (PNEPs) include the Maurer's clefts resident proteins SBP1, MAHRP1, REX1 and REX2 (Haase *et al.*, 2009).

Processing of the PEXEL motif facilitates export

Several studies have now revealed that the PEXEL sequence found in proteins destined for export is already cleaved after the leucine residue and N-acetylated within the ER prior to transport through the Golgi. If the R or L residues of the PEXEL motif are mutated to alanine, cleavage and export is prevented. In cases where PEXEL E>A mutations have been made, cleavage and

N-acetylation still occur but are insufficient for export (Boddey *et al.*, 2009). Similarly, if a linker region is not present between the PEXEL motif and GFP, parasites are unable to traffic GFP past the PV. However, the importance of N-acetylation remains unclear. In L>A mutants, cleavage occurs at the ER signal sequence and the end is acetylated, suggesting N-acetylation may be a common process in parasites and not specific to PEXEL processing (Boddey *et al.*, 2009).

The protease that cleaves the PEXEL motif has recently been identified (Fig. 9.2) (Boddey *et al.*, 2010; Russo *et al.*, 2010). The observation that HIV-1 aspartic acid protease cleaves a similar sequence to PEXEL and some HIV protease inhibitors prevent PEXEL cleavage, whilst leaving the signal sequence intact led Boddey *et al.* (2010) to investigate whether Plasmepsin V (PMV), an ER resident aspartic acid protease, is responsible for PEXEL processing. In parasites that harbour a mutated form of PMV, host cell homeostasis is impaired and the normally exported protein PfHRPII accumulates in its unprocessed form in the PV. In contrast, proteins that normally enter the secretory pathway that go to the PV and are then internalized are unaffected (Russo *et al.*, 2010). Moreover, both PMV isolated from the parasite and N-terminally truncated recombinant PMV purified from *E. coli* can cleave PEXEL motifs (Boddey *et al.*, 2010; Russo *et al.*, 2010). However, cleavage by signal peptidase to reveal the same N-terminal xE/Q as PEXEL-cleaved proteins is insufficient for export, signalling that PMV has a further role in directing the cleaved product into the export pathway (Fig. 9.2) (Boddey *et al.*, 2010). In keeping with its postulated essential role in protein export, PMV is refractory to deletion in *P. berghei*.

The PTEX machinery facilitates translocation of exported proteins across the PVM

Despite the identification of the PEXEL motif, the mechanism by which proteins destined for export cross the PVM to gain access to the erythrocyte cytosol has, until recently, been a complete mystery. Over the years, three different pathways have been proposed (Foley and Tilley, 1998). The first of these describes a one step process, in which vesicles containing the proteins to be exported fuse at regions where the parasite plasma membrane (PPM) and PVM are fused, releasing contents directly into the erythrocyte cytosol. So far, however, electron microscopy has not found any evidence of positions where the membranes are fused, although there are regions in which the membranes are in close proximity. Moreover, analysis of the topology of PfEXP1 in the ER and the PVM does not support this hypothesis (Gunther *et al.*, 1991).

Secondly, it has been suggested that double-membrane vesicles carrying protein cargo to be exported could fuse with the PPM, thereby releasing the inner vesicle into the PV. New vesicles could then form from the PVM and bud off into the erythrocyte cytosol. Double-membrane vesicle structures can be seen within the parasite at ring stages of development by transmission electron microscopy. In some cases, the outer membrane appears continuous with the ER. Other vesicles are seen in which the outer membrane appears to fuse with the PPM. It has been proposed that these allow the delivery of membranes to the vacuole membrane to accommodate the growing parasite providing the ability to generate vesicles from the PVM to export proteins into the erythrocyte cytosol. However, vesicles have never been seen within the PV and biochemical evidence is lacking to back up this hypothesis.

In the third postulated mechanism, vesicles fuse to the PPM to release protein cargo into the PV. The proteins to be exported are then recruited to and passed through a translocation channel in the PVM, to enter the erythrocyte cytosol. This is a common mechanism of translocation of proteins across membranes in many systems, including the ER and mitochondria, and requires an energy source and proteins to be in an unfolded state. Consistent with this hypothesis, export of GBP130 requires ATP and removal of ATP or erythrocytic proteins causes the accumulation of GBP130 in the PV (Ansorge *et al.*, 1996). Furthermore, trafficking of proteins across the PVM requires them to be unfolded (Gehde *et al.*, 2009). Nonetheless, homology searches for components from other translocon systems have never revealed any *Plasmodium* protein orthologues.

In an attempt to identify a malaria translocon at the PVM, de Koning-Ward *et al.* (2009) utilized an alternative approach, scrutinizing proteins present in a *P. falciparum* ring-stage detergent resistant membrane proteome enriched in PVM proteins (Sanders *et al.*, 2007) to determine whether any satisfied criteria predicted for translocon components. In this study, proteins were considered to be candidate translocon components if they were unique to *Plasmodium* as other *Apicomplexa* do not contain PEXEL proteins, expressed during the blood and liver stages when the parasite resides inside an intracellular vacuole, and refractory to gene deletion since some of the exported proteins are essential for parasite survival. Furthermore, the translocon components must also localize to the PVM and at the apical end of merozoites in order to be inserted into the PVM quickly after invasion. Importantly, components of the complex must bind PEXEL protein cargo specifically, and incorporate a power source such as an ATPase to translocate proteins across the membrane.

Within this detergent-resistant membrane proteome, the AAA$^+$-ATPase HSP101 and the hypothetical protein '*Plasmodium* translocon of exported protein' that has a molecular weight of 150 kDa (subsequently termed PTEX150) were identified. Both proteins are found throughout the *Plasmodium* genus and are essential to blood-stage survival (de Koning-Ward *et al.*, 2009; K. Mifsud, S. Haase and T. de Koning-Ward, unpublished), and although AAA$^+$-ATPases are present in other translocon systems, neither HSP101 nor PTEX150 has orthologues outside of the *Plasmodium* sp. Immunofluorescence assays using endogenous haemagglutinin (HA) epitope-tagged PTEX150 and HSP101 revealed that PTEX150 and HSP101 co-localize at the apical end of merozoites and in punctate foci in the PVM of ring-stage parasites (de Koning-Ward *et al.*, 2009). Moreover, the interaction of PTEX150 and HSP101 was found to be reciprocal in pull-downs using antibodies directed against the HA-epitope tag. Other interacting partners specific to the complex that were identified in these experiments were exported protein 2 (EXP2), an unknown protein termed PTEX88, and the thioredoxin-like protein, TRX2. All these proteins are specific to *Plasmodium*, have similar mRNA expression profiles, and have ER signal sequences, suggesting they traffic to the PV. EXP2 co-localizes to foci in the ring-stage with PTEX150 and HSP101, and reciprocal pull-down experiments using EXP2 antibodies have confirmed that EXP2 interacts with these proteins (de Koning-Ward *et al.*, 2009). Finally, this five-membered complex has been specifically shown to interact with native soluble PEXEL cargo destined for export as well as soluble PEXEL-containing GFP chimeric proteins.

The exact details of the interactions between the components of the translocon and their functions are not yet understood. However, through biochemical experiments it is known that the components reside on the vacuolar side of the PVM. The only likely membrane-associated protein in the complex is EXP2 and, since it has some structural homology to the membrane-inserting toxin Haemolysin E, has been hypothesized to form the putative channel (Fig. 9.2). HSP101 is predicted to form a hexamer in a similar manner to other ClpB/HSP100 family members. Analysis of the mass-spectrometry data of the PTEX150 pull-downs suggests that its stoichiometry with HSP101 is 1:1, but whether these two proteins exist as two stacked hexameric rings or whether HSP101 forms an inner ring through which the proteins pass with PTEX150 forming a ring around it remains to be established. TRX2, via its thioredoxin-like activity, may help to regulate the formation of the channel or help unfold cargo containing disulphide bonds to enable transport across the PVM. The function of PTEX88 is completely unknown as it bears little homology to other known proteins. On the cytosolic side of the membrane, it is anticipated that host and/or other parasite-derived proteins aid in protein translocation or refold proteins after passage across the membrane, particularly since removal of the erythrocyte cytosol prevents protein export, whereas the addition of cytosol from uninfected RBCs restores transport (Ansorge, 1996). A host HSP70-like protein, HSC70, and the DnaJ-containing exported protein PF08_0137 could be pulled-down with PTEX150-HA, suggesting they associate with the complex (Fig. 9.2) (de Koning-Ward *et al.*, 2009). In keeping with this, HSP70 is only found in a membrane-bound form in infected RBCs (Banumathy *et al.*, 2002) along with RESA and PF08_0137, 16 other predicted *P. falciparum* exported proteins possess a DnaJ/HSP40 domain (Sergeant *et al.*, 2006). HSP40 interacts and regulates the activity of HSP70 and thus these DnaJ domain proteins may potentially be involved in the refolding of exported molecules at the translocon and/or act as a co-chaperone with human HSP70 (de Koning-Ward *et al.*, 2009). Interestingly, 4 out of the 11 DnaJ proteins that could be disrupted by Maier *et al.* (2008) affected the display of PfEMP1 or led to alterations in membrane rigidity, whilst a recent study localized two PfHSP40s to novel extra-parasitic structures termed J-dots that have been proposed to be involved in trafficking of exported proteins (Külzer *et al.*, 2010).

Although the identity of the putative translocon has now been revealed, functional proof of PTEX in exporting cargo is still required. This is greatly hindered, however, by the fact that these proteins are essential to survival and thus attempts to block export to prove PTEX function are likely to be detrimental to parasite survival. This is further confounded by the fact that technologies to regulate gene expression in *Plasmodium* are either not available or robust enough (de Koning-Ward and Gilson, 2009) and thus delineating the functional role of these components in protein export, parasite virulence and survival becomes even more complicated.

Trafficking pathways

Despite the discovery of sequences that direct protein export, the protease that cleaves the PEXEL motif and the translocon at the PVM there are, nonetheless, key knowledge gaps still limiting our understanding of protein trafficking in malaria parasites. For example, how do proteins with a cleaved PEXEL motif (and thus non-specific xE N-terminus) exit the ER and migrate to the PV, and how are they specifically recognized by the translocon as cargo for export, whilst other non-exported proteins that traffic via the secretory pathway to the PV are not? Indeed, the lack of sequence specificity at the N-terminus generated after PEXEL cleavage makes the hypothetical 'barcode model' of trafficking to the translocon the least attractive export model of the three mechanisms postulated by Crabb *et al.* (2010). Rather, the 'chaperone model', in which chaperones or small molecular moieties bind after PEXEL cleavage to flag the protein for export once secreted into the PV, or the 'regional model' of export is much more likely. For the latter, PEXEL cleavage is postulated to direct proteins destined for export down a distinct trafficking pathway that targets them to a specific region of the parasite membrane, facilitating deposition into a restricted vacuole compartment in which the translocon resides. In this case, the translocon does not have to specifically recognize PEXEL-cleaved proteins because non-exported proteins are excluded from this compartment. Either of the chaperone or regional models are feasible trafficking mechanisms and now that it has been revealed that Plasmepsin V is the PEXEL cleaving protease at the ER, this should greatly aid in deciphering post-cleavage events leading to PEXEL proteins exiting the ER and arriving at the translocon.

Another important question to address with respect to host cell remodelling is whether multiple trafficking pathways exist for the diverse range of exported cargo (such as soluble, transmembrane or PEXEL-negative exported proteins) or whether the translocon acts as a common portal for translocation across the PVM. Moreover, once these exported proteins access the erythrocyte cytosol, how are they then transported to their final destination within the host cell? Since PEXEL motifs are present in both soluble and transmembrane domain-containing proteins, this suggests that both types of cargo are trafficked through the same pathway. Soluble proteins such as KAHRP and PfEMP3 are thought to diffuse across the erythrocyte cytoplasm and associate with the cytoplasmic surface of the Maurer's clefts before they redistribute to the erythrocyte cytoskeleton. Whilst both proteins harbour a N-terminal signal sequence and PEXEL motif to facilitate export into the erythrocyte cytosol (Hiller *et al.*, 2004; Marti *et al.*, 2004), PfEMP3 also requires a lysine-rich region and two predicted coiled-coil regions within the N-terminus to allow the attachment to the Maurer's clefts and its final association with the erythrocyte cytoskeleton (Knuepfer *et al.*, 2005a). Solubility assays and photobleaching experiments suggest that PfEMP3-GFP chimeras are trafficked in a complex (Knuepfer *et al.*, 2005a) rather than in vesicles as previously described (Trelka *et al.*, 2000).

But how then do transmembrane proteins pass through the translocon and reach their final destination? In the case of the STEVOR transmembrane proteins, trafficking occurs via the Maurer's clefts (Kaviratne *et al.*, 2002), prior to association with the erythrocyte membrane in late stage schizonts (Lavazec *et al.*, 2006). Their trafficking involves a signal sequence to mediate entry into the secretory pathway, a PEXEL motif to promote export into the erythrocyte and the transmembrane domain is required for final sorting to the Maurer's clefts (Przyborski *et al.*, 2005). Thus either the transmembrane classes of proteins are maintained in a soluble form by chaperones to allow passage through the translocon (Dixon *et al.*, 2008) in the first instance or, alternatively they are arrested during their translocation across the PVM via the hydrophobic transmembrane domain, with the translocon facilitating lateral insertion of the protein into this membrane. Once embedded in the PVM, these proteins could be loaded into nascent Maurer's clefts or vesicle-like structures (Haase and de Koning-Ward, 2010). Several observations of uncoated vesicle-like structures in the RBC cytosol have previously been reported (Taraschi *et al.*, 2003; Wickert *et al.*, 2003) although the function of these vesicles still remains to be investigated. Recent whole cell tomogram studies by Hanssen *et al.* (2010) demonstrate the presence of a number of 25-nm-diameter vesicles closely associated with the Maurer's clefts and underneath the erythrocyte membrane, so these could potentially be involved in trafficking cargo to the erythrocyte membrane or moving membrane material from one Maurer's cleft to another (Hanssen *et al.*, 2010; Tilley and Hanssen, 2008). In addition, a second population of 80 nm diameter spherical structures with an electron-dense coat, termed electron-dense vesicles (EDV), have been identified in close proximity to the erythrocyte membrane, and thus it is tempting to speculate that EDVs play a possible role in delivering integral membrane proteins to the RBC membrane (Hanssen *et al.*, 2010).

In the case of PfEMP1, which lacks an N-terminal hydrophobic signal sequence, the transmembrane domain has been proposed to mediate entry into the secretory pathway (Knuepfer *et al.*, 2005b). Although two different non-canonical PEXEL-like motifs have been identified within PfEMP1 and have been shown to mediate export into the RBC (Hiller *et al.*, 2004; Marti *et al.*, 2004), a recent study revealed the first PEXEL motif is sufficient to promote export of miniPfEMP1 proteins into the RBC cytosol and that additional information in the semi-conserved head structure (which includes the second PEXEL motif) in co-operation with the transmembrane domain and ATS is required for correct display at the erythrocyte surface (Melcher *et al.*, 2010). The mechanisms of PfEMP1 export are still not fully understood, however, different models have been proposed: that PfEMP1 exists first in a soluble state to facilitate trafficking across the PVM, bound to a protein complex during its passage through the RBC cytosol (Knuepfer *et al.*, 2005b; Papakrivos *et al.*, 2005) and then becoming 'more insoluble' once inserted into the Maurer's clefts. From here it gets trafficked to the erythrocyte membrane, with cholesterol-rich domains and serum lipoproteins promoting the delivery and/or presentation of PfEMP1 at the erythrocyte surface (Frankland *et al.*, 2006; Frankland *et al.*, 2007). On the other hand, the association with the Maurer's clefts could also argue for loading of PfEMP1 (as well as other integral membrane proteins destined for the Maurer's clefts) into nascent clefts as they form from the PVM and relocation towards the erythrocyte periphery as described above. Other studies suggest that PfEMP1 is transferred within transport vesicles (Trelka *et al.*, 2000). However, PfEMP1 does not appear to be cycled between the surface and Maurer's clefts as trypsinized infected erythrocytes do not replenish their surface PfEMP1 (Kriek *et al.*, 2003). PfEMP1 trafficking is indeed complex, with several accessory proteins involved. This includes KAHRP and PfEMP3, which transiently associate with the Maurer's clefts prior to their final destination at the erythrocyte membrane, as well as the Maurer's clefts resident proteins SBP1 and MAHRP1. Furthermore, using a large scale gene knockout approach of exported proteins, Maier *et al.*

(2008) identified six additional novel proteins that are required for export of PfEMP1; three of them appear to interfere with the early steps of PfEMP1 transport, whilst the other three affect PfEMP1 trafficking from the Maurer's clefts to the erythrocyte membrane (Maier *et al.*, 2008).

What is known then about the trafficking pathways of exported proteins such as RESA and some of its some paralogues that contain a 'relaxed' PEXEL motif (RxLxxE) and a deeply recessed hydrophobic signal sequence (Sergeant *et al.*, 2006), or those that lack an obvious PEXEL motif? For RESA, the timing of expression has been shown to be crucial for correct trafficking of RESA-GFP chimera to the dense granules in merozoites and ultimately into the erythrocyte cytosol (Rug *et al.*, 2004), however, the requirements of this unusual PEXEL motif still remain to be determined. The picture for the PNEPs is complicated, as they appear to require different signals to promote export. REX1 export is mediated by a recessed signal peptide plus an additional 10 amino acids and the presence of a predicted coiled-coil motif leads to the correct localization of soluble REX1 at the Maurer's clefts (Dixon *et al.*, 2008), whereas in REX2 the first 10 amino acids and the transmembrane domain are sufficient to promote targeting to the Maurer's clefts (Haase *et al.*, 2009). For SBP1 and MAHRP1, their transmembrane domains promote entry into the secretory pathway (Saridaki *et al.*, 2009; Spycher *et al.*, 2006) in addition to different domains acting independently, including a 15 amino acid domain with some similarity to the PEXEL motif in MAHRP-1 (Spycher *et al.*, 2006) and N-terminal domains with a high negative net charge in SBP-1 (Saridaki *et al.*, 2009). However, the N-termini of SBP1 and MAHRP1 could be interchanged with each other as could SBP1 and REX2, which has led to the hypothesis that a more generic factor such as the difference in the isoelectric point of the N- versus the C-terminus may play a role in the export of PNEPs (Haase *et al.*, 2009; Saridaki *et al.*, 2009; Spycher *et al.*, 2006). The trafficking requirements of MAHRP2 are again different from proteins destined for the Maurer's clefts. This protein does not contain a signal sequence, nor a PEXEL motif, however the first 15 residues, the his-rich N-terminal region and the predicted hydrophobic domain are all required for correct trafficking (Pachlatko *et al.*, 2010).

Concluding remarks

The proteins exported by *Plasmodium* parasites are enormous in number, form and function, allowing this deadly pathogen to carve out a living in one of the smallest and impoverished but most abundant cells in the vertebrate host. Parasites have figured out not only how to gain access to the nutrient rich host serum by increasing the permeability of their erythrocytic host but also how to stow away in the microvasculature of various organs to deny detection and destruction in the spleen. The impressive structures built by the parasite, namely the TVN, Maurer's clefts, variously shaped vesicles and knobs, facilitate the movement, assembly and display of exported parasite protein complexes, often in conjunction with host proteins. All this is achieved in the absence of a host secretory pathway, with the parasite instead evolving its own unique trafficking pathway by necessity. Despite recent breakthroughs in understanding how parasites export their protein cargoes many questions remain, such as how PTEX discriminates its cargo for export from other secreted proteins when the PEXEL motif responsible for translocation into the host cell cytosol is removed within the ER. Furthermore, other exported proteins completely lack a PEXEL motif and despite the identification of numerous sequences required for their export, it still remains to be established if this class of protein utilize the same mechanism to access the host cell. Thus, whether all exported proteins progress on a similar journey that utilize a common PTEX pathway or if there are multiple paths that enable proteins to traverse the PVM remains to be deciphered. Only with further work will we begin to appreciate just how much of the phenomenal success of malaria parasite rests on their ability to project so much of themselves into their host cells.

References

Adisa, A., Rug, M., Klonis, N., Foley, M., Cowman, A.F., and Tilley, L. (2003). The signal sequence of exported protein-1 directs the green fluorescent protein to the parasitophorous vacuole of transfected malaria parasites. J. Biol. Chem. *278*, 6532–6542.

An, X., and Mohandas, N. (2010). Red cell membrane and malaria. Transfus. Clin. Biol. *17*, 197–199.

Ansorge, I., Benting, J., Bhakdi, S., and Lingelbach, K. (1996). Protein sorting in *Plasmodium falciparum*-infected red blood cells permeabilized with the pore-forming protein streptolysin O. Biochem. J. *315*, 307–314.

Banumathy, G., Singh, V., and Tatu, U. (2002). Host chaperones are recruited in membrane-bound complexes by *Plasmodium falciparum*. J. Biol. Chem. *277*, 3902–3912.

Baruch, D.I., Pasloske, B.L., Singh, H.B., Bi, X., Ma, X.C., Feldman, M., Taraschi, T.F., and Howard, R.J. (1995). Cloning the *P. falciparum* gene encoding PfEMP1, a malarial variant antigen and adherence receptor on the surface of parasitized human erythrocytes. Cell *82*, 77–87.

Baum, J., Richard, D., Healer, J., Rug, M., Krnajski, Z., Gilberger, T.W., Green, J.L., Holder, A.A., and Cowman, A.F. (2006). A conserved molecular motor drives cell invasion and gliding motility across malaria life cycle stages and other apicomplexan parasites. J. Biol. Chem. *281*, 5197–5208.

Benedetti, C.E., Kobarg, J., Pertinhez, T.A., Gatti, R.M., de Souza, O.N., Spisni, A., and Meneghini, R. (2003). *Plasmodium falciparum* histidine-rich protein II binds to actin, phosphatidylinositol 4, 5-bisphosphate and erythrocyte ghosts in a pH-dependent manner and undergoes coil-to-helix transitions in anionic micelles. Mol. Biochem. Parasitol. *128*, 157–166.

Bennett, B.J., Mohandas, N., and Coppel, R.L. (1997). Defining the minimal domain of the *Plasmodium falciparum* protein MESA involved in the interaction with the red cell membrane skeletal protein 4.1. J. Biol. Chem. *272*, 15299–15306.

Bhattacharjee, S., Hiller, N.L., Liolios, K., Win, J., Kanneganti, T.D., Young, C., Kamoun, S., and Haldar, K. (2006). The malarial host-targeting signal is conserved in the Irish potato famine pathogen. PLoS Pathog. *2*, 453–465.

Black, C.G., Proellocks, N.I., Kats, L.M., Cooke, B.M., Mohandas, N., and Coppel, R.L. (2008). In vivo studies support the role of trafficking and cytoskeletal-binding motifs in the interaction of MESA with the membrane skeleton of *Plasmodium falciparum*-infected red blood cells. Mol. Biochem. Parasitol. *160*, 143–147.

Blisnick, T., Morales Betoulle, M.E., Barale, J.C., Uzureau, P., Berry, L., Desroses, S., Fujioka, H., Mattei, D., and Braun Breton, C. (2000). Pfsbp1, a Maurer's cleft *Plasmodium falciparum* protein, is associated with the erythrocyte skeleton. Mol. Biochem. Parasitol. *111*, 107–121.

Blisnick, T., Vincensini, L., Barale, J.C., Namane, A., and Braun Breton, C. (2005). LANCL1, an erythrocyte protein recruited to the Maurer's clefts during *Plasmodium falciparum* development. Mol. Biochem. Parasitol. *141*, 39–47.

Blythe, J.E., Surentheran, T., and Preiser, P.R. (2004). STEVOR – a multifunctional protein? Mol. Biochem. Parasitol. *134*, 11–15.

Boddey, J.A., Hodder, A.N., Gunther, S., Gilson, P.R., Patsiouras, H., Kapp, E.A., Pearce, J.A., de Koning-Ward, T.F., Simpson, R.J., Crabb, B.S., *et al.* (2010). An aspartyl protease directs malaria effector proteins to the host cell. Nature *463*, 627–631.

Boddey, J.A., Moritz, R.L., Simpson, R.J., and Cowman, A.F. (2009). Role of the Plasmodium export element in trafficking parasite proteins to the infected erythrocyte. Traffic *10*, 285–299.

Bonnefoy, S., Mattei, D., Dubremetz, J.F., Guillotte, M., Jouin, H., Ozaki, L.S., Sibilli, L., and Mercereau-Puijalon, O. (1988). *Plasmodium falciparum*: molecular analysis of a putative protective antigen, the thermostable 96-kDa protein. Exp. Parasitol. *65*, 69–83.

Bouyer, G., Egee, S., and Thomas, S.L. (2006). Three types of spontaneously active anionic channels in malaria-infected human red blood cells. Blood Cells Mol. Dis. *36*, 248–254.

Bultrini, E., Brick, K., Mukherjee, S., Zhang, Y., Silvestrini, F., Alano, P., and Pizzi, E. (2009). Revisiting the *Plasmodium falciparum* RIFIN family: from comparative genomics to 3D-model prediction. BMC Genomics *10*, 445.

Chang, H.H., Falick, A.M., Carlton, P.M., Sedat, J.W., DeRisi, J.L., and Marletta, M.A. (2008). N-terminal processing of proteins exported by malaria parasites. Mol. Biochem. Parasitol. *160*, 107–115.

Cheng, Q., Cloonan, N., Fischer, K., Thompson, J., Waine, G., Lanzer, M., and Saul, A. (1998). stevor and rif are *Plasmodium falciparum* multicopy gene families which potentially encode variant antigens. Mol. Biochem. Parasitol. *97*, 161–176.

Cooke, B.M., Buckingham, D.W., Glenister, F.K., Fernandez, K.M., Bannister, L.H., Marti, M., Mohandas, N., and Coppel, R.L. (2006). A Maurer's cleft-associated protein is essential for expression of the major malaria virulence antigen on the surface of infected red blood cells. J. Cell Biol. *172*, 899–908.

Coppel, R.L. (1992). Repeat structures in a *Plasmodium falciparum* protein (MESA) that binds human erythrocyte protein 4.1. Mol. Biochem. Parasitol. *50*, 335–347.

Coppel, R.L., Lustigman, S., Murray, L., and Anders, R.F. (1988). MESA is a *Plasmodium falciparum* phosphoprotein associated with the erythrocyte membrane skeleton. Mol. Biochem. Parasitol. *31*, 223–231.

Cowman, A.F., Coppel, R.L., Saint, R.B., Favaloro, J., Crewther, P.E., Stahl, H.D., Bianco, A.E., Brown, G.V., Anders, R.F., and Kemp, D.J. (1984). The ring-infected erythrocyte surface antigen (RESA) polypeptide of *Plasmodium falciparum* contains two separate blocks of tandem repeats encoding antigenic epitopes that are naturally immunogenic in man. Mol. Biol. Med. *2*, 207–221.

Crabb, B.S., Cooke, B.M., Reeder, J.C., Waller, R.F., Caruana, S.R., Davern, K.M., Wickham, M.E., Brown, G.V., Coppel, R.L., and Cowman, A.F. (1997). Targeted gene disruption shows that knobs enable malaria-infected red cells to cytoadhere under physiological shear stress. Cell *89*, 287–296.

Crabb, B.S., de Koning-Ward, T.F., and Gilson, P.R. (2010). Protein export in *Plasmodium* parasites: from the endoplasmic reticulum to the vacuolar export machine. Int. J. Parasitol. *40*, 509–513.

Culvenor, J.G., Day, K.P., and Anders, R.F. (1991). *Plasmodium falciparum* ring-infected erythrocyte surface antigen is released from merozoite dense granules after erythrocyte invasion. Infect. Immun. *59*, 1183–1187.

Culvenor, J.G., Langford, C.J., Crewther, P.E., Saint, R.B., Coppel, R.L., Kemp, D.J., Anders, R.F., and Brown, G.V. (1987). *Plasmodium falciparum*: identification and localization of a knob protein antigen expressed by a cDNA clone. Exp. Parasitol. *63*, 58–67.

Da Silva, E., Foley, M., Dluzewski, A.R., Murray, L.J., Anders, R.F., and Tilley, L. (1994). The *Plasmodium falciparum* protein RESA interacts with the erythrocyte cytoskeleton and modifies erythrocyte thermal stability. Mol. Biochem. Parasitol. *66*, 59–69.

de Koning-Ward, T.F., and Gilson, P.R. (2009). Keeping it simple: an easy method for manipulating the expression levels of malaria proteins. Trends Parasitol. *25*, 4–7.

de Koning-Ward, T.F., Gilson, P.R., Boddey, J.A., Rug, M., Smith, B.J., Papenfuss, A.T., Sanders, P.R., Lundie, R.J., Maier, A.G., Cowman, A.F., *et al.* (2009). A newly discovered protein export machine in malaria parasites. Nature *459*, 945–949.

Decherf, G., Egee, S., Staines, H.M., Ellory, J.C., and Thomas, S.L. (2004). Anionic channels in malaria-infected human red blood cells. Blood Cells Mol. Dis. *32*, 366–371.

Deitsch, K.W., and Wellems, T.E. (1996). Membrane modifications in erythrocytes parasitized by *Plasmodium falciparum*. Mol. Biochem. Parasitol. *76*, 1–10.

Dhermy, D., Schrevel, J., and Lecomte, M.C. (2007). Spectrin-based skeleton in red blood cells and malaria. Curr. Opin. Hematol. *14*, 198–202.

Dixon, M.W., Hawthorne, P.L., Spielmann, T., Anderson, K.L., Trenholme, K.R., and Gardiner, D.L. (2008). Targeting of the ring exported protein 1 to the Maurer's clefts is mediated by a two-phase process. Traffic *9*, 1316–1326.

Dluzewski, A.R., Fryer, P.R., Griffiths, S., Wilson, R.J., and Gratzer, W.B. (1989). Red cell membrane protein distribution during malarial invasion. J. Cell Sci. *92*, 691–699.

Dluzewski, A.R., Mitchell, G.H., Fryer, P.R., Griffiths, S., Wilson, R.J., and Gratzer, W.B. (1992). Origins of the parasitophorous vacuole membrane of the malaria parasite, *Plasmodium falciparum*, in human red blood cells. J. Cell Sci. *102*, 527–532.

Elmendorf, H.G., and Haldar, K. (1994). *Plasmodium falciparum* exports the Golgi marker sphingomyelin synthase into a tubovesicular network in the cytoplasm of mature erythrocytes. J. Cell Biol. *124*, 449–462.

Favaloro, J.M., Coppel, R.L., Corcoran, L.M., Foote, S.J., Brown, G.V., Anders, R.F., and Kemp, D.J. (1986). Structure of the RESA gene of *Plasmodium falciparum*. Nucleic Acids Res. *14*, 8265–8277.

Fernandez, V., Hommel, M., Chen, Q., Hagblom, P., and Wahlgren, M. (1999). Small, clonally variant antigens expressed on the surface of the *Plasmodium falciparum*-infected erythrocyte are encoded by the rif gene family and are the target of human immune responses. J. Exp. Med. *190*, 1393–1404.

Foley, M., Murray, L.J., and Anders, R.F. (1990). The ring-infected erythrocyte surface antigen protein of *Plasmodium falciparum* is phosphorylated upon association with the host cell membrane. Mol. Biochem. Parasitol. *38*, 69–75.

Foley, M., and Tilley, L. (1998). Protein trafficking in malaria-infected erythrocytes. Int. J. Parasitol. *28*, 1671–1680.

Foley, M., Tilley, L., Sawyer, W.H., and Anders, R.F. (1991). The ring-infected erythrocyte surface antigen of *Plasmodium falciparum* associates with spectrin in the erythrocyte membrane. Mol. Biochem. Parasitol. *46*, 137–147.

Frankland, S., Adisa, A., Horrocks, P., Taraschi, T.F., Schneider, T., Elliott, S.R., Rogerson, S.J., Knuepfer, E., Cowman, A.F., Newbold, C.I., *et al.* (2006). Delivery of the malaria virulence protein PfEMP1 to the erythrocyte surface requires cholesterol-rich domains. Eukaryot. Cell *5*, 849–860.

Frankland, S., Elliott, S.R., Yosaatmadja, F., Beeson, J.G., Rogerson, S.J., Adisa, A., and Tilley, L. (2007). Serum lipoproteins promote efficient presentation of the malaria virulence protein PfEMP1 at the erythrocyte surface. Eukaryot. Cell *6*, 1584–1594.

Gehde, N., Hinrichs, C., Montilla, I., Charpian, S., Lingelbach, K., and Przyborski, J.M. (2009). Protein unfolding is an essential requirement for transport across the parasitophorous vacuolar membrane of *Plasmodium falciparum*. Mol. Microbiol. *71*, 613–628.

Glenister, F.K., Coppel, R.L., Cowman, A.F., Mohandas, N., and Cooke, B.M. (2002). Contribution of parasite proteins to altered mechanical properties of malaria-infected red blood cells. Blood *99*, 1060–1063.

Glenister, F.K., Fernandez, K.M., Kats, L.M., Hanssen, E., Mohandas, N., Coppel, R.L., and Cooke, B.M. (2009). Functional alteration of red blood cells by a megadalton protein of *Plasmodium falciparum*. Blood *113*, 919–928.

Gunther, K., Tummler, M., Arnold, H.H., Ridley, R., Goman, M., Scaife, J.G., and Lingelbach, K. (1991). An exported protein of *Plasmodium falciparum* is synthesized as an integral membrane protein. Mol. Biochem. Parasitol. *46*, 149–157.

Haase, S., and de Koning-Ward, T.F. (2010). New insights into protein export in malaria parasites. Cell. Microbiol. *12*, 580–587.

Haase, S., Herrmann, S., Gruring, C., Heiber, A., Jansen, P.W., Langer, C., Treeck, M., Cabrera, A., Bruns, C., Struck, N.S., *et al.* (2009). Sequence requirements for the export of the *Plasmodium falciparum* Maurer's clefts protein REX2. Mol. Microbiol. *71*, 1003–1017.

Hanssen, E., Carlton, P., Deed, S., Klonis, N., Sedat, J., Derisi, J., and Tilley, L. (2010). Whole cell imaging reveals novel modular features of the exomembrane system of the malaria parasite, *Plasmodium falciparum*. Int. J. Parasitol. *40*, 123–134.

Hanssen, E., Hawthorne, P., Dixon, M.W., Trenholme, K.R., McMillan, P.J., Spielmann, T., Gardiner, D.L., and Tilley, L. (2008a). Targeted mutagenesis of the ring-exported protein-1 of *Plasmodium falciparum* disrupts the architecture of Maurer's cleft organelles. Mol. Microbiol. *69*, 938–953.

Hanssen, E., Sougrat, R., Frankland, S., Deed, S., Klonis, N., Lippincott-Schwartz, J., and Tilley, L. (2008b). Electron tomography of the Maurer's cleft organelles of *Plasmodium falciparum*-infected erythrocytes reveals novel structural features. Mol. Microbiol. *67*, 703–718.

Harris, P.K., Yeoh, S., Dluzewski, A.R., O'Donnell, R.A., Withers-Martinez, C., Hackett, F., Bannister, L.H., Mitchell, G.H., and Blackman, M.J. (2005). Molecular identification of a malaria merozoite surface sheddase. PLoS Pathog. *1*, 241–251.

Hawthorne, P.L., Trenholme, K.R., Skinner-Adams, T.S., Spielmann, T., Fischer, K., Dixon, M.W., Ortega, M.R., Anderson, K.L., Kemp, D.J., and Gardiner, D.L. (2004). A novel *Plasmodium falciparum* ring stage protein, REX, is located in Maurer's clefts. Mol. Biochem. Parasitol. *136*, 181–189.

Hiller, N.L., Bhattacharjee, S., van Ooij, C., Liolios, K., Harrison, T., Lopez-Estrano, C., and Haldar, K. (2004). A host-targeting signal in virulence proteins reveals a secretome in malarial infection. Science *306*, 1934–1937.

Hinterberg, K., Scherf, A., Gysin, J., Toyoshima, T., Aikawa, M., Mazie, J.C., da Silva, L.P., and Mattei, D. (1994). *Plasmodium falciparum*: the Pf332 antigen is secreted from the parasite by a brefeldin A-dependent pathway and is translocated to the erythrocyte membrane via the Maurer's clefts. Exp. Parasitol. *79*, 279–291.

Hodder, A.N., Maier, A.G., Rug, M., Brown, M., Hommel, M., Pantic, I., Puig-de-Morales-Marinkovic, M., Smith, B., Triglia, T., Beeson, J., *et al.* (2009). Analysis of structure and function of the giant protein Pf332 in *Plasmodium falciparum*. Mol. Microbiol. *71*, 48–65.

Horrocks, P., Pinches, R.A., Chakravorty, S.J., Papakrivos, J., Christodoulou, Z., Kyes, S.A., Urban, B.C., Ferguson, D.J., and Newbold, C.I. (2005). PfEMP1 expression is reduced on the surface of knobless *Plasmodium falciparum* infected erythrocytes. J. Cell Sci. *118*, 2507–2518.

Howard, R.F., Stanley, H.A., and Reese, R.T. (1988). Characterization of a high-molecular-weight phosphoprotein synthesized by the human malarial parasite *Plasmodium falciparum*. Gene *64*, 65–75.

Howard, R.J., Uni, S., Aikawa, M., Aley, S.B., Leech, J.H., Lew, A.M., Wellems, T.E., Rener, J., and Taylor, D.W. (1986). Secretion of a malarial histidine-rich protein (Pf HRP II) from *Plasmodium falciparum*-infected erythrocytes. J. Cell Biol. *103*, 1269–1277.

Iqbal, J., Siddique, A.B., Ahlborg, N., Perlmann, P., and Berzins, K. (1995). Cytoadherence-related homologous motifs in *Plasmodium falciparum* antigen Pf155/RESA and erythrocyte band 3 protein. Parasitology *110*, 503–511.

Jemmely, N.Y., Niang, M., and Preiser, P.R. (2010). Small variant surface antigens and *Plasmodium* evasion of immunity. Future Microbiol. *5*, 663–682.

Joannin, N., Abhiman, S., Sonnhammer, E.L., and Wahlgren, M. (2008). Sub-grouping and sub-functionalization of the RIFIN multi-copy protein family. BMC Genomics *9*, 19.

Johnson, D., Gunther, K., Ansorge, I., Benting, J., Kent, A., Bannister, L., Ridley, R., and Lingelbach, K. (1994). Characterization of membrane proteins exported from *Plasmodium falciparum* into the host erythrocyte. Parasitology *109*, 1–9.

Kaviratne, M., Khan, S.M., Jarra, W., and Preiser, P.R. (2002). Small variant STEVOR antigen is uniquely located within Maurer's clefts in *Plasmodium falciparum*-infected red blood cells. Eukaryot. Cell *1*, 926–935.

Khattab, A., and Klinkert, M.Q. (2006). Maurer's clefts-restricted localization, orientation and export of a *Plasmodium falciparum* RIFIN. Traffic *7*, 1654–1665.

Kilejian, A. (1979). Characterization of a protein correlated with the production of knob-like protrusions on membranes of erythrocytes infected with *Plasmodium falciparum*. Proc. Natl. Acad. Sci. U.S.A. *76*, 4650–4653.

Knuepfer, E., Rug, M., Klonis, N., Tilley, L., and Cowman, A.F. (2005a). Trafficking determinants for PfEMP3 export and assembly under the *Plasmodium falciparum*-infected red blood cell membrane. Mol. Microbiol. *58*, 1039–1053.

Knuepfer, E., Rug, M., Klonis, N., Tilley, L., and Cowman, A.F. (2005b). Trafficking of the major virulence factor to the surface of transfected *P. falciparum*-infected erythrocytes. Blood *105*, 4078–4087.

Kraemer, S.M., and Smith, J.D. (2006). A family affair: var genes, PfEMP1 binding, and malaria disease. Curr. Opin. Microbiol. *9*, 374–380.

Kriek, N., Tilley, L., Horrocks, P., Pinches, R., Elford, B.C., Ferguson, D.J., Lingelbach, K., and Newbold, C.I. (2003). Characterization of the pathway for transport of the cytoadherence-mediating protein, PfEMP1, to the host cell surface in malaria parasite-infected erythrocytes. Mol. Microbiol. *50*, 1215–1227.

Külzer, S., Rug, M., Brinkmann, K., Cannon, P., Cowman, A., Lingelbach, K., Blatch, G.L., Maier, A.G., and Przyborski, J.M. (2010). Parasite-encoded Hsp40 proteins define novel mobile structures in the cytosol of the *P. falciparum*-infected erythrocyte. Cell. Microbiol. *12*, 1398–1420.

Kun, J.F., Waller, K.L., and Coppel, R.L. (1999). *Plasmodium falciparum*: structural and functional domains of the mature-parasite-infected erythrocyte surface antigen. Exp. Parasitol. *91*, 258–267.

Kyes, S.A., Rowe, J.A., Kriek, N., and Newbold, C.I. (1999). Rifins: a second family of clonally variant proteins expressed on the surface of red cells infected with *Plasmodium falciparum*. Proc. Natl. Acad. Sci. U.S.A. *96*, 9333–9338.

Lanzer, M., Wickert, H., Krohne, G., Vincensini, L., and Braun Breton, C. (2006). Maurer's clefts: a novel multi-functional organelle in the cytoplasm of *Plasmodium falciparum*-infected erythrocytes. Int. J. Parasitol. *36*, 23–36.

Lavazec, C., Sanyal, S., and Templeton, T.J. (2006). Hypervariability within the Rifin, Stevor and

Pfmc-2TM superfamilies in *Plasmodium falciparum*. Nucleic Acids Res. *34*, 6696–6707.

Lavazec, C., Sanyal, S., and Templeton, T.J. (2007). Expression switching in the stevor and Pfmc-2TM superfamilies in *Plasmodium falciparum*. Mol. Microbiol. *64*, 1621–1634.

Lopez-Estrano, C., Bhattacharjee, S., Harrison, T., and Haldar, K. (2003). Cooperative domains define a unique host cell-targeting signal in *Plasmodium falciparum*-infected erythrocytes. Proc. Natl. Acad. Sci. U.S.A. *100*, 12402–12407.

Luse, S.A., and Miller, L.H. (1971). *Plasmodium falciparum* malaria. Ultrastructure of parasitized erythrocytes in cardiac vessels. Am. J. Trop. Med. Hyg. *20*, 655–660.

Lustigman, S., Anders, R.F., Brown, G.V., and Coppel, R.L. (1990). The mature-parasite-infected erythrocyte surface antigen (MESA) of *Plasmodium falciparum* associates with the erythrocyte membrane skeletal protein, band 4.1. Mol. Biochem. Parasitol. *38*, 261–270.

Magowan, C., Coppel, R.L., Lau, A.O., Moronne, M.M., Tchernia, G., and Mohandas, N. (1995). Role of the *Plasmodium falciparum* mature-parasite-infected erythrocyte surface antigen (MESA/PfEMP-2) in malarial infection of erythrocytes. Blood *86*, 3196–3204.

Maier, A.G., Rug, M., O'Neill, M.T., Beeson, J.G., Marti, M., Reeder, J., and Cowman, A.F. (2007). Skeleton-binding protein 1 functions at the parasitophorous vacuole membrane to traffic PfEMP1 to the *Plasmodium falciparum*-infected erythrocyte surface. Blood *109*, 1289–1297.

Maier, A.G., Rug, M., O'Neill, M.T., Brown, M., Chakravorty, S., Szestak, T., Chesson, J., Wu, Y., Hughes, K., Coppel, R.L., *et al.* (2008). Exported proteins required for virulence and rigidity of *Plasmodium falciparum*-infected human erythrocytes. Cell *134*, 48–61.

Marti, M., Good, R.T., Rug, M., Knuepfer, E., and Cowman, A.F. (2004). Targeting malaria virulence and remodeling proteins to the host erythrocyte. Science *306*, 1930–1933.

Mashima, R., Tilley, L., Siomos, M.A., Papalexis, V., Raftery, M.J., and Stocker, R. (2002). *Plasmodium falciparum* histidine-rich protein-2 (PfHRP2) modulates the redox activity of ferri-protoporphyrin IX (FePPIX): peroxidase-like activity of the PfHRP2–FePPIX complex. J. Biol. Chem. *277*, 14514–14520.

Mattei, D., and Scherf, A. (1992a). The Pf332 gene codes for a megadalton protein of *Plasmodium falciparum* asexual blood stages. Mem. Inst. Oswaldo Cruz *87*, 163–168.

Mattei, D., and Scherf, A. (1992b). The Pf332 gene of *Plasmodium falciparum* codes for a giant protein that is translocated from the parasite to the membrane of infected erythrocytes. Gene *110*, 71–79.

Melcher, M., Muhle, R.A., Henrich, P.P., Kraemer, S.M., Avril, M., Vigan-Womas, I., Mercereau-Puijalon, O., Smith, J.D., and Fidock, D.A. (2010). Identification of a role for the PfEMP1 semi-conserved head structure in protein trafficking to the surface of *Plasmodium falciparum* infected red blood cells. Cell. Microbiol. *12*, 1446–1462.

Merckx, A., Bouyer, G., Thomas, S.L., Langsley, G., and Egee, S. (2009). Anion channels in *Plasmodium-falciparum*-infected erythrocytes and protein kinase A. Trends Parasitol. *25*, 139–144.

Miller, L.H., Baruch, D.I., Marsh, K., and Doumbo, O.K. (2002). The pathogenic basis of malaria. Nature *415*, 673–679.

Moll, K., Chene, A., Ribacke, U., Kaneko, O., Nilsson, S., Winter, G., Haeggstrom, M., Pan, W., Berzins, K., Wahlgren, M., *et al.* (2007). A novel DBL-domain of the *P. falciparum* 332 molecule possibly involved in erythrocyte adhesion. PLoS One 2, 1–11.

Mphande, F.A., Ribacke, U., Kaneko, O., Kironde, F., Winter, G., and Wahlgren, M. (2008). SURFIN4.1, a schizont-merozoite associated protein in the SURFIN family of *Plasmodium falciparum*. Malar. J. 7, 116.

Murphy, S.C., Samuel, B.U., Harrison, T., Speicher, K.D., Speicher, D.W., Reid, M.E., Prohaska, R., Low, P.S., Tanner, M.J., Mohandas, N., *et al.* (2004). Erythrocyte detergent-resistant membrane proteins: their characterization and selective uptake during malarial infection. Blood *103*, 1920–1928.

Nash, G.B., O'Brien, E., Gordon-Smith, E.C., and Dormandy, J.A. (1989). Abnormalities in the mechanical properties of red blood cells caused by *Plasmodium falciparum*. Blood 74, 855–861.

Niang, M., Yan Yam, X., and Preiser, P.R. (2009). The *Plasmodium falciparum* STEVOR multigene family mediates antigenic variation of the infected erythrocyte. PLoS Pathog. *5*, 1–12.

Oh, S.S., Voigt, S., Fisher, D., Yi, S.J., LeRoy, P.J., Derick, L.H., Liu, S., and Chishti, A.H. (2000). *Plasmodium falciparum* erythrocyte membrane protein 1 is anchored to the actin–spectrin junction and knob-associated histidine-rich protein in the erythrocyte skeleton. Mol. Biochem. Parasitol. *108*, 237–247.

Pachlatko, E., Rusch, S., Muller, A., Hemphill, A., Tilley, L., Hanssen, E., and Beck, H.P. (2010). MAHRP2, an exported protein of *Plasmodium falciparum*, is an essential component of Maurer's cleft tethers. Mol. Microbiol. 77, 1136–1152.

Papakrivos, J., Newbold, C.I., and Lingelbach, K. (2005). A potential novel mechanism for the insertion of a membrane protein revealed by a biochemical analysis of the *Plasmodium falciparum* cytoadherence molecule PfEMP-1. Mol. Microbiol. *55*, 1272–1284.

Papalexis, V., Siomos, M.A., Campanale, N., Guo, X., Kocak, G., Foley, M., and Tilley, L. (2001). Histidine-rich protein 2 of the malaria parasite, *Plasmodium falciparum*, is involved in detoxification of the by-products of haemoglobin degradation. Mol. Biochem. Parasitol. *115*, 77–86.

Parker, P.D., Tilley, L., and Klonis, N. (2004). *Plasmodium falciparum* induces reorganization of host membrane proteins during intraerythrocytic growth. Blood *103*, 2404–2406.

Pasloske, B.L., Baruch, D.I., Ma, C., Taraschi, T.F., Gormley, J.A., and Howard, R.J. (1994). PfEMP3 and HRP1: co-expressed genes localized to chromosome 2 of *Plasmodium falciparum*. Gene *144*, 131–136.

Pasloske, B.L., Baruch, D.I., van Schravendijk, M.R., Handunnetti, S.M., Aikawa, M., Fujioka, H., Taraschi, T.F., Gormley, J.A., and Howard, R.J. (1993). Cloning and characterization of a *Plasmodium falciparum* gene encoding a novel high-molecular weight host membrane-associated protein, PfEMP3. Mol. Biochem. Parasitol. *59*, 59–72.

Pei, X., An, X., Guo, X., Tarnawski, M., Coppel, R., and Mohandas, N. (2005). Structural and functional studies of interaction between *Plasmodium falciparum* knob-associated histidine-rich protein (KAHRP) and erythrocyte spectrin. J. Biol. Chem. *280*, 31166–31171.

Pei, X., Guo, X., Coppel, R., Bhattacharjee, S., Haldar, K., Gratzer, W., Mohandas, N., and An, X. (2007a). The ring-infected erythrocyte surface antigen (RESA) of *Plasmodium falciparum* stabilizes spectrin tetramers and suppresses further invasion. Blood *110*, 1036–1042.

Pei, X., Guo, X., Coppel, R., Mohandas, N., and An, X. (2007b). *Plasmodium falciparum* erythrocyte membrane protein 3 (PfEMP3) destabilizes erythrocyte membrane skeleton. J. Biol. Chem. *282*, 26754–26758.

Perkins, M.E. (1984). Surface proteins of *Plasmodium falciparum* merozoites binding to the erythrocyte receptor, glycophorin. J. Exp. Med. *160*, 788–798.

Perlmann, H., Berzins, K., Wahlgren, M., Carlsson, J., Bjorkman, A., Patarroyo, M.E., and Perlmann, P. (1984). Antibodies in malarial sera to parasite antigens in the membrane of erythrocytes infected with early asexual stages of *Plasmodium falciparum*. J. Exp. Med. *159*, 1686–1704.

Petter, M., Bonow, I., and Klinkert, M.Q. (2008). Diverse expression patterns of subgroups of the rif multigene family during *Plasmodium falciparum* gametocytogenesis. PLoS One 3, 1–13.

Petter, M., Haeggstrom, M., Khattab, A., Fernandez, V., Klinkert, M.Q., and Wahlgren, M. (2007). Variant proteins of the *Plasmodium falciparum* RIFIN family show distinct subcellular localization and developmental expression patterns. Mol. Biochem. Parasitol. *156*, 51–61.

Pologe, L.G., Pavlovec, A., Shio, H., and Ravetch, J.V. (1987). Primary structure and subcellular localization of the knob-associated histidine-rich protein of *Plasmodium falciparum*. Proc. Natl. Acad. Sci. U.S.A. *84*, 7139–7143.

Przyborski, J.M. (2008). The Maurer's clefts of *Plasmodium falciparum*: parasite-induced islands within an intracellular ocean. Trends Parasitol. *24*, 285–288.

Przyborski, J.M., Miller, S.K., Pfahler, J.M., Henrich, P.P., Rohrbach, P., Crabb, B.S., and Lanzer, M. (2005). Trafficking of STEVOR to the Maurer's clefts in *Plasmodium falciparum*-infected erythrocytes. EMBO J. *24*, 2306–2317.

Rug, M., Prescott, S.W., Fernandez, K.M., Cooke, B.M., and Cowman, A.F. (2006). The role of KAHRP domains in knob formation and cytoadherence of *P. falciparum*-infected human erythrocytes. Blood *108*, 370–378.

Rug, M., Wickham, M.E., Foley, M., Cowman, A.F., and Tilley, L. (2004). Correct promoter control is needed for trafficking of the ring-infected erythrocyte surface antigen to the host cytosol in transfected malaria parasites. Infect. Immun. *72*, 6095–6105.

Russo, I., Babbitt, S., Muralidharan, V., Butler, T., Oksman, A., and Goldberg, D.E. (2010). Plasmepsin V licenses *Plasmodium* proteins for export into the host erythrocyte. Nature *463*, 632–636.

Safeukui, I., Correas, J.M., Brousse, V., Hirt, D., Deplaine, G., Mule, S., Lesurtel, M., Goasguen, N., Sauvanet, A., Couvelard, A., *et al.* (2008). Retention of *Plasmodium falciparum* ring-infected erythrocytes in the slow, open microcirculation of the human spleen. Blood *112*, 2520–2528.

Sam-Yellowe, T.Y. (2009). The role of the Maurer's clefts in protein transport in *Plasmodium falciparum*. Trends Parasitol. *25*, 277–284.

Sam-Yellowe, T.Y., Florens, L., Johnson, J.R., Wang, T., Drazba, J.A., Le Roch, K.G., Zhou, Y., Batalov, S., Carucci, D.J., Winzeler, E.A., *et al.* (2004). A *Plasmodium* gene family encoding Maurer's cleft membrane proteins: structural properties and expression profiling. Genome Res. *14*, 1052–1059.

Sanders, P.R., Cantin, G.T., Greenbaum, D.C., Gilson, P.R., Nebl, T., Moritz, R.L., Yates, J.R., 3rd, Hodder, A.N., and Crabb, B.S. (2007). Identification of protein complexes in detergent-resistant membranes of *Plasmodium falciparum* schizonts. Mol. Biochem. Parasitol. *154*, 148–157.

Sargeant, T.J., Marti, M., Caler, E., Carlton, J.M., Simpson, K., Speed, T.P., and Cowman, A.F. (2006). Lineage-specific expansion of proteins exported to erythrocytes in malaria parasites. Genome Biol. *7*, R12.

Saridaki, T., Frohlich, K.S., Braun-Breton, C., and Lanzer, M. (2009). Export of PfSBP1 to the *Plasmodium falciparum* Maurer's clefts. Traffic *10*, 137–152.

Scherf, A., Lopez-Rubio, J.J., and Riviere, L. (2008). Antigenic variation in *Plasmodium falciparum*. Annu. Rev. Microbiol. *62*, 445–470.

Sharp, S., Lavstsen, T., Fivelman, Q.L., Saeed, M., McRobert, L., Templeton, T.J., Jensen, A.T., Baker, D.A., Theander, T.G., and Sutherland, C.J. (2006). Programmed transcription of the var gene family, but not of stevor, in *Plasmodium falciparum* gametocytes. Eukaryot. Cell *5*, 1206–1214.

Silva, M.D., Cooke, B.M., Guillotte, M., Buckingham, D.W., Sauzet, J.P., Le Scanf, C., Contamin, H., David, P., Mercereau-Puijalon, O., and Bonnefoy, S. (2005). A role for the *Plasmodium falciparum* RESA protein in resistance against heat shock demonstrated using gene disruption. Mol. Microbiol. *56*, 990–1003.

Smith, J.D., Subramanian, G., Gamain, B., Baruch, D.I., and Miller, L.H. (2000). Classification of adhesive domains in the *Plasmodium falciparum* erythrocyte membrane protein 1 family. Mol. Biochem. Parasitol. *110*, 293–310.

Spielmann, T., and Beck, H.P. (2000). Analysis of stage-specific transcription in *Plasmodium falciparum* reveals a set of genes exclusively transcribed in ring stage parasites. Mol. Biochem. Parasitol. *111*, 453–458.

Spielmann, T., Hawthorne, P.L., Dixon, M.W., Hannemann, M., Klotz, K., Kemp, D.J., Klonis, N., Tilley, L., Trenholme, K.R., and Gardiner, D.L. (2006). A cluster of ring stage-specific genes linked to a locus implicated in cytoadherence in *Plasmodium falciparum* codes for PEXEL-negative and PEXEL-positive proteins exported into the host cell. Mol. Biol. Cell. *17*, 3613–3624.

Spycher, C., Klonis, N., Spielmann, T., Kump, E., Steiger, S., Tilley, L., and Beck, H.P. (2003). MAHRP-1, a novel *Plasmodium falciparum* histidine-rich protein, binds ferriprotoporphyrin IX and localizes to the Maurer's clefts. J. Biol. Chem. *278*, 35373–35383.

Spycher, C., Rug, M., Klonis, N., Ferguson, D.J., Cowman, A.F., Beck, H.P., and Tilley, L. (2006). Genesis of and trafficking to the Maurer's clefts of *Plasmodium falciparum*-infected erythrocytes. Mol. Cell. Biol. *26*, 4074–4085.

Spycher, C., Rug, M., Pachlatko, E., Hanssen, E., Ferguson, D., Cowman, A.F., Tilley, L., and Beck, H.P. (2008). The Maurer's cleft protein MAHRP1 is essential for trafficking of PfEMP1 to the surface of *Plasmodium falciparum*-infected erythrocytes. Mol. Microbiol. *68*, 1300–1314.

Staines, H.M., Ashmore, S., Felgate, H., Moore, J., Powell, T., and Ellory, J.C. (2006). Solute transport via the new permeability pathways in *Plasmodium falciparum*-infected human red blood cells is not consistent with a simple single-channel model. Blood *108*, 3187–3194.

Staines, H.M., Ellory, J.C., and Chibale, K. (2005). The new permeability pathways: targets and selective routes for the development of new antimalarial agents. Comb. Chem. High Throughput Screen *8*, 81–88.

Staines, H.M., Powell, T., Thomas, S.L., and Ellory, J.C. (2004). *Plasmodium falciparum*-induced channels. Int. J. Parasitol. *34*, 665–673.

Staines, H.M., Rae, C., and Kirk, K. (2000). Increased permeability of the malaria-infected erythrocyte to organic cations. Biochim. Biophys. Acta *1463*, 88–98.

Su, X.Z., Heatwole, V.M., Wertheimer, S.P., Guinet, F., Herrfeldt, J.A., Peterson, D.S., Ravetch, J.A., and Wellems, T.E. (1995). The large diverse gene family var encodes proteins involved in cytoadherence and antigenic variation of *Plasmodium falciparum*-infected erythrocytes. Cell *82*, 89–100.

Tamez, P.A., Bhattacharjee, S., van Ooij, C., Hiller, N.L., Llinas, M., Balu, B., Adams, J.H., and Haldar, K. (2008). An erythrocyte vesicle protein exported by the malaria parasite promotes tubovesicular lipid import from the host cell surface. PLoS Pathog. *4*, 1–12.

Taraschi, T.F., O'Donnell, M., Martinez, S., Schneider, T., Trelka, D., Fowler, V.M., Tilley, L., and Moriyama, Y. (2003). Generation of an erythrocyte vesicle transport system by *Plasmodium falciparum* malaria parasites. Blood *102*, 3420–3426.

Tilley, L., and Hanssen, E. (2008). A 3D view of the host cell compartment in *P. falciparum*-infected erythrocytes. Transfus. Clin. Biol. *15*, 72–81.

Tilley, L., Sougrat, R., Lithgow, T., and Hanssen, E. (2008). The twists and turns of Maurer's cleft trafficking in *P. falciparum*-infected erythrocytes. Traffic *9*, 187–197.

Trelka, D.P., Schneider, T.G., Reeder, J.C., and Taraschi, T.F. (2000). Evidence for vesicle-mediated trafficking of parasite proteins to the host cell cytosol and erythrocyte surface membrane in *Plasmodium falciparum* infected erythrocytes. Mol. Biochem. Parasitol. *106*, 131–145.

Tsarukyanova, I., Drazba, J.A., Fujioka, H., Yadav, S.P., and Sam-Yellowe, T.Y. (2009). Proteins of the *Plasmodium falciparum* two transmembrane Maurer's cleft protein family, PfMC-2TM, and the 130 kDa Maurer's cleft protein define different domains of the infected erythrocyte intramembranous network. Parasitol. Res. *104*, 875–891.

Waller, K.L., Cooke, B.M., Nunomura, W., Mohandas, N., and Coppel, R.L. (1999). Mapping the binding domains involved in the interaction between the *Plasmodium falciparum* knob-associated histidine-rich protein (KAHRP) and the cytoadherence ligand *P. falciparum* erythrocyte membrane protein 1 (PfEMP1). J. Biol. Chem. *274*, 23808–23813.

Waller, K.L., Nunomura, W., An, X., Cooke, B.M., Mohandas, N., and Coppel, R.L. (2003). Mature parasite-infected erythrocyte surface antigen (MESA) of *Plasmodium falciparum* binds to the 30-kDa domain of protein 4.1 in malaria-infected red blood cells. Blood *102*, 1911–1914.

Waller, K.L., Nunomura, W., Cooke, B.M., Mohandas, N., and Coppel, R.L. (2002). Mapping the domains of the cytoadherence ligand *Plasmodium falciparum* erythrocyte membrane protein 1 (PfEMP1) that bind to the knob-associated histidine-rich protein (KAHRP). Mol. Biochem. Parasitol. *119*, 125–129.

Waller, K.L., Stubberfield, L.M., Dubljevic, V., Nunomura, W., An, X., Mason, A.J., Mohandas, N., Cooke, B.M., and Coppel, R.L. (2007). Interactions of *Plasmodium falciparum* erythrocyte membrane protein 3 with the red blood cell membrane skeleton. Biochim. Biophys. Acta *1768*, 2145–2156.

Waller, R.F., Keeling, P.J., Donald, R.G., Striepen, B., Handman, E., Lang-Unnasch, N., Cowman, A.F., Besra, G.S., Roos, D.S., and McFadden, G.I. (1998). Nuclear-encoded proteins target to the plastid in *Toxoplasma gondii* and *Plasmodium falciparum*. Proc. Natl. Acad. Sci. U.S.A. *95*, 12352–12357.

Waller, R.F., Reed, M.B., Cowman, A.F., and McFadden, G.I. (2000). Protein trafficking to the plastid of *Plasmodium falciparum* is via the secretory pathway. EMBO J. *19*, 1794–1802.

Waterkeyn, J.G., Wickham, M.E., Davern, K.M., Cooke, B.M., Coppel, R.L., Reeder, J.C., Culvenor, J.G., Waller, R.F., and Cowman, A.F. (2000). Targeted mutagenesis of *Plasmodium falciparum* erythrocyte membrane protein 3 (PfEMP3) disrupts cytoadherence of malaria-infected red blood cells. EMBO J. *19*, 2813–2823.

WHO. (2010). World Malaria report. http://www.who.int/malaria/world_malaria_report_2010/world_malaria_report_2010.pdf.

Wickert, H., Wissing, F., Andrews, K.T., Stich, A., Krohne, G., and Lanzer, M. (2003). Evidence for trafficking of PfEMP1 to the surface of *P. falciparum*-infected erythrocytes via a complex membrane network. Eur. J. Cell Biol. *82*, 271–284.

Wickham, M.E., Rug, M., Ralph, S.A., Klonis, N., McFadden, G.I., Tilley, L., and Cowman, A.F. (2001). Trafficking and assembly of the cytoadherence complex in *Plasmodium falciparum*-infected human erythrocytes. EMBO J. *20*, 5636–5649.

Winter, G., Kawai, S., Haeggstrom, M., Kaneko, O., von Euler, A., Kawazu, S., Palm, D., Fernandez, V., and Wahlgren, M. (2005). SURFIN is a polymorphic antigen expressed on *Plasmodium falciparum* merozoites and infected erythrocytes. J. Exp. Med. *201*, 1853–1863.

Dissecting Mosquito–Parasite Interactions through Molecular Biology and Biochemistry: Genomic, Proteomic and Glycomic Analyses

10

Lindsay A. Parish, Lindsey S. Garver, David R. Colquhoun, Ceereena Ubaida Mohien, Elizabeth Weissbrod and Rhoel R. Dinglasan

Abstract

Our understanding of the malaria parasite-mosquito vector host interactions has grown significantly in the post-genomic era. The sequencing of the *Anopheles gambiae* genome and functional genomics has revolutionized our approaches at examining the role of the mosquito vector host in malaria transmission. It is hoped that 13 other anopheline species (including the major vectors for *Plasmodium vivax*) will be completely sequenced within the next decade. Now, more than ever, there is a heightened research focus on Vector Biology and the transmission stages of the malaria parasite through the mosquito, from gametocytes to sporozoites, with the underlying aim of developing novel transmission-blocking vaccines/interventions. This chapter provides an overview of what is currently known at the genomic, proteomic and glycomic levels, about the role of different molecules, tissue compartments and molecular pathways in the mosquito vector and the ensuing strategies that have evolved, which can potentially impact malaria parasite transmission.

Introduction

The complex parasite life cycle includes developmental stages within both the mammalian host and its insect vector, the *Anopheles* mosquito (Fig. 10.1). Immediately following ingestion of the blood meal, *Plasmodium* sexual stages (male and female gametocytes) undergo gametogenesis. During this rapid process (8–10 min), female gametocytes transform into female gametes, called macrogametes, that involves egress from the red blood cell and an increase in size. Male gametocytes undergo a fascinating transformation within the same time period that includes an increase in the size of the nucleus, displacement of the cytoplasm to the periphery of the cell and then finally an emergence of eight highly motile, flagellar axonemes or microgametes (recently reviewed in Sinden, 2009). Collectively these events are referred to as exflagellation, which occur 10–15 min post activation in the blood meal. The exflagellating males search for macrogametes within the blood meal bolus inside the mosquito midgut. Upon finding female macrogametes, fertilization occurs and results in the formation of a motile zygote, called an ookinete. *Plasmodium* ookinetes that have formed in the mosquito midgut lumen migrate to the periphery where they are thought to recognize the peritrophic matrix (PM), a chitinous and proteinaceous sac which envelops the ingested blood meal (Sieber *et al.*, 1991; Shen and Jacobs-Lorena, 1998; Dinglasan *et al.*, 2009). The ookinetes pass through the PM into the ectoperitrophic space where they are thought to recognize and adhere to molecules that are present on the apical face of the mosquito midgut epithelial cell or what is termed as the brush border microvilli. After binding to the midgut surface, the ookinete subsequently traverses the single cell layer of the midgut epithelium. Once reaching the basal lamina of the midgut epithelium, these ookinetes transform into oocysts just beneath the basal labyrinth, which mature over 7–15 days

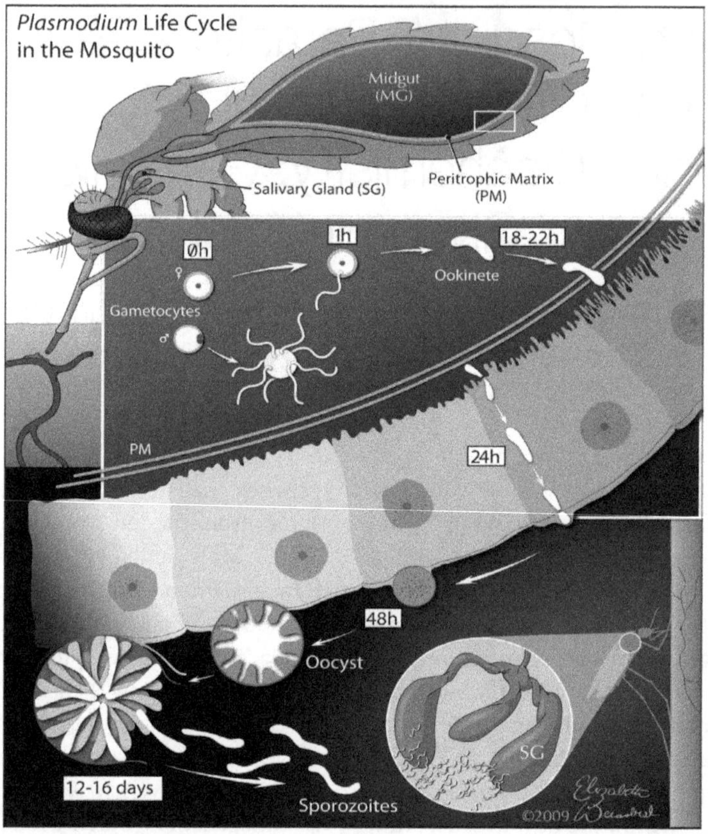

Figure 10.1 Generalized schematic of the obligatory sporogonic development of human malaria parasites in anopheline mosquito vectors. The sexual stage male and female gametocytes are taken up into the mosquito midgut during blood feeding. Within minutes, these gametocytes transform into male and female gametes. Gamete fertilization produces a motile zygote called an ookinete. At around 24 h post-blood feeding, the ookinetes have already penetrated the peritrophic matrix (PM) and the midgut apical plasma membrane, actively traversing the midgut epithelial cell to form an oocyst just beneath the basal lamina. Within a single oocyst thousands of sporozoites develop and at around 12–16 days post blood feeding, the sporozoites are released into the haemocoel of the mosquito. The mosquito's circulatory system brings the sporozoites from the abdomen of the mosquito to the thorax, allowing the sporozoites to contact and invade the mosquito's salivary glands (see Fig. 10.3 for additional details).

depending on the species, culminating with the release of thousands of sporozoites into the mosquito haemocoel. The sporozoites use the rapid anterograde flow of the mosquito haemolymph to reach the salivary glands (Akaki and Dvorak, 2005; Hillyer, 2007; Glenn *et al.*, 2010). Sporozoites adhere to the basal membrane of this tissue prior to active invasion and traversal of the salivary gland epithelial cell (reviewed in Mueller *et al.*, 2010). Finally, the majority of the sporozoites assemble themselves into parallel bundles within the secretory cavity with a few entering the salivary duct in single file. At this point, the mosquito is capable of transmitting the sporozoites to a vertebrate host when the mosquito takes a subsequent blood meal (Frishknecht *et al.*, 2004; Amino *et al.*, 2005, 2006; Frishknecht *et al.*, 2010). The trek of *Plasmodium* from the midgut lumen to the salivary glands, involves multiple interactions between the parasite and *Anopheles* macromolecules. This section reviews the known mosquito host molecules which have been shown to mediate 'contact' with the parasite, and the current genetic, proteomic and glycomic approaches that have been or can be used to characterize these interactions.

Mosquito–parasite interactions

In order to understand the complex relationship of the parasite with different tissue compartments in the mosquito vector, it is necessary to consider the similarities and differences between the different tissues *Plasmodium* parasites interact with. This section describes briefly the tissue and cellular architecture of the midgut, haemolymph and the salivary glands of the mosquito, in order to provide a foundation for understanding what is currently known about the initial interaction of parasite molecules with these compartments.

Midgut architecture

The sugar-fed, steady-state mosquito midgut is composed primarily of a single layer of polarized columnar epithelial cells (Fig. 10.2) with the apical membrane exposed to the midgut lumen and the basal side surrounded by the basal lamina, which is bathed in haemolymph. Numerous finger-like projections of microvilli extend from the apical membrane and serve to secrete digestive enzymes and aid in nutrient absorption. The basal membrane is characterized by elaborate convolutions which facilitate water and ion transport and is often referred to as the basal labyrinth. During blood feeding, a second membrane is formed, which is called the peritrophic matrix (PM). The Type 1 PM is a sac that completely envelops the blood meal, protecting the mosquito midgut apical plasma membrane from potentially harmful by-products of blood digestion. Components of the PM are stored in secretory vesicles within midgut epithelial cells and are released into the lumen of the midgut upon blood feeding (Devenport *et al.*, 2004, 2005). The PM has an underlying structure composed of long chitin fibrils, which are repeating β-linked *N*-acetylglucosamine disaccharides (GlcNAcβ1–4GlcNAc). This framework allows for the attachment and assembly of glycoproteins onto the PM, giving its characteristic thickness and function as a sieve involved, in part by sequestering haem that is released following digestion of the blood (Devenport *et al.*, 2006). PM formation normally requires ingestion of a blood meal. However, the formation of the PM can also be artificially induced without a blood meal by feeding mosquitoes protein-free latex beads (Dinglasan *et al.*, 2009). Latex beads are mixed with a buffer containing ATP which is added as a phagostimulant. The proteome of the latex-induced PM has been recently characterized by multidimensional protein identification technology (MudPIT; Dinglasan *et al.*, 2009) and, as expected, was found to be composed of chitin-binding proteins, known as peritrophins (Abraham *et al.*, 2003; Dinglasan *et al.*, 2009) and intercalating digestive enzymes, some of which appear to be chimaeras, with one domain involved in binding to the chitin fibrils and the other in enzymatic reactions. However, although PM

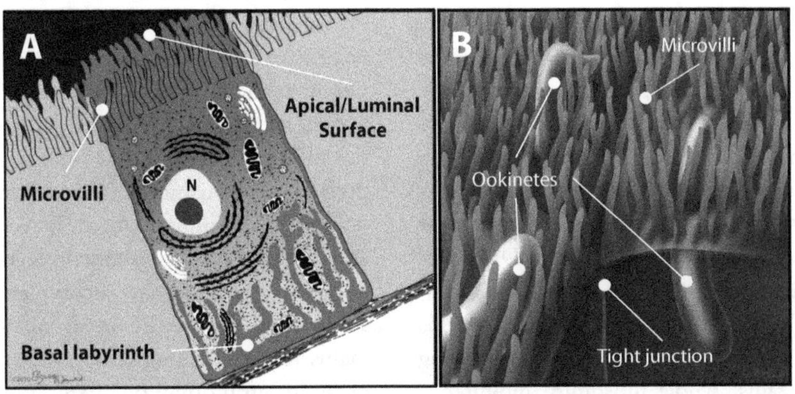

Figure 10.2 Ookinete recognition, adhesion and invasion of the mosquito epithelium. (A) The midgut epithelium is composed of individual polarized epithelial cells with both apical (luminal) and basal domains. N, nucleus. (B) Ookinetes adhere to and glide along the apical midgut surface, interacting with glycoproteins and other glycoconjugates on microvilli as attachment ligands. Ookinetes appear to migrate and invade the plasma membrane of the midgut epithelial cells in close proximity to tight junctions between cells.

secretion can be induced artificially in *Anopheles* mosquitoes after feeding, the protein-free latex meal appears to be a weak stimulator of digestive enzyme expression and secretion (Billingsley and Rudin, 1992; Lemos *et al.*, 1996). Given this limitation in sample preparation, it is likely that the current PM proteome remains incomplete.

Proteins in the apical midgut membrane can be enriched through the preparation of brush border membrane vesicles (BBMVs). To isolate BBMVs, the midgut tissue is homogenized and the cells are then hypotonically lysed in a buffer containing mannitol. After multiple centrifugation steps, a pellet containing the BBMV is obtained. The resulting BBMVs are characterized by alkaline phosphatase (Hua *et al.*, 2009) and aminopeptidase (Okuda *et al.*, 2005) activity and have aided in the exploration of the biological roles of BBMV glycoproteins across different insect systems by enzymatic and binding assays (Carroll and Ellar, 1993; Lee *et al.*, 2006). Examination by electron microscopy reveals that BBMVs appear as spherical vesicles of varying size and are devoid of organelles (Abdul-Rauf and Ellar, 1999). However, common co-purifying contaminants in BBMV preps include mitochondrial and cytoplasmic proteins (Forcella *et al.*, 2006; Popova-Butler and Dean, 2009) and therefore, bioinformatic analyses of proteins identified from such samples by mass spectrometry require several rounds of post-spectra acquisition curation and scrutiny.

Midgut interactions

After penetrating the peritrophic matrix, ookinetes contact the apical membrane of the midgut epithelium which is abundantly covered with a microvilli-associated glycocalyx and various glycoproteins. Recent evidence suggests that *Plasmodium* parasites interact with multiple *Anopheles* midgut proteins and glycans in order to traverse the midgut epithelium (reviewed in Dinglasan and Jacobs-Lorena, 2005; Dinglasan and Jacobs-Lorena, 2008; Table 10.1). In support of this hypothesis, it was shown early on that generating antibodies against whole mosquito midgut can block the development of *Plasmodium falciparum* and *Plasmodium vivax* (Lal *et al.*, 2001). However, the identities of the key protein epitopes on the midgut were not identified in this study. More recently, polyclonal antibodies raised against a specific protein target, the *Anopheles gambiae* alanyl aminopeptidase (AgAPN1) appear to significantly block both *P. falciparum* and *Plasmodium berghei* development in two different vectors, *Anopheles gambiae* and *Anopheles stephensi* (Dinglasan *et al.*, 2007; Mathias *et al.*, 2012). Similarly, salivary gland and midgut peptide 1 (SM1), a twelve amino acid peptide that has been shown to adhere to the apical midgut, can block up to 70% of the development of *P. berghei* but not *P. falciparum* parasites (Ghosh *et al.*, 2001). The blocking effects of SM1 and AgAPN1 antibodies were found to be additive and result in 95% inhibition of parasite development (Dinglasan *et al.*, 2007). These results suggest that the SM1 midgut binding peptide and APN1 are distinct ligands recognized by *P. berghei* ookinetes. Moreover, it appears that *P. berghei* and *P. falciparum* utilize different adhesion pathways as SM1 fails to block *P. falciparum* development (Dinglasan *et al.*, 2007). However, at present the exact SM1 binding ligand(s) on midgut surface remains unknown. Antibodies raised against another *An. gambiae* midgut protein, carboxypeptidase B 1 (CPBAg1), reduced both parasite development within the midgut and carboxypeptidase B activity (Lavazec *et al.*, 2006). The precise mechanism of how APN1 and CPBAg1 antibodies inhibit ookinetes from traversing the midgut remains to be elucidated. A list of mosquito midgut molecules that have been successfully targeted molecularly and immunologically are found in Table 10.1.

To date, six *An. gambiae* midgut proteins (AgAPN1, ANXB10B, ANXB10C, ANXB9, CpbAg1, and SCRBQ2) have been demonstrated experimentally to facilitate ookinete invasion of the midgut. However, our understanding of the precise mechanisms of invasion, including how ookinetes coordinate multiple protein–protein and protein–glycan interactions on the apical midgut surface, remains incomplete. One recent study has proposed the involvement of microdomains in the plasma membrane, known as lipid rafts, in promoting ookinete midgut invasion (Parish *et al.*, 2011). Lipid rafts are dynamic, mobile microdomains which are enriched in cholesterol, sphingolipids, and proteins involved in the cellular processes of signal transduction,

Table 10.1 A summary list of anopheline proteins that have been implicated as ligands for midgut attachment and invasion by *Plasmodium* parasites. The list is not intended to be comprehensive. The genes listed in this table represent those molecules that have been targeted either by antibodies or RNAi-mediated silencing and, importantly, result in a *decrease* in parasite oocyst numbers in the midgut following treatment

Name	Accession	Description	Overall findings	Reference
ANT	AGAP006782	Adenine nucleotide translocator	Knockdown of ANT in results in a significant decrease in *P. berghei* oocysts	Oliveira *et al.* (2011)
ANXB9	AGAP003790	Annexin	Antibodies against ANXB9 block 30.0–38.4% of *P. berghei* development in the midgut[a]	Kotsyfakis *et al.* (2005)
ANXB10B	AGAP003721	Annexin	Antibodies against ANXB10B blocked 36.3–39.6% of *P. berghei* development in the midgut[a]	Kotsyfakis *et al.* (2005)
ANXB10C	AGAP003722	Annexin	Antibodies against ANXB10C blocked 28.2–43.7% of P. berghei development in the midgut[a]	Kotsyfakis *et al.* (2005)
AgAPN1	AGAP004809	Aminopeptidase N	Antibodies against AgAPN1 significantly inhibit both *P. falciparum* and *P. berghei* development in the midgut[a]	Dinglasan *et al.* (2007)
ArgK	AGAP005627	Arginine kinase	Knockdown of ARGK in results in a significant decrease of both *P. berghei* and *P. falciparum* oocysts	Jaramillo-Gutierrez *et al.* (2009)
AgCatalase	AGAP004904	Catalase	Knockdown of *AgCatalase* results in a significant decrease of *P. berghei* oocysts 1, 2 and 11 days post infection	Molina-Cruz *et al.* (2008)
CpbAg1	AGAP006209	Carboxypeptidase B	Antibodies against CpbAg1 significantly inhibit both *P. falciparum* and *P. berghei* development in the midgut[a]	Lavazec *et al.* (2007)
CTL4	AGAP005335	C-type Lectin 4	Knockdown of *CTL4* results in a significant decrease of *P. berghei* oocysts but has no effect on *P. falciparum* oocyst numbers	Osta *et al.* (2004), Cohuet *et al.* (2006)
CTLMA2	AGAP005334	C-type lectin (mannose binding)	Knockdown of *CTLMA2* results in 53% melanization of invading *P. berghei* ookinetes	Osta *et al.* (2004)
Duox	AGAP009978	Dual oxidase	Knockdown of *Duox* results in a substantial decrease in *P. berghei* oocysts[a]	Kumar *et al.* (2010)
GSTT1	AGAP000761	Glutathione S-transferase (theta class 1)	Knockdown of *GSTT1* in results in a significant reduction in *P. berghei* oocyst but has no effect *on P. falciparum* oocyst numbers	Jaramillo-Gutierrez *et al.* (2009)
GSTT2	AGAP000888	Glutathione S-transferase (theta class 2)	Knockdown of *GSTT2* in results in a significant reduction in P. berghei oocysts but results in an increase in *P. falciparum* oocysts	Jaramillo-Gutierrez *et al.* (2009)
Hsc-3	AGAP004192	Heat-Shock Cognate 3	Knockdown of *Hsc-3* in results in a significant reduction in *P. falciparum* oocysts but has no effect on the numbers of *P. berghei* oocyst	Jaramillo-Gutierrez *et al.* (2009)
IMPer	AGAP013327	Immunomodulatory peroxidase	Knockdown of *IMPer* by RNAi results in a 9.2 fold reduction in *P. berghei* oocysts[a]	Kumar, S. *et al.*, 2010.
LANB2	AGAP007629	Laminin camma 1	Knockdown of *LANB2* in results in a significant reduction in *P. berghei* oocyst	Arrighi, R.B. *et al.*, 2005.
AgOXR1	AGAP001751	Oxidation resistance 1	Knockdown of *AgOXR1* in results in a 7-fold decrease in the median number of oocyst 2 days post infection and a significant decrease 6 days post infection. However, knockdown of *AgOXR1* has no effect of the median number of *P. falciparum* oocyst 7–8 days post infection	Jaramillo-Gutierrez *et al.* (2010), Brandt *et al.* (2008)
PRS1	AGAP006102	*Plasmodium*-responsive salivary 1	Knockdown of *PRS1* in results in a significant decrease in *P. berghei* oocysts glands	Chertemps *et al.* (2010)

Table 10.1 Continued

Name	Accession	Description	Overall findings	Reference
SCRBQ2	AGAP010133	Scavenger receptor; Croquemort homologue	Knockdown of *SCRBQ2* results in a 62.5% inhibition of *P. berghei* oocyst formation[a]	González-Lázaro *et al.* (2009)
SOCS	AGAP011042	Suppressor of cytokine signalling	Knockdown of *SOCS* results in a significant inhibition of *P. berghei* oocyst formation 8 days post infection	Gupta *et al.* (2009)
SRPN2	AGAP006911	Serine protease inhibitor	Knockdown of SRPN2 by RNAi results in a 97% decrease in *P. berghei* oocysts	Michel *et al.* (2005)

[a]Data suggest that the protein is found on the surface of the mosquito midgut. Localization of the other proteins have not been confirmed.

cell adhesion, and vesicle trafficking (reviewed in Simons *et al.*, 2010). Numerous pathogens throughout nature appear to exploit lipid rafts as the clustering of host receptors in a concentrated region of the plasma membrane is thought to allow for the strengthening of multivalent protein–protein and protein–glycan interactions between the pathogen and the host cell which are essential for attachment and invasion of the host cell to occur (Riethmuller *et al.*, 2006). Recently, the detergent resistant membrane (DRM) proteome (which contains lipid rafts) of *An. gambiae* midguts was characterized and altogether, 1452 proteins were identified in DRMs including five of the six (AgAPN1, ANXB10B, ANXB9, CpbAg1, and SCRBQ2) known ookinete-interacting proteins (Parish *et al.*, 2011). These results imply that lipid rafts may be utilized by ookinetes during the invasion process. Thus far, the cognate mosquito receptors for each of the parasite ligands have not been identified implying that uncharacterized midgut proteins involved in the invasion process remain to be discovered. Further characterization of the identified DRM proteins may aid in revealing new proteins which are utilized by ookinetes during invasion of the midgut epithelium.

In order to traverse the midgut epithelium, ookinetes penetrate the apical cell membrane and frequently pass through several neighbouring midgut epithelial cells before reaching the basal lamina. This invasion process triggers gross changes in cell morphology including loss of microvilli from the cell surface, DNA fragmentation, nitric oxide synthase expression, and results in apoptosis of the invaded cells (Han *et al.*, 2000; Kumar *et al.*, 2002). Eventually, these damaged cells are pinched off from the epithelial layer and are expelled into the midgut lumen. The 'time bomb model' of ookinete invasion of midgut epithelial cells (reviewed in Kumar and Barillas-Mury, 2005) proposes that parasite invasion elicits a cascade of events that culminates in cell death. To successfully reach the basal lamina, ookinetes must exit the midgut epithelial layer before the damaged cells produce nitric oxide which is potentially toxic to the invading parasite (Kumar *et al.*, 2004).

The haemocoel and haemolymph

Mosquitoes possess an open circulatory system that disseminates haemolymph and nutrients throughout the haemocoel. Haemolymph flow throughout the mosquito body is bidirectional and is controlled by the mosquito heart (Glenn *et al.*, 2010). There are several types of mosquito blood cells (haemocytes), hormones, nutrients and other factors (e.g. products of blood digestion) that accumulate or are present in the circulating haemolymph. The three major types of haemocytes that are present in *Anopheles gambiae* adult mosquitoes are granulocytes, oenocytoids and prohaemocytes (Castillo *et al.*, 2006). Among the many functions of haemocytes is a role in the defence of the mosquito against pathogens or invading microorganisms (Pinto *et al.*, 2009). This defensive action can be direct, e.g. phagocytosis for granulocytes, in particular, or indirect, e.g. the secretion of soluble antimicrobial peptides or components of the phenoloxidase cascade mediating melanization and blood coagulation.

Haemocoel interactions

Once invading ookinetes cross the midgut epithelium and become exposed to the basal labyrinth, soluble immune components from haemocytes such as thioester-containing protein 1 (TEP1), leucine-rich repeat immune protein 1 (LRIM1), and *Anopheles Plasmodium*-responsive leucine-rich repeat 1 (APL1) are able to cross the labyrinth and coat ookinetes (reviewed in Blandin *et al.*, 2008). TEP1 is a complement-like protein which is structurally and functionally similar to the vertebrate complement factor C3 (Baxter *et al.*, 2007; Levashina *et al.*, 2001). Haemocytes constitutively secrete TEP1 into the haemolymph where it is proteolytically cleaved and then capable of opsonizing pathogens such as Gram-positive and Gram-negative bacteria, thereby enhancing their phagocytosis (Levashina *et al.*, 2001). Similarly, TEP1 has been shown to bind to the surface of ookinetes but appears to cause parasite lysis rather than induce phagocytosis at this stage as haemocytes, the major phagocytic cells, cannot penetrate the basal lamina (Blandin *et al.*, 2004). The importance of TEP1 in the mosquito immune response is shown when TEP1 is knocked down by RNA interference. This results in a four to five fold increase of *P. berghei* oocysts in susceptible and refractory strains of *An. gambiae* mosquitoes (Levashina *et al.*, 2001). Additionally, LRIM1 and APL1 form a protein complex with TEP1. The formation of this complex is necessary for TEP1 stabilization as knockdown of either LRIM1 or APL1 depletes TEP1 from the haemolymph (Fraiture *et al.*, 2009).

Another mosquito immune response to pathogens is melanization, a process by which pathogens are coated with a hardened proteinaceous capsule (Collins *et al.*, 1986; Beerntsen *et al.* ,1989). Melanization is initiated when multiple Clip domain serine proteases convert prophenoloxidase into phenoloxidase. Phenoloxidase subsequently oxidizes compounds containing a phenol group into quinines which then in turn form melanin (Jiang *et al.*, 1998). Melanization of invading ookinetes is seldom observed in *An. gambiae* strains that are susceptible to *P. berghei* infection (Collins *et al.*, 1986). In contrast, in the L35 *An. gambiae* strain, which is refractory to infection with *P. berghei* and many other *Plasmodium* species, over 95% of ookinetes are melanized. When TEP1 is knocked down in refractory mosquitoes by RNA interference, ookinete melanization is no longer observed and refractory females become susceptible to *P. berghei* infection (Blandin *et al.*, 2004). These data indicate that mosquito immunity is a major determinant of mosquito susceptibility to *Plasmodium* infection. Interestingly, some *P. falciparum* strains of African origin are not melanized and survive well in the refractory L35 strain (Collins *et al.*, 1996). Laminin, a primary component of the basal lamina, appears to closely associate with the maturing oocyst and has been hypothesized to aid in inhibiting melanization of the parasite (Arrighi *et al.*, 2005; Warburg *et al.*, 2006). The mechanism for this proposed protective effect is unknown.

After the oocyst ruptures, thousands of sporozoites are released into the mosquito haemolymph and are circulated throughout the haemocoel. Sporozoites are now accessible to the mosquito cellular immune response which is mediated predominantly by haemocytes. At this point, the parasite population suffers a drastic bottleneck as only 10–20% of sporozoites successfully reach and invade the salivary glands. The remaining 80–90% of sporozoites appear to be either phagocytosed by haemocytes or degraded by an unknown mechanism in the haemolymph (Hillyer *et al.*, 2003, 2007). A proportion of the sporozoites are found to accumulate in non-target tissues such as the flight muscle (Akaki and Dvorak, 2005). It was proposed that sporozoites use a chemotactic gradient present in the haemolymph to home in on the salivary glands (Akaki and Dvorak, 2005) but given the velocity of the haemolymph, the relative importance of this gradient on the success of sporozoite homing and invasion of the salivary glands remains unclear.

The salivary gland

The adult mosquito has a pair of salivary glands located in the thorax. Each gland consists of one medial and two lateral lobes (Fig. 10.3A). Each lobe, in turn, consists of a single epithelial layer surrounding a common duct and can be subdivided into proximal and distal regions.

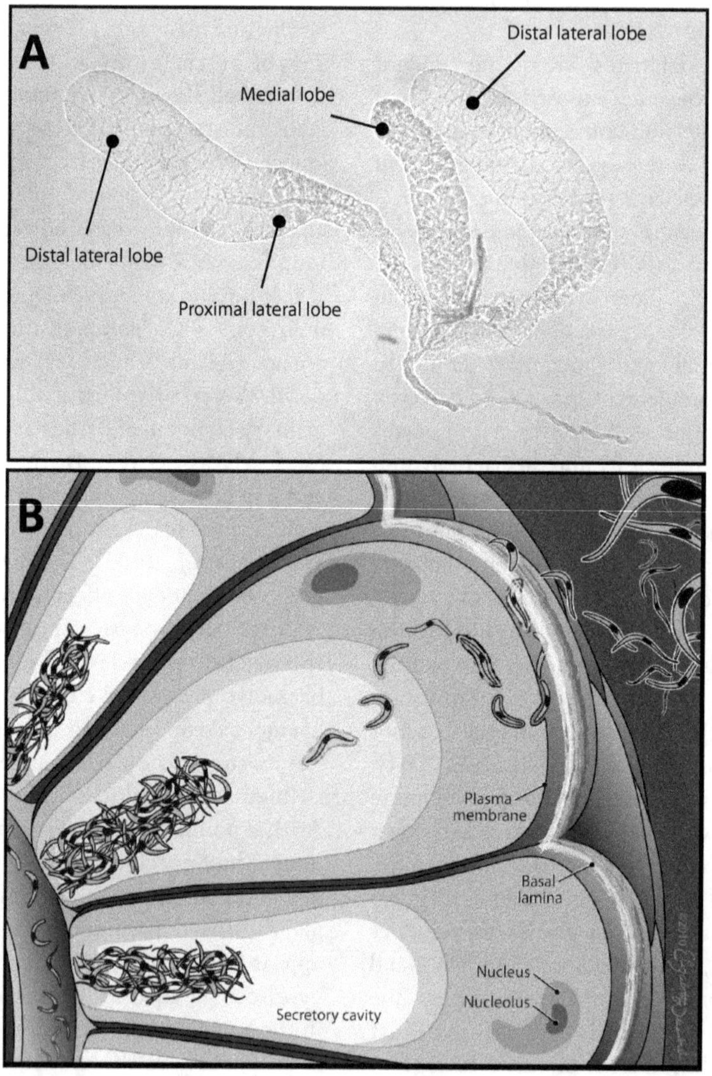

Figure 10.3 Generalized schematic of sporozoite invasion of the distal lobes of the mosquito salivary glands. Sporozoites are brought to the salivary glands by the anterograde flow of the mosquito haemolymph. Upon contact with the distal lateral lobes of the salivary glands (A), the sporozoites quickly adhere to the basal membrane of this tissue prior to active invasion and traversal of the salivary gland epithelial cell (B). The majority of the sporozoites assemble themselves into parallel bundles within the secretory cavity. A few sporozoites are able to enter the salivary duct in single file ready to be delivered to a new vertebrate host once the mosquito takes another blood meal. Micrograph courtesy of R. Dinglasan.

Salivary gland interactions

Sporozoites circulating in the haemolymph come into contact with many different cells and in some cases find themselves closely associated and irreversibly attached to inappropriate tissues. However, the only tissue that sporozoites invade is the mosquito salivary glands (Fig. 10.3), suggesting that there are ligands on the basal side of salivary glands which enable the parasite to specifically target this organ (reviewed in Ghosh and Jacobs-Lorena, 2009; Hegge *et al.*, 2010; Table 10.1). Sporozoites can invade the medial lobes and the distal portion of the lateral lobes of the salivary glands (Sterling *et al.*, 1973; Rossignol *et al.*, 1984), but are more commonly found in the distal lateral lobes. Like ookinete invasion, salivary gland invasion by sporozoites appears to be a multistep process (Pimenta *et al.*, 1994) and has been shown to be partly mediated by glycosaminoglycans (Armistead *et al.*, 2011). Sporozoites recognize and attach to the basal lamina extracellular matrix (ECM) of the salivary glands across their entire length, but presumably invasion proceeds following motile orientation of the anterior or apical tip of the sporozoite towards the salivary gland epithelial cell beneath the ECM. During penetration of the salivary gland basal ECM, the parasite interacts with the host cell membrane and forms a transient parasitophorous vacuole-like membrane as it progresses through the cell. The exact mechanism of formation and dissolution of this membrane remains unexplained but electron microscopy studies indicated that it is akin to a moving tight junction (Pimenta *et al.*, 1994; Rodriguez and Hernandez-Hernandez, 2004). As sporozoites exit the salivary epithelial cell, they are no longer surrounded by a membrane and accumulate in the central secretory cavity.

In order to discover potential *Plasmodium* salivary gland ligands, monoclonal antibodies against salivary gland proteins were produced and then assessed for the capability to prevent invasion by sporozoites. One monoclonal antibody recognized a 100-kDa protein and when injected into the mosquito haemocoel causes a 73% decrease in sporozoites invasion of the salivary glands (Brennan *et al.*, 2000). This monoclonal antibody was later shown to recognize the *An. gambiae* salivary protein saglin (Okulate *et al.*, 2007). Independent experiments with the SM1 peptide, which inhibits ookinete invasion of the midgut and sporozoite invasion of the salivary glands, also corroborate saglin as a receptor for sporozoites. Pull-down assays using a UV cross-linker identified saglin as one of the SM1 ligands in the salivary glands. Two other SGS (Salivary Gland Surface) proteins, gSG1 and gSG1b, that share sequence homology to saglin were also cross-linked to SM1 (Ghosh *et al.*, 2009). However, the localization of gSG1 and gSG1b to the salivary basal surface and their function in sporozoite invasion of the salivary glands has not been confirmed. The role of saglin in salivary gland invasion is further supported by RNA interference experiments which result in up to 98% inhibition of salivary gland invasion by sporozoites as compared to controls. Moreover, a monoclonal antibody to SM1 was produced and found to recognize the *Plasmodium* sporozoite protein TRAP. Recombinant TRAP protein fragments were shown to bind mosquito salivary glands and this interaction can be inhibited with the addition of SM1. Additionally, recombinant TRAP has been demonstrated to bind to recombinant saglin *in vitro* (Ghosh *et al.*, 2009). Data suggest that Saglin plays a significant role in sporozoite invasion of the salivary glands.

Functional genomics and transcriptomics

In 2002 the genome sequence of *Anopheles gambiae*, the mosquito responsible for the majority of human malaria transmission in the world, became available, changing the landscape of medical entomology (Holt *et al.*, 2002). What followed was a boom of molecular genetics, spurring use of techniques in functional genomics, transcriptomics and reverse genetics in an organism that crucially affects human health but had a paltry repertoire of molecular tools. In the last few years, the use of microarrays to globally assess how mosquitoes respond to parasites has evolved from a description of induction/repression to a targeted screening method aimed at identifying candidate genes for RNAi. In the coming years, next generation sequencing is expected to add an unprecedented depth and refinement to the existing functional genomics methods.

Microarrays

Arrays have shown that mosquitoes respond differently to different parasites and that these mosquitoes respond variably to antigens in the lumen (infected blood) and active invasion (Vlachou *et al.*, 2005; Dong *et al.*, 2006a). More Anopheles genes are regulated by *P. berghei* than *P. falciparum* infection and more genes were regulated by the presence of *P. berghei* in the bolus than were regulated by active invasion, particularly in the body (tissue remaining after midgut dissections) (Dong *et al.*, 2006a). Invasion-responsive genes come from functional classes of cytoskeletal arrangement, apoptosis, immunity, redox-metabolism, detoxification and others (Vlachou *et al.*, 2005 and Dong *et al.*, 2006a).

Haemocytes robustly sense and respond to *P. berghei* infection at both midgut invasion and sporozoite migration, regulating many of the immune defence genes employed by the mosquito including components of the IMD and JNK and melanization pathways and several pattern recognition receptors (PRRs). Little overlap (5.3% of regulated genes) is observed between haemocyte responses to the two stages, indicating this cell type inherently changes over time or is very flexible in responding to different parasite life stages (Baton *et al.*, 2009). SAGE analysis of salivary glands of infected versus non-infected mosquitoes shows that transcriptomic responses to sporozoites include the up-regulation of immune genes encoding cecropin2, serpin6, defensin1 and GNBP as well as genes important for transport and metabolism (Rosinski-Chupin *et al.*, 2007). Together, these analyses demonstrate that mosquitoes are responding to parasites all throughout their association, in multiple tissue types and with transcripts encoding a wide variety of proteins.

As the materials became more affordable and genome annotation became more thorough, arrays became more specific and used in more hypothesis-driven applications. For example, an array was developed to hone in on transcription of detoxification genes during and after invasion and in two tissues, the midgut and the fat body (Felix *et al.*, 2010). By observing such patterns, one can speculate on which genes are functionally related or in common pathways. Another custom array analysis uses exon specific probes to identify alternative splicing under variable conditions. *Dscam* is an Ig-domain-containing gene that theoretically encodes over 36,000 isoforms. By probing variable exons, researchers were able to determine that some exons are used more frequently during assaults by *P. berghei* and *P. falciparum*, suggesting Dscam is differentially spliced accordingly (Dong *et al.*, 2006b).

RNAi

In the last decade, reverse genetics techniques have dominated the way vector biologists determine a gene's function and has resulted in a battery of genes related to the insect's interactions with *Plasmodium*. For a review of RNAi in mosquitoes, please see Catteruccia and Levashina, 2009. The major discoveries made using RNAi have been in the subfields of ligand–parasite interactions and innate immune defence (for review, see Christophides *et al.*, 2004; Osta *et al.*, 2004b; Baton *et al.*, 2008).

In insects, there are three known immune pathways, the Toll, the IMD (with a separate JNK branch), and the Jak/Stat pathways. These pathways link detection of microbes with induced transcription of immune effector genes and, through knockdown of positive and negative regulatory components, all pathways have been implicated in the defence against at least one species of *Plasmodium*. Among them, the Toll pathway seems to have some specificity for *P. berghei* while the IMD pathway and the Jak/Stat pathway have reported activity against both species and the JNK pathway is necessary for defence against *P. berghei* although the relation of this phenotype to that of the IMD pathway is unexplored (Meister *et al.*, 2005; Frolet *et al.*, 2006; Garver *et al.*, 2009; Gupta *et al.*, 2009; Jaromillo-Gutierrez *et al.*, 2010).

Microarrays coupled to gene silencing have determined that the IMD pathway causes regulation of a battery of immune genes such as *cecropin3* (*CEC3*), *gambicin1* (*GAM1*), several CLIP domain serine proteases, *thioester-containing protein 1* (*TEP1*), *fibrinogen immunolectin 9* (*FBN9*), and others (Meister *et al.*, 2005, Garver *et al.*, 2009). Similar arrays show the Toll pathway regulates a separate but overlapping gene set (Garver *et al.*, 2009). Double silencing experiments show TEP1 acting downstream of the Toll and IMD pathways while APL1C is downstream of Toll and NOS is acting downstream of the Jak-Stat pathway (Frolet *et al.*, 2006; Riehle *et al.*, 2008; Garver *et al.*, 2009; Gupta *et al.*, 2009).

TEP1, FBN family members, immunoglobulin family members (Dscam and others), leucine-rich repeat-containing proteins (LRIM/APL families), C-type lectins (CTLs), beta-glucan binding proteins (GNBPs) and scavenger receptors have all been implicated as players during *Plasmodium* infections (Blandin *et al.*, 2004; Osta *et al.*, 2004; Dong *et al.*, 2006a,b, 2009a; Garver *et al.*, 2008; Riehle *et al.*, 2008; Warr *et al.*, 2008; Gonzalez-Lazaro *et al.*, 2009). Some (like *GNBPs*) encode characteristic PRRs likely playing a detection role, though the epitopes recognized are unknown. Others, such as TEP1 have pathogen-binding and other functional domains suggesting effector activity. Further analysis has led to excellent characterization of TEP1 and its binding partners as executors of parasite lysis (Blandin *et al.*, 2004; Fraiture *et al.*, 2009; Povelones *et al.*, 2009) making TEP1 a leading example of how deeper knowledge of *Anopheles–Plasmodium* interactions was gained by synergizing functional genomics with a multitude of other techniques.

Intriguingly, gene silencing does not give identical results when different mosquitoes or parasite species are examined. In field studies, natural *P. falciparum* populations carried by natural *An. gambiae* populations are unaffected by *LRIM1* silencing (Cohuet *et al.*, 2006) and APL family member silencing in natural *An. gambiae* describes *APL1A*'s efficiency against *P. falciparum* but inefficiency against *P. berghei* while the opposite is true for *APL1C* (Mitri *et al.*, 2009). In contrast, silencing of *WASP*, *ApoII/I*, *ArgK* and a tetraspanin have similar effects on both *P. berghei* and *P. falciparum* in a G3 strain of *An. gambiae* (Mendes *et al.*, 2008; Jaramillo-Gutierrez *et al.*, 2009). The favoured hypothesis is that parasites adapt to the immune responses of their natural vectors. A well adapted parasite has evolved strategies that allow it to evade the immune system of its vector, resulting in a highly compatible interaction that triggers very little activation of the mosquito immune system (Jaramillo-Gutierrez *et al.*, 2009).

Melanization

The melanization cascade is a series of proteolytic events leading to cleavage of prophenoloxidase (PPO) into phenoloxidase (PO) which leads to protein crosslinking and melanization of parasites and is primarily made up of CLIP-domain serine proteases, serine protease homologues and inhibitors, serpins. A comprehensive analysis of CLIPA and CLIPB genes showed that some are antagonistic while others are agonistic. Strain specificity and the distinction between melanization mediating either direct parasite killing or disposal of dead parasites add to the complexity of this vector response (Volz *et al.*, 2006; reviewed by Barillas-Mury, 2007). Of the serpins, SRPN2 drastically affects *P. berghei* (but not *P. falciparum*) ookinete survival and melanization (Michel *et al.*, 2005, 2006) while knock-down of *SPRN6* plays vector-specific roles: it prevents lysis of *P. berghei* ookinetes in *An. stephensi*, has no effect on *An. gambiae* G3 females and increases the number of parasites that complete midgut invasion in the *An. gambiae* L35 strain (Abraham *et al.*, 2005). SRPN6 silencing also increases the number of sporozoites that invade the salivary glands (Pinto *et al.*, 2008).

Reactive oxygen species (ROS)

Earlier studies of *P. berghei* in *An. stephensi* and *An. gambiae* implicate ROS in a series of H_2O_2-dependent biochemical events in ookinete-invaded midgut cells that leads to apoptosis, cell extrusion and parasite death (reviewed by Kumar and Barillas-Mury, 2005). Recent RNAi-based studies have shown that lack of detoxification enzymes, such as catalase, or oxidation resistance 1 (OXR1), which regulates catalase expression, increases local levels of H_2O_2 and reduces survival of *P. berghei*. *P. berghei* ookinete invasion down-regulates catalase expression in midgut epithelial cells, indicating that ROS detoxification enzymes are in a fine balance, as ROS are final effectors of antimicrobial responses and must accumulate to mount an effective defence response, but damage to the host mounting the response needs to be prevented (Molina-Cruz *et al.*, 2008; Jaramillo-Gutierrez *et al.*, 2010).

Other interactions determined through functional genomics

Undoubtedly, functional genomics analyses of vector–parasite interactions are dominated by the sub-field of mosquito immune response to malaria parasites (reviewed in Hillyer, 2010). This is likely a combination of having an excellent model (*Drosophila*) for candidate gene selection, the ability to knockdown genes with limited affect on other physiological processes, and a direct phenotypic read-out of oocyst (or sporozoite) numbers. But, other interactions have been identified using this same method.

Knockdown of mosquito laminin gamma 1, a component of the basal lamina, leads to a reduction in *P. berghei* oocysts. Thus, laminin may trigger the ookinete to oocyst developmental transition or may allow the parasite to disguise itself by covering the oocyst surface (Arrighi *et al.*, 2005). Also as part of the midgut response, frizzled2 (Fz2) and cell division cycle 42 (cdc42) were shown to affect the actin polymerization within epithelial cells containing a dead or dying ookinete (Shiao *et al.*, 2006). This mechanism is linked to PPO activation but not parasite killing and harkens back to array data showing up-regulation of many cytoskeletal genes during invasion. And, recently, knockdown of a midgut peroxidase gene (*IMPer1*) and dual oxidase gene (*duox*) unveiled their involvement in the formation of a dinitrotyrosine network that forms on the luminal side of the midgut shortly after blood ingestion. This network slows the diffusion of microbial elicitors to the epithelium, dampening responses to bacteria and parasites (Kumar *et al.*, 2010). Finally, in a more systemic way, miRNAs and the proteins that generate them, Dicer1 and Argonaut1 have been implicated in parasite survival though the mechanism remains to be elucidated (Winter *et al.*, 2007).

New trends

In addition to the previously mentioned trends of investigating compatible pairs and combing RNAi with other techniques to determine biological function and interaction of genes that affect parasite survival, there are newly emerging directions for the functional genomics of malaria vectors. The role of endogenous bacteria is a recent revelation in mosquito–parasite interactions. New research shows these are actually tripartite interactions that include the insect's gut flora: treating mosquitoes with antibiotics has revealed that some of the responses presumed to be anti-*Plasmodium* are actually bystander-effects from an anti-bacterial response (Dong *et al.*, 2009b, Meister *et al.*, 2009). Determination of unique versus shared anti-pathogen responses is one future direction of the field.

Methodologically, the field is moving towards next generation sequencing techniques that are expected to provide a new wave of information. Funding has recently been approved for a project that was submitted in August 2008, which was entitled, 'Genome Analysis of Vectorial Capacity In Major *Anopheles* Vectors of Malaria Parasites'. This effort is spearheaded by the *Anopheles* Genomes Cluster Committee and the overall aim of the project is to use next generation sequencing and a overarching comparative genomics approach to sequence thirteen additional anopheline genomes in the near term (http://www.vectorbase.org/sections/Other/addtl_org_includes/Anopheles_Genomes_4Aug08.pdf). Deep sequencing does not require prior knowledge of sequence for probe design which not only provides an opportunity to discover un-annotated genes and alternative splicing but allows sequencing of the whole system (mosquito, parasite, bacteria, cells from blood meal), rather than just the known targets. Additionally, it does not require competitive hybridization to observe quantitative differences between samples which means analysis of highly abundant transcripts will not suffer saturation and analysis of scarce transcripts will not be obscured by noise. (For a more in-depth view of deep sequencing for disease vectors, please see Gibbons *et al.*, 2009). Already, several groups have begun the process of analysing the transcriptomes of a few non-African anopheline vectors such as *An. stephensi* (Dixit *et al.*, 2009); *An. darlingi* (Calvo *et al.*, 2009) and *An. albimanus* (Martinez-Barnetche, 2012).

Proteomics

Proteomics is defined as the analysis of the proteome, or the entire complement of proteins, present in a tissue, organism or system (Wilkins *et al.*, 1996). The overarching goals of proteomics studies are typically threefold: to confidently identify as many proteins as possible from a tissue or organism; to identify proteins and pathways that are altered in response to challenge or perturbation; and, to identify candidate targets for directed research, such as gene silencing or vaccine development. Proteomic high throughput techniques can therefore be considered primarily as screening tools to generate vast amounts of biologically relevant information, which can then be utilized to target specific proteins or pathways.

Within *Anopheles gambiae* proteomics studies these goals are the same, with the primary aim of identifying candidate targets for inhibiting the roles of these insects as vectors in diseases. Initial work has focused on identifying proteins present in organs, tissues or fluids, and mapping these to the genome (Amenya *et al.*, 2010; Paskewitz and Shi, 2005; He *et al.*, 2007; Dinglasan *et al.*, 2009). Additional studies have investigated the differential expression of proteins in response to infection of the mosquito with *Plasmodium* parasites (Choumet *et al.*, 2007; Lefevre *et al.*, 2007). In particular, the mosquito midgut, and to an extent the salivary glands, are of increased interest due to their potential as a source of transmission-blocking targets (Dinglasan and Jacobs-Lorena, 2008).

Challenges for mosquito proteomics

One of the major challenges in proteomic analysis of *Anopheles* and other mosquito species is the low yield of proteins from preparations, and thus the need for large numbers of insects and subsequent dissection and enrichment of target organs, fluids or tissues. Initial research into many relevant areas of *Anopheles* proteomics was advanced from the knowledge obtained from Lepidoptera, since protein yields are significantly higher in these species. Likewise, studying larval stages of *Anopheles* can be complicated by low yield, and is not as critical in the understanding of vector–parasite interactions, since only adult female mosquitoes blood feed and thus, transmit pathogens via this process. Therefore, the majority of proteomic studies in *Anopheles* are conducted in the mature adult (see Table 10.2). Also of relevance, proteomic studies in *Aedes aegypti* show similar experimental designs, strategy and outcomes in terms of proteome coverage, although it is a completely different mosquito species. *Aedes* was originally the model organism for understanding malaria transmission biology by virtue of its interaction with the bird malaria parasite, *P. gallinaceum*. It is likely that insight gained from the earlier studies of *Aedes* may still have applicability and utility in shedding light on the *Anopheles–Plasmodium* system. However, given the availability of *Anopheles* colonies and a sequenced genome, the rationale for continued use of the *Aedes* system for understanding *Anopheles–Plasmodium* interactions has become less persuasive.

The most critical step in a proteomics experiment is sample preparation. The selection of appropriate buffer, detergents and enrichment steps is complex and somewhat tissue and technique specific. The dissection and isolation of only the organs of interest will improve the ability to interpret subsequent data, and as a crude fractionation technique is excellent for reducing sample complexity. Generic sample preparation considerations are discussed in detail elsewhere (Gorg *et al.*, 2004; Bodzon-Kulakowska *et al.*, 2007; Gundry *et al.*, 2009). Other procedural and practical questions to consider when designing proteomics and mass spectrometry (MS)-based experiments include the number of insects to use for analysis; the yield of subcellular preparations can be small, and a large '*n*' may be needed to observe the lower abundant proteins

Table 10.2 Summary of the methods and findings from selected proteomic studies of *Anopheles gambiae* and *Aedes aegypti*

Species	Subproteome	Methods Utilized [a]	Proteins	Reference
An. gambiae	Haemolymph	SDS-PAGE, LC-MS/MS	26	Paskewitz and Shi (2005)
An. gambiae	Salivary glands	2-DE, LC-MS/MS iTRAQ	122	Choumet *et al.* (2007)
An. gambiae	Cast cuticles	SDS-PAGE, LC-MS/MS	69–119	He *et al.* (2007)
An. gambiae	Head	2-D DIGE, PMF	12	Lefevre *et al.* (2007)
Ae. aegypti	Male reproductive glands	LC-MS/MS	63	Sirot *et al.* (2008)
An. gambiae	Adult midgut PM	LC-MS/MS	209	Dinglasan *et al.* (2009)
Ae. aegypti	Brush border membrane vesicles (BBMV)	2-DE, LC-MS/MS	36–89	Popova-Butler and Dean (2009)
An. gambiae	Eggshell	LC-MS/MS	44	Amenya *et al.* (2010)

[a]LC-MS/MS, liquid chromatography–tandem mass spectrometry; 2-DE, two-dimensional electrophoresis; DIGE, differential in-gel electrophoresis; PMF, peptide mass fingerprinting; iTRAQ, isobaric tags for the relative and absolute quantification of proteins.

that are involved in some complex processes. Furthermore, temporal and experimental conditions may be critical, including diet, age, method of harvesting, and short- and long-term storage conditions.

Following sample preparation, there are two common approaches for proteome analysis; initial separation and (optional) quantification by two-dimensional gel electrophoresis (2-DE and DIGE) (O'Farrell, 1975; Unlu *et al.*, 1997; Viswanathan *et al.*, 2006), or tryptic digestion followed by multidimensional fractionation and subsequent tandem mass spectrometry (MS/MS) analysis; the most common technique used in the previously cited *Anopheles* studies is MudPIT (Washburn *et al.*, 2001), a combination of strong cation exchange and reversed-phase (C18) peptide separation prior to MS.

Analytical techniques

In 2-DE, proteins are separated by isoelectric point and then molecular weight in gels; protein spots can be visualized by staining, excised, and then processed for identification using MS methods. Using fluorescent dyes (e.g. DIGE, GE Healthcare), multiple samples can be analysed on the same gel and differentially expressed spots selected for MS identification. The drawback of this approach is that it is time- and labour-intensive to pick and identify all the spots in a gel, so typically only a handful of spots are processed from MS analysis and assigned a protein identity (see, for example, Paskewitz and Shi, 2005; Lefevre *et al.*, 2007). Using MudPIT, many more proteins in the sample can be identified, although instruments are obviously limited by speed and sensitivity. With complex fractionation, thousands of proteins can be identified in a single experiment. Adding quantitative methods such as iTRAQ (Ross *et al.*, 2004) allows for comparison of protein abundance in samples, permitting the simultaneous identification and quantification of proteins. The limitation of the MS-only approach is that large volumes of potentially biologically unimportant data are generated, and finding relevant and meaningful data may be more difficult. Additionally, the dynamic range of instruments and separations may prevent the identification of a low abundance target of interest, biasing identifications towards more abundant proteins. In both gel and MS-only approaches, orthogonal validation techniques (e.g. western blots) are currently required to confirm the presence and quantification of identified proteins.

The recent advances in instrument capabilities have allowed a dramatic increase in the ability to identify large numbers of proteins with increased speed, resolution and confidence (Olsen *et al.*, 2005, 2009; Vollmer and van de Goor, 2009). However, instrument improvements in the near future are unlikely to yield similarly large increases in our ability to identify more proteins, and interrogate their structure and modifications in more detail. Therefore, improved sample preparation, alternative fractionation techniques and more targeted approaches are needed to improve the proteome coverage for *An. gambiae*. For instance, isolation of subcellular fractions and specific tissues or fractions thereof (e.g. brush border membrane vesicles from midgut) are useful for analysing specific host–parasite interaction sites; however, the purity of the enrichment should always be assessed and potential cellular contamination accounted for. In addition to sample preparation, sample fractionation strategies can reveal a greater number, and a different population, of proteins. For example, using liquid phase isoelectric focusing (Horth *et al.*, 2006), peptides separated by pH gradient can be directly injected into the MS without clean-up, and proteome coverage is comparable but different to other fractionation protocols (Elschenbroich *et al.*, 2009).

The techniques involved in analysing large numbers of proteins in a system are advancing rapidly. Using techniques developed in clinical fields, and applying them to understanding *An. gambiae* biological processes has already yielded some promising results. As techniques (and more importantly, genomic databases and annotation) improve, the general mechanisms of host–parasite interactions will be better understood and specific mechanisms targeted for further study and analysis. By integrating a systems biology approach (Hood, 2003) of merging genomic, proteomics, and metabolomic data, drug and vaccine interactions may be not only advanced but better understood.

Glycomics

This section covers a discipline that is not very well known despite the considerable number of significant contributions to several fields from cancer biology to infectious disease (von Itzstein *et al.*, 2008; Blow, 2009), and as such, requires a few concrete as well as evolving definitions. Glycobiology is the study of macromolecules, i.e. complex carbohydrates/oligosaccharides or glycans that are involved in many different cellular processes. For example, roles for glycans have been demonstrated in cell signalling, cell–cell interaction, and structural organization, to name a few (reviewed extensively in Varki *et al.*, 1999). Glycobiology also includes the study of plant and animal proteins that preferentially and specifically bind glycans, i.e. lectins. Glycomics, on the other hand, is the global study of the range of glycan structures present in cells, tissues or whole organisms across various development or disease states and how they are involved in these various important cellular processes. Glycoproteomics includes the identification, study and characterization of proteins that are co- and/or post-translationally modified by glycans. Although they are not research approaches that have been extensively applied to the analysis of malaria transmission biology, this section will highlight what is known in terms of mosquito–parasite glycobiology and glycoproteomics and provide some views on where and how glycomic analyses can be applied to this dynamic in the near future.

Mosquito–parasite glycobiology

It is only in the past two decades that a small number of studies have begun outlining the critical role of glycans in the *Plasmodium* parasite–mosquito interface. It has been shown that a panel of legume lectins are capable of binding distinct oligosaccharides on the lumenal face of the mosquito midgut (Rudin and Hecker, 1989; Zieler *et al.*, 1999; Wilkins and Billingsley, 2001). More importantly, in an *ex vivo* assay it was shown that *Plasmodium* ookinetes binding to the luminal face of the midgut was inhibited by periodate treatment, which deglycosylated the midgut glycoproteins (Zieler *et al.*, 1999). This was then followed with experiments showing that several legume lectins (e.g. jacalin and wheat germ agglutinin) significantly inhibited ookinete adhesion to the midgut via glycan ligands (Zieler *et al.*, 2000). In fact, ookinete recognition of glycans appears to be quite specific, since not all lectins inhibit parasite adhesion and of the panel of legume lectins used, only the lectin jacalin, which has specificity for O-linked *N*-acetylgalactosamine (GalNAc) has a significant inhibitory effect. A few years later, MG96, an anti-glycan monoclonal antibody (MAb) that recognizes glycan epitopes on the midgut was shown to completely inhibit *Plasmodium* development (Dinglasan *et al.*, 2003), presumably by blocking adhesion which necessarily precedes active invasion of the midgut epithelial cell. Follow up studies using competition ELISA and glycan array screening demonstrated that the cognate glycan epitope for this antibody is O-GalNAc, which is recognized by jacalin (Dinglasan *et al.*, 2005). Moreover, it was found that *Plasmodium* ookinetes directly bind to GalNAc-BSA neoglycoproteins arrayed on a slide (Billingsley, 1994). Such highly corroborative data from independent groups strongly suggest that indeed O-glycans somehow mediate parasite interaction with the midgut. Given, that O-GalNAc is a post-translational modification present on proteins, the two immediate outstanding questions that came out of these studies were (1) What are the midgut microvillar glycoproteins with this modification? and (2) 'What ookinete molecules are involved in midgut glycan recognition?

Parasite lectin-like molecules

Current evidence suggests that several malaria parasite ookinete molecules have glycan recognition or lectin-like properties (for a review of lectins, see Sharon and Lis, 2004). *Plasmodium*, like its apicomplexan relatives, uses adhesive gliding motility for active penetration of host cells (Siden-Kiamos *et al.*, 2006a,b). Although the substrate for this motility in insects has not yet been fully defined, it is assumed to be a combination of glycoconjugates (proteins and lipids modified by glycans) on vertebrate cells in the blood meal, the mosquito PM, and on the midgut epithelial cell apical membrane. Two parasite micronemal adhesive proteins, the membrane bound circumsporozoite- and thrombospondin-related adhesion protein-related protein (CTRP) and the secreted vWF A-domain-like-related protein (WARP), are predominantly expressed on *Plasmodium* ookinetes and have both been shown to be important for parasite invasion of the midgut (reviewed in Vlachou *et al.*, 2006; Sinden, 2009). The vWF A-domain is conserved across the thrombospondin- related anonymous protein (TRAP) family of proteins and repeated in varying numbers (Li *et al.*, 2004; Yuda *et al.*, 1999; Yuda *et al.*, 2001). TRAP, CTRP, and WARP have been shown by ELISA to have binding affinity for heparin (McCormick *et al.*, 1999; Li *et al.*, 2004). Heparin is a highly sulphated glycosaminoglycan and is structurally related to heparan sulphate, which has been found across invertebrates. Glycosaminoglycans are anionic polysaccharides that modify proteoglycans and glycoproteins. Surface plasmon resonance analyses also confirmed that TRAP, as an entire molecule, has higher binding affinity for heparin than the vWF A-domains alone (Akhouri *et al.*, 2004). This is most likely due to the contribution of the TSP-1 repeat domain on TRAP to heparin affinity. The vWF A-domains were also found to have two putative ligand-binding surfaces, a metal-binding site and a putative heparin binding site. In mosquitoes, TRAP, vWF A-domain, and TSP-1 knockout mutant parasites were incapable of invasion of mosquito salivary glands, but gliding motility remained unaffected (Matuschewski *et al.*, 2002). It is quite likely that several critical parasitic proteins are involved in the orchestrated actions of gliding, attachment, and invasion.

A search of the current *Plasmodium* genome database (http://plasmodb.org/) suggests that *P. falciparum* has about 43 genes, *P. vivax* has ~39 genes, and both *P. yoelii* and *P. berghei* have ~17 genes that have been annotated to contain lectin domains, with predicted affinities for neutral oligosaccharides. Recently, six genes belonging to the *Limulus* clotting factor C, Coch-5b2, and Lgl1 (LCCL)-lectin adhesive-like protein family (LAP/CCp) were shown to be conserved across different *Plasmodium* species (Trueman *et al.*, 2004; Pradel *et al.*, 2004, 2006). These protein domains have definitive C-type lectin-like properties (Dessens *et al.*, 2004) in other systems and are expressed specifically in female gametocytes (Khan *et al.*, 2005; Raine *et al.*, 2007) as a protein family complex (Simon *et al.*, 2009). LAP1, LAP2, LAP4, LAP5 and LAP6 have been shown to have varying effects on parasite development in the mosquito (Claudianos *et al.*, 2002; Raine *et al.*, 2007; Ecker *et al.*, 2008). Given the protein domain composition, LAP/CCp proteins may have glycan binding properties; however this has yet to be demonstrated. Another protein with a predicted C-terminal glycan binding domain, is the putative secreted ookinete protein 7 (PSOP7) (Ecker *et al.*, 2008). PSOP7 is not the only ookinete molecule with a predicted lectin domain. However, it should be noted that only independent analysis of PSOP7 revealed a predicted glycan binding domain, suggesting that the current number of lectin-like genes for different *Plasmodium* species is underestimated. It was found that *Δpsop7* ookinetes were inhibited in their ability to bind to the midgut wall, at 24 h post-blood feeding although they were observed to have normal motility and morphology *in vitro*. Together, these observations suggest that *Plasmodium* ookinetes secrete molecules that have binding affinity to both anionic polysaccharides (glycosaminoglycans) as well as neutral oligosaccharides (GalNAc).

Mosquito glycans, glycoproteins and proteoglycans

The thick chitinous PM is the first mosquito membrane that the ookinete encounters and not surprisingly, some species of the parasite which develop slowly in the midgut lumen (e.g. *P. falciparum*) secrete a chitinase that prevents the PM from being a physical barrier to accessing the midgut epithelium (Vinetz *et al.*, 1999). Other species, such as *P. vivax* (Ramasamy *et al.*, 1997), are hypothesized to develop much more quickly and can therefore access the midgut epithelium before the PM has had time to thicken. There is some evidence to suggest that N-acetylglucosamine (GlcNAc) and mannose residues on PM-associated glycoproteins are ligands for parasite adhesion since feeding GlcNAc along with an infective blood meal to mosquitoes significantly impairs parasite invasion (Billingsley *et al.*, 1994; Basseri *et al.*, 2008) and Wheat germ agglutinin, which has binding affinity for GlcNAc can inhibit (to some degree) parasite adhesion (Zeiler *et al.*, 1999, 2000). The evidence for mannose is much more indirect and is based on the following arguments: (a) several ookinete genes encode for concanavalin A (Con A)-like molecules; (b) it is known that Con A has binding preference for terminal mannose residues; and (c) the mosquito midgut and PM contains N-linked glycans that are of the high mannose type (Wilkins and Billingsley, 2001). However, since biochemical evidence supporting mannose binding affinity by any of the molecules studied to date is missing, this recognition specificity remains highly speculative.

In the previous section a screen of the first generation Glycominds™ glycan microarray with the anti-glycan MG96 MAb identified O-GalNAc as a requisite component of the glycan epitope (Dinglasan *et al.*, 2005). This terminal GalNAc can exist as a single moiety directly modifying a Ser/Thr residue on the polypeptide backbone of a glycoprotein. However, the glycan structures (glycoforms) represented on this mammalian derived glycan array do not represent the complete repertoire of glycoforms on the mosquito midgut. Therefore, since a single O-linked GalNAc (Tn antigen) or the O-linked galactoseβ1–3GalNAc (T antigen) are highly conserved across all eukaryotes as the canonical modification in mucins, our understanding of potential O-linked glycan ligands for ookinetes remains severely limited to what we know about the minimum jacalin and MG96 binding epitope (O-GalNAc). Mucins are glycoproteins that are heavily modified by O-GalNAc that are conserved across metazoans and are present in anopheline midguts (Devenport *et al.*, 2005). Glycoproteins that are not 'mucins' per se can also contain mucin-like domains, which are rich in Pro, Ser and Thr and, thus, predicted to be heavily O-glycosylated. Using jacalin affinity chromatography and protein identification by mass spectrometry, a shortlist of heavily O-GalNAc modified midgut glycoproteins have recently been identified (Dinglasan *et al.*, 2007a). These included three alanyl aminopeptidases, a maltase-like enzyme, an O-glycosyl hydrolase and a Glu-Asp symporter. Of these, only one protein, an *An. gambiae* midgut specific, glycosylphosphotidyl inositol-anchored alanyl aminopeptidase (AgAPN1) has been further characterized (Dinglasan *et al.*, 2007a). Interestingly, targeting the aglycosylated N-terminal region of AgAPN1 with antibodies significantly inhibits parasite development in the mosquito, suggesting that while jacalin can presumably bind to this highly abundant glycoprotein preventing physical adhesion by the ookinete, the polypeptide itself also plays a role in the overall midgut invasion process. The other O-glycosylated glycoproteins in the list have not yet been characterized, namely ookinete invasion of the midgut.

The presence of heparin-like glycosaminoglycans in the mosquito was recently confirmed (Sinnis *et al.*, 2007; Dinglasan *et al.*, 2007b). Heparin itself was not observed in either study, but heparan sulphate and chondroitin sulphate were found in both the midgut and salivary glands of the mosquito. There also appeared to be polarized distribution of heparan and chondroitin sulphate, with the former residing predominantly in the basal membrane of both the midgut and salivary glands, and the latter is found on the apical face of the midgut and therefore may represent the proposed binding ligand for CTRP and WARP. The two studies differ in their interpretation of the disaccharide composition of the midgut chondroitin sulphate. Sinnis *et al.* (2007) did not observe the presence of a 6-sulfated chondroitin disaccharide unit, as we had proposed and concluded that midgut chondroitin sulphate was predominantly unsulphated with only a few 4-sulfated GalNAc forms. Such a conclusion implied that the midgut chondroitin could not be a ligand for CTRP or WARP as described previously. Note that we had observed that labelled 6-sulphated chondroitin disaccharides do indeed bind to *P. falciparum* ookinetes *in vitro* (Dinglasan *et al.*, 2007b) and used single chain variable fragment antibodies (scFv) which require 6-sulphate for binding in our analyses. Moreover, our more recent composition analysis via capillary electrophoresis laser-induced fluorescence analysis confirmed the presence of 6-Sulfated GalNAc units (unpublished observations) further corroborating the hypothesis that midgut chondroitin may act as a binding ligand for ookinetes. It now remains to be seen whether surface plasmon resonance analysis of CTRP and WARP, as described earlier will indeed confirm chondroitin sulphate binding affinity. Unfortunately, to date no midgut or salivary gland proteoglycans (proteins that are modified by glycosaminoglycans) have been fully characterized in the mosquito although it has been over a decade since it has been shown that lectins can also inhibit sporozoite invasion of the salivary glands (Barreau *et al.*, 1995).

There are no MAbs against mosquito-specific chondroitin or heparan sulphate glycosaminoglycans and scFvs isolated from phage are limited in their utility for such assays because of purification issues. Since glycans are not encoded by DNA, targeting these glycoforms directly by genetic approaches is impossible. However, modification of polypeptides and the biosynthesis of mono-, oligo- and polysaccharides requires the action of enzymes called glycosyltransferases (reviewed in Varki *et al.*, 1999), which can be targeted by genetic approaches, for example RNA interference (RNAi). This approach was used to block the biosynthesis of chondroitin sulphate on the apical midgut membrane and in so doing significantly inhibited *P. falciparum* development in the mosquito (Dinglasan *et al.*, 2007b). This was possible since the single peptide-O-xylosyltransferase (AgOXT1); the first transferase involved in glycosaminoglycan biosynthesis in the mosquito was targeted. Recently, following this same approach, it was shown that RNAi-mediated disruption of heparan sulphate biosynthesis (via AgOXT1 targeting) in the salivary glands reduced *P. falciparum* sporozoite invasion between 59% and 79% (Armistead *et al.*, 2011). However, since the protein saglin (or another basal lamina protein) remained unaffected by the above RNAi approach, sporozoites were still successful in invading the salivary glands. Importantly, the demonstration of the partial involvement of salivary gland basal heparan sulphate supports the earlier biochemical characterization of TRAP binding affinity for glycosaminoglycans (McCormick *et al.*, 1999; Matuschewski *et al.*, 2002; Akhouri *et al.*, 2004; Li *et al.*, 2004). Although these studies were successful in achieving transferase silencing, efforts at targeting glycosyltransferases involved in O-GalNAc modification of midgut glycoproteins may be more difficult, given the presence of several O-GalNAc transferases, a poor understanding of their functional specificity and the possibility of functional redundancy.

Outlook and future trends

There has traditionally been limited interest in pursuing glycobiological approaches in the malaria community, since much of the analytical platforms required for such analyses is capital heavy and highly specialized (e.g. mass spectrometry). Moreover, analysis of protein–glycan interactions between the parasite and mosquito tissues is limited by the lack of an appropriate cell line model for the midgut or salivary glands (Wilkins and Billingsley, 2010). However, this trend is rapidly changing with several interdisciplinary groups coming together to address some of the outstanding issues, technological hurdles and critical biological questions. Aside from the mosquito–parasite interactions described here, other groups are looking at the asexual stages of *Plasmodium* as well (von Itzstein *et al.*, 2008; Nebl *et al.*, 2005; Tamborrini *et al.*, 2010; Boyle *et al.*, 2010; Kamena *et al.*, 2008). The proven ability to use new chemistries and mass spectrometry platforms (Bielik and Zaia, 2010) for acquiring and analysing the whole glycome, glycoproteome, and proteoglycome of cells, tissues and even organisms (Nishimura *et al.*, 2004; Miura *et al.*, 2008; Furukawa *et al.*, 2008; Hsu and Mahal, 2009; Jang-Lee *et al.*, 2006; Lepenies and Seeberger, 2010) in other fields will soon find applicability in addressing malaria and vector biology. For a review of a few of the glycobiology-related bioinformatic approaches, see 'Glycan/glycome bioinformatics', below.

Bioinformatics

The sequencing of the *P. falciparum*, *An. gambiae*, and human genomes (Gershon, 2002) has facilitated the identification of genes and pathways involved in the transmission, pathogenicity and drug responses of the malaria parasite, and has provided a greater understanding of the complex human host–parasite, insect–vector interactions. The number of data generated by the sequencing project and other transcriptomics, proteomics, glycomics, pharmacogenomics and vaccinomics projects derived from the genomics project causes problems for traditional methods of data storage, data analysis and data interpretation. Therefore, it has become necessary to supplement current bioinformatics resources and tools with new approaches to accommodate, integrate and analyse malaria derived 'omics' information.

Sequencing data repositories such as the NCBI's Malaria genetics and genomics, the TDR database for malaria, MR4, as well as the NIH funded programs such as *VectorBase* (http://www.vectorbase.org/index.php) and *PlasmoDB* (http://PlasmoDB.org) have facilitated data accessibility with enhanced usability and data concurrency (Table 10.3). These databases provide relational database features, which allow data to be mined for different questions and for different biological perspectives. Moreover, they provide entire genome sequences and computationally predicted, manually curated gene models, as well as ESTs, protein feature predictions, and functional annotation at different levels of biological function. The features incorporating BLAST, such as degenerate searches, phylogenetic comparisons, gene/protein expression analysis, polymorphism data, small molecule information, and the provision of publications related to the known gene, allow users to answer many questions that arise in research contexts.

Protein bioinformatics

The capabilities of early two dimensional studies of malaria parasite proteins are greatly extended by mass spectrometry derived proteomics (Sims and Hyde, 2006). MS identified peak lists are searched against a known protein FASTA database by using various search algorithms. Probabilistic algorithms such as *Mascot* (Perkins *et al.*, 1999), *OMSSA* (Geer *et al.*, 2004), *MyriMatch* (see Table 10.3) reports a probability based ion score between observed match and experimental data and database sequence as a random event. While *X!Tandem* (Craig and Beavis, 2004), *Sequest* (Eng *et al.*, 1994), *Spectrum Mill* (Agilent Technologies) and *Inspect* (see Table 10.3) uses a heuristic algorithm, which correlates the spectrum to calculate a score based on the similarities between experimental mass spectrum with the

Table 10.3 List of useful genomic and proteomic bioinformatics resources.

Databases	
NCBI malaria genetics and genomics	http://www.ncbi.nlm.nih.gov/projects/Malaria
TDR database	http://apps.who.int/tdr/svc/diseases/malaria
MR4	http://www.mr4.org/
VectorBase	http://www.vectorbase.org
Ensembl	http://metazoa.ensembl.org/Anopheles_gambiae
PlasmoDB	http://PlasmoDB.org
Protein identification search tools	
Mascot	http://www.matrixscience.com/
Spectrum Mill (Agilent)	http://www.chem.agilent.com/
X!Tandem	http://www.thegpm.org/tandem/
OMSSA	http://pubchem.ncbi.nlm.nih.gov/omssa/
Sequest	http://fields.scripps.edu/sequest/
Inspect	http://proteomics.ucsd.edu/
MyriMatch	http://fenchurch.mc.vanderbilt.edu/lab/software.php
PepArML	https://edwardslab.bmcb.georgetown.edu/pymsio/
Data analysis software	
Maspectras 2	http://genome.tugraz.at/maspectras/maspectras_description.shtml
Scaffold	http://www.proteomesoftware.com/Proteome_software_prod_Scaffold.html
Trans-Proteomic Pipeline	http://tools.proteomecenter.org/software.php
ProteinCenter	http://www.proxeon.com

derived theoretical spectrum. Search engine identified peptides above a significance threshold are accepted, clustered, annotated and analysed by software programs such as *Scaffold* (Searle, 2009), *ProteinCenter* (Proxeon), *Trans-Proteomic Pipeline* (Institute for Systems Biology, Seattle) and *Maspectras 2* (Ubaida Mohien *et al.*, 2010). Beyond the common workflow described here, data from a given mass spectrometry experiment can be re-mined and maximized by various software methods. These include search against EST databases, post-translational modification, and most importantly for parasite biology, sequence variation away from database sequences. Methods such as these were detailed in (Tharakan *et al.*, 2010).

The meta search engine *PepArML* uses machine learning techniques to allow better analysis of proteomic mass spectrometry data than that delivered by a single search engine and produces a combined E-value, allowing for the possibility for a common statistical standard required for a combined search score (Edwards and Tseng, 2009). Since the accuracy of each search algorithm is affected differently by the characteristics of each data set, using complementary and consensus search methods, and combining search results from different search engines is a good approach to enhance peptide identification confidence. Hence this method improves sensitivity (the true positive hits identified irrespective of score) and specificity (the significance of true positive hits in relation to the other hits), and it thereby improves peptide identification (Alves *et al.*, 2008) and significantly benefits protein identification (Kapp *et al.*, 2005). Nevertheless, the effectiveness of the resulting analysis does depend to some degree upon the strengths and weaknesses of different search engines, and their combinations, and also upon correctly specified search parameters like mass tolerance (instrument dependent and calibration), enzymatic digestion (tryptic-semi tryptic), search space (protein sequence database), etc., and the appropriateness of the search engine(s) for these experimental conditions.

Glycan/glycome bioinformatics

Although the fields of mosquito and parasite glycobiology are in their infancy, commonly used glycobiology bioinformatics tools are also applicable to this subfield of general glycobiology. These include tools for glycosylation analysis, glycomics, glycan biomarker prediction, glycan structure analysis, glyco-gene expression analysis and glycan structure mining tools. In terms of the current glycobiology-related DBs, KEGG Glycan (http://www.genome.jp/kegg/glycan/), KEGG chemical function (KCF) (Hattori *et al.*, 2003) as well as the suite of tools and databases available from Glycosciences.de (http://www.glycosciences.de/), the Complex Carbohydrate Research Centre (http://www.ccrc.uga.edu/) and especially the Consortium for Functional Glycomics (http://www.functionalglycomics.org/), all provide well structured glycan structures to reduce the ambiguity arising in data capture between derived spectrum and the prediction of any type of glycan structure. More recently, glycobiology bioinformatics has focused much research in glycomics on further deconvoluting these ambiguities. Programs like *Cartoonist* (Goldberg *et al.*, 2005), *GlycoPep ID* (Irungu *et al.*, 2005), *GlycoMod* (Cooper *et al.*, 2001a), *GlycoSuiteDB* (Cooper *et al.*, 2001b), *BOLD*, *O-glycan DB* (Cooper *et al.*, 1999), *GlycoPeakFinder* (Maass *et al.*, 2007), *GlyDB* (Ren *et al.*, 2007), *GlycoMiner* (Ozohanics *et al.*, 2008), *GlycoMaster* (Shan *et al.*, 2008), *GlycoWorkbench* (Ceroni *et al.*, 2008) *GlycoCT* (Herget, *et al.*, 2008), *Branch-and-Bound* (Peltoniemi *et al.*, 2009) and the dynamic program-based *GLYCH* algorithm, which lists all possible glycan fragments (Tang *et al.*, 2005) are existing tools for glycan and glycopeptide structure mining and prediction (see previous section) and can complement the genomics and proteomics that are being evaluated in malarial research programmes.

Summary

This chapter provided an overview of what is currently known at the genomic, proteomic and glycomic levels, about the role of different molecules, tissues and molecular pathways in the mosquito vector that mediate malaria parasite transmission. Although there is still much that we need to discover and learn about malaria transmission biology, over the past decade alone, it is clear that much of the prevailing scientific gaps of knowledge that existed in the decades preceding the advent of the *An. gambiae* and *Plasmodium* genomes have been addressed vigorously and to some degree diminished considerably. We now look forward to an exciting time in malaria research, as we continue to observe the evolution of our existing technologies and the development of new approaches for dissecting the intricate mosquito vector–parasite interaction. In this regard, the 'Omics revolution has clearly opened new avenues for investigation and the resulting data will help lay the groundwork for future 'Systems Biology' or 'Multi-Omics' (genomic, transcriptomic, proteomic, and glycomic) analyses of this unique host-parasite system. With the renewed focus on the importance of the vector in parasite transmission, there is a sense of optimism that research efforts in this context may one day contribute significantly towards the global malaria elimination and eradication effort.

References

Abdul-Rauf, M., and Ellar, D.J. (1999). Isolation and characterization of brush border membrane vesicles from whole *Aedes aegypti* larvae. J. Invertebr. Pathol. *73*, 45–51.

Abraham, E.G., Donnelly-Doman, M., Fujioka, H., Ghosh, A., Moreira, L., and Jacobs-Lorena, M. (2005). Driving midgut-specific expression and secretion of a foreign protein in transgenic mosquitoes with AgAper1 regulatory elements. Insect Mol. Biol. *14*, 271–279.

Abraham, E.G., Pinto, S.B., Ghosh, A., Vanlandingham, D.L., Budd, A., Higgs, S., Kafatos, F.C., Jacobs-Lorena, M., and Michel, K. (2005). An immune-responsive serpin, SRPN6, mediates mosquito defense against malaria parasites. Proc. Natl. Acad. Sci. U.S.A. *102*, 16327–16332.

Akaki, M., and Dvorak, J.A. (2005). A chemotactic response facilitates mosquito salivary gland infection by malaria sporozoites. J. Exp. Biol. *208*, 3211–3218.

Akhouri, R.R., Bhattacharyya, A., Pattnaik, P., Malhotra, P., and Sharma, A. (2004). Structural and functional dissection of the adhesive domains of *Plasmodium falciparum* thrombospondin-related anonymous protein (TRAP). Biochem. J. *379*, 815–822.

Amenya, D.A., Chou, W., Li, J., Yan, G., Gershon, P., James, A.A., and Marinotti, O. (2010). Proteomics reveals novel components of the *Anopheles gambiae* eggshell. J. Insect Physiol. Epub ahead of print.

Amino, R., Menard, R., and Frischknecht, F. (2005). In vivo imaging of malaria parasites – recent advances and future directions. Curr. Opin. Microbiol. *8*, 407–414.

Amino, R., Thiberge, S., Shorte, S., Frischknecht, F., and Menard, R. (2006). Quantitative imaging of *Plasmodium* sporozoites in the mammalian host. C. R. Biol. *329*, 858–862.

Arrighi, R.B., Lycett, G., Mahairaki, V., Siden-Kiamos, I., and Louis, C. (2005). Laminin and the malaria parasite's journey through the mosquito midgut. J. Exp. Biol. *208*, 2497–2502.

Armistead, J.S., Wilson, I.B., van Kuppevelt, T.H., and Dinglasan, R.R. (2011). A role for heparan sulfate proteoglycans in *Plasmodium falciparum* sporozoite invasion of anopheline mosquito salivary glands. Biochem. J. *438*, 475–483.

Barillas-Mury, C. (2007). CLIP proteases and *Plasmodium* melanization in *Anopheles gambiae*. Trends Parasitol. *23*, 297–299.

Barreau, C., Touray, M., Pimenta, P.F., Miller, L.H., and Vernick, K.D. (1995). *Plasmodium gallinaceum*: sporozoite invasion of *Aedes aegypti* salivary glands is inhibited by anti-gland antibodies and by lectins. Exp. Parasitol. *81*, 332–343.

Basseri, H.R., Doosti, S., Akbarzadeh, K., Nateghpour, M., Whitten, M.M., and Ladoni, H. (2008). Competency of *Anopheles stephensi mysorensis* strain for *Plasmodium vivax* and the role of inhibitory carbohydrates to block its sporogonic cycle. Malar J. *7*, 131.

Baton, L., Garver, L., Xi, Z., and Dimopoulos, G. (2008). Functional genomics studies on the innate immunity of disease vectors. Insect Science *15*, 15–15–27.

Baton, L.A., Robertson, A., Warr, E., Strand, M.R., and Dimopoulos, G. (2009). Genome-wide transcriptomic profiling of *Anopheles gambiae* hemocytes reveals pathogen-specific signatures upon bacterial challenge and *Plasmodium berghei* infection. BMC Genomics *10*, 257.

Baxter, R.H., Chang, C.I., Chelliah, Y., Blandin, S., Levashina, E.A., and Deisenhofer, J. (2007). Structural basis for conserved complement factor-like function in the antimalarial protein TEP1. Proc. Natl. Acad. Sci. U.S.A. *104*, 11615–11620.

Beerntsen, B.T., Luckhart, S., and Christensen, B.M. (1989). *Brugia malayi* and *Brugia pahangi*: inherent difference in immune activation in the mosquitoes Armigeres subalbatus and *Aedes aegypti*. J. Parasitol. *75*, 76–81.

Bielik, A.M., and Zaia, J. (2010). Historical overview of glycoanalysis. Methods Mol. Biol. *600*, 9–30.

Billingsley, P.F. (1994). Approaches to vector control: new and trusted. 2. Molecular targets in the insect midgut. Trans. R. Soc. Trop. Med. Hyg. *88*, 136–140.

Billingsley, P.F., and Rudin, W. (1992). The role of the mosquito peritrophic membrane in blood meal digestion and infectivity of *Plasmodium* species. J. Parasitol. *78*, 430–440.

Blandin, S., Shiao, S.H., Moita, L.F., Janse, C.J., Waters, A.P., Kafatos, F.C., and Levashina, E.A. (2004). Complement-like protein TEP1 is a determinant of vectorial capacity in the malaria vector *Anopheles gambiae*. Cell *116*, 661–670.

Blandin, S.A., Marois, E., and Levashina, E.A. (2008). Antimalarial responses in *Anopheles gambiae*: from a complement-like protein to a complement-like pathway. Cell. Host Microbe 3, 364–374.

Blow, N. (2009). Glycobiology: A spoonful of sugar. Nature *457*, 617–620.

Bodzon-Kulakowska, A., Bierczynska-Krzysik, A., Dylag, T., Drabik, A., Suder, P., Noga, M., Jarzebinska, J., and Silberring, J. (2007). Methods for samples preparation in proteomic research. J. Chromatogr. B. Analyt. Technol. Biomed. Life. Sci. *849*, 1–31.

Boyle, M.J., Richards, J.S., Gilson, P.R., Chai, W., and Beeson, J.G. (2010). Interactions with heparin-like molecules during erythrocyte invasion by *Plasmodium falciparum* merozoites. Blood *115*, 4559–4568.

Brandt, S.M., Jaramillo-Gutierrez, G., Kumar, S., Barillas-Mury, C., and Schneider, D.S. (2008). Use of a *Drosophila* model to identify genes regulating *Plasmodium* growth in the mosquito. Genetics *180*, 1671–1678.

Brennan, J.D., Kent, M., Dhar, R., Fujioka, H., and Kumar, N. (2000). *Anopheles gambiae* salivary gland proteins as putative targets for blocking transmission of malaria parasites. Proc. Natl. Acad. Sci. U.S.A. *97*, 13859–13864.

Calvo, E., Pham, V.M., Marinotti, O., Andersen, J.F., and Ribeiro, J.M. (2009). The salivary gland transcriptome of the neotropical malaria vector Anopheles darlingi reveals accelerated evolution of genes relevant to hematophagy. BMC Genomics *10*, 57.

Carroll, J., and Ellar, D.J. (1993). An analysis of *Bacillus thuringiensis* delta-endotoxin action on insect-midgut-membrane permeability using a light-scattering assay. Eur. J. Biochem. *214*, 771–778.

Castillo, J.C., Robertson, A.E., and Strand, M.R. (2006). Characterization of hemocytes from the mosquitoes *Anopheles gambiae* and *Aedes aegypti*. Insect Biochem. Mol. Biol. *36*, 891–903.

Catteruccia, F., and Levashina, E.A. (2009). RNAi in the malaria vector, *Anopheles gambiae*. Methods Mol. Biol. *555*, 63–75.

Ceroni, A., Maass, K., Geyer, H., Geyer, R., Dell, A., and Haslam, S.M. (2008). GlycoWorkbench: a tool for the computer-assisted annotation of mass spectra of glycans. J. Proteome Res. *7*, 1650–1659.

Chertemps, T., Mitri, C., Perrot, S., Sautereau, J., Jacques, J.C., Thiery, I., Bourgouin, C., Rosinski-Chupin, I. (2010). *Anopheles gambiae* PRS1 modulates *Plasmodium* development at both midgut and salivary gland steps. PLoS One. 5, e11538.

Choumet, V., Carmi-Leroy, A., Laurent, C., Lenormand, P., Rousselle, J.C., Namane, A., Roth, C., and Brey, P.T. (2007). The salivary glands and saliva of *Anopheles gambiae* as an essential step in the *Plasmodium* life cycle: a global proteomic study. Proteomics 7, 3384–3394.

Christophides, G.K., Vlachou, D., and Kafatos, F.C. (2004). Comparative and functional genomics of the innate immune system in the malaria vector *Anopheles gambiae*. Immunol. Rev. *198*, 127–148.

Claudianos, C., Dessens, J.T., Trueman, H.E., Arai, M., Mendoza, J., Butcher, G.A., Crompton, T., and Sinden, R.E. (2002). A malaria scavenger receptor-like protein Alves, G., Wu, W.W., Wang, G., Shen, R.F., and Yu, Y.K. (2008). Enhancing peptide identification confidence by combining search methods. J. Proteome Res. 7, 3102–3113.

Cohuet, A., Osta, M.A., Morlais, I., Awono-Ambene, P.H., Michel, K., Simard, F., Christophides, G.K., Fontenille, D., and Kafatos, F.C. (2006). *Anopheles* and *Plasmodium*: from laboratory models to natural systems in the field. EMBO Rep. 7, 1285–1289.

Collins, F.H., Sakai, R.K., Vernick, K.D., Paskewitz, S., Seeley, D.C., Miller, L.H., Collins, W.E., Campbell, C.C., and Gwadz, R.W. (1986). Genetic selection of a *Plasmodium*-refractory strain of the malaria vector *Anopheles gambiae*. Science *234*, 607–610.

Cooper, C.A., Gasteiger, E., and Packer, N.H. (2001a). GlycoMod – a software tool for determining glycosylation compositions from mass spectrometric data. Proteomics *1*, 340–349.

Cooper, C.A., Harrison, M.J., Wilkins, M.R., and Packer, N.H. (2001b). GlycoSuiteDB: a new curated relational database of glycoprotein glycan structures and their biological sources. Nucleic Acids Res. *29*, 332–335.

Cooper, C.A., Wilkins, M.R., Williams, K.L., and Packer, N.H. (1999). BOLD – a biological O-linked glycan database. Electrophoresis *20*, 3589–3598.

Craig, R., and Beavis, R.C. (2004). TANDEM: matching proteins with tandem mass spectra. Bioinformatics *20*, 1466–1467.

Dessens, J.T., Sinden, R.E., and Claudianos, C. (2004). LCCL proteins of apicomplexan parasites. Trends Parasitol. *20*, 102–108.

Devenport, M., Alvarenga, P.H., Shao, L., Fujioka, H., Bianconi, M.L., Oliveira, P.L., and Jacobs-Lorena, M. (2006). Identification of the *Aedes aegypti* peritrophic matrix protein AeIMUCI as a heme-binding protein. Biochemistry *45*, 9540–9549.

Devenport, M., Fujioka, H., Donnelly-Doman, M., Shen, Z., and Jacobs-Lorena, M. (2005). Storage and secretion of Ag-Aper14, a novel peritrophic matrix protein, and Ag-Muc1 from the mosquito *Anopheles gambiae*. Cell Tissue Res. *320*, 175–185.

Devenport, M., Fujioka, H., and Jacobs-Lorena, M. (2004). Storage and secretion of the peritrophic matrix protein Ag-Aper1 and trypsin in the midgut of *Anopheles gambiae*. Insect Mol. Biol. *13*, 349–358.

Dinglasan, R.R., Alaganan, A., Ghosh, A.K., Saito, A., van Kuppevelt, T.H., and Jacobs-Lorena, M. (2007). *Plasmodium falciparum* ookinetes require mosquito midgut chondroitin sulfate proteoglycans for cell invasion. Proc. Natl. Acad. Sci. U.S.A. *104*, 15882–15887.

Dinglasan, R.R., Devenport, M., Florens, L., Johnson, J.R., McHugh, C.A., Donnelly-Doman, M., Carucci, D.J., Yates, J.R., 3rd, and Jacobs-Lorena, M. (2009). The *Anopheles gambiae* adult midgut peritrophic matrix proteome. Insect Biochem. Mol. Biol. *39*, 125–134.

Dinglasan, R.R., and Jacobs-Lorena, M. (2008). Flipping the paradigm on malaria transmission-blocking vaccines. Trends Parasitol. *24*, 364–370.

Dinglasan, R.R., Kalume, D.E., Kanzok, S.M., Ghosh, A.K., Muratova, O., Pandey, A., and Jacobs-Lorena, M. (2007). Disruption of *Plasmodium falciparum* development by antibodies against a conserved mosquito midgut antigen. Proc. Natl. Acad. Sci. U.S.A. *104*, 13461–13466.

Dinglasan, R.R., Valenzuela, J.G., and Azad, A.F. (2005). Sugar epitopes as potential universal disease transmission blocking targets. Insect Biochem. Mol. Biol. 35, 1–10.

Dixit, R., Sharma, A., Mourya, D.T., Kamaraju, R., Patole, M.S., and Shouche, Y.S. (2009). Salivary gland transcriptome analysis during *Plasmodium* infection in malaria vector *Anopheles stephensi*. Int. J. Infect. Dis. *13*, 636–646.

Dong, Y., Aguilar, R., Xi, Z., Warr, E., Mongin, E., and Dimopoulos, G. (2006a). *Anopheles gambiae* immune responses to human and rodent Plasmodium parasite species. PLoS Pathog. *2*, e52.

Dong, Y., and Dimopoulos, G. (2009). Anopheles fibrinogen-related proteins provide expanded pattern recognition capacity against bacteria and malaria parasites. J. Biol. Chem. *284*, 9835–9844.

Dong, Y., Taylor, H.E., and Dimopoulos, G. (2006b). AgDscam, a hypervariable immunoglobulin domain-containing receptor of the *Anopheles gambiae* innate immune system. PLoS Biol. *4*, e229.

Ecker, A., Bushell, E.S., Tewari, R., and Sinden, R.E. (2008). Reverse genetics screen identifies six proteins important for malaria development in the mosquito. Mol. Microbiol. *70*, 209–220.

Ecker, A., Pinto, S.B., Baker, K.W., Kafatos, F.C., and Sinden, R.E. (2007). *Plasmodium berghei*: plasmodium perforin-like protein 5 is required for mosquito midgut invasion in *Anopheles stephensi*. Exp. Parasitol. *116*, 504–508.

Edwards, N., Wu, X., and Tseng, C.W. (2009). An unsupervised, model-free, machine-learning combiner for peptide identifications from tandem mass spectra. Clin. Proteomics *5*, 23–36.

Elschenbroich, S., Ignatchenko, V., Sharma, P., Schmitt-Ulms, G., Gramolini, A.O., and Kislinger, T. (2009). Peptide separations by on-line MudPIT compared to isoelectric focusing in an off-gel format: application to a membrane-enriched fraction from C2C12 mouse skeletal muscle cells. J. Proteome Res. *8*, 4860–4869.

Eng, J.K., McCormack, A.J., and Yates, J.R.I. (1994). An approach to correlate tandem mass spectral data of peptides with amino acid sequences in a protein database. Am. Soc. Mass Spectrom. *5*, 976–989.

Felix, R.C., Muller, P., Ribeiro, V., Ranson, H., and Silveira, H. (2010). Plasmodium infection alters *Anopheles gambiae* detoxification gene expression. BMC Genomics *11*, 312.

Forcella, M., Berra, E., Giacchini, R., and Parenti, P. (2006). Leucine transport in brush border membrane vesicles from freshwater insect larvae. Arch. Insect Biochem. Physiol. *63*, 110–122.

Fraiture, M., Baxter, R.H., Steinert, S., Chelliah, Y., Frolet, C., Quispe-Tintaya, W., Hoffmann, J.A., Blandin, S.A., and Levashina, E.A. (2009). Two mosquito LRR proteins function as complement control factors in the TEP1-mediated killing of *Plasmodium*. Cell. Host Microbe *5*, 273–284.

Frischknecht, F. (2010). Imaging parasites at different scales. Cell. Host Microbe *8*, 16–19.

Frischknecht, F., Baldacci, P., Martin, B., Zimmer, C., Thiberge, S., Olivo-Marin, J.C., Shorte, S.L., and Menard, R. (2004). Imaging movement of malaria parasites during transmission by Anopheles mosquitoes. Cell. Microbiol. *6*, 687–694.

Frolet, C., Thoma, M., Blandin, S., Hoffmann, J.A., and Levashina, E.A. (2006). Boosting NF-kappaB-dependent basal immunity of *Anopheles gambiae* aborts development of *Plasmodium berghei*. Immunity *25*, 677–685.

Furukawa, J., Shinohara, Y., Kuramoto, H., Miura, Y., Shimaoka, H., Kurogochi, M., Nakano, M., and Nishimura, S. (2008). Comprehensive approach to structural and functional glycomics based on chemoselective glycoblotting and sequential tag conversion. Anal. Chem. *80*, 1094–1101.

Garver, L.S., Dong, Y., and Dimopoulos, G. (2009). Caspar controls resistance to *Plasmodium falciparum* in diverse anopheline species. PLoS Pathog. *5*, e1000335.

Garver, L.S., Xi, Z., and Dimopoulos, G. (2008). Immunoglobulin superfamily members play an important role in the mosquito immune system. Dev. Comp. Immunol. *32*, 519–531.

Geer, L.Y., Markey, S.P., Kowalak, J.A., Wagner, L., Xu, M., Maynard, D.M., Yang, X., Shi, W., and Bryant, S.H. (2004). Open mass spectrometry search algorithm. J. Proteome Res. *3*, 958–964.

Gershon, D. (2002). Malaria research tools up for the future. Nature *419*, 4–5.

Ghosh, A.K., Devenport, M., Jethwaney, D., Kalume, D.E., Pandey, A., Anderson, V.E., Sultan, A.A., Kumar, N., and Jacobs-Lorena, M. (2009). Malaria parasite invasion of the mosquito salivary gland requires interaction between the *Plasmodium* TRAP and the *Anopheles* saglin proteins. PLoS Pathog. *5*, e1000265.

Ghosh, A.K., and Jacobs-Lorena, M. (2009). Plasmodium sporozoite invasion of the mosquito salivary gland. Curr. Opin. Microbiol. *12*, 394–400.

Ghosh, A.K., Ribolla, P.E., and Jacobs-Lorena, M. (2001). Targeting *Plasmodium* ligands on mosquito salivary glands and midgut with a phage display peptide library. Proc. Natl. Acad. Sci. U.S.A. *98*, 13278–13281.

Ghosh, A., Srinivasan, P., Abraham, E.G., Fujioka, H., and Jacobs-Lorena, M. (2003). Molecular strategies to study *Plasmodium*–mosquito interactions. Trends Parasitol. *19*, 94–101.

Gibbons, J.G., Janson, E.M., Hittinger, C.T., Johnston, M., Abbot, P., and Rokas, A. (2009). Benchmarking next-generation transcriptome sequencing for functional and evolutionary genomics. Mol. Biol. Evol. *26*, 2731–2744.

Glenn, J.D., King, J.G., and Hillyer, J.F. (2010). Structural mechanics of the mosquito heart and its function in bidirectional hemolymph transport. J. Exp. Biol. *213*, 541–550.

Goldberg, D., Sutton-Smith, M., Paulson, J., and Dell, A. (2005). Automatic annotation of matrix-assisted laser desorption/ionization N-glycan spectra. Proteomics *5*, 865–875.

Gonzalez-Lazaro, M., Dinglasan, R.R., Hernandez-Hernandez Fde, L., Rodriguez, M.H., Laclaustra, M., Jacobs-Lorena, M., and Flores-Romo, L. (2009). *Anopheles gambiae* Croquemort SCRBQ2, expression profile in the mosquito and its potential interaction with the malaria parasite *Plasmodium berghei*. Insect Biochem. Mol. Biol. *39*, 395–402.

Gorg, A., Weiss, W., and Dunn, M.J. (2004). Current two-dimensional electrophoresis technology for proteomics. Proteomics *4*, 3665–3685.

Gundry, R.L., White, M.Y., Murray, C.I., Kane, L.A., Fu, Q., Stanley, B.A., and Van Eyk, J.E. (2009). Preparation of proteins and peptides for mass spectrometry analysis in a bottom-up proteomics workflow. Curr. Protoc. Mol. Biol. *Chapter 10*, Unit10.25.

Gupta, L., Molina-Cruz, A., Kumar, S., Rodrigues, J., Dixit, R., Zamora, R.E., and Barillas-Mury, C. (2009). The STAT pathway mediates late-phase immunity against Plasmodium in the mosquito *Anopheles gambiae*. Cell. Host Microbe *5*, 498–507.

Han, Y.S., Thompson, J., Kafatos, F.C., and Barillas-Mury, C. (2000). Molecular interactions between *Anopheles stephensi* midgut cells and *Plasmodium berghei*: the time bomb theory of ookinete invasion of mosquitoes. EMBO J. *19*, 6030–6040.

Hattori, M., Okuno, Y., Goto, S., and Kanehisa, M. (2003). Development of a chemical structure comparison method for integrated analysis of chemical and genomic information in the metabolic pathways. J. Am. Chem. Soc. *125*, 11853–11865.

He, N., Botelho, J.M., McNall, R.J., Belozerov, V., Dunn, W.A., Mize, T., Orlando, R., and Willis, J.H. (2007). Proteomic analysis of cast cuticles from *Anopheles gambiae* by tandem mass spectrometry. Insect Biochem. Mol. Biol. *37*, 135–146.

Hegge, S., Munter, S., Steinbuchel, M., Heiss, K., Engel, U., Matuschewski, K., and Frischknecht, F. (2010). Multistep adhesion of *Plasmodium* sporozoites. FASEB J. *24*, 2222–2234.

Herget, S., Ranzinger, R., Maass, K., and Lieth, C.W. (2008). GlycoCT-a unifying sequence format for carbohydrates. Carbohydr. Res. *343*, 2162–2171.

Hillyer, J.F. (2010). Mosquito Immunity. In Invertebrate Immunity, K. Soderhall, ed. (New York: Springer Science+Business Media, LLC.), pp. 218–238.

Hillyer, J.F., Barreau, C., and Vernick, K.D. (2007). Efficiency of salivary gland invasion by malaria sporozoites is controlled by rapid sporozoite destruction in the mosquito haemocoel. Int. J. Parasitol. *37*, 673–681.

Hillyer, J.F., Schmidt, S.L., and Christensen, B.M. (2003). Hemocyte-mediated phagocytosis and melanization in the mosquito *Armigeres subalbatus* following immune challenge by bacteria. Cell Tissue Res. *313*, 117–127.

Holt, R.A., Subramanian, G.M., Halpern, A., Sutton, G.G., Charlab, R., Nusskern, D.R., Wincker, P., Clark, A.G., Ribeiro, J.M., Wides, R., *et al.* (2002). The genome sequence of the malaria mosquito *Anopheles gambiae*. Science *298*, 129–149.

Hood, L. (2003). Systems biology: integrating technology, biology, and computation. Mech. Ageing Dev. *124*, 9–16.

Horth, P., Miller, C.A., Preckel, T., and Wenz, C. (2006). Efficient fractionation and improved protein identification by peptide OFFGEL electrophoresis. Mol. Cell. Proteomics *5*, 1968–1974.

Hsu, K.L., and Mahal, L.K. (2009). Sweet tasting chips: microarray-based analysis of glycans. Curr. Opin. Chem. Biol. *13*, 427–432.

Hua, G., Zhang, R., Bayyareddy, K., and Adang, M.J. (2009). *Anopheles gambiae* alkaline phosphatase is a functional receptor of *Bacillus thuringiensis* jegathesan Cry11Ba toxin. Biochemistry *48*, 9785–9793.

Irungu, J., Go, E.P., Dalpathado, D.S., and Desaire, H. (2007). Simplification of mass spectral analysis of acidic glycopeptides using GlycoPep ID. Anal. Chem. *79*, 3065–3074.

Jang-Lee, J., North, S.J., Sutton-Smith, M., Goldberg, D., Panico, M., Morris, H., Haslam, S., and Dell, A. (2006). Glycomic profiling of cells and tissues by mass spectrometry: fingerprinting and sequencing methodologies. Methods Enzymol. *415*, 59–86.

Jaramillo-Gutierrez, G., Molina-Cruz, A., Kumar, S., and Barillas-Mury, C. (2010). The *Anopheles gambiae* oxidation resistance 1 (OXR1) gene regulates expression of enzymes that detoxify reactive oxygen species. PLoS One *5*, e11168.

Jaramillo-Gutierrez, G., Rodrigues, J., Ndikuyeze, G., Povelones, M., Molina-Cruz, A., and Barillas-Mury, C. (2009). Mosquito immune responses and compatibility between Plasmodium parasites and anopheline mosquitoes. BMC Microbiol. *9*, 154.

Jiang, H., Wang, Y., and Kanost, M.R. (1998). Pro-phenol oxidase activating proteinase from an insect, *Manduca sexta*: a bacteria-inducible protein similar to *Drosophila easter*. Proc. Natl. Acad. Sci. U.S.A. *95*, 12220–12225.

Khan, S.M., Franke-Fayard, B., Mair, G.R., Lasonder, E., Janse, C.J., Mann, M., and Waters, A.P. (2005). Proteome analysis of separated male and female gametocytes reveals novel sex-specific *Plasmodium* biology. Cell *121*, 675–687.

Kamena, F., Tamborrini, M., Liu, X., Kwon, Y.U., Thompson, F., Pluschke, G., and Seeberger, P.H. (2008). Synthetic GPI array to study antitoxic malaria response. Nat. Chem. Biol. *4*, 238–240.

Kapp, E.A., Schutz, F., Connolly, L.M., Chakel, J.A., Meza, J.E., Miller, C.A., Fenyo, D., Eng, J.K., Adkins, J.N., Omenn, G.S., and Simpson, R.J. (2005). An evaluation, comparison, and accurate benchmarking of several publicly available MS/MS search algorithms: sensitivity and specificity analysis. Proteomics *5*, 3475–3490.

Korochkina, S., Barreau, C., Pradel, G., Jeffery, E., Li, J., Natarajan, R., Shabanowitz, J., Hunt, D., Frevert, U., and Vernick, K.D. (2006). A mosquito-specific protein family includes candidate receptors for malaria sporozoite invasion of salivary glands. Cell. Microbiol. *8*, 163–175.

Kotsyfakis, M., Ehret-Sabatier, L., Siden-Kiamos, I., Mendoza, J., Sinden, R.E., and Louis, C. (2005). *Plasmodium berghei* ookinetes bind to *Anopheles gambiae* and *Drosophila melanogaster* annexins. Mol. Microbiol. *57*, 171–179.

Kotsyfakis, M., Vontas, J., Siden-Kiamos, I., and Louis, C. (2005). The annexin gene family in the malaria mosquito *Anopheles gambiae*. Insect Mol. Biol. *14*, 555–562.

Kumar, S., Gupta, L., Han, Y.S., and Barillas-Mury, C. (2004). Inducible peroxidases mediate nitration of anopheles midgut cells undergoing apoptosis in response to Plasmodium invasion. J. Biol. Chem. *279*, 53475–53482.

Kumar, S., and Barillas-Mury, C. (2005). Ookinete-induced midgut peroxidases detonate the time bomb in anopheline mosquitoes. Insect Biochem. Mol. Biol. *35*, 721–727.

Kumar, S., Molina-Cruz, A., Gupta, L., Rodrigues, J., and Barillas-Mury, C. (2010). A peroxidase/dual oxidase system modulates midgut epithelial immunity in *Anopheles gambiae*. Science *327*, 1644–1648.

Lal, A.A., Patterson, P.S., Sacci, J.B., Vaughan, J.A., Paul, C., Collins, W.E., Wirtz, R.A., and Azad, A.F. (2001). Anti-mosquito midgut antibodies block development of *Plasmodium falciparum* and *Plasmodium vivax* in multiple species of Anopheles mosquitoes and reduce vector fecundity and survivorship. Proc. Natl. Acad. Sci. U.S.A. *98*, 5228–5233.

Lavazec, C., Boudin, C., Lacroix, R., Bonnet, S., Diop, A., Thiberge, S., Boisson, B., Tahar, R., and Bourgouin, C. (2007). Carboxypeptidases B of *Anopheles gambiae* as targets for a *Plasmodium falciparum* transmission-blocking vaccine. Infect. Immun. 75, 1635–1642.

Lee, M.K., Miles, P., and Chen, J.S. (2006). Brush border membrane binding properties of *Bacillus thuringiensis* Vip3A toxin to *Heliothis virescens* and Helicoverpa zea midguts. Biochem. Biophys. Res. Commun. *339*, 1043–1047.

Lefevre, T., Thomas, F., Schwartz, A., Levashina, E., Blandin, S., Brizard, J.P., Le Bourligu, L., Demettre, E., Renaud, F., and Biron, D.G. (2007). Malaria Plasmodium agent induces alteration in the head proteome of their *Anopheles* mosquito host. Proteomics 7, 1908–1915.

Lemos, F.J., Cornel, A.J., and Jacobs-Lorena, M. (1996). Trypsin and aminopeptidase gene expression is affected by age and food composition in *Anopheles gambiae*. Insect Biochem. Mol. Biol. *26*, 651–658.

Lepenies, B., and Seeberger, P.H. (2010). The promise of glycomics, glycan arrays and carbohydrate-based vaccines. Immunopharmacol. Immunotoxicol. *32*, 196–207.

Levashina, E.A., Moita, L.F., Blandin, S., Vriend, G., Lagueux, M., and Kafatos, F.C. (2001). Conserved role of a complement-like protein in phagocytosis revealed by dsRNA knockout in cultured cells of the mosquito, *Anopheles gambiae*. Cell *104*, 709–718.

Li, F., Templeton, T.J., Popov, V., Comer, J.E., Tsuboi, T., Torii, M., and Vinetz, J.M. (2004). Plasmodium ookinete-secreted proteins secreted through a common micronemal pathway are targets of blocking malaria transmission. J. Biol. Chem. *279*, 26635–26644.

Maass, K., Ranzinger, R., Geyer, H., von der Lieth, C.W., and Geyer, R. (2007). 'Glyco-peakfinder' – de novo composition analysis of glycoconjugates. Proteomics 7, 4435–4444.

Martinez-Barnetche, J., Gomez-Barreto, R.E., Ovilla-Munoz, M., Tellez-Sosa, J., Garcia-Lopez, D.E., Dinglasan, R.R., Ubaida Mohien, C., Maccallum, R.M., Redmond, S.N., Gibbons, J.G., *et al.* (2012) Transcriptome of the adult female malaria mosquito vector anopheles albimanus. BMC Genomics *13*, 207.

Mathias, D.K., Plieskatt, J.L., Armistead, J.S., Bethony, J.M., Abdul-Majid, K.B., McMillan, A., Angov, E., Aryee, M.J., Zhan, B., Gillespie, P., *et al.* (2012) Expression, immunogenicity, histopathology, and potency of a mosquito-based malaria transmission-blocking recombinant vaccine. Infect Immun *80*, 1606–1614.

Matuschewski, K., Nunes, A.C., Nussenzweig, V., and Menard, R. (2002). Plasmodium sporozoite invasion into insect and mammalian cells is directed by the same dual binding system. EMBO J. *21*, 1597–1606.

McCormick, C.J., Tuckwell, D.S., Crisanti, A., Humphries, M.J., and Hollingdale, M.R. (1999). Identification of heparin as a ligand for the A-domain of *Plasmodium falciparum* thrombospondin-related adhesion protein. Mol. Biochem. Parasitol. *100*, 111–124.

Meister, S., Kanzok, S.M., Zheng, X.L., Luna, C., Li, T.R., Hoa, N.T., Clayton, J.R., White, K.P., Kafatos, F.C., Christophides, G.K., and Zheng, L. (2005). Immune signaling pathways regulating bacterial and malaria parasite infection of the mosquito *Anopheles gambiae*. Proc. Natl. Acad. Sci. U.S.A. *102*, 11420–11425.

Mendes, A.M., Schlegelmilch, T., Cohuet, A., Awono-Ambene, P., De Iorio, M., Fontenille, D., Morlais, I., Christophides, G.K., Kafatos, F.C., and Vlachou, D. (2008). Conserved mosquito/parasite interactions affect development of *Plasmodium falciparum* in Africa. PLoS Pathog. *4*, e1000069.

Michel, K., Budd, A., Pinto, S., Gibson, T.J., and Kafatos, F.C. (2005). *Anopheles gambiae* SRPN2 facilitates midgut invasion by the malaria parasite *Plasmodium berghei*. EMBO Rep. *6*, 891–897.

Michel, K., Suwanchaichinda, C., Morlais, I., Lambrechts, L., Cohuet, A., Awono-Ambene, P.H., Simard, F., Fontenille, D., Kanost, M.R., and Kafatos, F.C. (2006). Increased melanizing activity in *Anopheles gambiae* does not affect development of *Plasmodium falciparum*. Proc. Natl. Acad. Sci. U.S.A. *103*, 16858–16863.

Mitri, C., Jacques, J.C., Thiery, I., Riehle, M.M., Xu, J., Bischoff, E., Morlais, I., Nsango, S.E., Vernick, K.D., and Bourgouin, C. (2009). Fine pathogen discrimination within the APL1 gene family protects *Anopheles gambiae* against human and rodent malaria species. PLoS Pathog. *5*, e1000576.

Miura, Y., Shinohara, Y., Furukawa, J., Nagahori, N., and Nishimura, S. (2007). Rapid and simple solid-phase esterification of sialic acid residues for quantitative glycomics by mass spectrometry. Chemistry *13*, 4797–4804.

Molina-Cruz, A., DeJong, R.J., Charles, B., Gupta, L., Kumar, S., Jaramillo-Gutierrez, G., and Barillas-Mury, C. (2008). Reactive oxygen species modulate *Anopheles gambiae* immunity against bacteria and Plasmodium. J. Biol. Chem. *283*, 3217–3123.

Mueller, A.K., Kohlhepp, F., Hammerschmidt, C., and Michel, K. (2010). Invasion of mosquito salivary glands by malaria parasites: prerequisites and defense strategies. Int. J. Parasitol. *40*, 1229–1235.

Nebl, T., De Veer, M.J., and Schofield, L. (2005). Stimulation of innate immune responses by malarial glycosylphosphatidylinositol via pattern recognition receptors. Parasitology *130* (Suppl.), S45–62.

Nishimura, S., Niikura, K., Kurogochi, M., Matsushita, T., Fumoto, M., Hinou, H., Kamitani, R., Nakagawa, H., Deguchi, K., Miura, N., Monde, K., and Kondo, H. (2004). High-throughput protein glycomics: combined use of chemoselective glycoblotting and MALDI-TOF/TOF mass spectrometry. Angew. Chem. Int. Ed Engl. *44*, 91–96.

O'Farrell, P.H. (1975). High resolution two-dimensional electrophoresis of proteins. J. Biol. Chem. *250*, 4007–4021.

Okuda, K., Caroci, A., Ribolla, P., Marinotti, O., de Bianchi, A.G., and Bijovsky, A.T. (2005). Morphological and enzymatic analysis of the midgut of Anopheles darlingi during blood digestion. J. Insect Physiol. *51*, 769–776.

Okulate, M.A., Kalume, D.E., Reddy, R., Kristiansen, T., Bhattacharyya, M., Chaerkady, R., Pandey, A., and Kumar, N. (2007). Identification and molecular characterization of a novel protein Saglin as a target of monoclonal antibodies affecting salivary gland infectivity of Plasmodium sporozoites. Insect Mol. Biol. *16*, 711–722.

Oliveira, J.H., Gonçalves, R.L., Oliveira, G.A., Oliveira, P.L., Oliveira, M.F., and Barillas-Mury, C. (2011). Energy metabolism affects susceptibility of *Anopheles gambiae* mosquitoes to Plasmodium infection. Insect Biochem. Mol. Biol. PMID:21320598.

Olsen, J.V., de Godoy, L.M., Li, G., Macek, B., Mortensen, P., Pesch, R., Makarov, A., Lange, O., Horning, S., and Mann, M. (2005). Parts per million mass accuracy on an Orbitrap mass spectrometer via lock mass injection into a C-trap. Mol. Cell. Proteomics *4*, 2010–2021.

Olsen, J.V., Schwartz, J.C., Griep-Raming, J., Nielsen, M.L., Damoc, E., Denisov, E., Lange, O., Remes, P., Taylor, D., Splendore, M., *et al.* (2009). A dual pressure linear ion trap Orbitrap instrument with very high sequencing speed. Mol. Cell. Proteomics *8*, 2759–2769.

Osta, M.A., Christophides, G.K., and Kafatos, F.C. (2004a). Effects of mosquito genes on *Plasmodium* development. Science *303*, 2030–2032.

Osta, M.A., Christophides, G.K., Vlachou, D., and Kafatos, F.C. (2004b). Innate immunity in the malaria vector *Anopheles gambiae*: comparative and functional genomics. J. Exp. Biol. *207*, 2551–2563.

Ozohanics, O., Krenyacz, J., Ludanyi, K., Pollreisz, F., Vekey, K., and Drahos, L. (2008). GlycoMiner: a new software tool to elucidate glycopeptide composition. Rapid Commun. Mass Spectrom. *22*, 3245–3254.

Parish, L.A., Colquhoun, D.R., Ubaida Mohien, C., Lyashkov, A.E., Graham, D.R., and Dinglasan, R.R. (2011). Ookinete-interacting proteins on the microvillar surface are partitioned into detergent resistant membranes of *Anopheles gambiae* midguts. J. Proteome Res. *10*, 5150–5162.

Paskewitz, S.M., and Shi, L. (2005). The hemolymph proteome of *Anopheles gambiae*. Insect Biochem. Mol. Biol. 35, 815–824.

Peltoniemi, H., Joenvaara, S., and Renkonen, R. (2009). De novo glycan structure search with the CID MS/MS spectra of native N-glycopeptides. Glycobiology *19*, 707–714.

Perkins, D.N., Pappin, D.J., Creasy, D.M., and Cottrell, J.S. (1999). Probability-based protein identification by searching sequence databases using mass spectrometry data. Electrophoresis *20*, 3551–3567.

Pimenta, P.F., Touray, M., and Miller, L. (1994). The journey of malaria sporozoites in the mosquito salivary gland. J. Eukaryot. Microbiol. *41*, 608–624.

Pinto, S.B., Kafatos, F.C., and Michel, K. (2008). The parasite invasion marker SRPN6 reduces sporozoite numbers in salivary glands of *Anopheles gambiae*. Cell. Microbiol. *10*, 891–898.

Pinto, S.B., Lombardo, F., Koutsos, A.C., Waterhouse, R.M., McKay, K., An, C., Ramakrishnan, C., Kafatos, F.C., and Michel, K. (2009). Discovery of *Plasmodium* modulators by genome-wide analysis of circulating hemocytes in *Anopheles gambiae*. Proc. Natl. Acad. Sci. U.S.A. *106*, 21270–21275.

Popova-Butler, A., and Dean, D.H. (2009). Proteomic analysis of the mosquito *Aedes aegypti* midgut brush border membrane vesicles. J. Insect Physiol. *55*, 264–272.

Povelones, M., Waterhouse, R.M., Kafatos, F.C., and Christophides, G.K. (2009). Leucine-rich repeat protein complex activates mosquito complement in defense against Plasmodium parasites. Science *324*, 258–261.

Pradel, G., Hayton, K., Aravind, L., Iyer, L.M., Abrahamsen, M.S., Bonawitz, A., Mejia, C., and Templeton, T.J. (2004). A multidomain adhesion protein family expressed in *Plasmodium falciparum* is essential for transmission to the mosquito. J. Exp. Med. *199*, 1533–1544.

Pradel, G., Wagner, C., Mejia, C., and Templeton, T.J. (2006). *Plasmodium falciparum*: co-dependent expression and co-localization of the PfCCp multi-adhesion domain proteins. Exp. Parasitol. *112*, 263–268.

Raine, J.D., Ecker, A., Mendoza, J., Tewari, R., Stanway, R.R., and Sinden, R.E. (2007). Female inheritance of malarial lap genes is essential for mosquito transmission. PLoS Pathog. 3, e30.

Ramasamy, M.S., Kulasekera, R., Wanniarachchi, I.C., Srikrishnaraj, K.A., and Ramasamy, R. (1997). Interactions of human malaria parasites, *Plasmodium vivax* and *P. falciparum*, with the midgut of *Anopheles* mosquitoes. Med. Vet. Entomol. *11*, 290–296.

Ren, J.M., Rejtar, T., Li, L., and Karger, B.L. (2007). N-Glycan structure annotation of glycopeptides using a linearized glycan structure database (GlyDB). J. Proteome Res. *6*, 3162–3173.

Riehle, M.M., Xu, J., Lazzaro, B.P., Rottschaefer, S.M., Coulibaly, B., Sacko, M., Niare, O., Morlais, I., Traore, S.F., and Vernick, K.D. (2008). *Anopheles gambiae* APL1 is a family of variable LRR proteins required for Rel1-mediated protection from the malaria parasite, *Plasmodium berghei*. PLoS One 3, e3672.

Riethmuller, J., Riehle, A., Grassme, H., and Gulbins, E. (2006). Membrane rafts in host–pathogen interactions. Biochim. Biophys. Acta *1758* 2139–2147.

Rodriguez, M.H., and Hernandez-Hernandez Fde, L. (2004). Insect-malaria parasites interactions: the salivary gland. Insect Biochem. Mol. Biol. *34*, 615–624.

Rosinski-Chupin, I., Briolay, J., Brouilly, P., Perrot, S., Gomez, S.M., Chertemps, T., Roth, C.W., Keime, C., Gandrillon, O., Couble, P., and Brey, P.T. (2007). SAGE analysis of mosquito salivary gland transcriptomes during *Plasmodium* invasion. Cell. Microbiol. *9*, 708–724.

Ross, P.L., Huang, Y.N., Marchese, J.N., Williamson, B., Parker, K., Hattan, S., Khainovski, N., Pillai, S., Dey, S., Daniels, S., *et al.* (2004). Multiplexed protein quantitation in Saccharomyces cerevisiae using amine-reactive isobaric tagging reagents. Mol. Cell. Proteomics 3, 1154–1169.

Rossignol, P.A., Ribeiro, J.M., and Spielman, A. (1984). Increased intradermal probing time in sporozoite-infected mosquitoes. Am. J. Trop. Med. Hyg. 33, 17–20.

Rudin, W., and Hecker, H. (1989). Lectin-binding sites in the midgut of the mosquitoes *Anopheles stephensi* Liston and *Aedes aegypti* L. (Diptera: Culicidae). Parasitol. Res. 75, 268–279.

Searle, B.C. (2010). Scaffold: a bioinformatic tool for validating MS/MS-based proteomic studies. Proteomics *10*, 1265–1269.

Shahabuddin, M., Toyoshima, T., Aikawa, M., and Kaslow, D.C. (1993). Transmission-blocking activity of a chitinase inhibitor and activation of malarial parasite chitinase by mosquito protease. Proc. Natl. Acad. Sci. U.S.A. *90*, 4266–4270.

Shan, B., Ma, B., Zhang, K., and Lajoie, G. (2008). Complexities and algorithms for glycan sequencing using tandem mass spectrometry. J. Bioinform Comput. Biol. *6*, 77–91.

Sharon, N., and Lis, H. (2004). History of lectins: from hemagglutinins to biological recognition molecules. Glycobiology *14*, 53R–62R.

Shen, Z., and Jacobs-Lorena, M. (1998). A type I peritrophic matrix protein from the malaria vector *Anopheles gambiae* binds to chitin. Cloning, expression, and characterization. J. Biol. Chem. *273*, 17665–17670.

Shiao, S.H., Whitten, M.M., Zachary, D., Hoffmann, J.A., and Levashina, E.A. (2006). Fz2 and cdc42 mediate melanization and actin polymerization but are dispensable for Plasmodium killing in the mosquito midgut. PLoS Pathog. 2, e133.

Siden-Kiamos, I., Ecker, A., Nyback, S., Louis, C., Sinden, R.E., and Billker, O. (2006a). *Plasmodium berghei* calcium-dependent protein kinase 3 is required for ookinete gliding motility and mosquito midgut invasion. Mol. Microbiol. *60*, 1355–1363.

Siden-Kiamos, I., Pinder, J.C., and Louis, C. (2006b). Involvement of actin and myosins in *Plasmodium berghei* ookinete motility. Mol. Biochem. Parasitol. *150*, 308–317.

Sieber, K.P., Huber, M., Kaslow, D., Banks, S.M., Torii, M., Aikawa, M., and Miller, L.H. (1991). The peritrophic membrane as a barrier: its penetration by Plasmodium gallinaceum and the effect of a monoclonal antibody to ookinetes. Exp. Parasitol. 72, 145–156.

Simon, N., Scholz, S.M., Moreira, C.K., Templeton, T.J., Kuehn, A., Dude, M.A., and Pradel, G. (2009). Sexual stage adhesion proteins form multi-protein complexes in the malaria parasite *Plasmodium falciparum*. J. Biol. Chem. *284*, 14537–14546.

Simons, K., and Gerl, M.J. (2010). Revitalizing membrane rafts: new tools and insights. Nat. Rev. Mol. Cell Biol. *11*, 688–699.

Sinden, R.E. (2009). Malaria, sexual development and transmission: retrospect and prospect. Parasitology *136*, 1427–1434.

Sinnis, P., Coppi, A., Toida, T., Toyoda, H., Kinoshita-Toyoda, A., Xie, J., Kemp, M.M., and Linhardt, R.J. (2007). Mosquito heparan sulfate and its potential role in malaria infection and transmission. J. Biol. Chem. *282*, 25376–25384.

Sterling, C.R., Aikawa, M., and Vanderberg, J.P. (1973). The passage of *Plasmodium berghei* sporozoites through the salivary glands of *Anopheles stephensi*: an electron microscope study. J. Parasitol. 59, 593–605.

Tamborrini, M., Liu, X., Mugasa, J.P., Kwon, Y.U., Kamena, F., Seeberger, P.H., and Pluschke, G. (2010). Synthetic glycosylphosphatidylinositol microarray reveals differential antibody levels and fine specificities in children with mild and severe malaria. Bioorg. Med. Chem. *18*, 3747–3752.

Tang, H., Mechref, Y., and Novotny, M.V. (2005). Automated interpretation of MS/MS spectra of oligosaccharides. Bioinformatics *21* (Suppl. 1), i431–9.

Tharakan, R., Edwards, N., and Graham, D.R. (2010). Data maximization by multipass analysis of protein mass spectra. Proteomics *10*, 1160–1171.

Trueman, H.E., Raine, J.D., Florens, L., Dessens, J.T., Mendoza, J., Johnson, J., Waller, C.C., Delrieu, I., Holders, A.A., Langhorne, J., *et al.* (2004). Functional characterization of an LCCL-lectin domain containing protein family in *Plasmodium berghei*. J. Parasitol. *90*, 1062–1071.

Ubaida Mohien, C., Hartler, J., Breitwieser, F., Rix, U., Remsing Rix, L., Winter, G.E., Thallinger, G.G., Bennett, K.L., Superti-Furga, G., Trajanoski, Z., and Colinge, J. (2010). MASPECTRAS 2: An Integration and Analysis Platform for Proteomic Data. Proteomics *10*, 2719–2722.

Unlu, M., Morgan, M.E., and Minden, J.S. (1997). Difference gel electrophoresis: a single gel method for detecting changes in protein extracts. Electrophoresis *18*, 2071–2077.

Varki, A., Cummings, R.D., Esko, J.D., Freeze, H.H., Stanley, P., Bertozzi, C.R., Hart, G.W., and Etzler, M.E. (2009). Essentials of Glycobiology.784.

Vinetz, J.M., Dave, S.K., Specht, C.A., Brameld, K.A., Xu, B., Hayward, R., and Fidock, D.A. (1999). The chitinase PfCHT1 from the human malaria parasite *Plasmodium falciparum* lacks proenzyme and chitin-binding domains and displays unique substrate preferences. Proc. Natl. Acad. Sci. U.S.A. *96*, 14061–14066.

Viswanathan, S., Unlu, M., and Minden, J.S. (2006). Two-dimensional difference gel electrophoresis. Nat. Protoc. *1*, 1351–1358.

Vlachou, D., Schlegelmilch, T., Christophides, G.K., and Kafatos, F.C. (2005). Functional genomic analysis of midgut epithelial responses in Anopheles during Plasmodium invasion. Curr. Biol. *15*, 1185–1195.

Vlachou, D., Schlegelmilch, T., Runn, E., Mendes, A., and Kafatos, F.C. (2006). The developmental migration of Plasmodium in mosquitoes. Curr. Opin. Genet. Dev. *16*, 384–391.

Vollmer, M., and van de Goor, T. (2009). HPLC-Chip/MS technology in proteomic profiling. Methods Mol. Biol. *544*, 3–15.

Volz, J., Muller, H.M., Zdanowicz, A., Kafatos, F.C., and Osta, M.A. (2006). A genetic module regulates the melanization response of *Anopheles* to *Plasmodium*. Cell. Microbiol. *8*, 1392–1405.

von Itzstein, M., Plebanski, M., Cooke, B.M., and Coppel, R.L. (2008). Hot, sweet and sticky: the glycobiology of *Plasmodium falciparum*. Trends Parasitol. *24*, 210–218.

Warburg, A., Shtern, A., Cohen, N., and Dahan, N. (2007). Laminin and a Plasmodium ookinete surface protein inhibit melanotic encapsulation of Sephadex beads in the hemocoel of mosquitoes. Microbes Infect. *9*, 192–199.

Warr, E., Das, S., Dong, Y., and Dimopoulos, G. (2008). The Gram-negative bacteria-binding protein gene family: its role in the innate immune system of *Anopheles gambiae* and in anti-*Plasmodium* defence. Insect Mol. Biol. *17*, 39–51.

Washburn, M.P., Wolters, D., and Yates, J.R., 3rd. (2001). Large-scale analysis of the yeast proteome by multidimensional protein identification technology. Nat. Biotechnol. *19*, 242–247.

Wilkins, M.R., Pasquali, C., Appel, R.D., Ou, K., Golaz, O., Sanchez, J.C., Yan, J.X., Gooley, A.A., Hughes, G., Humphery-Smith, I., Williams, K.L., and Hochstrasser, D.F. (1996). From proteins to proteomes: large scale protein identification by two-dimensional electrophoresis and amino acid analysis. Biotechnology (N. Y.) *14*, 61–65.

Wilkins, S., and Billingsley, P.F. (2010). Mosquito cell line glycoproteins: an unsuitable model system for the *Plasmodium* ookinete–mosquito midgut interaction? Parasit. Vectors 3, 22.

Wilkins, S., and Billingsley, P.F. (2001). Partial characterization of oligosaccharides expressed on midgut microvillar glycoproteins of the mosquito, *Anopheles stephensi* Liston. Insect Biochem. Mol. Biol. *31*, 937–948.

Winter, F., Edaye, S., Huttenhofer, A., and Brunel, C. (2007). *Anopheles gambiae* miRNAs as actors of defence reaction against *Plasmodium* invasion. Nucleic Acids Res. *35*, 6953–6962.

Yuda, M., Sakaida, H., and Chinzei, Y. (1999). Targeted disruption of the *plasmodium berghei* CTRP gene reveals its essential role in malaria infection of the vector mosquito. J. Exp. Med. *190*, 1711–1716.

Yuda, M., Yano, K., Tsuboi, T., Torii, M., and Chinzei, Y. (2001). von Willebrand Factor A domain-related protein, a novel microneme protein of the malaria ookinete highly conserved throughout Plasmodium parasites. Mol. Biochem. Parasitol. *116*, 65–72.

Zieler, H., Garon, C.F., Fischer, E.R., and Shahabuddin, M. (2000). A tubular network associated with the brush-border surface of the *Aedes aegypti* midgut: implications for pathogen transmission by mosquitoes. J. Exp. Biol. *203*, 1599–1611.

Zieler, H., Nawrocki, J.P., and Shahabuddin, M. (1999). Plasmodium gallinaceum ookinetes adhere specifically to the midgut epithelium of *Aedes aegypti* by interaction with a carbohydrate ligand. J. Exp. Biol. *202*, 485–495.

The Malariologist's Molecular Toolbox

11

Alexander G. Maier

> If the only tool you have is a hammer,
> you tend to see every problem as a nail.
>
> Abraham Maslow (1908–1970)

Abstract

This chapter will summarize many of the recent advances in experimental techniques that have greatly increased our ability to address fundamental questions of parasite biology. Many of these advances are improvements on basic transfection techniques through new generation vectors and regulatable expression systems; however, advances in the analysis of gene expression, microscopy and the analysis of field samples have also proven very fruitful.

Introduction

In this day and age it seems superfluous to justify the use of molecular techniques for anything. However, the study of *Plasmodium* possesses particular challenges for the molecular biologist. Due to these challenges in the laboratory, the pressing urgency of finding a cure, and the practical limitations in the field, the fight against malaria has historically been focusing on empirical approaches. Only in the last 20 years molecular techniques have caught on and are used on a more wide spread basis. There are several reasons why dissecting molecular mechanisms is especially important for the study of *Plasmodium*: The large number of parasitic cells present in a single infection provides a high evolutionary potential for variations and mutations and it is therefore essential to understand the working mechanism of any drug or vaccine. Moreover, the use of animal models for *Plasmodium* species infecting humans is limited due to biological, logistical, financial and most importantly ethical reasons. Other non-human *Plasmodium* species currently offer the advantage of dissecting *in vivo* function of proteins, access to all life-cycle stages, and the utilization of better transfection systems. It is here where the power of comparative genomics of *Plasmodium* becomes most apparent.

Our understanding of *Plasmodium* biology has grown considerably in the past few years due to whole-genome approaches. Complete genome sequences of *P. falciparum*, *P. vivax*, *P. knowlesi*, *P. berghei*, *P. yoelii* (Carlton, 2003; Carlton *et al.*, 2002, 2008; Gardner *et al.*, 2002; Hall, 2005) are now available. Genome sequencing of these and several more species, isolates and strains on the way provide us with a stock-take of available genes, which allows us to compare between species and to more established eukaryotic model organisms. The typical *Plasmodium* genome consists of around 5000 genes distributed across 14 chromosomes with an AT-content that ranges from ~65% (*P. vivax*) to ~80% (*P. falciparum*). A total of 77% of proteins are conserved across the different *Plasmodium* species.

In addition these genomes have been functionally subdivided according to transcription profiles (Ben Mamoun *et al.*, 2001; Bozdech *et al.*, 2003; Cowman, 2001; Hayward *et al.*, 2000; Le Roch *et al.*, 2002, 2003), protein expression (Florens *et al.*, 2002; Hall *et al.*, 2005; Johnson *et al.*, 2004; Lasonder *et al.*, 2002; Sam-Yellowe, 2004), protein interaction networks (Lacount *et al.*, 2005) and other bioinformatic parameters (Adisa *et al.*, 2003; Date and Stoeckert, 2006; Foth *et al.*, 2003; Hiller *et al.*, 2004; Lopez-Estraño *et al.*, 2003; Marti *et al.*, 2004; Myrick *et al.*, 2003; Sijwali and Rosenthal, 2004; Suthram *et al.*, 2005; Waller *et* al., 2000). This led to the identification of stage-specific or concurrently expressed genes. The challenge remains to convert this genome information into information about biological activity.

P. falciparum was the first parasitic disease where the sequence of the host, the parasite and the vector was known (Gardner *et al.*, 2002; Holt *et al.*, 2002; Venter *et al.*, 2001). With this privileged knowledge comes the obligation to use the information and turn it into usable applications not only to understand these organisms, but also to fight malaria. The intriguing and enigmatic biology of *Plasmodium* is also reflected in the fact that almost half of the genes have no homologues in commonly studied model organisms, and lack clear cellular function (Gardner *et al.*, 2002). Once these functions are elucidated this uniqueness might prove an Achilles' heel of the organism and could allow for the development of effective intervention strategies; however, it does not make the task of deciphering the function of gene products any easier.

As the malaria field entered the post-genomic era the elucidation of gene function and the identification of essential genes became a major focus. Consequently, the repertoire of genetic tools available for functional genomics approaches has increased enormously. Gene disruption and allelic replacement helped to elucidate key parasite strategies like host cell invasion, immune-evasion, pathogenesis, drug resistance and transmission.

However, several of the techniques are very demanding, resulting in various degrees of reproducibility, and have not been disseminated at the desired level. Larger scale systematic studies have for the most part been missing, since random approaches of mutagenesis and selection are lacking and gene targeting is inefficient and cumbersome. The next few paragraphs will outline the stumbling blocks that hinder the adaption of such methods, and some current attempts that have been made to overcome them.

Current challenges

Limitations that are imposed by current techniques to clone DNA constructs and to efficiently delivery these DNA constructs into the parasite cell are affecting our ability to dissect gene function.

Cloning

Gene sequences from *Plasmodium* are not very compatible with the workhorse of molecular biology, *E. coli*. Plasmid scrambling and sequence deletions are very frequently encountered. Addressing these challenges seems to be a prerequisite to significantly move the field forward. These problems emerge due to the AT-richness and the repetitiveness of the *Plasmodium* genomes. Some phages infecting Gram-negative bacteria like *E. coli* also contain very AT-rich genomes and consequently *E. coli* has developed various mechanisms to resist bacteriophage infection. One of them is the detection and degradation of AT-rich DNA sequences conferred by enzymes and recombinases. Deletion of these genes allows the stabilization of exogenous AT-rich DNA; however, it comes with the price of losing fidelity in the replication of these DNA stretches. Therefore there is not one universal bacterial strain that is optimal for the cloning of *Plasmodium* DNA, but depending on the particular sequence different strains might effectively stabilize the sequence. In our hands the genetic makeup of PMC103 (Doherty *et al.*, 1993) or Stbl4® (Invitrogen) have yielded good results. Dropping the growth temperature for the bacteria from 37°C to 27°C has proven to be beneficial on occasions.

AT-rich DNA is also very susceptible to nicking during UV-visualization and has its challenges for the amplification and sequencing reactions. Ways of overcoming this is minimizing the UV exposure (e.g. by using a blue-light transilluminator to visualize DNA) or to drop the elongation temperature for a polymerase chain reaction (PCR) to 65–68°C (Su *et al.*, 1996). Alternative methods for ligation-independent cloning have been explored only to various degrees. One approach to a more efficient way of cloning has taken advantage of recombinase-based cloning, namely the GATEWAY® system (Invitrogen) (Skinner-Adams *et al.*, 2003a; Tonkin *et al.*, 2004). This cloning system is based on the site-specific recombination properties of bacteriophage lambda (Landy, 1989). The cloning of PCR products into the gateway vectors is promoted by an enzyme mix consisting of the lambda integrase and the integration host factor from *E. coli*. Both PCR product and vector contain recombination specific sites (attB and attP, respectively), which after successful recombination form an attL site. In a subsequent step this attL site can integrate into attR-containing end vectors with the aid of another enzyme mix that in addition to the integrase and integration host factor also contains lambda excisionase.

The advantage of this method is that preparation and purification of DNA products and their resulting loss are kept to a minimum. Another advantage is that DNA products can be easily 'shuttled' into other vectors that contain the appropriate recombination sites (e.g. bacterial expression vectors for protein purification). The system also allows the simultaneous recombination of several components into one vector (e.g. promoter, coding region, tag) in one reaction, since several sequence specific attL/attR combinations can be used. However, this strategy does not seem to perform with the same consistency in *Plasmodium* as it does in other organisms, maybe due to the AT-rich att sites.

Another ligation-independent method is the single-stranded annealing (SSA) recombination system that joins DNA pieces which have at least 15 base pairs of homology with the help of the 3′–5′ exonuclease activity of a vaccinia virus DNA polymerase (Hamilton *et al.*, 2007). This activity excises nucleotides from the 3′ end of the strand exposing overlapping overhangs, which than can anneal to the complementary vector sequence. The strand breaks are being joined by the *E. coli* repair mechanism. Various commercially available products take advantage of this method [e.g. In-Fusion® (Clontech)/CloneEZ® (Genscript)]. Again the advantage is to minimize handling of the DNA without the use of restriction enzymes, phosphatases, or purification procedures. However, the verdict is still out on the suitability of this system for *Plasmodium*.

A method based both on ligand-dependant and –independent cloning at the same time is the Mini Tn-5-mediated shuttle transposon mutagenesis in *P. berghei* (Sakamoto *et al.*, 2005), which will be discussed in the section outlining 'Tools for Forward Genetics'.

The common problem of all of these systems is that amplification of plasmids takes place in *E. coli* with its associated problems. Therefore diligent cloning procedures have to be applied, approaches have to be flexible in order to cater for the specific requirements of the particular sequence to be cloned and proper analysis of the plasmid are essential.

Transfection

The ability to genetically manipulate cells ultimately depends on the transfer of exogenous DNA into the cell. Due to our inability to effectively culture some *Plasmodium* species for longer time periods, the genetic manipulation of *Plasmodium* is mainly restricted to a few species (Table 11.1). The most advanced systems have been established for *P. berghei* as a representative of *Plasmodium* causing malaria in rodents and *P. falciparum* for the species causing primate malaria. Comparison of the two systems highlights the aspects of the procedure that can potentially be varied and optimized (Table 11.2). The standard mode of DNA delivery into the *Plasmodium* cells is electroporation. Alternative methods have been tried, but have not found wide acceptance within the field (e.g. Mamoun *et al.*, 1999b; Skinner-Adams *et al.*, 2003b).

Blood-stage *P. berghei* parasites for the culturing and purification of mature schizonts are sourced from infected rats or mice, since *P. berghei* cannot be cultured *in vitro* without difficulty for more than one cycle. *P. berghei* cells are transfected in the schizonts/merozoites stage, which are viable *in vitro* for several hours. Depending on the intended outcome, circular (for episomal replication in the parasite) or linearized DNA can be used. Plasmid DNA is linearized within the targeting sequence to drive integration into the genome via a single-crossover event or, alternatively, 5′ and 3′ targeting sequences flanking a selectable marker are used for double cross-over recombination events. Immediately after electroporation the parasites are injected into mice.

A quantum leap in *P. berghei* transfection was achieved by using nucleofection rather than electroporation (Janse *et al.*, 2006a; Sakamoto *et al.*, 2005). Nucleofection (Amaxa®) is a modified form of electroporation using a mixture of electrical parameters in combination with cell type-specific

Table 11.1 Molecular tools

Host	Species	Key experiments	Reference
Rodents	*P. berghei*	Stable transfection (episome)	van Dijk *et al.* (1995)
		Targeted integration into genome	van Dijk *et al.* (1996)
		GFP and luciferase reporter genes	de Koning-Ward *et al.* (1998, 1999)
		Additional positive selectable marker (hDHFR)	de Koning-Ward *et al.* (2000)
		Complementation of gene disruption	Sultan *et al.* (2001)
		Conditional mutagenesis using site-specific recombination (Flp/FRT)	Carvalho *et al.* (2004)
		Nucleofection	Janse *et al.* (2006c)
		Negative selectable marker	Braks *et al.* (2006)
		Systematic 'large scale' reverse genetics screen	Ecker *et al.* (2008a), Pfander *et al.* (2011)
	P. yoelii	Gene targeting	Mota *et al.* (2001)
		Selectable markers/nucleofection	Jongco *et al.* (2006)
	P. chabaudi	GFP transfection	Reece and Thompson (2008), Spence *et al.* (2011)
Primates	*P. falciparum*	Stable transfection (episome)	Wu *et al.* (1995)
		Targeted integration into genome	Wu *et al.* (1996), Crabb *et al.* (1997b)
		Promoter analysis	Crabb and Cowman (1996), Dechering *et al.* (1999)
		GFP transfection	VanWye and Haldar, 1997)
		Targeted gene disruption	Crabb *et al.* (1997a)
		Additional positive selectable markers	Mamoun *et al.* (1999a), Wang *et al.* (2002)
		Negative selectable marker	Duraisingh *et al.* (2002), Maier *et al.* (2006)
		Double knockout	Baum *et al.* (2005)
		Systematic 'large scale' reverse genetics screen	Maier *et al.* (2008)
		Systems to modulate expression	Agop-Nersesian *et al.* (2008), Armstrong and Goldberg (2007), Epp *et al.* (2008), Gardiner *et al.* (2006), Gehde *et al.* (2009), Meissner *et al.* (2005), Saridaki *et al.* (2008), Waller *et al.* (2003)
	P. knowlesi	Stable transfection (episome)	van der Wel *et al.* (1997)
		Double cross-over gene disruption	Kocken *et al.* (2002)
		Selectable markers and single cross-over integration	Wel *et al.* (2004)
	P. cynomolgi	Stable transfection (episome)	Kocken *et al.* (1999)
Birds	*P. gallinaceum*	Transient transfection	Goonewardene *et al.* (1993)

reagents, which enables the direct delivery of the DNA to the nucleus. The transfection efficiency jumped from ~10^{-6} to ~10^{-2}, opening the door to a whole new set of genetic approaches. Using nucleofection therefore requires less time to select for a population, fewer parasites, fewer numbers of laboratory animals, a region of homology as little as 200 base pairs, and less DNA. Based on this a protocol for small-scale in-vitro culture and transfection was recently developed (Somsak *et al.*, 2011). Despite several attempts the same quantum leap in transfection efficiency remains

Table 11.2 Molecular tools

	Plasmodium berghei	*Plasmodium falciparum*
Stage to transfect	Merozoites	Ring stages
Construct	Linear	Circular
Amount of DNA	5 μg	50–150 μg
Efficiency	$10^{-2}/10^{-3}$ (Nucleofaction)	10^{-9} (Nucleofaction)
	$10^{-6}/10^{-9}$ (GenePulser)	10^{-9} (GenePulser)
Methods	Free merozoites	Preloading or infected red blood cells
Selection Time	~ 6–12 days	~ 18–60 days
Selection	Drug treatment of infected animals	Drug treatment of infected human red blood cells grown *in vitro*

elusive in *P. falciparum*. Nucleofection is also used for several of the other *Plasmodium* species. In general their transfection protocol is more similar to the conditions for *P. berghei* than to those for *P. falciparum* (e.g. use of mature stage parasites, except for *P. gallinaceum*, where a zygote/gamete fraction is electroporated) (Goonewardene *et al.*, 1993) (see also Table 11.1).

For *P. falciparum* another approach has to be taken, since the merozoites in this species are only viable for 5–10 min. Transfection of isolated *P. falciparum* merozoites is also possible; however, given the difficulties of obtaining these stages in the correct time frame, the advantages have been so far annihilated by the disadvantages. However, new methods to isolate larger numbers of *P. falciparum* merozoites (Boyle *et al.*, 2010) might prompt researcher to revisit the transfection of this stage. In the meantime the stage used for *P. falciparum* transfection is the ring stage, which is relatively resilient. This comes with the drawback that the DNA does not only have to cross the plasma membrane of the parasite before it can be delivered to the nucleus, but also the plasma membrane of the host cell and the membrane of the parasitophorous vacuole and space. This represents a major physical barrier preventing efficient DNA delivery. The efficiency of transfection for *P. falciparum* has been estimated at ~ 1 in 10^6 cells (O'Donnell *et al.*, 2002). Two main techniques of DNA delivery into *P. falciparum* have been developed: pre-loading of DNA into uninfected erythrocytes and the subsequent infection of these cells with *P. falciparum* parasites (Deitsch *et al.*, 2001) and the direct transfection of ring-stage infected erythrocytes (Wu *et al.*, 1995, modified by Crabb *et al.*, 2004). Skinner-Adams *et al.* (2003b) observed a slightly higher *in vitro* efficacy of the preloading method, and in our hands both methods provide comparable outcomes (unpublished). The delivery mode of DNA into the erythrocyte in both cases is *via* electroporation, but the mechanism of DNA uptake into the parasites is unknown. It has been speculated that even in the case of direct transfection of infected erythrocytes, the DNA is only preloaded into the erythrocyte and then taken up into the parasite during the invasion process. However, the application of drug pressure to these parasites after four hours of transfection yields a population of drug-resistant parasites, indicating that even without the process of reinvasion, plasmids conferring drug resistance had been delivered to the parasites and resulted in the efficient translation of the drug resistance protein. A recent paper by Allen and Kirk (2010) reports significantly reduced selection times for transfected *P. falciparum* parasites grown in suspension in comparison to standard static conditions. This highlights the influence and importance of culture conditions for the success of *Plasmodium* transfections.

Another major difference between the transfection of *P. berghei* and *P. falciparum* is the physical state of the DNA. While circular plasmid DNA can be used for the transfection of both *P. berghei* and *P. falciparum* cells, after which the circular DNA propagates as episomes, *P. berghei* also accepts linear DNA, which is then incorporated into the genome by homologous recombination. The use of linear DNA for transfection allows at

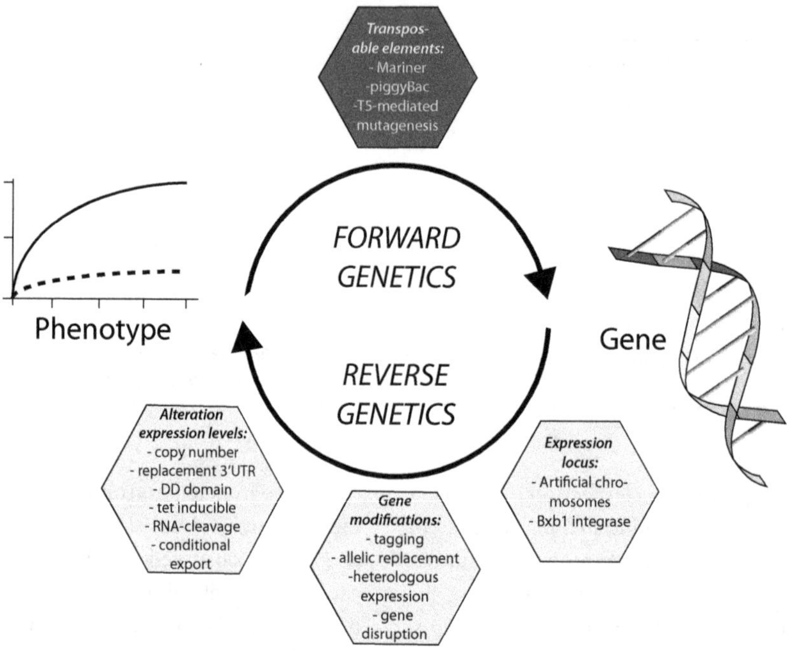

Figure 11.1 Schematic outline of forward and reverse genetics screens in *Plasmodium*. Forward genetics screens aim to identify the gene(s) responsible for a particular phenotype. Examples for tools employed in forward genetics screens are shown in the dark grey hexagon. Reverse genetics screens on the contrary have a particular gene as a starting point and try to unveil the phenotype that is associated with this gene. The light grey hexagons depict the tools employed in reverse genetics screens.

least theoretically the generation of transfection constructs via gene synthesis. Circular plasmid DNA for transfection presents a major hurdle for a large-scale systematic transformation approach in *P. falciparum*.

The current sub-optimal DNA delivery for *P. falciparum* represents a severe bottleneck to obtain functional genomic data on a larger scale. This fact is compounded by an inefficient mechanism of the *Plasmodium* parasite to integrate exogenous DNA through homologous recombination, which is essential for knockout studies. Consequently, these kind of studies are currently inefficient, long and tedious.

Genetic tools

There are two major avenues by which molecular tools can be employed to obtain data on gene function (Fig. 11.1):

1 forward genetics – the screening of a heterogeneous population of cells for a certain phenotype, and the consequent identification of the involved gene;
2 reverse genetics – assessing the function of a particular gene product by deletion, modification, and modulation of the associated gene.

Tools for reverse genetics

Maintaining exogenous DNA

Especially for the reverse genetics approach the introduction of heterogenous DNA into the parasite is necessary and – as outlined above – this very often takes the form of plasmid DNA (Fig. 11.2A). A typical stably maintained transfection plasmid consists of the following components:

- a 'backbone' encoding a gene cassette for positive selection in bacteria (usually conferring ampicillin resistance) and an origin of replication;

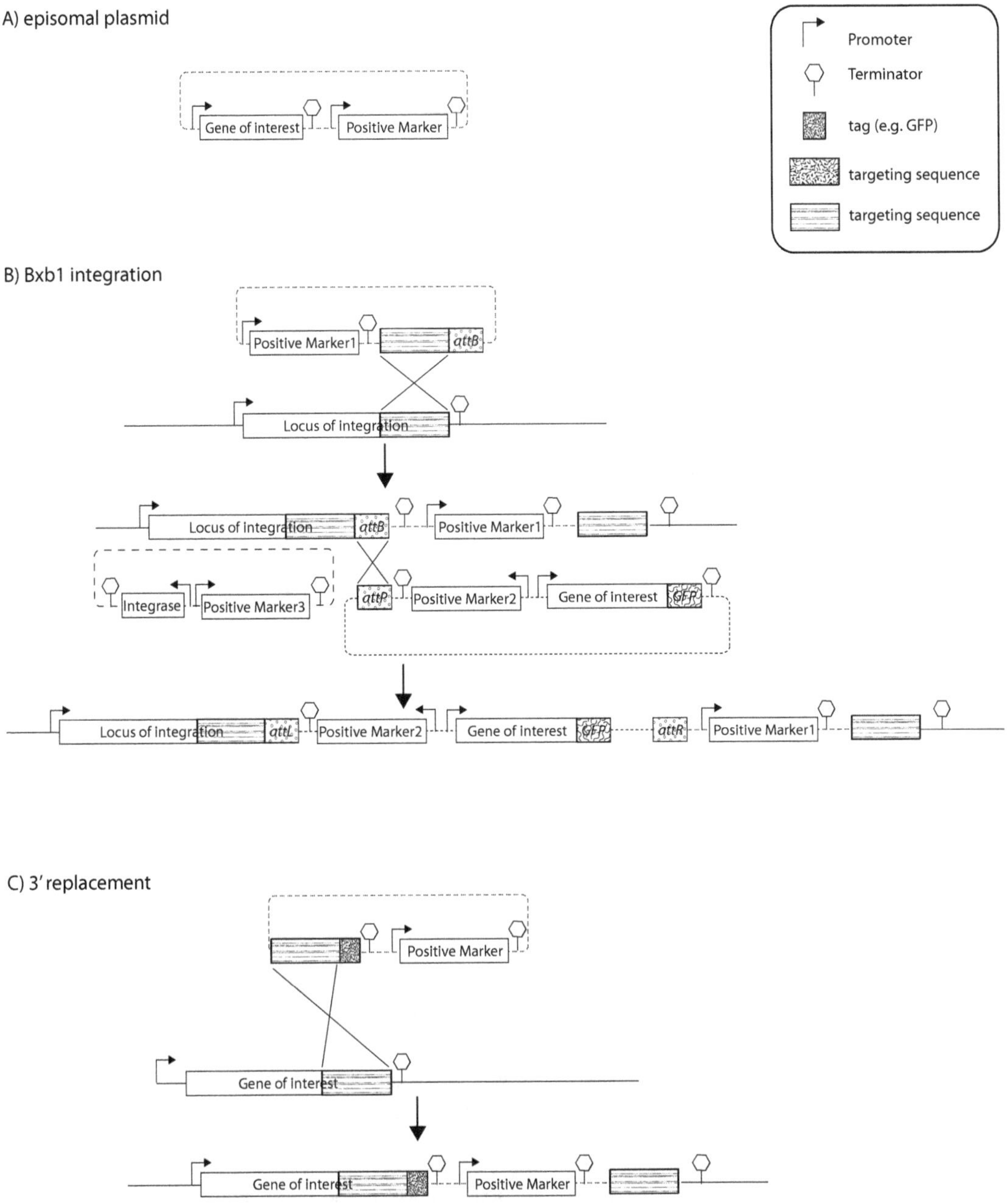

Figure 11.2 Episomal expression and potential homologous recombination events, that lead to integration into the *Plasmodium* genome. (A) Episomal plasmids containing a gene cassette for the expression of the gene of interest and a positive selection cassette (broken line represents the plasmid backbone). Reasons for expressing a transgene from an episomal plasmid include tagging, modification, and overexpression. (B) Site-specific integration via the Bxb1 mediated recombination. In a first step a single recombination event introduces an attB site into a genomic locus. Cells harbouring this modified locus are then transfected with a second plasmid containing a transgene cassette of interest and an attP site. Recombination between the attB and attP site is initiated by the Bxb1 integrase expressed from a third plasmid. The resulting locus contains the transgene and a positive selectable marker cassette flanked by attL and attR sites. The attL and attR sites are composites and result from recombination between the attB and attP sites. (C) 3′ replacements are used for tagging and allelic replacement experiments. A single cross-over homologous recombination event results in the integration of the plasmid into the genomic locus flanked by a duplication of the targeting sequence.

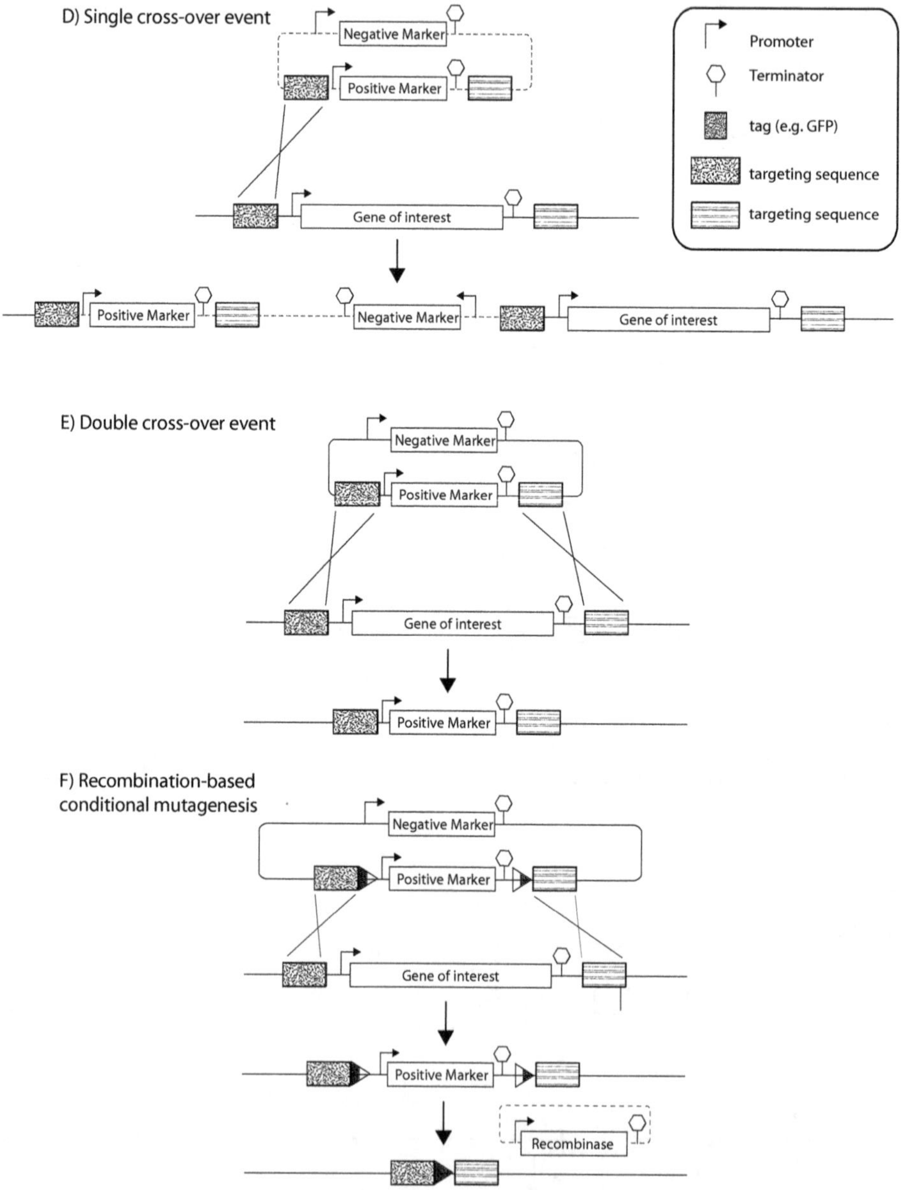

Figure 11.2 Continued. (D) Single cross-over events can be used for gene disruptions and incorporation of additional sequence information into a genomic locus. Similar to the 3′ replacement this event results in a duplication of the targeting sequence and the incorporation of the plasmid (or even multiple copies of the plasmid sequence). The negative maker, which is not utilized to achieve this event, is only shown to illustrate the difference to double cross-over events. (E) Double cross-over homologous recombination events are widely used for gene disruption/deletions. Upon positive and negative selection the sequence between the two targeting regions is replaced with a selectable marker cassette and the plasmid sequence containing the negative selectable maker is lost. (F) Recombination-based conditional mutagenesis system (Cre/lox or Flp/frt). This example depicts the application of this system for recycling of selectable markers. Recombination sites (lox or frt) are placed between the targeting sequences and the selectable marker cassette. Upon replacement of the endogenous gene locus via double cross-over recombination, the selectable marker cassette can be removed through the activation of an exogenously expressed recombinase (Cre or Flp). Instead of the selectable marker cassette only a remnant of the recombination sites are left between the targeting sequences in the disrupted locus. The selectable marker cassette becomes available again to be re-used for additional modifications of the *Plasmodium* genome in these transgenic cells.

- a 'positive selection cassette' for selection of transfected *Plasmodium*;
- either an 'expression cassette' for a transgene and/or a 'targeting sequence' for integration into the genome;
- optionally, features like a 'negative selection cassette' are added.

As a result of these requirements plasmids tend to exceed 6–7 kilobases.

Most plasmids have evolved over generations of projects and have to be seen as a consequence of that rather than a rationalized purpose-built product, and there might be potential to improve/streamline certain aspects.

Positive selectable markers

For the stable maintenance of transfection plasmids and for integration purposes selectable marker genes are used. The most commonly used selectable marker genes for *Plasmodium* are *Toxoplasma gondii* dihydrofolate reductase-thymidylate synthetase (*Tg dhfr-ts*) (Donald and Roos, 1993) and human dihydrofolate reductase (*hDHFR*) (de Koning-Ward *et al.*, 2000). These genes encode proteins that confer resistance to antifolate drugs (pyrimethamine, methotrexate, WR99210). There are other selectable markers like blasticidin-S-deaminase (*BSD*), neomycin phosphotransferase (*NPT*) (Mamoun *et al.*, 1999a) and puromycin-N-acetyltransferase (*PAC*) (de Koning-Ward *et al.*, 2001); however, their use is less reliable. Therefore the set of available selectable marker genes is quite limited, despite the fact that the transfectable stages of *Plasmodium* are haploid. The stable maintenance of a plasmid can be a means by itself (e.g. episomal expression of a transgene) or a pre-requisite for modifying an endogenous genetic locus (e.g. deletion or truncation of a gene).

Challenges to maintaining exogenous DNA

Interestingly, bacterial plasmids transfected into human- (*P. falciparum*) or rodent- (*P. berghei*) infecting *Plasmodium* replicate, but do not segregate evenly into daughter cells (van Dijk *et al.*, 1997). That leads to the rapid loss of plasmid upon removal of the selective drug pressure in erythrocytic stages (>95% within 21 days) (Fidock and Wellems, 1997) (O'Donnell *et al.*, 2002).

Upon transfection the monomeric, supercoiled plasmid DNA is converted into concatameric and high copy number forms within the parasite (O'Donnell *et al.*, 2001b). These concatamers are a result of the 'rolling-circle' plasmid replication process and might be a requirement that allows a more efficient segregation of plasmids. The prerequisite to convert the monomeric plasmid into a multi-copy head-to-tail concatamer might also explain the delay in observing transfected parasites (O'Donnell *et al.*, 2001b). The size of the concatamer varies in response to drug pressure, since increased demand for protein conferring the resistance is met by increased copy numbers of drug resistance cassette (see section on 'Strategies to alter expression levels') (Mamoun *et al.*, 1999a).

The unequal segregation and concatameric presence of plasmid has major experimental implications:

- The maintenance of drug pressure in the stages occurring in the mosquito or liver is a major challenge and basically precludes the episomal expression of transgenes in sexual stage parasites. However, the loss of plasmid in the absence of drug pressure can be exploited for the enrichment/selection of integration events (see section on 'Gene deletion/disruption').
- Plasmid copy numbers between individual cells vary and influence protein levels. This makes gene regulation, localization, and functional studies with episomes difficult.

Improved maintenance of exogenous DNA

The maintenance of exogenous DNA for transgene expression can be improved by either facilitating the equal segregation of plasmids (through the inclusion of Rep20 or centromeric sequences into the plasmid) or through the integration of the transgenic gene cassette into the chromosomal context of the parasite (through the use of the Bxb1 integrase system).

Rep20 sequences

A *P. falciparum*-specific solution to uneven segregation of plasmids is the incorporation of a 21 base pair repeat sequence found in subtelomeric regions of *P. falciparum* chromosomes, known as Rep20 (O'Donnell *et al.*, 2002). This sequence allows the tethering of the plasmid to chromosomes and therefore improves equal segregation. Cell lines transfected with constructs that contain the Rep20 sequence are established considerably quicker. However, Rep20 sequences interfere with gene targeting approaches and are only recommended for plasmids that should be maintained episomally.

Centromeric sequences

One exciting new development and an elegant way of addressing some of the shortcomings of episomal transfection is the construction of a *Plasmodium* artificial chromosome (PAC) in *P. berghei* (Iwanaga *et al.*, 2010). As outlined above the maintenance of plasmids in *Plasmodium* is quite unstable. Iwanaga *et al.* cloned a 1.2 kilobase centromere of the *P. berghei* chromosome 5 into a plasmid. This plasmid was evenly and stably segregated during nuclear division, meiosis in zygotes, and mitosis in mosquito oocysts – at this stage ~90% of the parasites still contained the plasmid, versus 2% for the control plasmid without the centromere. When *Plasmodium* telomere sequences were added to a linearized centromere-containing plasmid, it was found that these sequences were elongated over several rounds of replication until they reached the classical *Plasmodium* telomere size of 1.2 kilobases. In addition, the transfection efficiency via electroporation increased 10–100 fold in comparison to the circular plasmid and 1000-fold compared to classical plasmids.

Whether this approach will also work in *P. falciparum* remains to be explored. Interestingly, the exchange of the *P. berghei* centromere with one from *P. falciparum* resulted in lower segregation rates in *P. berghei* (Iwanaga *et al.*, 2010).

Their high transfection efficiency in *P. berghei*, low copy number, and stable maintenance makes artificial chromosomes good expression vectors. They might become vectors of choice in studies that look at the localization of tagged proteins, the regulation of gene transcription, and the complementation of protein function in defective cell lines. They could replace the need for homologous integration for such studies. Moreover, they will provide a valuable tool for forward genetic studies, where genes involved in the particular (very often polygenic) phenotype will be identified.

Bacteriophage N15 recombineering

To trounce the limitations imposed by the low efficiency of homologous recombination and the instability of the AT-rich sequences in *E. coli* a system for cloning and high frequency insertion was developed for *P. berghei* (Pfander *et al.*, 2011). Here genomic *P. berghei* DNA is converted into a bacteriophage N15-based vector library. The genomic DNA inserts provide long homologous DNA regions, which facilitate increased recombination frequency. A bacterial marker is inserted into the genomic DNA of the bacteriophage vector to delete or otherwise manipulate the gene of interest. In a second step the bacterial marker is then replaced *in vitro* with a selection cassette for *P. berghei* with the aid of a Gateway LR clonase reaction. Finally the manipulated insert is cut out from the vector backbone using unique restriction sites and transfected into *P. berghei* cells using standard procedures. This method holds great promises to finally conduct genetic modification experiments on a genome-wide scale in *Plasmodium*.

Mycobacteriophage Bxb1 integrase

The Bxb1 integrase systems permits the efficient and site-specific integration of genetic information into *P. falciparum* chromosomes (Fig. 11.2B) (Nkrumah *et al.*, 2006). This system mediates integration into a defined locus, unlike the piggyBac system, which enables random mutagenesis by using frequently occurring transposition target sites in the parasite genome (see 'PiggyBac', below). Cell lines harbouring the defined locus have to be engineered before a transposition event can take place. Hence the primary use of the Bxb1 integrase method is transgene expression from a chromosomal location.

Bxb1 integrase catalyses recombination events between an incoming attP and a previously-inserted chromosomal attB site. It is a mycobacteriophage integrase-based system defined by its high efficiency of recombination. AttP and attB sites are relatively short (~ 50 bp) and GC-rich. The *Plasmodium* genomes do not bear any region of detectable homology to them, which in turn allows for specific integration. AttP and attB sites are DNA sequences of inverted symmetry flanking a conserved 5′ GT dinucleotide. The fact that they are non-palindromic determines the polarity of integration. The maximum length of DNA that can be integrated via this system is not yet known. The maximum length of integration attempted by Nkrumah *et al.* (2006) was 10.2 kilobases (of that 5.4 kilobases were encoding the transgene), but this might not be the upper limit as the mycobacteriophage genome is 50–55 kilobases in size. It is more likely that cloning such long *P. falciparum* fragments will be the major limitation.

This system consists of three components. The first is a *P. falciparum* cell line containing an attB site. Nkrumah *et al.* (2006) have chosen the glutaredoxin-like gene cg6 for insertion of the attB site, since its disruption does not result in any observable growth defect in asexual blood stages. The second component is a plasmid providing an attP site for recombination, the transgene cassette for expression, and a drug resistance cassette for selection. A preference of integration of plasmids containing two or three attP sites has been shown. The third component is a plasmid for the transient expression of the integrase, which facilitates the recombination event.

Upon expression of the Bxb1 integrase, integration of the attP-containing plasmid into the attB site occurs. Bxb1 integrase only mediates integration, which results in asymmetric attL and attR sites. AttL and attR sites are refractory to Bxb1 integrase excision. This prevents the spontaneous reversion of events and results in a very stable locus. However, plasmid excision and the resulting reconstitution of the attB locus might be desirable for complementation experiments (to confirm a transgene phenotype) or for the recovery of marker genes. For these purposes the gene product gp47 from the mycobacteriophage could be utilized, which has been reported to mediate excision from attL and attR sites (Ghosh *et al.*, 2006).

One application of this system is the rapid and specific integration of transgenes into the *P. falciparum* genome. Genetically uniform recombinant parasites which express a single-copy gene can be generated within 2 to 4 weeks. The recombination is genetically stable in absence of drug and can therefore be carried throughout the life cycle. Combined with other systems it could be an efficient tool for complementation experiments.

It has to be pointed out that the integrase of Bxb1 is a serine integrase and therefore distinct from tyrosine integrases. Examples of tyrosine integrases/recombinases include lambda recombinase used in the GATEWAY® cloning system (see section on efficient cloning), Cre recombinase from the *E. coli* P1 phage, or the Flp recombinase from *Saccharomyces cerevisiae* (see 'Recombinase-based conditional mutagenesis systems'). They mediate both integration and excision events between two identical target sites; however, excision is thermodynamically favoured. The recombination events mediated by these recombinases are considered less stable. In contrast to the tyrosine integrases, serine integrases do not require bacterial host factors, divalent cations, or DNA supercoiling for their function (Ghosh *et al.*, 2003; Groth and Calos, 2004).

Major limitations of the Bxb1 system are the use of another selectable marker in addition to the one used for the establishment of the transgene plasmid, and the pre-requisite of an engineered host cell line, which requires time to engineer, and might preclude some primary cell lines from being analysed.

Strategies to modify malaria genes

A wide variety of techniques have been developed that allow the alteration of malaria genes. Depending on the experimental question, adding protein tags, modifying, replacing or deleting genes might provide answers.

Gene tagging

The expression of fusion proteins, from both episome and 3′ replacement integrants, into the endogenous locus (Fig. 11.2A and C) is probably the most common application of transfection technique in *Plasmodium*. Tags include fluorophores like green fluorescent protein for localization and transport studies, and epitope tags for localization and pull-down experiments. The technically easiest way is to express it at the C-terminus.

The biggest advantage of proteins tagged with GFP or its variations is that it allows for live cell imaging (Ploemen *et al.*, 2009; Rankin *et al.*, 2010). Dynamic processes in the cell can be visualized. In addition transgenic cells expressing GFP can be used for flow cytometry assays (e.g. Janse *et al.*, 2006b; Wilson *et al.*, 2010). GFPs can also be used as epitope tags and detected with antibodies in immunofluorescence assays to assess localization or interaction with other molecules. GFP can also be used for pull-down studies with antibody–coupled beads (GFP-trap Chromotek®). Problems here are the large size of the tag (27 kDa) and the potential of interfering with protein function. Another shortcoming is that if the GFP tag is destined to cellular compartments which prevent the secondary beta barrel structure from forming, fluorescence is only faint or lacking completely.

Epitope tags like myc-, haemagglutin (HA), or Streptag II®, on the other hand, are much smaller in size (1–1.2 kDa). These tags can be used for localization studies, since reliable and commercially available antibodies against these epitopes exist. In order to enhance the antigenicity of the tags they can be expressed in triple repeats (e.g. 3 × HA). These tags can also be used as affinity tags to isolate and analyse protein complexes from cells. Combining two different tags allows for the gentle and specific purification of protein complexes. This tandem affinity purification was used for example by Takebe *et al.* (2007), who expressed translation elongation factor PfEF-1beta C-terminally fused to a tandem tag consisting of protein C epitope, a protease cleavage site, and two protein A domains. Using these tags subsequently for pull-down experiments, not only could PfEF-1beta be isolated from crude extract of transgenic parasites, but also PFEF-1alpha and two unknown proteins, PfEF-1gamma and PfEF-1delta.

Especially with the advances in mass-spectroscopy in recent years, these affinity tags can provide a tool for the systematic analysis of protein–protein interactions in the malaria parasites.

Genetically modified parasites used for imaging

Many advances in the imaging of *Plasmodium* have been made over the last decade. These developments have been driven both by the new developments in instrumentation and by progress in our ability to genetically manipulate the parasite. A comprehensive review of the newly available tools is beyond the scope of this chapter; however, to provide a snapshot of the field, a few developments will be highlighted (for a comprehensive overview see, for example, Amino *et al.*, 2005; Frischknecht, 2010).

Fusion proteins linked to fluorescent proteins have enabled us to follow the fate of proteins during the morphological changes taking place in the parasite, and have lead to a much better understanding of the localization, interactions, and dynamics of *Plasmodium* proteins (Tilley *et al.*, 2007). The adaptation of new fluorescence proteins, which for example are photoactivatable, holds great promise for further refinement. Dixon *et al.* (Dixon *et al.*, 2009) developed a system, which allows the detection of erythrocytic stages, which have committed to develop into gametocyte, by fusing GFP to the early sexual blood-stage protein Pfs16. Another valuable application is the use of fluorescently labelled fusion proteins as probes to measure intracellular environments like pH (Klonis *et al.*, 2007) and Ca^{2+} concentrations (Rohrbach, 2009). Fluorescence/luminescence can be used to track and image labelled parasites in live mice (Franke-Fayard *et al.*, 2005, 2006) or mosquitoes (Frischknecht *et al.*, 2004) and their movement in affected organs (Mwakingwe *et al.*, 2009; Rankin *et al.*, 2010). These cell lines are also an alternative to radioisotopic methods to determine viability in drug assays (Sanchez *et al.*, 2004, 2007).

Allelic replacement

Allelic replacement allows for the exchange and analysis of altered protein molecules. Instead of adding a tag to the C-terminus of a gene, part of genes can be exchanged/altered by 3′ replacement (Fig. 11.2C). One of the first applications in *P. falciparum* was to prove the involvement of mutations in the *dhps* gene and the relative contribution of each mutation to sulfadoxine resistance. Triglia *et al.* (1998) introduced specific mutations into the endogenous locus. Another example involved the merozoite surface protein 1 (MSP-1). The C-terminus of MSP-1 shows limited polymorphism between *P. falciparum* strains and is associated with immunity against *P. falciparum*. O'Donnell *et al.* (2000) replaced the C-terminus of *P. falciparum* with the MSP-1 C-terminus of the rodent malaria *P. chabaudi*. The authors could show that this highly divergent sequence functionally complements the *P. falciparum* sequence and that therefore the *P. falciparum* MSP-1 C-terminal sequence is not constrained by function. Furthermore similar transgenic parasites were used to show that antibodies specific for this domain are a major component of inhibitory immune responses in both *P. falciparum*-immune humans and *P. chabaudi*/*P. berghei*-immune mice (De Koning-Ward *et al.*, 2003; O'Donnell *et al.*, 2001a). In another study the epidermal growth factor-like (EGF) domains of *P. falciparum* MSP-1 were replaced by the highly divergent EGF domains of *P. berghei* MSP-8, showing that the EGF domains of MSP-1 can accommodate diverse sequences; however, there are certain structural constraints imposed by the immune pressure in the field (Drew *et al.*, 2004).

Heterologous expression in *P. falciparum*

To overcome the limitation in functional analysis of *P. vivax* genes imposed by the lack of an effective continuous *in vitro* culture system, Sá *et al.* (2006) expressed the *P. vivax* ortholog of the *P. falciparum* chloroquine resistance transporter in *P. falciparum*. Heterologously expressed *P. vivax* CRT-O localized to the digestive vacuole and conferred a chloroquine resistance phenotype in *P. falciparum*.

However, this approach is still limited by our ability to manipulate *P. falciparum*. In addition, *P. falciparum* homologues might interfere with the expressed *P. vivax* molecules, making the dissection of phenotypes very challenging.

Gene deletion/disruption

The deletion or functional disruption of a gene is the most direct method to understand its function. Unfortunately it is also one of the most cumbersome. Disruption of a gene is traditionally achieved by introducing a selectable marker gene cassette into its open reading frame. Incorporation takes place via homologous recombination and specific genes are targeted by flanking the selection cassette with gene specific sequences (Fig. 11.2D and E). Gene disruptions obtained through double cross-over events are preferred, since they lead to irrevocable deletion of gene sequence, which in turn allows for the transgenic cells to be maintained in the absence of selective drug pressure (Fig. 11.2E). Gene disruptions obtained through single cross-over on the other hand still contain the complete gene sequence (although disrupted by the selection cassette) and have the potential to revert back to wild-type status in the absence of drug pressure (Fig. 11.3D). Maintaining drug pressure might interfere with assays that assess gene function and might make a comparison with the parental wild-type cells difficult.

The main challenge is that in these experiments the low transfection efficiency is potentiated by the low rate of recombination. The fact that circular plasmid DNA is transfected and maintained as concatamers in *P. falciparum* creates additional problems: it creates a platform for recombination events between the concatamers. Secondly, in order to obtain a cell line with the desired integration event, stable episomes have to be removed. This is done by enriching transfected cell lines for the presence of integrants by exposing them to cycles of on and off drug selection over a period of months. The adaptation of negative selectable markers, which provide a selection pressure against free episomes, have shortened the time frame necessary to select for a cell line with a double cross-over integration event.

Negative selection plasmids contain a 'suicide gene'. This suicide gene encodes an enzyme which converts a non-toxic pro-drug into a potent drug and kills the cell expressing the enzyme by interfering with the parasite's *de novo* pyrimidine biosynthesis and nucleic acid synthesis. Two negative selection systems have been adapted for use in *Plasmodium*: thymidine kinase from the herpes simplex virus (HSV-TK) (Duraisingh *et al.*, 2002) and deaminase/uracil phosphoribosyl transferase from the yeast *Saccharomyces cerevisiae*. HSV-TK converts the

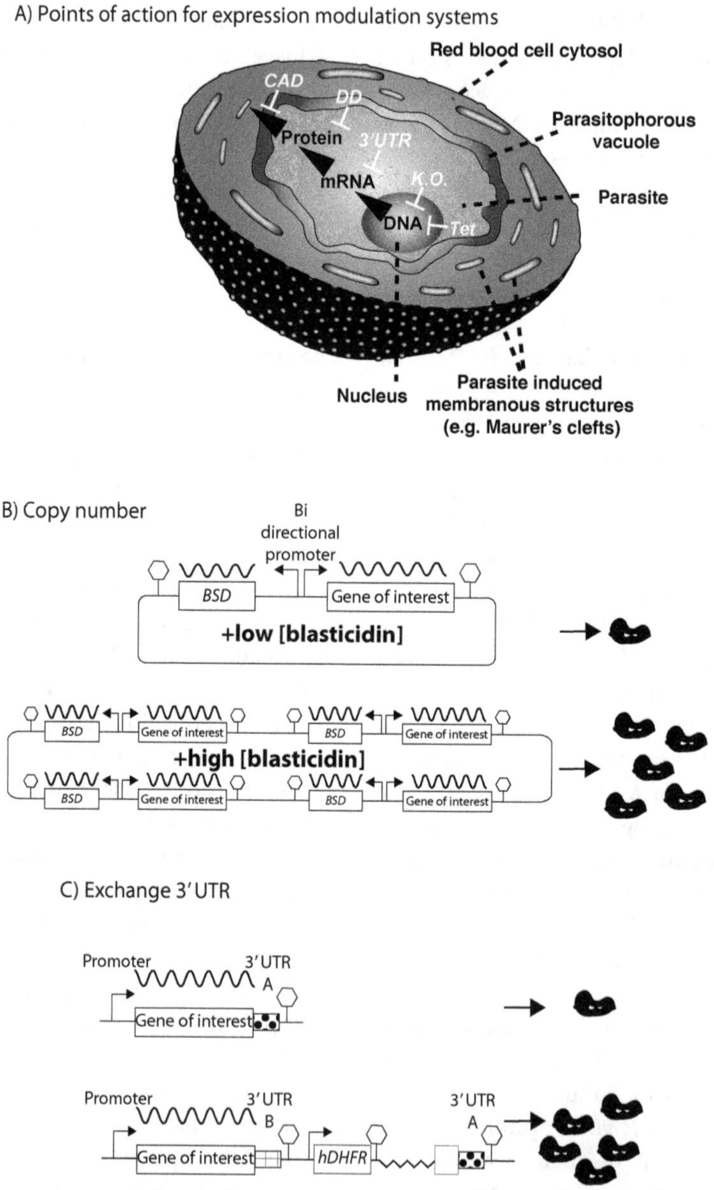

Figure 11.3 Systems to influence functional gene-expression in *Plasmodium*. (A) Point of action when applying molecular tools to influence gene-expression within the *Plasmodium* infected red blood cell. Gene disruptions/deletions (K.O.) and Tetracycline-inducible transcriptional regulation (Tet) influence the transcription of a gene. Different 3′UTR have an impact on mRNA stability and therefore expression levels. The destabilization domain (DD) influences protein stability at the post-translational level. The conditional aggregation domain (CAD) prevents the protein from reaching the host cell cytoplasm and its parasite-induced structures (e.g. Maurer's clefts). (B) Variations in the amount of drug used for selection results in different copy number of the transgenic plasmid. The level of transgene-expression correlates with gene copy number. (C) Stability of mRNA is influenced by the 3′ UTR. Increased half-life of mRNA results in increased protein expression.

harmless (to the parasite) drug gancyclovir into a toxic metabolite that inhibits DNA synthesis. HSV-TK has the disadvantage that it can show a 'bystander effect', in which upon the death of the suicide gene-containing cell, activated drug is released into its surroundings, thereby killing

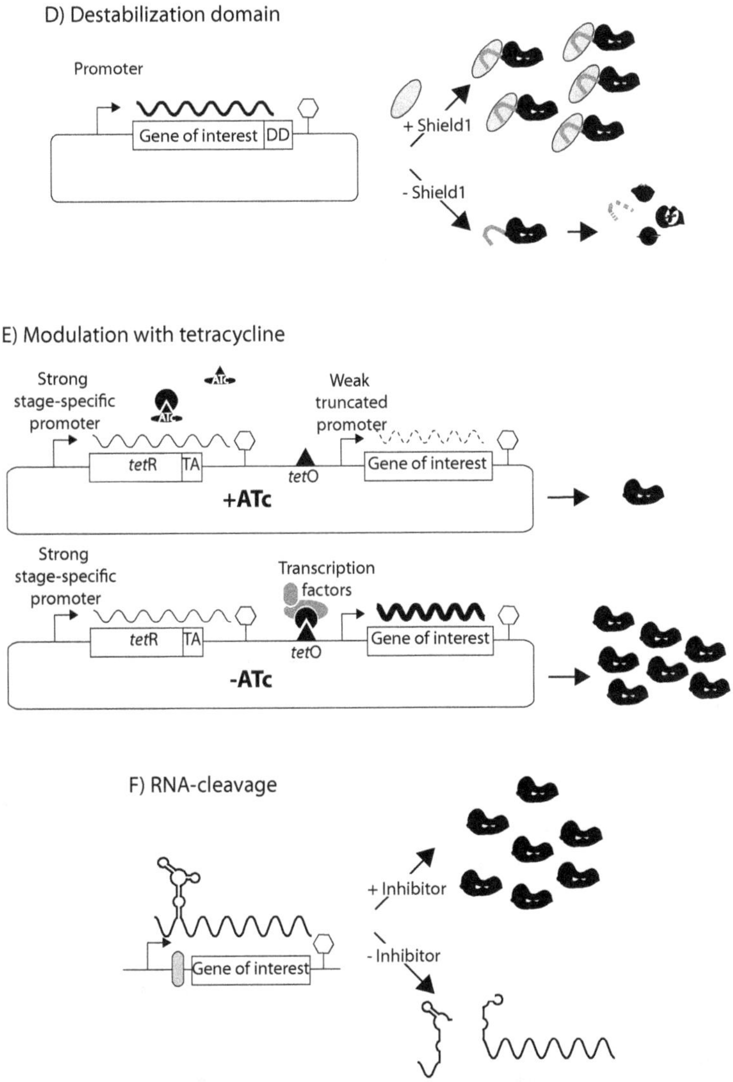

Figure 11.3 Continued. (D) Chimeric proteins containing the destabilization domain are degraded via the ubiquitin-proteasome system. The addition of the membrane permeable small molecule ligand Shield1 prevents the tag from being recognized and leads to the stabilization of the chimeric protein. (E) Mechanism of the tetracycline–inducible system: a gene of interest is expressed under a weak truncated promoter. A second expression cassette encodes for transactivator (tetR/TA). In the presence of anhydrotetracycline (ATc) transactivator does not bind to the tet-operator sequence upstream of the weak truncated promoter. Upon removal of ATc transactivators bind to the tet-operator sequences, which leads to the recruitment of transcription factors and a significant increase in transcription of the gene of interest. (F) Ribozymes are places near the transcriptional start site of a gene of interest. Owing to the autocatalysis of the ribozyme the mRNA is cleaved and translation prevented. Addition of specific ribozyme inhibitors, stabilize the mRNA, which leads to an increase in protein expression.

neighbouring cells which might not contain the suicide gene. The HSV-TK system in *P. falciparum* also shows a relatively high rate of spontaneous compensatory mutation in secondary sites, which probably allows for a resistance mechanism to take place and renders selection useless (Duraisingh *et al.*, 2002). Duraisingh *et al.* (2002) also looked at cytosine deaminase from *Escherichia coli*, but were unable to obtain double cross-over recombinations using this

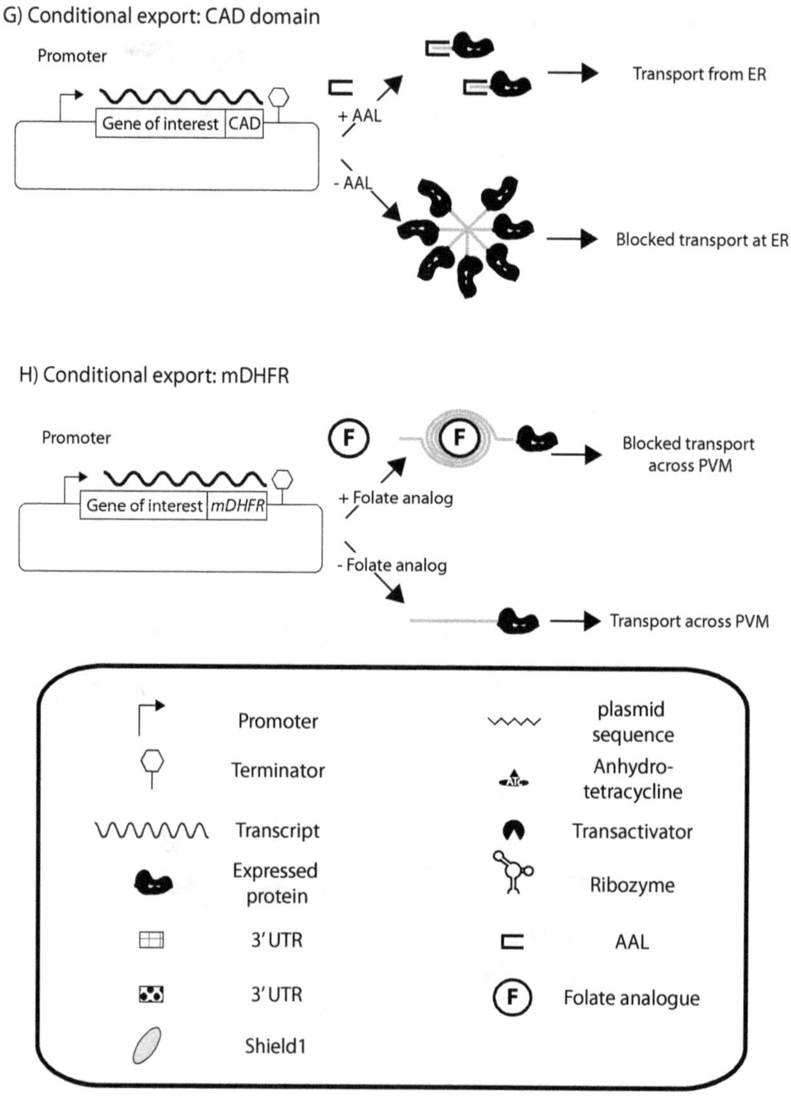

Figure 11.3 Continued. (G) Protein chimeras harbouring the CAD domain aggregate in the endoplasmic reticulum (ER) and are prevented from being transported out of the ER. Addition of anti-aggregation ligand (AAL) prevents the formation of aggregates due to the CAD domain and normal export of the chimeric protein from the ER into the parasite resumes. (H) Parasite proteins destined for the export into the red blood cell cytoplasm have to be unfolded in order to be transported across the parasitophorous vacuole membrane (PVM). In the presence of folate analogues mouse dihydrofolate reductase (mDHFR) forms a tertiary structure that cannot be unfolded by the cell and therefore prevents the entry of a protein that is fused to mDHFR into the red blood cell cytoplasm. In the absence of folate analogues normal unfolding of the fusion protein takes place resulting in the translocation across the PVM into the red blood cell cytoplasm. Illustrations have been inspired by de Koning-Ward and Gilson (2008) and modified with permission.

system. The second system is the cytosine deaminase/uracil phosphoribosyl transferase – a fusion gene from the yeast *Saccharomyces cerevisiae*, which has been adapted for both *P. berghei* (Braks *et al.*, 2006) and *P. falciparum* (Maier *et al.*, 2006). This enzyme combination makes a selection 1000-fold more potent than cytosine deaminase alone and decreases the change of spontaneous resistant mutants. The fusion protein converts 5-fluorocytosine into 5-fluorouracil, which inhibits thymidylate synthase as well as RNA synthesis. Although not 100%

perfect, this system increases the probability of obtaining parasites with a double cross-over integration event significantly.

Due to the lengthy selection process there is an increased risk that secondary mutations will occur in a transfected cell line and that an observed phenotype might be caused by these secondary alterations rather than the specific gene that was targeted. Given this and the advances in molecular techniques, Goldberg *et al.* (2010) proposed standards which should be observed when data on genetically manipulated *Plasmodium* is presented. These standards include basing a conclusion on at least two independent mutants, or verifying it (where technically possible) via a genetic complementation of a knockout mutant. In addition, unambiguous evidence for the genetic events should be presented (ideally in the form of Southern blots).

Building on the increased repertoire of molecular tools, the last couple of years saw a few systematic larger scale knockout libraries for *Plasmodium* being generated and phenotypically analysed. Maier *et al.* (2008) attempted to knock out 83 genes that are putatively involved in remodelling of the *P. falciparum*-infected red blood cell (Maier *et al.*, 2008). Fifty-three of these attempts were successful, and the functional screen of the knockout cell lines revealed eight genes that are required for the export and correct display of the virulence factor PfEMP1. In addition, several gene products were identified that have an influence on the rigidity of the plasma membrane of infected erythrocytes.

A similar approach was conducted in *P. berghei* to look at putatively secreted proteins expressed by the ookinete (Ecker *et al.*, 2008b). Twenty genes encoding such proteins were identified and knockouts were attempted, of which 12 were successful. Their analysis revealed six gene products that are important for the development of the parasite in the mosquito.

Overall more than 100 parasite gene disruptions have been reported since transfection technology was first established in *Plasmodium* (van Dijk *et al.*, 1995; Wu *et al.*, 1995). This represents little more than 2% of the whole genome, and there is great capacity to discover many more interesting phenotypes through this approach.

Another important event was the establishment of a database that accumulates genotypical and phenotypical data of genetically modified rodent malaria species (RMgm database, www.pberghei.eu; Janse *et al.*, 2010). It provides a comprehensive listing of not only successful gene knockouts, but also of attempts that were not successful (and therefore might indicate essentiality of the gene), and of other genetically modified *Plasmodium*. As molecular tools disseminate and the collection of available genetic mutants grows, such databases are becoming more and more important to avoid unnecessary duplication and provide focus of research efforts.

Strategies to dissect essential genes

The identification of essential genes is a major priority, since they provide a starting point to explore novel important cellular functions in the parasite and identify new targets for anti-malarial drugs. Since only blood-stage parasites can be transformed, genes essential for blood-stage growth cannot be analysed via the current knockout procedures. Proof of essentiality for blood-stage genes is hence difficult, but the Mini-Tn5-mediated transposon mutagenesis system in *P. berghei* might offer a possibility to provide this information (see later in the chapter).

RNA interference

Knock-down of gene expression by the use of RNA interference (RNAi) has been successfully applied in other protozoan parasites to elucidate gene function on a genome-wide scale (Subramaniam *et al.*, 2006). However, several RNA-based gene silencing approaches in *Plasmodium* were unsuccessful and a bioinformatic comparative genomics search revealed the absence of RNAi machinery (Baum *et al.*, 2009).

Recombinase-based conditional mutagenesis systems

Expression of site-specific recombinases is a standard tool when it comes to the genetic manipulation of mice and other organisms (Fig. 11.2F) (Birling *et al.*, 2009; Bischof and Basler, 2008). Carvalho *et al.* (2004) modified the Flp/FRT system from *Saccharomyces cerevisiae* to be used as a conditional mutagenesis technique in *P. berghei*. The flip-recombinase (Flp) mediates reversible integration and excision between two identical target sites, but excision is thermodynamically favoured. Recognition sites for the Flp recombinase were introduce via double cross-over homologous recombination in such a way that they flanked the regulatory sequences (Promoter or 3′ UTR) of the locus to be modulated. Site-specific recombination was then induced after cross-fertilization in the mosquito with a clone that expressed Flp recombinase. Using this technique, specific genes can be deleted in the mosquito stages and the resulting parasite can then be analysed for loss of function (Combe, 2009).

O'Neill *et al.* (2010) tested both Flp/FRT and Cre/LOX (derived from bacteriophage P1) for an ability to mediate the removal of genes that were previously inserted into the genome from blood-stage parasites (Fig. 11.2F). Instead of cross-fertilization, two other modes of recombinase delivery were tested: transfection of an episome encoding the recombinase, and the conditional expression of the recombinase under the influence of a tet-inducible promoter. Flp recombinase could catalyse removal of specific gene sequences; however, it was found to be very inefficient. One reason could be that Flp recombinase activity is the highest at 27°C, a temperature compatible with mosquito stages, but not with blood stages. Cre on the other hand could efficiently remove LOX-flanked gene sequences. It worked so efficiently that the residual expression under the tet-inducible promoter was enough to constitutively induce recombination. This limits its use in a conditional knockout system. The plasmid containing the Cre recombinase can be lost upon the removal of drug selection and the only remnant of recombinase activity is the 34 base pair sequence created through the removal of the target sequence.

These systems cannot only be used for the conditional removal of endogenous genes, but also for the removal of selectable marker genes after drug selection has taken place. Removal is useful to allow the reuse of the same selectable marker gene in an already-transfected parasite to perform reconstitution experiments, or in cases where the presence of a resistance gene is not desired (e.g. in genetically attenuate live parasite vaccine) (VanBuskirk *et al.*, 2009).

Strategies to alter expression levels

Since our current molecular toolbox for *Plasmodium* does not contain a tool to disrupt essential genes for blood stages by homologous recombination, the regulation of transgene expression is one potential avenue to assess gene function of essential genes. It is hoped that either over- or under-expression of the gene will lead to an observable change in phenotype, which in turn can give a clue about gene function. Several systems of various complexity have been adapted in the last few years (Fig. 11.3A).

These different techniques capable of influencing the protein levels at the site of function can be employed for different purposes. A main application would be to generate conditional knockout mutants for essential genes. The endogenous gene could either be removed in the presence of an induced episomal copy or replaced by an inducible copy. The phenotypical analysis could then take place by comparing cells producing wild-type levels of the protein and cells with a reduced amount of functional protein. For some proteins it might be possible to deduce their function by creating dominant-negative mutants: if over-expression of the protein in question interferes with normal protein function the altered phenotype might give a clue as to its function. Some of these techniques might also prove to be useful for complementation studies of generated knockout mutants. It is early days for most of the outlined techniques, which probably require significant tweaking in order to become easy and reliable tools, and experience will further highlight the strong points and shortcomings of each technique.

Alteration of gene copy number

The simplest way of over-expressing a gene is to transfect a plasmid containing a second copy into wild-type cells. Gardiner *et al.* (2006) used this approach to over-express leucyl aminopeptidase and found that these cells are more resistant to the inhibitor bestatin. As discussed previously plasmids are very often maintained as multi-copy concatamers. Expression strength and timing can be modulated by the choice of promoter selected for the transgene expression. Disadvantages of this method include that it is only usable if the expression level is high enough to see an altered phenotype, but low enough to ensure the survival of the cell. A danger is also that the condition of over-expression is present during the time of selection for a uniform transgenic parasite population. During this time the parasite can potentially adapt or compensate for the increased expression level, so that by the time the cells can be analysed no clear phenotype alteration can be observed any more.

Epp *et al.* (2008) have developed a system for regulated transgene expression which uses a single bidirectional promoter to drive both the expression of a selectable marker and a transgene (Fig. 11.3B). The selectable marker encodes blasticidin-S-deaminase (BSD), which confers resistance to blasticidin S. In response of increasing amounts of blasticidin S the plasmid copy number is altered, thereby modulating the amount of transgene expression. Increasing the amount of blasticidin from 2μg/ml to 20μg/ml led to a 10-fold increase in renilla luciferase activity, showing a transcriptional co-regulation. The bidirectional promoter allows also for a smaller construct that is more stable in *E. coli.* The usability of this system depends on the promoter employed; the bidirectional promoter (which is from the *var* intron PFC0005w) is only suitable for genes expressed in the second half of the 48 h asexual blood-stage life cycle. Epp *et al.* also evaluated a promoter combination that leads to a co-regulated more constitutive expression (*PSTI/CAM*), although the promoters were orientated in a head to head fashion.

Influencing mRNA stability

Reduced protein levels can be achieved by replacing the endogenous 3′ untranslated region (UTR) with another 3′ UTR via 3′ replacement (Fig. 11.3C). This technique is based on the influence of the 3′ UTR on the stability of mRNA and, hence, protein expression levels. A truncation of the 3′ UTR of the gene for the Pf chloroquine resistance transporter (PfCRT) just prior to the putative polyadenylation site resulted in a 10-fold decrease of reporter protein level and a 30–40% decrease in protein expression levels (Waller *et al.*, 2003). Replacement of the endogenous 3′UTR of the sodium-proton transporter PfNHE by the truncated PfCRT 3′ UTR resulted in 50% PfNHE protein expression (Nkrumah *et al.*, 2009). This technique is ideal to look at proteins involved in drug resistance, since the cell lines can be generated in the absence of drug pressure. Presumably the technique might not be so successful when evolutionary pressure cannot be removed during the generation of the transgenic cell lines.

Destabilisation domain (DD)

Lately a ligand-controlled destabilization domain based on the mutated FKBP protein has been adapted to *P. falciparum* (Fig. 11.3D) (Armstrong and Goldberg, 2007). Proteins fused to this 12 kDa tag are destabilized and rapidly degraded by the proteasome. However, the fusion protein becomes stable in the cell upon the reversible binding of a small, membrane-permeable ligand (Shield1). This ligand–tag interaction has been optimized by both screening for efficient FKBP mutants and using analogues of rapamycin ligands.

Armstrong and Goldberg used this system to modulate the expression of falcipain-2 in a falcipain 2 knockout background (Armstrong and Goldberg, 2007). Falcipain 2 is an integral membrane cysteine protease located in the food vacuole and involved in haemoglobin degradation (Sijwali and Rosenthal, 2004). They could show that stabilization is proportional to the amount of added Shield1 within a range of 10–500nM. They also reported favourable kinetics for this system; observable stabilization took place within 3h of adding Shield1 and removal of Shield1 led to a protein degradation within 6 h.

Russo *et al.* (Russo *et al.*, 2010; Russo *et al.*, 2009) used DD fusion to identify the role of calpain in cell cycle progression. Calpain proved to be refractory to gene disruption by double cross-over, gene truncation by single cross-over recombination, or allelic replacement. By integrating a GFP–DD fusion into the endogenous calpain locus via single cross-over homologous recombination, calpain levels could be knocked down by removal of Shield1, and cells showed an arrest in ring stages, highlighting the potential of this system. More recently, a further optimized variation was used to decipher the role of a plant-like kinase of *P. falciparum* in its invasion process (Dvorin *et al.*, 2010). This system holds great promises in the regulation of protein quantity and consequently the identification of protein function.

Although the DD can be added to both the N- and C-terminus, the tag at the N-terminus showed a higher impact on protein modulation (Banaszynski *et al.*, 2006). Unfortunately, due to the presence of signal sequences/export elements present at the N-terminus of *Plasmodium* proteins, the C-terminus might be the preferred option for many proteins. Also, the addition of Shield during the selection phase is not optimal, since variation of Shield concentration due to turnover could lead to potential fluctuations in protein expression levels.

Several labs have experienced a deregulation of this system over time or when expression levels drop below a level essential for the parasite's survival (e.g. (Dvorin *et al.*, 2010). Along this line a considerable variation in the reduction of expression levels between cells were observed (Armstrong and Goldberg, 2007), which might make the interpretation of some phenotypes difficult. Fusion of the 12-kDa DD could potentially interfere with protein function and not every protein might be amendable to this technique. The current high price for commercially available Shield-1 is also a dent in the usefulness of this molecular tool.

Tetracycline-inducible system

Gene regulation via tetracycline-responsive promoters have been widely used in mammalian and other systems (Gossen *et al.*, 1995; Meissner *et al.*, 2002; Wirtz and Clayton, 1995). Its adaptation to *Plasmodium* uses a TET-off version and is situated on one plasmid (Fig. 11.3E) (Meissner *et al.*, 2005). The system consists of a gene for the tet repressor protein (*tetR*) fused to a transactivator domain (*TA*) under the influence of a strong stage-specific promoter (e.g. *MSP2*). Also on the plasmid is the gene of interest under the control of a bipartite promoter consisting of the tet-operator sequence (7×42 base pair repeats) and a truncated promoter sequence with a low level basal activity (*calmodulin*). In the 'on' state *trans* activator protein is transcribed and binds to the tet operator sequence, recruiting transcription factors, which in turn initiate strong transcription of the gene of interest. Upon the addition of anhydrotetracycline, tet complexes with the tetR/TA-transactivator protein, prevent it from binding to the operator and only allow for a low level of basic transcription.

The kinetics of removing ATc from the cells and consequently the induction of gene expression are probably much slower than the repression induced by the addition of ATc. The large plasmid size represents a major drawback of this system. In addition there is considerable variation in expression levels between cells (de Koning-Ward and Gilson, 2008) and gene regulation seems to be overwritten, once a critically low level of expression is reached.

Inducible RNA cleavage

One approach which is still in its infancy is the use of self-cleaving ribozymes to regulate gene-expression (Fig. 11.3F). Introduction of a *cis*-acting hammerhead ribozyme near the translational start site results in the cleavage and degradation of the mRNA. Addition of the ribozyme inhibitor toyocamycin ablates the autocatalysis and leads to the stabilization of the mRNA, and its successful translation. This system of ectopic gene regulation has recently been developed for mammalian cells (Yen *et al.*, 2004), and a proof of principle study has been conducted for the apicomplexan parasites *Toxoplasma gondii* and *Plasmodium falciparum* (Agop-Nersesian *et al.*, 2008). Its biggest advantage is that it does not rely on the expression of any other trans-acting factors. The ribozyme is self-sufficient and contains all components needed for catalysis within its 85 nucleotides. Placement under the control of the endogenous promoter of the gene of interest ensures correct timing and levels of expression, and vectors are relatively simple. Unfortunately the nucleoside analogue toyocamycin has to be phosphorylated by an adenosine kinase, which is absent in *Plasmodium falciparum*. Furthermore, in *Toxoplasma*, activated toyocamycin displays cell toxicity. Therefore another specific non-toxic inhibitor for this autocatalytic ribozyme is needed, which to date has not been identified. Should such an inhibitor become available this system holds great promises to provide an easy and rapid way to down-regulate gene expression.

Conditional export system based on CAD

To analyse the function of gene products that are exported, a conditional export system was adapted to its use in *P. falciparum* (Fig. 11.3G) (Saridaki *et al.*, 2008). It is based on a conditional aggregation domain (CAD). When expressed as a fusion protein with the protein to be analysed, the CAD domain self-aggregates in the endoplasmic reticulum, preventing it from being transported any further. Upon the addition of small, cell-permeable ligand molecules (termed anti-aggregation ligand or AP21998) this aggregation can be reversed, and transport of the fusion protein through the constitutive secretory pathway resumes. This technique allows not only the conditional disruption of the function of exported gene products, but also allows the resolution of transport events in their temporal sequence. Moreover, transient interactions with molecules along the export pathway can be identified. The arrest of proteins in the ER seems to have no negative influence on the export of other exported proteins, and the de-aggregation of the arrested fusion protein is dose-dependent. Parasite age did not affect protein transport regulation. The amount of added anti-aggregation ligand had no effect on parasite growth at concentrations used to arrest transport. Saridaki *et al.* (2008) analysed the controlled export of a soluble protein (truncated form of STEVOR) and a transmembrane protein (full-length SBP-1). By adding brefeldin A, an inhibitor of protein transport from the ER to the Golgi apparatus (Klausner *et al.*, 1992), induced export was inhibited. This confirmed the previous findings that exported proteins destined for the host cell enter the parasite's secretory pathway (Charpian and Przyborski, 2008; Marti, 2005; Van Ooij and Haldar, 2007). Since restoration of export also took place in the presence of the protein synthesis inhibitor cycloheximide, it was concluded that the secretory machinery does not require additional de novo synthesis of further proteins at the trophozoite stage, and is fully operational. However, stages younger than 26 h post invasion were not looked at.

It remains to be seen whether the slow rate of observed protein export is an intrinsic property or whether it is due to the CAD system. Export could potentially be delayed by inefficient recognition of the fusion protein by the export machinery or by additional time required for the unfolding of stabilized tertiary structures of the CAD fusion protein. Nonetheless, the system represents an exciting tool to look at transport mechanism and protein function of exported *Plasmodium* proteins.

Conditional export using mDHFR

Another tool to dissect protein transport and translocation into cellular subcompartments is mouse/human dihydrofolate reductase (Fig. 11.3H). It is expressed as a fusion protein with the protein or signal sequence of interest. Upon addition of a folate analogue the folding of mDHFR stabilizes. This stabilization interferes with the transport processes that require the unfolding of the protein. In the past this technique has provided important clues into the folding state of transported proteins for cellular organelles like mitochondria, chloroplasts, and glycosomes (Eilers and Schatz, 1986; Häusler *et al.*, 1996; Soll and Schleiff, 2004). Gehde *et al.* (2009) used this technique to find out about the folding state of *P. falciparum* proteins that are trafficked across the parasitophorous vacuole membrane. Translocation of the fusion protein across this membrane takes place efficiently, but stabilization of mDHFR in its folded state results in a retention within the lumen of the parasitophorous vacuole. This indicates that unfolding of proteins destined for the red blood cell cytoplasm is necessary to cross the parasitophorous vacuole membrane. The transport block was reversible and dose-dependent. The system can be used to look at translocation into cellular organelles like the apicoplast or Maurer's clefts. Whether this system can also be utilizes as a conditional functional knockout system for exported proteins in general remains to be seen. One limitation is that folate analogues are toxic to wild-type *Plasmodium* and therefore the studies have to be done in a background of resistant parasites.

Tools for forward genetics

In order to identify mutants that possess a phenotype of interest, genetics screens can be employed. The components of a genetic screen are a large library of genetically distinct cell lines ('mutants'), a selection procedure for a certain phenotype, and a test to analyse the selected mutants and identify the associated genetic variation. Phenotypes of interests in *Plasmodium* include drug resistance, invasion, virulence, or cytoadhesion. This approach is especially powerful for phenotypes that involve several genes.

Classically, mutant libraries are created by exposing the cells to either radiation or chemical mutagenesis. Alternatively, a heterogenous library of cells can be generated by transfecting genomic libraries into a cell line lacking a certain phenotype and selecting for cells that have restored the phenotype (e.g. drug resistance). The advantage of these approaches is the ability to perform a saturation screen to uncover every single gene involved in a particular phenotype. Rounds of screening and mapping are done until no new genes are found. In order to reach saturation these libraries must be of a size several times the total number of genes in the genome. These approaches require the generation of a huge number of cell lines and take considerable effort at the analysis stage. The identification of gene products, which are essential for the survival of the cell requires an even greater effort, if possible at all. Given the logistical challenges of *Plasmodium* cell culture and the bottleneck to analyse these mutants in insect stages, this approach has not been pursued intensively.

Transposable elements

The use of transposable genetic elements, which integrate into the genome, is a popular alternative to chemical or radiation mutagenesis. One advantage is that the integration event does not only disrupt gene function, but also provides a marker to indicate where the mutation event has occurred. Moreover some of the screens allow for the systematic identification of essential gene products (see Mini-Tn5-mediated shuttle transposon mutagenesis).

The transposable elements used for genetic screens typically consist of a selection cassette (often a resistance or reporter gene) flanked by sequence repeats. These inverted sequences are transposon-specific and determine the integration locus in the genome of the organism to be mutagenized. The second element is an open reading frame encoding a transposase that initiates the transposition of the selectable marker in the genome. Often the transposase mediates integration as well as excision from the genome, and the conditions that favour one over the other differ between the systems. To be used for a systematic genome-wide mutagenesis the following attributes are desirable: an unbiased, even distribution of integration events along the genome; a frequency of integration events that allows several integration events per transfection experiment; a high efficiency of transfection; and a very specific integration and selection mechanism that avoids *in vivo* recombination and allows for the specific recovery of desired integration events. No endogenous transposable elements have been identified in *Plasmodium*.

Transposon-mediated mutagenesis can do much more than just disrupt gene function. Transposons can be used to identify gene regulatory elements. Deficient gene cassettes surrounding a reporter or selectable marker gene are used to screen for integration events that complement the gene cassette and restore full gene function. Consequently the full expression of the reporter/selectable marker gene allows for the selection and analysis of the integration event. To identify (for example) enhancer elements, a reporter gene under the influence of a minimal promoter is placed onto the transposon. Upon transposition into a locus under the influence of an enhancer element the promoter activity is enhanced, leading to higher expression of the reporter gene. Likewise, to screen for promoter or polyadenylation signals, reporter gene cassettes lacking these elements are transposed. It is noteworthy that the insertion of a reporter gene to identify promoter sequences usually leads to the disruption of the endogenous gene under the control of the identified promoter. This limits the output to promoter regions of non-essential genes. On the other hand this limitation does not prevent this approach from being used to identify transcriptionally inactive genes in certain life cycle stages.

The exploitation of the transposon-mediated mutagenesis for the identification of regions of the genome which are active under particular circumstances is called trapping, since such experiments trap and highlight a desired integration event. At the moment only few regulatory elements like promoters and terminators are characterized in malaria parasites (e.g. Crabb and Cowman, 1996; Dechering *et al.*, 1999; Fernandez-Becerra, 2003). However, information about promoter length, stage specificity, and bi-directionality would help establishing new molecular tool and make existing ones more versatile. Efficient reporter gene systems are established (e.g. Militello *et al.*, 2004). Given the sparse available information on gene regulatory elements, surprisingly little effort has been spent on employing transposons to learn more about these DNA sequences.

As outlined above the challenges of maintaining *Plasmodium* cells in the laboratory and the genome size of *Plasmodium* limits the analysis to the population level, since analysing individual clones obtained from transposition or library transfections is logistically demanding. Another possibility to alleviate this restriction of a whole-genome mutagenesis is conducting more focused screens, where potential candidate genes have been identified *in silico* by comparing the available sequence, expression, and proteomics data, such as data centrally compiled and easily accessible via PlasmoDB (Aurrecoechea *et al.*, 2009; www.plasmodb.org). The available genome information, for example, allows the identification of regulatory units of co-transcribed genes, or prediction of potential gene function. In other words, virtual genetics allows the definition of gene sets. Transposon-mediated mutagenesis suffers from the same limitation as all the other systems as its success still relies on the efficacy of transformation.

Mariner element

First attempts to establish a transposon-mediated mutagenesis system used the mariner element from *Drosophila* (Mamoun *et al.*, 2000). This system performs well in other protozoan parasites like *Leishmania* (Gueiros-Filho and Beverley, 1997) or trypanosomes (Leal *et al.*, 2004). However, due to the very low frequency of transformation, uncertainty about the transposase activity, and only few insertions in the same locus, this system did not catch on.

PiggyBac

The piggyBac-based transposition system allows for transposon-based random mutagenesis, gene delivery, or systematic gene knockout studies in *P. falciparum*. It was first discovered in insects (Fraser *et al.*, 1985) and adapted for use in *Plasmodium* by Balu *et al.* (2005) (Fig. 11.4A). PiggyBac is a DNA transposon that works by a 'cut and paste' mechanism and consists of an open reading frame encoding a 64 kDa transposase flanked by short inverted repeats of 35bp (5′) and 63 bp (3′) length. PiggyBac elements can carry up to 10–14 kilobases of foreign DNA without sacrificing too much efficiency (Ding *et al.*, 2005). It integrates at TTAA sites in the genome. Balu *et al.* predicted 311155 of these sites in the *P. falciparum* genome resulting in more than 20 insertion sites per average *P. falciparum* gene. Of these sites 40.5% are found in the EST database, making an integration event in a *P. falciparum* transcriptional unit (thereby disrupting gene expression) likely. PiggyBac integrates non-preferentially into the overall genome (Thibault *et al.*, 2004; Balu *et al.*, 2005), but has a greater tendency to insert into 5′ UTR and 5′ segments of genes. It is believed that this is due to a less condensed state of the chromatin structure in these regions, which allows better access for the transposase. This preference facilitates studies of the 5′ regions of genes like promoter traps.

One of its biggest advantages of the piggyBac system is that it can be separated into a *cis* element (inverted terminal repeats surrounding a drug resistance/reporter gene cassette) and *trans* element (transiently expressed helper plasmid with transposase), which adds to the versatility of the system. By using different promoters for the expression of the transposase, the time point of transposition can be determined. For example, gene products can be analysed for their importance in insect stages by transfecting blood stages with a helper plasmid expressing the transposase under the influence of an insect stage specific promoter. Any effect of the gene alteration in blood stages is thereby eliminated.

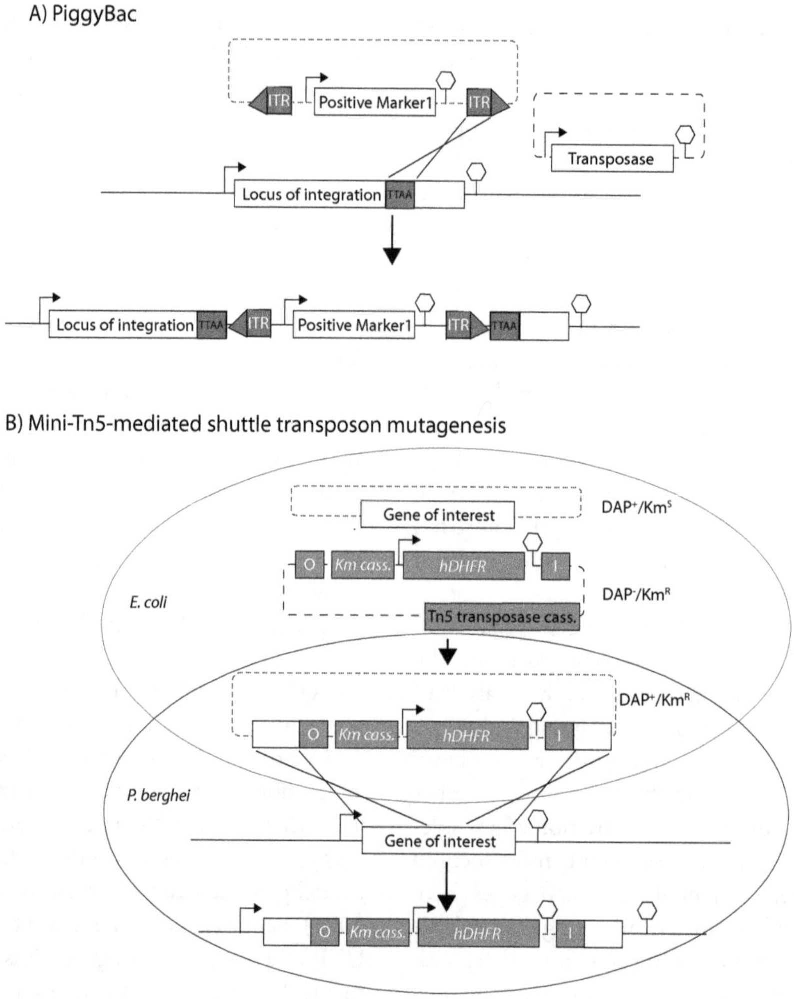

Figure 11.4 Transposon- and recombinase-mediated integration into the *Plasmodium* genome. (A) PiggyBac transformation of *Plasmodium*: a plasmid containing a positive selectable marker cassette flanked by inverted terminal repeat regions (ITR) is co-transfected with a plasmid expressing the piggyBac transposase. The transposase mediates random integration of the selectable marker cassette into a TTAA containing locus in the *Plasmodium* genome. (B) Transposition with the mini-Tn5 mediated shuttle transposon mutagenesis system takes place in bacteria and is used for the generation of transfection constructs. The gene sequence to be disrupted is cloned into a plasmid and introduced into a bacterial strain auxotroph for diaminopimelic acid (DAP^+). A second bacterial strain contains a plasmid encoding for an hDHFR cassette, a kanamycin resistance cassette and flanking region from the Tn5 transposon (O and I). Transposition is induced by bacterial conjugation and bacterial clones containing transposition events are selected for by DAP auxotrophy and Kanamycin resistance (DAP^+/Km^R). Colonies containing plasmids where the coding region of the gene of interest is disrupted are identified by PCR. The resulting plasmid is then used to transfect *Plasmodium* and integrated into the genome by a traditional double cross-over recombination event.

To adopt the piggyBac system for use in *P. falciparum* Balu *et al.* (2005) placed the *cis* element on a circular plasmid that also contained an origin of replication (*ori*) sequence and an antibiotic resistance cassette, which allows for the maintenance of the plasmid in *P. falciparum*, but also for the production of the plasmid in bacteria. The introduction of both the *cis* and *trans* elements led to the transposition of the drug selection cassette carried within piggyBac into the genome with high efficiency and a frequency of 1 in 10^{-3} schizonts (Balu *et al.*, 2005).

The transformation of few parasites in a 96-well microtitre plate format allowed for the adaptation of this system for large-scale genetic screens (Balu and Adams, 2006).

As for the goal of making a mutant library, it is estimated that 15,000 random mutations achieved by the piggyBac system would saturate the *P. falciparum* genome by 50%. Higher saturations are more difficult to achieve, due to the occurrence of redundant mutations.

The usefulness of the piggyBac system is not restricted to gene disruption experiments; it can also be used for rapidly obtaining cell lines for stable transgene expression. Integration into the genome via double cross-over recombination takes place as soon as transcription of the transposase is initiated. Another advantage of the piggyBac system is that potentially any strain or clone of *P. falciparum* can be transformed without relying on a genetically engineered host strain. This system can potentially be adapted to *P. berghei* and *P. yoelii*, adding the advantages of these systems, i.e. higher transfection efficiencies and the ability to transfect with linear DNA. Its major limitation is the inability to identify essential genes for blood stages or genes that are detrimental in these stages and might be lost during selection procedures. A remote, but very cumbersome solution would be to initiate the transposition in sexual stages and observe the effect of the transposition in blood stages.

Recently the piggyBac system was also adapted for use in *P. berghei* (Fonager *et al.*, 2011). Here the system inserted with a 16–18 times higher efficiency and more frequent integration into coding sequences compared to *P. falciparum*. This could allow for conducting saturating genome-wide mutagenesis experiments.

In addition the piggyBac system has also been used for the heterologous expression of *P. vivax* genes in *P. falciparum*; wild-type and mutant dihydrofolate reductase-thymidylate synthase from *P. vivax* were expressed episomally in *P. falciparum* to assess their role in antifolate resistance (Auliff *et al.*, 2010). This not only allowed to functionally characterize a *P. vivax* gene via reverse genetics, but also established a rapid screening system for studying drug resistance mechanisms in *P. vivax*.

Mini-Tn5-mediated shuttle transposon mutagenesis

Instead of manipulating the genome this tool provides an efficient means of generating knockout fragments, which can be used to transfect *P. berghei* cells (Fig. 11.4B) (Sakamoto *et al.*, 2005). Strictly speaking it represents a cloning method, but since it represents a new application in forward genetic screens it is presented here. For this method *Plasmodium* genes are cloned into a plasmid and transferred to bacteria. These genes are then disrupted via a bacterial Tn5-transposon system; transposition of the transposon into the *P. berghei* DNA is induced by bacterial conjugation. The donor cells contain the pUT plasmid encoding the mini-Tn5Pb1 transposon, the recipient cells contain the target construct. Selection for conjugation is done by kanamycin (resistance provided by the donor cells) and auxotrophy for diaminopimelic acid (provided by recipient cells). Insertion takes place almost randomly via a cut and paste process (Berg *et al.*, 1983). The transposition results in plasmids containing the mini-Tn5Pb1-interrupted *P. berghei* fragments, which are identified by polymerase chain reaction (PCR) and digested with ApaI to release a linear fragment. This disrupted fragment is then used for parasite transfection to replace the wild-type locus of the *Plasmodium* gene in the parasite genome by double-crossover homologous recombination. Currently the use of a linear fragment prevents its use in *P. falciparum*.

The Tn5 based shuttle mutagenesis simplifies the construction of gene knockouts since it requires a minimum number of simple bacterial manipulations, is more efficient than classical DNA cloning methods, provides a choice of distinct insertions, and can be automated, which is very attractive in the light of generating knockout screens and libraries.

It also provides the possibility to systematically, although indirectly, identify essential blood-stage genes in *Plasmodium*. Identical amounts of knockout constructs are mixed and used to be transfected. The resulting pool of *P. berghei* knockout cells are subjected to PCR analysis with Tn5 specific primer and gene replacement specific primers. The absence of a specific disruption would be indicative to the essentiality of the gene.

Since it is relatively easy to detect the presence or absence of a knockout cell line via PCR, essentiality of genes in certain life cycle stages can be established. Batches of ~ 10 knockout cell lines could be maintained through the different life-cycle stages and the composition of the mixture analysed at every stage via PCR. Such a strategy is only suitable for haploid stages and hence excludes the diploid zygote, ookinete, and oocyst stages, since they would result in false positives. Microarray analysis might allow for the detection of compensatory mechanisms (Sakamoto *et al.*, 2005).

Field application of molecular tools

Using these molecular tools in the laboratory has provided us with valuable insights into the biology of the malaria parasites and is continuing to do so. However, this is only one piece of the puzzle; as in all models there are certain limitations imposed on the extrapolation of laboratory data to the field. The transcriptional silencing or loss of genes important for transmission or maintenance of the parasite in the host under culture conditions is only one example. Field isolates are notoriously difficult to adapt to *in vitro* culture conditions, and the small amount of DNA available from these isolates represents a severe bottleneck in verifying laboratory findings in the field. One solution is the amplification of the DNA via molecular biological methods. Whole-genome amplification has been achieved via several PCR-based methods, but these methods are prone to uneven, biased amplification (Pinard *et al.*, 2006). An alternative is multiple-displacement amplification (Wang *et al.*, 2009). Here, random hexamer primers are annealed to the template, and DNA is synthesized by the high-fidelity enzyme φ29 (or Bst polymerase) at a constant temperature. The high processivity, high fidelity, and a 3′–5′ proofreading capability of φ29 result in longer products with fewer errors. Femtograms of starting material can be amplified to micrograms of DNA. This technique might also prove useful for single-cell analysis in the laboratory.

Conclusion-a peek into the crystal ball…

The expansion and refinement of molecular tools we now have available for *Plasmodium* have set the scene for a more systematic and high-throughput analysis of these important parasites and their fascinating biology. However, despite the recent significant developments, expanding the repertoire of molecular tools remains a project with many major building sites. Even fundamental information on optimal culture conditions or DNA delivery is sparse and there is plenty of room for improvement. In case of *P. falciparum* the low transfection efficiency is the major bottleneck for further adapting and developing new tools. Other *Plasmodium* species, on the other hand, are at the brink of being exploited for genetic discoveries and are just lacking the critical mass of researchers to progress further.

Our ability of genetically manipulating *Plasmodium* is still lagging behind other organisms and a major effort is required to close this gap. Adapting techniques currently used in other organisms, like genetic bar codes utilized in functional genomics studies in bacteria (Chiang *et al.*, 1999; Hensel *et al.*, 1995) and yeast (Kumar and Snyder, 2001; Shoemaker *et al.*, 1996), the variety in selectable markers/marker recycling standard in other organisms, or semi-automated culturing systems, might help in this effort. Maybe the combination of already existing systems will allow us to develop more efficient and robust methods for complementation and loss-of-function studies.

Major improvements and breakthroughs have expanded our molecular toolbox significantly in recent time and we should expect that these efforts will bear fruit in our fight against this disease.

Acknowledgements

This chapter is based on the work of countless people, but by nature it can only provide a biased snapshot and I apologize for any omissions. Our research is supported by the Australian Research Council and the National Health and Medical Research Council of Australia. Support was also received from the Hugh D T Williamson Foundation, which is managed by ANZ Trustees. I would like to thank Dr Melanie Rug for the critical reading of the manuscript and helpful comments.

References

Adisa, A., Rug, M., Klonis, N., Foley, M., Cowman, A.F., and Tilley, L. (2003). The signal sequence of exported protein-1 directs the green fluorescent protein to the parasitophorous vacuole of transfected malaria parasites. J. Biol. Chem. *278*, 6532–6542.

Agop-Nersesian, C., Pfahler, J., Lanzer, M., and Meissner, M. (2008). Functional expression of ribozymes in Apicomplexa: towards exogenous control of gene expression by inducible RNA-cleavage. Int. J. Parasitol. *38*, 673–681.

Allen, R.J.W., and Kirk, K. (2010). *Plasmodium falciparum* culture: the benefits of shaking. Mol. Biochem. Parasitol. *169*, 63–65.

Amino, R., Menard, R., and Frischknecht, F. (2005). imaging of malaria parasites – recent advances and future directions. Curr. Opin. Microbiol. *8*, 407–414.

Armstrong, C.M., and Goldberg, D.E. (2007). An FKBP destabilization domain modulates protein levels in *Plasmodium falciparum*. Nat. Meth. *4*, 1007–1009.

Auliff, A.M., Adams, J.H., O'Neil, M.T., and Cheng, Q. (2010). Defining the role of mutations in *Plasmodium vivax* dihydrofolate reductase-thymidylate synthase gene using an episomal *Plasmodium falciparum* transfection system. Antimicrob. Agents Chemother. *54*, 3927–3932.

Aurrecoechea, C., Brestelli, J., Brunk, B.P., Dommer, J., Fischer, S., Gajria, B., Gao, X., Gingle, A., Grant, G., Harb, O.S., *et al.* (2009). PlasmoDB: a functional genomic database for malaria parasites. Nucleic Acids Res. 37, D539–543.

Balu, B., and Adams, J.H. (2006). Functional genomics of *Plasmodium falciparum* through transposon-mediated mutagenesis. Cell. Microbiol. *8*, 1529–1536.

Balu, B., Shoue, D.A., Fraser, M.J., and Adams, J.H. (2005). High-efficiency transformation of *Plasmodium falciparum* by the lepidopteran transposable element piggyBac. Proc. Natl. Acad. Sci. U.S.A. *102*, 16391–16396.

Banaszynski, L., Chen, L., Maynardsmith, L., Ooi, A., and Wandless, T. (2006). A rapid, reversible, and tunable method to regulate protein function in living cells using synthetic small molecules. Cell *126*, 995–1004.

Baum, J., Maier, A.G., Good, R.T., Simpson, K.M., and Cowman, A.F. (2005). Invasion by *P. falciparum* merozoites suggests a hierarchy of molecular interactions. PLoS Pathog. *1*, e37.

Baum, J., Papenfuss, A.T., Mair, G.R., Janse, C.J., Vlachou, D., Waters, A.P., Cowman, A.F., Crabb, B.S., and De Koning-Ward, T.F. (2009). Molecular genetics and comparative genomics reveal RNAi is not functional in malaria parasites. Nucleic Acids Res. *37*, 3788–3798.

Ben Mamoun, C., Gluzman, I.Y., Hott, C., MacMillan, S.K., Amarakone, A.S., Anderson, D.L., Carlton, J.M., Dame, J.B., Chakrabarti, D., Martin, R.K., *et al.* (2001). Co-ordinated programme of gene expression during asexual intraerythrocytic development of the human malaria parasite *Plasmodium falciparum* revealed by microarray analysis. Mol. Microbiol. 39, 26–36.

Berg, D.E., Schmandt, M.A., and Lowe, J.B. (1983). Specificity of transposon Tn5 insertion. Genetics *105*, 813–828.

Birling, M.-C., Gofflot, F., and Warot, X. (2009). Site-specific recombinases for manipulation of the mouse genome. Methods Mol. Biol. *561*, 245–263.

Bischof, J., and Basler, K. (2008). Recombinases and their use in gene activation, gene inactivation, and transgenesis. Methods Mol. Biol. *420*, 175–195.

Boyle, M.J., Wilson, D.W., Richards, J.S., Riglar, D.T., Tetteh, K.K., Conway, D.J., Ralph, S.A., Baum, J., and Beeson, J.G. (2010). Isolation of viable *Plasmodium falciparum* merozoites to define erythrocyte invasion events and advance vaccine and drug development. Proc. Natl. Acad. Sci. U.S.A. *107*, 14378–14383.

Bozdech, Z., Llinás, M., Pulliam, B.L., Wong, E.D., Zhu, J., and Derisi, J.L. (2003). The transcriptome of the intraerythrocytic developmental cycle of *Plasmodium falciparum*. PLoS Biol. *1*, E5.

Braks, J.A.M., Franke-Fayard, B., Kroeze, H., Janse, C.J., and Waters, A.P. (2006). Development and application of a positive-negative selectable marker system for use in reverse genetics in Plasmodium. Nucleic Acids Res. *34*, e39.

Carlton, J. (2003). The *Plasmodium vivax* genome sequencing project. Trends Parasitol. *19*, 227–231.

Carlton, J.M., Adams, J.H., Silva, J.C., Bidwell, S.L., Lorenzi, H., Caler, E., Crabtree, J., Angiuoli, S.V., Merino, E.F., Amedeo, P., *et al.* (2008). Comparative genomics of the neglected human malaria parasite *Plasmodium vivax*. Nature 455, 757–763.

Carlton, J.M., Angiuoli, S.V., Suh, B.B., Kooij, T.W., Pertea, M., Silva, J.C., Ermolaeva, M.D., Allen, J.E., Selengut, J.D., Koo, H.L., *et al.* (2002). Genome sequence and comparative analysis of the model rodent malaria parasite *Plasmodium yoelii yoelii*. Nature *419*, 512–519.

Carvalho, T.G., Thiberge, S., Sakamoto, H., and Ménard, R. (2004). Conditional mutagenesis using site-specific recombination in *Plasmodium berghei*. Proc. Natl. Acad. Sci. U.S.A. *101*, 14931–14936.

Charpian, S., and Przyborski, J.M. (2008). Protein transport across the parasitophorous vacuole of *Plasmodium falciparum*: into the great wide open. Traffic 9, 157–165.

Chiang, S.L., Mekalanos, J.J., and Holden, D.W. (1999). In vivo genetic analysis of bacterial virulence. Annu. Rev. Microbiol. 53, 129–154.

Combe, A., Giovannini, D., Carvalho, T., Spath, S., Boisson, B., Loussert, C., Thiberge, S., Lacroix, C., Gueirard, P., and Menard, R. (2009). Clonal conditional mutagenesis in malaria parasites. Cell Host Micro. 5, 386–396.

Cowman, A.F. (2001). Functional analysis of drug resistance in *Plasmodium falciparum* in the post-genomic era. Int. J. Parasitol. *31*, 871–878.

Crabb, B.S., Cooke, B.M., Reeder, J.C., Waller, R.F., Caruana, S.R., Davern, K.M., Wickham, M.E., Brown, G.V., Coppel, R.L., and Cowman, A.F. (1997a). Targeted gene disruption shows that knobs enable malaria-infected red cells to cytoadhere under physiological shear stress. Cell *89*, 287–296.

Crabb, B.S., and Cowman, A.F. (1996). Characterization of promoters and stable transfection by homologous and nonhomologous recombination in *Plasmodium falciparum*. Proc. Natl. Acad. Sci. U.S.A. *93*, 7289–7294.

Crabb, B.S., Rug, M., Thompson, J.K., Triglia, T., Maier, A.G., and Cowman, A.F. (2004). Transfection of the human malaria parasite *Plasmodium falciparum*. Methods Mol. Biol. *270*, 263–276.

Crabb, B.S., Triglia, T., Waterkeyn, J.G., and Cowman, A.F. (1997b). Stable transgene expression in *Plasmodium falciparum*. Mol. Biochem. Parasitol. *90*, 131–144.

Date, S.V., and Stoeckert, C.J. (2006). Computational modeling of the *Plasmodium falciparum* interactome reveals protein function on a genome-wide scale. Genome Res. *16*, 542–549.

de Koning-Ward, T.F., Fidock, D.A., Thathy, V., Menard, R., van Spaendonk, R.M., Waters, A.P., and Janse, C.J. (2000). The selectable marker human dihydrofolate reductase enables sequential genetic manipulation of the *Plasmodium berghei* genome. Mol. Biochem. Parasitol. *106*, 199–212.

de Koning-Ward, T.F., O'Donnell, R.A., Drew, D.R., Thomson, R., Speed, T.P., and Crabb, B.S. (2003). A new rodent model to assess blood-stage immunity to the *Plasmodium falciparum* antigen merozoite surface protein 119 reveals a protective role for invasion inhibitory antibodies. J. Exp. Med. *198*, 869–875.

de Koning-Ward, T.F., Sperança, M.A., Waters, A.P., and Janse, C.J. (1999). Analysis of stage specificity of promoters in *Plasmodium berghei* using luciferase as a reporter. Mol. Biochem. Parasitol. *100*, 141–146.

de Koning-Ward, T.F., Thomas, A.W., Waters, A.P., and Janse, C.J. (1998). Stable expression of green fluorescent protein in blood and mosquito stages of *Plasmodium berghei*. Mol. Biochem. Parasitol. *97*, 247–252.

de Koning-Ward, T.F., Waters, A.P., and Crabb, B.S. (2001). Puromycin-N-acetyltransferase as a selectable marker for use in *Plasmodium falciparum*. Mol. Biochem. Parasitol. *117*, 155–160.

de Koning-Ward, T., and Gilson, P. (2008). Keeping it simple: an easy method for manipulating the expression levels of malaria proteins. Trends Parasitol. *25*, 4–7.

Dechering, K.J., Kaan, A.M., Mbacham, W., Wirth, D.F., Eling, W., Konings, R.N., and Stunnenberg, H.G. (1999). Isolation and functional characterization of two distinct sexual-stage-specific promoters of the human malaria parasite *Plasmodium falciparum*. Mol. Cell Biol. *19*, 967–978.

Deitsch, K., Driskill, C., and Wellems, T. (2001). Transformation of malaria parasites by the spontaneous uptake and expression of DNA from human erythrocytes. Nucleic Acids Res. *29*, 850–853.

Ding, S., Wu, X., Li, G., Han, M., Zhuang, Y., and Xu, T. (2005). Efficient transposition of the piggyBac (PB) transposon in mammalian cells and mice. Cell *122*, 473–483.

Dixon, M.W.A., Peatey, C.L., Gardiner, D.L., and Trenholme, K.R. (2009). A green fluorescent protein-based assay for determining gametocyte production in *Plasmodium falciparum*. Mol. Biochem. Parasitol. *163*, 123–126.

Doherty, J.P., Lindeman, R., Trent, R.J., Graham, M.W., and Woodcock, D.M. (1993). Escherichia coli host strains SURE and SRB fail to preserve a palindrome cloned in lambda phage: improved alternate host strains. Gene *124*, 29–35.

Donald, R.G., and Roos, D.S. (1993). Stable molecular transformation of Toxoplasma gondii: a selectable dihydrofolate reductase-thymidylate synthase marker based on drug-resistance mutations in malaria. Proc. Natl. Acad. Sci. U.S.A. *90*, 11703–11707.

Drew, D.R., O'Donnell, R.A., Smith, B.J., and Crabb, B.S. (2004). A common cross-species function for the double epidermal growth factor-like modules of the highly divergent plasmodium surface proteins MSP-1 and MSP-8. J. Biol. Chem. *279*, 20147–20153.

Duraisingh, M.T., Triglia, T., and Cowman, A.F. (2002). Negative selection of *Plasmodium falciparum* reveals targeted gene deletion by double crossover recombination. Int. J. Parasitol. *32*, 81–89.

Dvorin, J.D., Martyn, D.C., Patel, S.D., Grimley, J.S., Collins, C.R., Hopp, C.S., Bright, A.T., Westenberger, S., Winzeler, E., Blackman, M.J., *et al.* (2010). A plant-like kinase in *Plasmodium falciparum* regulates parasite egress from erythrocytes. Science *328*, 910–912.

Ecker, A., Bushell, E., Tewari, R., and Sinden, R. (2008a). Reverse genetics screen identifies six proteins important for malaria development in the mosquito. Mol. Microbiol. *70*, 209–220.

Ecker, A., Bushell, E.S.C., Tewari, R., and Sinden, R.E. (2008b). Reverse genetics screen identifies six proteins important for malaria development in the mosquito. Mol. Microbiol. *70*, 209–220.

Eilers, M., and Schatz, G. (1986). Binding of a specific ligand inhibits import of a purified precursor protein into mitochondria. Nature *322*, 228–232.

Epp, C., Raskolnikov, D., and Deitsch, K.W. (2008). A regulatable transgene expression system for cultured *Plasmodium falciparum* parasites. Malar. J. *7*, 86.

Fernandez-Becerra, C. (2003). *Plasmodium falciparum*: new vector with bi-directional promoter activity to stably express transgenes. Exp. Parasitol. *103*, 88–91.

Fidock, D.A., and Wellems, T.E. (1997). Transformation with human dihydrofolate reductase renders malaria parasites insensitive to WR99210 but does not affect the intrinsic activity of proguanil. Proc. Natl. Acad. Sci. U.S.A. *94*, 10931–10936.

Florens, L., Washburn, M.P., Raine, J.D., Anthony, R.M., Grainger, M., Haynes, J.D., Moch, J.K., Muster, N., Sacci, J.B., Tabb, D.L., *et al.* (2002). A proteomic view of the *Plasmodium falciparum* life cycle. Nature *419*, 520–526.

Fonager, J., Franke-Fayard, B., Adams, J., Ramesar, J., Klop, O., Khan, S.M., Janse, C.J., and Waters, A.P. (2011). Development of the piggyBac transposable system for *Plasmodium berghei* and its application for random mutagenesis in malaria parasites. BMC Genomics *12*, 155.

Foth, B.J., Ralph, S.A., Tonkin, C.J., Struck, N.S., Fraunholz, M., Roos, D.S., Cowman, A.F., and

Mcfadden, G.I. (2003). Dissecting apicoplast targeting in the malaria parasite *Plasmodium falciparum*. Science *299*, 705–708.

Franke-Fayard, B., Janse, C.J., Cunha-Rodrigues, M., Ramesar, J., Büscher, P., Que, I., Löwik, C., Voshol, P.J., den Boer, M.A.M., van Duinen, S.G., *et al.* (2005). Murine malaria parasite sequestration: CD36 is the major receptor, but cerebral pathology is unlinked to sequestration. Proc. Natl. Acad. Sci. U.S.A. *102*, 11468–11473.

Franke-Fayard, B., Waters, A.P., and Janse, C.J. (2006). Real-time in vivo imaging of transgenic bioluminescent blood stages of rodent malaria parasites in mice. Nat. Protoc. *1*, 476–485.

Fraser, M.J., Brusca, J.S., Smith, G.E., and Summers, M.D. (1985). Transposon-mediated mutagenesis of a baculovirus. Virology *145*, 356–361.

Frischknecht, F. (2010). Imaging parasites at different scales. Cell Host Microbe *8*, 16–19.

Frischknecht, F., Baldacci, P., Martin, B., Zimmer, C., Thiberge, S., Olivo-Marin, J.-C., Shorte, S.L., and Ménard, R. (2004). Imaging movement of malaria parasites during transmission by Anopheles mosquitoes. Cell. Microbiol. *6*, 687–694.

Gardiner, D.L., Trenholme, K.R., Skinner-Adams, T.S., Stack, C.M., and Dalton, J.P. (2006). Overexpression of leucyl aminopeptidase in *Plasmodium falciparum* parasites. Target for the antimalarial activity of bestatin. J. Biol. Chem. *281*, 1741–1745.

Gardner, M.J., Hall, N., Fung, E., White, O., Berriman, M., Hyman, R.W., Carlton, J.M., Pain, A., Nelson, K.E., Bowman, S., *et al.* (2002). Genome sequence of the human malaria parasite *Plasmodium falciparum*. Nature *419*, 498–511.

Gehde, N., Hinrichs, C., Montilla, I., Charpian, S., Lingelbach, K., and Przyborski, J.M. (2009). Protein unfolding is an essential requirement for transport across the parasitophorous vacuolar membrane of *Plasmodium falciparum*. Mol. Microbiol. *71*, 613–628.

Ghosh, P., Kim, A.I., and Hatfull, G.F. (2003). The orientation of mycobacteriophage Bxb1 integration is solely dependent on the central dinucleotide of attP and attB. Mol. Cell *12*, 1101–1111.

Ghosh, P., Wasil, L.R., and Hatfull, G.F. (2006). Control of phage Bxb1 excision by a novel recombination directionality factor. PLoS Biol. *4*, e186.

Goonewardene, R., Daily, J., Kaslow, D., Sullivan, T.J., Duffy, P., Carter, R., Mendis, K., and Wirth, D. (1993). Transfection of the malaria parasite and expression of firefly luciferase. Proc. Natl. Acad. Sci. U.S.A. *90*, 5234–5236.

Gossen, M., Freundlieb, S., Bender, G., Müller, G., Hillen, W., and Bujard, H. (1995). Transcriptional activation by tetracyclines in mammalian cells. Science *268*, 1766–1769.

Groth, A.C., and Calos, M.P. (2004). Phage integrases: biology and applications. J. Mol. Biol. *335*, 667–678.

Gueiros-Filho, F.J., and Beverley, S.M. (1997). Trans-kingdom transposition of the *Drosophila* element mariner within the protozoan *Leishmania*. Science *276*, 1716–1719.

Hall, N. (2005). A comprehensive survey of the *Plasmodium* life cycle by genomic, transcriptomic, and proteomic analyses. Science *307*, 82–86.

Hall, N., Karras, M., Raine, J.D., Carlton, J.M., Kooij, T.W.A., Berriman, M., Florens, L., Janssen, C.S., Pain, A., Christophides, G.K., *et al.* (2005). A comprehensive survey of the Plasmodium life cycle by genomic, transcriptomic, and proteomic analyses. Science *307*, 82–86.

Hamilton, M.D., Nuara, A.A., Gammon, D.B., Buller, R.M., and Evans, D.H. (2007). Duplex strand joining reactions catalyzed by vaccinia virus DNA polymerase. Nucleic Acids Res. *35*, 143–151.

Häusler, T., Stierhof, Y.D., Wirtz, E., and Clayton, C. (1996). Import of a DHFR hybrid protein into glycosomes in vivo is not inhibited by the folate-analogue aminopterin. J. Cell Biol. *132*, 311–324.

Hayward, R.E., Derisi, J.L., Alfadhli, S., Kaslow, D.C., Brown, P.O., and Rathod, P.K. (2000). Shotgun DNA microarrays and stage-specific gene expression in *Plasmodium falciparum* malaria. Mol. Microbiol. *35*, 6–14.

Hensel, M., Shea, J.E., Gleeson, C., Jones, M.D., Dalton, E., and Holden, D.W. (1995). Simultaneous identification of bacterial virulence genes by negative selection. Science *269*, 400–403.

Hiller, N.L., Bhattacharjee, S., Van Ooij, C., Liolios, K., Harrison, T., Lopez-Estraño, C., and Haldar, K. (2004). A host-targeting signal in virulence proteins reveals a secretome in malarial infection. Science *306*, 1934–1937.

Holt, R.A., Subramanian, G.M., Halpern, A., Sutton, G.G., Charlab, R., Nusskern, D.R., Wincker, P., Clark, A.G., Ribeiro, J.M.C., Wides, R., *et al.* (2002). The genome sequence of the malaria mosquito *Anopheles gambiae*. Science *298*, 129–149.

Iwanaga, S., Khan, S.M., Kaneko, I., Christodoulou, Z., Newbold, C., Yuda, M., Janse, C.J., and Waters, A.P. (2010). Functional Identification of the Plasmodium Centromere and Generation of a Plasmodium Artificial Chromosome. Cell Host Microbe *7*, 245–255.

Janse, C.J., Franke-Fayard, B., Mair, G.R., Ramesar, J., Thiel, C., Engelmann, S., Matuschewski, K., van Gemert, G.J., Sauerwein, R.W., and Waters, A.P. (2006a). High efficiency transfection of *Plasmodium berghei* facilitates novel selection procedures. Mol. Biochem. Parasitol. *145*, 60–70.

Janse, C.J., Franke-Fayard, B., and Waters, A.P. (2006b). Selection by flow-sorting of genetically transformed, GFP-expressing blood stages of the rodent malaria parasite, *Plasmodium berghei*. Nat. Protoc. *1*, 614–623.

Janse, C.J., Kroeze, H., van Wigcheren, A., Mededovic, S., Fonager, J., Franke-Fayard, B., Waters, A.P., and Khan, S.M. (2011). A genotype and phenotype database of genetically modified malaria-parasites. Trends Parasitol. *27*, 31–39.

Janse, C.J., Ramesar, J., and Waters, A.P. (2006c). High-efficiency transfection and drug selection of genetically transformed blood stages of the rodent malaria parasite *Plasmodium berghei*. Nat. Protoc. *1*, 346–356.

Johnson, J.R., Florens, L., Carucci, D.J., and Yates, J.R. (2004). Proteomics in malaria. J. Proteome Res. *3*, 296–306.

Jongco, A.M., Ting, L.-M., Thathy, V., Mota, M.M., and Kim, K. (2006). Improved transfection and new selectable markers for the rodent malaria parasite *Plasmodium yoelii*. Mol. Biochem. Parasitol. *146*, 242–250.

Klausner, R.D., Donaldson, J.G., and Lippincott-Schwartz, J. (1992). Brefeldin A: insights into the control of membrane traffic and organelle structure. J. Cell Biol. *116*, 1071–1080.

Klonis, N., Tan, O., Jackson, K., Goldberg, D., Klemba, M., and Tilley, L. (2007). Evaluation of pH during cytostomal endocytosis and vacuolar catabolism of haemoglobin in *Plasmodium falciparum*. Biochem. J. *407*, 343–354.

Kocken, C.H., van der Wel, A., and Thomas, A.W. (1999). *Plasmodium cynomolgi*: transfection of blood-stage parasites using heterologous DNA constructs. Experimental Parasitology *93*, 58–60.

Kocken, C.H.M., Ozwara, H., van der Wel, A., Beetsma, A.L., Mwenda, J.M., and Thomas, A.W. (2002). *Plasmodium knowlesi* provides a rapid in vitro and in vivo transfection system that enables double-crossover gene knockout studies. Infect. Immun. *70*, 655–660.

Kumar, A., and Snyder, M. (2001). Emerging technologies in yeast genomics. Nat. Rev. Genet. *2*, 302–312.

Lacount, D.J., Vignali, M., Chettier, R., Phansalkar, A., Bell, R., Hesselberth, J.R., Schoenfeld, L.W., Ota, I., Sahasrabudhe, S., Kurschner, C., *et al.* (2005). A protein interaction network of the malaria parasite *Plasmodium falciparum*. Nature *438*, 103–107.

Landy, A. (1989). Dynamic, structural, and regulatory aspects of lambda site-specific recombination. Annu. Rev. Biochem. *58*, 913–949.

Lasonder, E., Ishihama, Y., Andersen, J.S., Vermunt, A.M.W., Pain, A., Sauerwein, R.W., Eling, W.M.C., Hall, N., Waters, A.P., Stunnenberg, H.G., *et al.* (2002). Analysis of the *Plasmodium falciparum* proteome by high-accuracy mass spectrometry. Nature *419*, 537–542.

Le Roch, K.G., Zhou, Y., Batalov, S., and Winzeler, E.A. (2002). Monitoring the chromosome 2 intraerythrocytic transcriptome of *Plasmodium falciparum* using oligonucleotide arrays. Am. J. Trop. Med. Hyg. *67*, 233–243.

Le Roch, K.G., Zhou, Y., Blair, P.L., Grainger, M., Moch, J.K., Haynes, J.D., De La Vega, P., Holder, A.A., Batalov, S., Carucci, D.J., *et al.* (2003). Discovery of gene function by expression profiling of the malaria parasite life cycle. Science *301*, 1503–1508.

Leal, S., Acosta-Serrano, A., Morris, J., and Cross, G.A.M. (2004). Transposon mutagenesis of Trypanosoma brucei identifies glycosylation mutants resistant to concanavalin A. J. Biol. Chem. *279*, 28979–28988.

Lopez-Estraño, C., Bhattacharjee, S., Harrison, T., and Haldar, K. (2003). Cooperative domains define a unique host cell-targeting signal in *Plasmodium falciparum*-infected erythrocytes. Proc. Natl. Acad. Sci. U.S.A. *100*, 12402–12407.

Maier, A., Braks, J., Waters, A., and Cowman, A. (2006). Negative selection using yeast cytosine deaminase/uracil phosphoribosyl transferase in *Plasmodium falciparum* for targeted gene deletion by double crossover recombination. Mol. Biochem. Parasitol. *150*, 118–121.

Maier, A., Rug, M., Oneill, M., Brown, M., Chakravorty, S., Szestak, T., Chesson, J., Wu, Y., Hughes, K., and Coppel, R. (2008). Exported proteins required for virulence and rigidity of *Plasmodium falciparum*-infected human erythrocytes. Cell *134*, 48–61.

Mamoun, C.B., Gluzman, I.Y., Beverley, S.M., and Goldberg, D.E. (2000). Transposition of the *Drosophila* element mariner within the human malaria parasite *Plasmodium falciparum*. Mol. Biochem. Parasitol. *110*, 405–407.

Mamoun, C.B., Gluzman, I.Y., Goyard, S., Beverley, S.M., and Goldberg, D.E. (1999a). A set of independent selectable markers for transfection of the human malaria parasite *Plasmodium falciparum*. Proc. Natl. Acad. Sci. U.S.A. *96*, 8716–8720.

Mamoun, C.B., Truong, R., Gluzman, I., Akopyants, N.S., Oksman, A., and Goldberg, D.E. (1999b). Transfer of genes into *Plasmodium falciparum* by polyamidoamine dendrimers. Mol. Biochem. Parasitol. *103*, 117–121.

Marti, M. (2005). Signal-mediated export of proteins from the malaria parasite to the host erythrocyte. J. Cell Biol. *171*, 587–592.

Marti, M., Good, R.T., Rug, M., Knuepfer, E., and Cowman, A.F. (2004). Targeting malaria virulence and remodeling proteins to the host erythrocyte. Science *306*, 1930–1933.

Meissner, M., Krejany, E., Gilson, P.R., De Koning-Ward, T.F., Soldati, D., and Crabb, B.S. (2005). Tetracycline analogue-regulated transgene expression in *Plasmodium falciparum* blood stages using Toxoplasma gondii transactivators. Proc. Natl. Acad. Sci. U.S.A. *102*, 2980–2985.

Meissner, M., Schlüter, D., and Soldati, D. (2002). Role of Toxoplasma gondii myosin A in powering parasite gliding and host cell invasion. Science *298*, 837–840.

Militello, K.T., Dodge, M., Bethke, L., and Wirth, D.F. (2004). Identification of regulatory elements in the *Plasmodium falciparum* genome. Mol. Biochem. Parasitol. *134*, 75–88.

Mota, M.M., Thathy, V., Nussenzweig, R.S., and Nussenzweig, V. (2001). Gene targeting in the rodent malaria parasite *Plasmodium yoelii*. Mol. Biochem. Parasitol. *113*, 271–278.

Mwakingwe, A., Ting, L.-M., Hochman, S., Chen, J., Sinnis, P., and Kim, K. (2009). Noninvasive real-time monitoring of liver-stage development of bioluminescent *Plasmodium* parasites. J. Infect. Dis. *200*, 1470–1478.

Myrick, A., Munasinghe, A., Patankar, S., and Wirth, D.F. (2003). Mapping of the *Plasmodium falciparum* multidrug resistance gene 5′-upstream region, and evidence of induction of transcript levels by antimalarial drugs in chloroquine sensitive parasites. Mol. Microbiol. *49*, 671–683.

Nkrumah, L.J., Muhle, R.A., Moura, P.A., Ghosh, P., Hatfull, G.F., Jacobs, W.R., and Fidock, D.A.

(2006). Efficient site-specific integration in *Plasmodium falciparum* chromosomes mediated by mycobacteriophage Bxb1 integrase. Nat. Meth. 3, 615.

Nkrumah, L.J., Riegelhaupt, P.M., Moura, P., Johnson, D.J., Patel, J., Hayton, K., Ferdig, M.T., Wellems, T.E., Akabas, M.H., and Fidock, D.A. (2009). Probing the multifactorial basis of *Plasmodium falciparum* quinine resistance: evidence for a strain-specific contribution of the sodium–proton exchanger PfNHE. Mol. Biochem. Parasitol. *165*, 122–131.

O'Donnell, R.A., de Koning-Ward, T.F., Burt, R.A., Bockarie, M., Reeder, J.C., Cowman, A.F., and Crabb, B.S. (2001a). Antibodies against merozoite surface protein (MSP)-1(19) are a major component of the invasion-inhibitory response in individuals immune to malaria. J. Exp. Med. *193*, 1403–1412.

O'Donnell, R.A., Freitas-Junior, L.H., Preiser, P.R., Williamson, D.H., Duraisingh, M., McElwain, T.F., Scherf, A., Cowman, A.F., and Crabb, B.S. (2002). A genetic screen for improved plasmid segregation reveals a role for Rep20 in the interaction of *Plasmodium falciparum* chromosomes. EMBO J. *21*, 1231–1239.

O'Donnell, R.A., Preiser, P.R., Williamson, D.H., Moore, P.W., Cowman, A.F., and Crabb, B.S. (2001b). An alteration in concatameric structure is associated with efficient segregation of plasmids in transfected *Plasmodium falciparum* parasites. Nucleic Acids Res. *29*, 716–724.

O'Donnell, R.A., Saul, A., Cowman, A.F., and Crabb, B.S. (2000). Functional conservation of the malaria vaccine antigen MSP-119across distantly related *Plasmodium* species. Nat. Med. *6*, 91–95.

O'Neill, M.T., Phuong, T., Healer, J., Richard, D., and Cowman, A.F. (2010). Gene deletion from *Plasmodium falciparum* using FLP and Cre recombinases: Implications for applied site-specific recombination. Int. J. Parasitol. *41*, 117–123.

Pfander, C., Anar, B., Schwach, F., Otto, T.D., Brochet, M., Volkmann, K., Quail, M.A., Pain, A., Rosen, B., Skarnes, W., Rayner, J.C., and Billker, O. (2011). A scalable pipeline for highly effective genetic modification of a malaria parasite. Nat. Methods *8*, 1078–1082.

Pinard, R., de Winter, A., Sarkis, G.J., Gerstein, M.B., Tartaro, K.R., Plant, R.N., Egholm, M., Rothberg, J.M., and Leamon, J.H. (2006). Assessment of whole genome amplification-induced bias through high-throughput, massively parallel whole genome sequencing. BMC Genomics 7, 216.

Ploemen, I.H.J., Prudêncio, M., Douradinha, B.G., Ramesar, J., Fonager, J., van Gemert, G.-J., Luty, A.J.F., Hermsen, C.C., Sauerwein, R.W., Baptista, F.G., *et al.* (2009). Visualisation and quantitative analysis of the rodent malaria liver stage by real time imaging. PLoS ONE 4, e7881.

Rankin, K.E., Graewe, S., Heussler, V.T., and Stanway, R.R. (2010). Imaging liver-stage malaria parasites. Cell. Microbiol. *12*, 569–579.

Reece, S.E., and Thompson, J. (2008). Transformation of the rodent malaria parasite *Plasmodium chabaudi* and generation of a stable fluorescent line PcGFPCON. Malar. J. 7, 183.

Rohrbach, P. (2009). Imaging ion flux and ion homeostasis in blood-stage malaria parasites. Biotechnol. J. *4*, 812–825.

Russo, I., Babbitt, S., Muralidharan, V., Butler, T., Oksman, A., and Goldberg, D.E. (2010). Plasmepsin V licenses *Plasmodium* proteins for export into the host erythrocyte. Nature *463*, 632–636.

Russo, I., Oksman, A., Vaupel, B., and Goldberg, D.E. (2009). A calpain unique to alveolates is essential in *Plasmodium falciparum* and its knockdown reveals an involvement in pre-S-phase development. Proc. Natl. Acad. Sci. U.S.A. *106*, 1554–1559.

Sá, J.M., Yamamoto, M.M., Fernandez-Becerra, C., de Azevedo, M.F., Papakrivos, J., Naudé, B., Wellems, T.E., and Del Portillo, H.A. (2006). Expression and function of pvcrt-o, a *Plasmodium vivax* ortholog of pfcrt, in *Plasmodium falciparum* and *Dictyostelium discoideum*. Mol. Biochem. Parasitol. *150*, 219–228.

Sakamoto, H., Thiberge, S., Akerman, S., Janse, C.J., Carvalho, T.G., and Ménard, R. (2005). Towards systematic identification of *Plasmodium* essential genes by transposon shuttle mutagenesis. Nucleic Acids Res. 33, e174.

Sam-Yellowe, T.Y. (2004). A *Plasmodium* gene family encoding Maurer's cleft membrane proteins: structural properties and expression profiling. Genome Res. *14*, 1052–1059.

Sanchez, B.A.M., Mota, M.M., Sultan, A.A., and Carvalho, L.H. (2004). *Plasmodium berghei* parasite transformed with green fluorescent protein for screening blood schizontocidal agents. Int. J. Parasitol. *34*, 485–490.

Sanchez, B.A.M., Varotti, F.P., Rodrigues, F.G., and Carvalho, L.H. (2007). Validation of a *Plasmodium falciparum* parasite transformed with green fluorescent protein for antimalarial drug screening. J. Microbiol. Methods *69*, 518–522.

Saridaki, T., Sanchez, C.P., Pfahler, J., and Lanzer, M. (2008). A conditional export system provides new insights into protein export in *Plasmodium falciparum*-infected erythrocytes. Cell. Microbiol. *10*, 2483–2495.

Shoemaker, D.D., Lashkari, D.A., Morris, D., Mittmann, M., and Davis, R.W. (1996). Quantitative phenotypic analysis of yeast deletion mutants using a highly parallel molecular bar-coding strategy. Nat. Genet. *14*, 450–456.

Sijwali, P.S., and Rosenthal, P.J. (2004). Gene disruption confirms a critical role for the cysteine protease falcipain-2 in hemoglobin hydrolysis by *Plasmodium falciparum*. Proc. Natl. Acad. Sci. U.S.A. *101*, 4384–4389.

Skinner-Adams, T.S., Hawthorne, P.L., Trenholme, K.R., and Gardiner, D.L. (2003a). GATEWAY vectors for *Plasmodium falciparum* transfection. Trends Parasitol. *19*, 17–18.

Skinner-Adams, T.S., Lawrie, P.M., Hawthorne, P.L., Gardiner, D.L., and Trenholme, K.R. (2003b). Comparison of *Plasmodium falciparum* transfection methods. Malar. J. 2, 19.

Soll, J., and Schleiff, E. (2004). Protein import into chloroplasts. Nat. Rev. Mol. Cell Biol. 5, 198–208.

Somsak, V., Srichairatanakool, S., Kamchonwongpaisan, S., Yuthavong, Y., and Uthaipibull, C. (2011). Small-scale in vitro culture and purification of *Plasmodium berghei* for transfection experiment. Mol. Biochem. Parasitol. *177*, 156–159.

Spence, P., Cunningham, D., Jarra, W., Lawton, J., Langhorne, J., and Thompson, J. (2011). Transformation of the rodent malaria parasite *Plasmodium chabaudi*. Nat. Protoc. *6*, 553–561.

Su, X.Z., Wu, Y., Sifri, C.D., and Wellems, T.E. (1996). Reduced extension temperatures required for PCR amplification of extremely A+T-rich DNA. Nucleic Acids Res. *24*, 1574–1575.

Subramaniam, C., Veazey, P., Redmond, S., Hayes-Sinclair, J., Chambers, E., Carrington, M., Gull, K., Matthews, K., Horn, D., and Field, M.C. (2006). Chromosome-wide analysis of gene function by RNA interference in the African trypanosome. Eukaryotic Cell *5*, 1539–1549.

Sultan, A.A., Thathy, V., de Koning-Ward, T.F., and Nussenzweig, V. (2001). Complementation of *Plasmodium berghei* TRAP knockout parasites using human dihydrofolate reductase gene as a selectable marker. Mol. Biochem. Parasitol. *113*, 151–156.

Suthram, S., Sittler, T., and Ideker, T. (2005). The Plasmodium protein network diverges from those of other eukaryotes. Nature *438*, 108–112.

Takebe, S., Witola, W.H., Schimanski, B., Günzl, A., and Ben Mamoun, C. (2007). Purification of components of the translation elongation factor complex of *Plasmodium falciparum* by tandem affinity purification. Eukaryotic Cell *6*, 584–591.

Thibault, S.T., Singer, M.A., Miyazaki, W.Y., Milash, B., Dompe, N.A., Singh, C.M., Buchholz, R., Demsky, M., Fawcett, R., Francis-Lang, H.L., *et al.* (2004). A complementary transposon tool kit for *Drosophila melanogaster* using P and piggyBac. Nat. Genet. *36*, 283–287.

Tilley, L., McFadden, G., Cowman, A., and Klonis, N. (2007). Illuminating *Plasmodium falciparum*-infected red blood cells. Trends Parasitol. *23*, 268–277.

Tonkin, C.J., van Dooren, G.G., Spurck, T.P., Struck, N.S., Good, R.T., Handman, E., Cowman, A.F., and McFadden, G.I. (2004). Localization of organellar proteins in *Plasmodium falciparum* using a novel set of transfection vectors and a new immunofluorescence fixation method. Mol. Biochem. Parasitol. *137*, 13–21.

Triglia, T., Wang, P., Sims, P.F., Hyde, J.E., and Cowman, A.F. (1998). Allelic exchange at the endogenous genomic locus in *Plasmodium falciparum* proves the role of dihydropteroate synthase in sulfadoxine-resistant malaria. EMBO J. *17*, 3807–3815.

van der Wel, A.M., Tomás, A.M., Kocken, C.H., Malhotra, P., Janse, C.J., Waters, A.P., and Thomas, A.W. (1997). Transfection of the primate malaria parasite *Plasmodium knowlesi* using entirely heterologous constructs. J. Exp. Med. *185*, 1499–1503.

van Dijk, M.R., Janse, C.J., and Waters, A.P. (1996). Expression of a Plasmodium gene introduced into subtelomeric regions of *Plasmodium berghei* chromosomes. Science *271*, 662–665.

van Dijk, M.R., Vinkenoog, R., Ramesar, J., Vervenne, R.A., Waters, A.P., and Janse, C.J. (1997). Replication, expression and segregation of plasmid-borne DNA in genetically transformed malaria parasites. Mol. Biochem. Parasitol. 86, 155–162.

van Dijk, M.R., Waters, A.P., and Janse, C.J. (1995). Stable transfection of malaria parasite blood stages. Science *268*, 1358–1362.

Van Ooij, C., and Haldar, K. (2007). Protein export from Plasmodium parasites. Cell. Microbiol. *9*, 573–582.

VanBuskirk, K.M., O'Neill, M.T., De La Vega, P., Maier, A.G., Krzych, U., Williams, J., Dowler, M.G., Sacci, J.B., Kangwanrangsan, N., Tsuboi, T., *et al.* (2009). Preerythrocytic, live-attenuated *Plasmodium falciparum* vaccine candidates by design. Proc. Natl. Acad. Sci. U.S.A. *106*, 13004–13009.

VanWye, J.D., and Haldar, K. (1997). Expression of green fluorescent protein in *Plasmodium falciparum*. Mol. Biochem. Parasitol. *87*, 225–229.

Venter, J.C., Adams, M.D., Myers, E.W., Li, P.W., Mural, R.J., Sutton, G.G., Smith, H.O., Yandell, M., Evans, C.A., Holt, R.A., *et al.* (2001). The sequence of the human genome. Science *291*, 1304–1351.

Waller, K.L., Muhle, R.A., Ursos, L.M., Horrocks, P., Verdier-Pinard, D., Sidhu, A.B.S., Fujioka, H., Roepe, P.D., and Fidock, D.A. (2003). Chloroquine resistance modulated in vitro by expression levels of the *Plasmodium falciparum* chloroquine resistance transporter. J. Biol. Chem. *278*, 33593–33601.

Waller, R.F., Reed, M.B., Cowman, A.F., and McFadden, G.I. (2000). Protein trafficking to the plastid of *Plasmodium falciparum* is via the secretory pathway. EMBO J. *19*, 1794–1802.

Wang, P., Wang, Q., Sims, P.F.G., and Hyde, J.E. (2002). Rapid positive selection of stable integrants following transfection of *Plasmodium falciparum*. Mol. Biochem. Parasitol. *123*, 1–10.

Wang, Y., Nair, S., Nosten, F., and Anderson, T.J.C. (2009). Multiple displacement amplification for malaria parasite DNA. J. Parasitol. *95*, 253–255.

Wel, A.v.d., Kocken, C.H.M., Pronk, T.C., Franke-Fayard, B., and Thomas, A.W. (2004). New selectable markers and single crossover integration for the highly versatile *Plasmodium knowlesi* transfection system. Mol. Biochem. Parasitol. *134*, 97–104.

Wilson, D.W., Crabb, B.S., and Beeson, J.G. (2010). Development of fluorescent *Plasmodium falciparum* for in vitro growth inhibition assays. Malar. J. *9*, 152.

Wirtz, E., and Clayton, C. (1995). Inducible gene expression in trypanosomes mediated by a prokaryotic repressor. Science *268*, 1179–1183.

Wu, Y., Kirkman, L.A., and Wellems, T.E. (1996). Transformation of *Plasmodium falciparum* malaria parasites by homologous integration of plasmids that confer resistance to pyrimethamine. Proc. Natl. Acad. Sci. U.S.A. *93*, 1130–1134.

Wu, Y., Sifri, C.D., Lei, H.H., Su, X.Z., and Wellems, T.E. (1995). Transfection of *Plasmodium falciparum* within human red blood cells. Proc. Natl. Acad. Sci. U.S.A. *92*, 973–977.

Yen, L., Svendsen, J., Lee, J.-S., Gray, J.T., Magnier, M., Baba, T., D'Amato, R.J., and Mulligan, R.C. (2004). Exogenous control of mammalian gene expression through modulation of RNA self-cleavage. Nature *431*, 471–476.

www.ingramcontent.com/pod-product-compliance
Lightning Source LLC
LaVergne TN
LVHW062119200926
848484LV00007B/1719